Advances in
MICROBIAL ECOLOGY

Volume 10

ADVANCES IN MICROBIAL ECOLOGY

A continuation Order Plan is available for this series. A continuation order will bring delivery of
each new volume immediately upon publication. Volumes are billed only upon actual shipment.
For further information please contact the publisher.

Advances in
MICROBIAL ECOLOGY

Volume 10

Edited by

K. C. Marshall
University of New South Wales
Kensington, New South Wales, Australia

PLENUM PRESS · NEW YORK AND LONDON

The Library of Congress cataloged the first volume of this title as follows:

Advances in microbial ecology. v. 1–
  New York, Plenum Press c1977–
  v. ill. 24 cm.
  Key title: Advances in microbial ecology, ISSN 0147-4863
  1. Microbial ecology — Collected works.
QR100.A36   576′.15   77-649698

ISBN 0-306-42710-9

© 1988 Plenum Press, New York
A Division of Plenum Publishing Corporation
233 Spring Street, New York, N.Y. 10013

Printed in the United States of America

Contributors

Jacqueline Aislabie, Department of Biology, University of Louisville, Louisville, Kentucky 40292

Lynne Boddy, Department of Microbiology, University College, Cardiff CF2 ITA, United Kingdom

P. Bossier, Laboratory of Microbial Ecology, University of Ghent, B-9000 Ghent, Belgium

Philip Bremer, Microbiology Department, University of Otago, Dunedin, New Zealand

Douglas G. Capone, Chesapeake Biological Laboratory, Center for Environmental and Estuarine Studies, University of Maryland, Solomons, Maryland 20688-0038

Ralf Conrad, Max-Planck-Institut für Chemie, D-6500 Mainz, Federal Republic of Germany. *Present address:* Fakultät für Biologie, Universität Konstanz, D-7750 Konstanz 1, Federal Republic of Germany

Simon Ford, Program in Social Ecology, University of California, Irvine, California 92717

M. Hofte, Laboratory of Microbial Ecology, University of Ghent, B-9000 Ghent, Belgium

Richard E. Lenski, Department of Ecology and Evolutionary Biology, University of California, Irvine, California 92717

Margaret W. Loutit, Microbiology Department, University of Otago, Dunedin, New Zealand

Birgit Nordbring-Hertz, Department of Microbial Ecology, University of Lund, S-223 62 Lund, Sweden

Betty H. Olson, Program in Social Ecology, University of California, Irvine, California 92717

Ronald S. Oremland, Water Resources Divison, United States Geological Survey, Menlo Park, California 94025

Aharon Oren, Division of Microbial and Molecular Ecology, Institute of Life Sciences, Hebrew University of Jerusalem, Jerusalem 91904, Israel

Christopher Pillidge, Department of Microbiology, University of Maryland, College Park, Maryland 20742

Karen G. Porter, Department of Zoology, University of Georgia, Athens, Georgia 30602

A. D. M. Rayner, School of Biological Sciences, University of Bath, Bath BA2 7AY, United Kingdom

Robert W. Sanders, Department of Zoology, University of Georgia, Athens, Georgia 30602

W. Verstraete, Laboratory of Microbial Ecology, University of Ghent, B-9000 Ghent, Belgium

Preface

The publication of Volume 10 of *Advances in Microbial Ecology* represents something of a milestone in the history of modern microbial ecology. *Advances in Microbial Ecology* was established by the International Committee on Microbial Ecology (ICOME) to provide a vehicle for in-depth, critical, and even provocative reviews to emphasize current trends in the rapidly expanding field of microbial ecology. Martin Alexander was the Founding Editor of the series and was responsible for editing the first five volumes. The next five volumes were edited by Kevin Marshall. *Advances in Microbial Ecology* has attained recognition as an authoritative source of information and inspiration for practicing and prospective microbial ecologists. The Editorial Board usually invites contributions from leading microbial ecologists, but also encourages individuals to submit outlines of unsolicited contributions to any member of the Board for consideration for publication in *Advances.*

Contributions to Volume 10 again cover a broad range of topics related to microbial ecology. Interactions between microorganisms are well represented with chapters on bacterium–virulent bacteriophage interactions by R. E. Lenski, on fungal communities in the decay of wood by A. D. M. Rayner and L. Boddy, on recognition in the nematode-nematophagous fungus system by B. Nordbring-Hertz, and on phago-trophic phytoflagellates by R. W. Sanders and K. G. Porter. Chapters with both an ecological and a biogeochemical flavor include those on atmospheric CO and H_2 by R. Conrad, on the use of "specific" inhibitors by R. S. Oremland and D. G. Capone, and on chromium in sediments by M. Loutit, J. Aislabie, P. Bremer, and C. Pillidge. Of topical interest is the contribution by S. Ford and B. H. Olson on methods for detecting genetically engineered microorganisms in the environment. P. Bossier, M. Hofte, and W. Verstraete discuss the ecological significance of sider-ophores in soil. A very specialized and extreme ecosystem, the Dead Sea, is considered in detail by A. Oren.

Howard Slater has completed his term as a member of the Editorial Board and, with the publication of this volume, we welcome Gwyn Jones to the Board.

K. C. Marshall, Editor
R. Atlas
J. G. Jones
B. B. Jørgensen

Contents

Chapter 3

Ecology and Recognition in the Nematode–Nematophagous Fungus System

Birgit Nordbring-Hertz

Chapter 4

Fungal Communities in the Decay of Wood

A. D. M. Rayner and Lynne Boddy

Chapter 5

Phagotrophic Phytoflagellates

Robert W. Sanders and Karen G. Porter

Chapter 6

The Microbial Ecology of the Dead Sea

Aharon Oren

Chapter 7

Biogeochemistry and Ecophysiology of Atmospheric CO and H$_2$

Ralf Conrad

Chapter 8

Use of "Specific" Inhibitors in Biogeochemistry and Microbial Ecology

Ronald S. Oremland and Douglas G. Capone

Chapter 9

Ecological Significance of Siderophores in Soil

P. Bossier, M. Hofte, and W. Verstraete

Chapter 10

Bacteria and Chromium in Marine Sediments

Margaret W. Loutit, Jacqueline Aislabie, Philip Bremer, and
Christopher Pillidge

Dynamics of Interactions between Bacteria and Virulent Bacteriophage

RICHARD E. LENSKI

1. Introduction

The interactions of bacteria and their viruses (bacteriophage) are, by and large, ones of trophic exploitation. In fact, "phage" is derived from the Greek word for "devour." Using the criterion of relative size, the interactions can be defined as parasitism (Bull and Slater, 1982). Because replication by most virulent phage necessarily results in bacterial death, these interactions could also be called predation. Certain interactions could even be termed mutualistic, as some temperate phage encode phenotypic characteristics that are of direct benefit to their hosts. Semantics aside, the fundamental ecological question that I will attempt to address in this chapter is: What role do bacteriophage infections play in limiting the abundance of bacteria?

Such a broad question cannot be answered using any single approach or line of evidence. Therefore, I have chosen to organize this chapter in a hierarchical manner, moving from mathematical models, through simple laboratory communities, and finally to much more complex communities in natural settings. But first it is necessary to review the basic biological features of the interactions between bacteria and phage, as revealed by the extraordinary advances in the areas of microbial genetics and molecular biology. This research provides a precise methodological and conceptual framework for examining ecological hypotheses, probably unrivaled for any other parasite–host interaction. The same features of

RICHARD E. LENSKI • Department of Ecology and Evolutionary Biology, University of California, Irvine, California 92717.

the phage–bacteria system that have been so valuable in nonecological research also contribute to its power in addressing fundamental ecological questions: specifically, ease of culture and sampling, high population densities, and short generation times. In fact, generations are so short that it becomes imperative to consider the effects of evolutionary change on population dynamics, even over the course of short-term experiments.

2. Molecular and Genetic Bases of the Interaction

There are a number of excellent references on the basic features of the interaction between bacteria and phage. Stent (1963) provides a historical perspective on the progress in elucidating the bases of the interaction, but this book is now somewhat dated. Luria *et al.* (1978) provide a more up-to-date introduction to phage biology. Mathews *et al.* (1983) and Hendrix *et al.* (1983) provide comprehensive summaries of recent research on virulent coliphage T4 and temperate coliphage Lambda, respectively, emphasizing the regulation and expression of phage genes.

2.1. The Course of Lytic Infections

2.1.1. Virulent and Temperate Phage

Bacteriophages are generally divided into two basic classes, virulent and temperate. For both, infection of a host bacterium commences with the adsorption of the phage to the bacterial surface, followed by the introduction of the the genome of the phage into the bacterium. For virulent phage, there follows a period of vegetative growth during which time its genome is replicated and encapsulated intracellularly, terminating with lysis of the bacterium and the release of infective progeny phage (Ellis and Delbruck, 1939). Temperate phage infection may also proceed via this lytic process, or the infecting phage may form a semistable association with the bacterium, a phenomenon known as lysogeny (Lwoff, 1953). In the event of this lysogenic response, the phage genome (referred to as a prophage) is replicated along with the bacterial genome, and is inherited by each of the daughter cells (referred to as lysogens). Lysogens are immune to reinfection by the same temperate phage, although they are not usually immune to virulent mutants of this phage. After one or more generations of lysogenic replication, the prophage may be stimulated to enter the lytic cycle (induction), or it may be lost from the genome of the bacterium during subsequent replication (segregation). The distinction between virulent and temperate phage is thus made on functional grounds, with virulent phage capable only of the lytic mode of replication and temperate phage capable of both the lytic and lysogenic modes of replication.

The existence of temperate phage raises a number of interesting questions that will not be addressed in this chapter. What selective pressures are responsible for the evolution of lysogeny (Levin and Lenski, 1983, 1985; Stewart and Levin, 1984)? What effect does prophage carriage have on the competitive ability of lysogens (Dykhuizen and Hartl, 1983)? What conditions determine the outcome of the facultative "decision" by a temperate phage to lyse or lysogenize its host (Echols, 1972)? What role do phage, especially temperate, play in the infectious transfer of bacterial genes (Reanney, 1976; Reanney et al., 1983)? Furthermore, I will not consider the actual phylogenies of bacterial viruses, nor the role that recombination may have played in this phylogenetic evolution (Bradley, 1967; Botstein, 1980; Reanney and Ackermann, 1982; Campbell and Botstein, 1983).

While the focus of this chapter is on the interaction of virulent phage and their hosts, the distinction between temperate and virulent is somewhat artificial. The repressor protein responsible for the maintenance of lysogeny is coded for by the genome of the temperate phage, but this response is also dependent on environmental and genetic factors acting on the host bacterium. Thus, the classification of a phage as temperate or virulent may depend on the particular host or environment in which the lysogeny criterion is tested. Nor can the distinction be made on phylogenetic grounds; with one or a few mutations, temperate phage can lose their ability to form lysogens and hence become virulent phage (Lwoff, 1953; Bronson and Levine, 1971; Ptashne et al., 1980). Even for phage clearly identifiable as temperate, the likelihood of the lysogenic response for any given infection is usually much lower than the likelihood of the lytic response.

Moreover, certain types of phage infections are not readily classified as either lytic or lysogenic. For example, the normally virulent coliphage T3 may replicate alongside the bacterial genome for several generations before lysing the infected cells (Fraser, 1957). This response differs from true lysogeny, however, in that this pseudolysogenic state is not maintained by a phage-encoded repressor protein (Kruger and Schroeder, 1981). Similarly, normally virulent phage of *Bacillus subtilis* may persist in living cells through the sporulation process (Barksdale and Arden, 1974). The distinction is even more problematic for the filamentous phages, which are released continuously from growing and dividing cells. The growth rate of bacteria is usually reduced by infection with these phage, but lysis does not normally occur (Marvin and Hohn, 1969). Associations between filamentous phage and their hosts are perhaps more analogous to "typical" host–parasite associations, in contrast to the dormant lysogenic state and the lethal lytic response.

2.1.2. Adsorption

Phage and bacteria encounter one another through random Brownian motion. The adsorption process commences with the binding of the phage adsorption organelle to highly specific receptor sites on the bacterial surface. Sites of attachment vary from bacterium to bacterium and from phage to phage. A wide variety of moieties at the cell surface may serve as receptors for particular phage, including proteins (Schwartz, 1980), lipopolysaccharide in Gram-negative bacteria (Wright *et al.,* 1980), and peptidoglycan and teichoic acid in Gram-positive bacteria (Archibald, 1980). It is not uncommon for two or more different phage to recognize the same receptor sites on a particular bacterium (Schwartz, 1980), and a few phage may even be capable of adsorbing to two or more different receptors (Morona and Henning, 1984, 1986).

Not surprisingly, the receptor sites on the bacterial surface serve particular functions, and have been exploited secondarily by bacteriophage. For example, many of the proteinaceous receptor sites are involved in the transport of specific nutrients, including sugars, amino acids, and vitamins (Braun and Hantke, 1977). Other receptors are associated with organelles of specific function, such as flagella and conjugative pili, the latter serving as adsorption sites for the so-called male-specific phages.

The attachment of the phage adsorption organelle to bacterial receptors is initially reversible (Goldberg, 1980). That is, the phage may be dissociated, for example, by diluting the reaction mixture or killing the bacteria, with the phage retaining their infectious capacity. The association eventually becomes irreversible, and this irreversibility is linked to the steps leading to the penetration of the genetic material of the phage into the bacterial cell.

The kinetics of the adsorption process can be studied by means of an approach pioneered by Krueger (1931) and Schlesinger (1932). Known densities of bacteria and phage are combined, usually with fewer phage than bacteria (i.e., a low multiplicity of infection, or MOI), so that complications arising from multiple phage adsorptions to a single bacterium can be ignored. If the rate of irreversible adsorption of phage to bacteria is sufficiently high, it is possible to measure directly the decline in free (i.e., unadsorbed) phage. At frequent intervals, a subsample of the culture is diluted and chloroformed (Adams, 1959). Dilution effectively stops the cell density-dependent process of phage adsorption, while chloroform kills bacteria and phage that have adsorbed irreversibly to the bacteria, but does not usually affect free phage. The duration of the experiment must be limited to preclude the formation of complete phage progeny within those cells that were infected first (see discussion of the eclipse period in Section 2.1.3). If the proportional decline in free phage is slight, then more complex procedures are necessary. For example, using anti-serum active against the phage, it is possible to remove free phage and thus measure directly the increase in infected bacteria (Adams, 1959).

The dynamics of phage adsorption are usually described empirically by a first-order rate constant, which is expressed as unit volume per unit time. This rate can be calculated from the slope of the exponential decline in concentration of free phage divided by the density of bacteria on which adsorption took place. Departures from first-order kinetics become apparent at very high densities of bacteria (Stent and Wollman, 1952). In such cases, the rate of phage adsorption reaches some maximum that is independent of bacterial density, as the kinetics are limited not by the formation of reversible attachments, but by the formation of irreversible associations. It has been found that the rate of phage adsorption to bacteria under favorable conditions can be very close to the rate of collision by Brownian motion (Schlesinger, 1932; Delbruck, 1940a; Schwartz, 1976). Thus, the attachment of an adsorption organelle of a phage to receptor sites on a bacterium can be an extremely efficient process.

The kinetics of adsorption can also be affected at higher multiplicities of phage. In extreme cases, one may observe the competition of phage for limiting receptor sites. A large number of adsorptions to an individual bacterium may also cause a phenomenon known as lysis-from-without, whereby the cell is ruptured and infection rendered nonproductive (Delbruck, 1940b). Infection by coliphage T5 results in the inactivation of remaining T5 receptors on the bacterial surface, thus restricting the chances for reinfection (Dunn and Duckworth, 1977).

The adsorption rate is, of course, also dependent on the medium in which the interaction takes place, and on the physiological state of the cells (Delbruck, 1940a). Certain phages adsorb to killed as well as to living bacteria; other phages, such a Lambda, require an energized cell membrane, and adsorb irreversibly only to healthy bacteria (Schwartz, 1976). More subtle variation in adsorption rate may also be significant. The receptor for coliphage Lambda is involved in the uptake of maltose, and the surface density of the receptor depends on the carbon source on which the bacteria are grown. On maltose, the density of receptors is uniformly high, whereas on glucose the concentration of receptors is much lower

(Schwartz, 1976) and highly variable from cell to cell (Howes, 1965; Ryter *et al.,* 1975). These effects are due to the substrate-dependent rate of induction of synthesis of the receptor (Ryter *et al.,* 1975).

2.1.3. Vegetative Growth and Lysis

Subsequent to the adsorption of a virulent phage and the penetration of its genetic material, the metabolic machinery of the bacterium comes under the control of the phage genome; to a greater or lesser extent, the bacterium is converted into a "factory" for the production of progeny phage. The duration of the period extending from adsorption to lysis of the cell is referred to as the latent period, and the number of phage progeny released upon lysis is referred to as the burst size.

These parameters can be estimated using the one-step growth experiment devised by Ellis and Delbruck (1939). At time zero, known densities of bacteria and phage are mixed; as with the adsorption rate experiment, bacteria are in excess to avoid the complications of multiple infections. After several minutes, the proportion of phage that have irreversibly adsorbed is determined (preferably near 100%), and the culture is diluted to prevent any further adsorption. This culture is then assayed at frequent intervals for the concentration of plaque-forming units (infected bacteria plus remaining free phage), which should remain unchanged over the duration of the latent period. After an elapsed time equivalent to the latent period, the first infected cells begin to lyse, releasing progeny phage and increasing the concentration of plaque-forming units. During the rise period, all infected cells eventually burst, and a new plateau is reached for the concentration of plaque-forming units, which remains constant because dilution has effectively stopped further phage adsorption and replication. The ratio of the concentration of plaque-forming units after the rise period to their concentration immediately after dilution is the average burst size; this ratio must be corrected for the fraction of phage that were not adsorbed.

Although the intracellular dynamics of phage growth during the latent period can be viewed as a "black box" from the perspective of population dynamics, it is frequently divided into two phases. Prior to the eclipse, an infected bacterium that is lysed artificially (e.g., with chloroform) contains no infective phage; subsequent to the eclipse, but still prior to natural lysis, an artificially lysed bacterium contains increasing numbers of phage progeny. This discontinuity occurs because phage are assembled not one-by-one, but rather *en masse* from precursor subunits.

While adsorption for some phage can occur even on dead bacteria, this would not seem to be the case for vegetative growth. However, coliphage T2 can, under appropriate conditions, actually replicate in and lyse otherwise nonviable cells (T. F. Anderson, 1948). This is because T2 relies almost exclusively on its own genetic information, shutting off all host-controlled macromolecular synthesis within a few minutes after the penetration of its DNA (Koerner and Snustad, 1979). In contrast, vegetative growth by most phage requires a living cell, and host metabolism may continue almost until lysis.

The course of vegetative growth also can be influenced by multiple infections, with the effects dependent on the time elapsed between successive events. For some phage, subsequent infections are usually nonproductive, such that later adsorbed phage are essentially killed. For others, like the T-even coliphages, a phenomenon known as lysis inhibition can occur, in which lysis is delayed (and total burst size increased) by reinfecting the bacteria (Doermann, 1948). It is also frequently possible to demonstrate priority effects, whereby phage of one type partially or completely exclude the production of progeny phage of another type (Delbruck and Luria, 1942). If exclusion is not complete, phage genomes may exhibit recombination, as first shown by Hershey and Rotman (1949).

2.2. Bacterial Defenses and Phage Counterdefenses

2.2.1. Resistance Mutations in Bacteria

The rate at which phage irreversibly adsorb to bacteria may be affected by mutations in either the bacteria or the phage. I have already noted that phage adsorption rates can be nearly as high as the maximum rates allowed by Brownian motion, a result undoubtedly due to selection acting on phage to increase the efficiency of their adsorption to specific receptor sites on bacteria. Similarly, bacteria can be selected that have reduced rates of adsorption by particular phage.

Mutant bacteria may be resistant to adsorption by a particular phage for any of several reasons: (1) the structure of the receptor sites may be altered; (2) the exposure of the receptor sites may be altered; (3) the density of the receptor sites may be reduced; or (4) the receptor sites may be lost altogether because of failure in their production or their incorporation into the cell envelope. This fourth class of mutants, altogether lacking particular receptor sites, is especially useful in the identification of receptor moieties through biochemical comparisons of sensitive and resistant cell envelopes (Schwartz, 1980). These mutants also have special evolutionary significance, as they present a challenge that is not readily overcome by mutations in the phage genome.

Because phage receptor sites perform functions of use to the bacterium, their alteration or loss may interfere with bacterial metabolism. I will return to evidence for this point later, and present just two specific examples now. *Escherichia coli* mutants that are resistant to phage Lambda are unable to grow on maltose at low concentrations, although at higher concentrations they are able to grow (Szmelcman and Hofnung, 1975). It appears that Lambda receptors are involved in the active transport of maltose, but that diffusion alone is sufficient for bacterial growth when this substrate is abundant. The coliphage T1 adsorbs to receptor sites that bind and transport iron complexes (Braun and Hantke, 1977), yet E. H. Anderson (1946) demonstrated that some T1-resistant mutants of *E. coli* are auxotrophic for the amino acid tryptophan. This odd result probably derives from the fact that certain T1-resistance mutations map immediately adjacent to the structural genes of the tryptophan operon (Bachmann and Low, 1980). This auxotrophy thus appears to be an indirect consequence of deletion mutations that confer resistance to phage T1.

The rate at which a bacterial strain mutates to resistance to a particular phage can be estimated by means of a fluctuation test developed in the classic paper by Luria and Delbruck (1943). A number of independent cultures of bacteria are grown to some final density, and the entire content of each culture is plated in the presence of excess phage. All bacteria that are sensitive to the phage are killed, whereas any resistant mutants that may be present can grow and are detected. The rate of mutation is estimated from the proportion of cultures that contain zero phage-resistant mutants, according to the Poisson distribution and taking into account the number of bacteria in each culture. It is possible to modify this estimation procedure to make use of the frequency of mutants (rather than just their presence or absence), but this requires an additional assumption concerning the relative growth rates of resistant and sensitive cells.

The phenotypic expression of phage resistance may require one or more generations after the original mutational event, as preexisting phage receptors are diluted sufficiently to allow the survival of progeny in the presence of excess phage (Kubitschek, 1970). The detection of these "latent" mutants requires that the excess phage be added only after plated bacteria have had several generations to grow.

In principle, use of these procedures could result in the isolation not only of mutants that are resistant to phage adsorption, but also of mutants that survive by virtue of their inability to support vegetative growth of the phage. In practice, however, almost all mutants isolated in this way are resistant to phage adsorption. Several factors may account for this. First, for many phage the adsorption process itself is lethal to the bacterium (Duckworth, 1970); this is especially likely when there are excess phage due to the phenomenon of lysis-from-without. Second, the loss of a bacterial function essential for phage replication is also likely to prevent the growth of the bacteria, hence precluding detection. In contrast, the loss of receptor function, though often deleterious, usually does not completely incapacitate bacteria, probably because most nutrients can be obtained via several routes. Third, most abortive phage infections are due to the acquisition, not the loss, of specific host gene functions. Such events are more likely to occur by transfer of genetic material than by spontaneous mutation.

Rates of mutation to phage resistance can vary considerably, depending on the particular bacterium and phage: as high as 10^{-3} per cell generation for phage that adsorb to the antigenic determinants of phase variation in *Salmonella* and *Shigella* (E. S. Anderson, 1957; Barksdale and Arden, 1974), but more typically on the order of 10^{-7} or less (Demerec and Fano, 1945). Because different phage often share receptor sites (or some component of the receptors), selection for resistance to one phage often results in resistance to other phage. For example, nearly all mutants of *E. coli* B that are resistant to phage T3 are resistant to phage T4, while about one-half of these mutants are also resistant to phage T7 (Demerec and Fano, 1945). T3 and T7 are closely related on morphological and other grounds, but are unrelated to T4 (Bradley, 1967; Kruger and Schroeder, 1981); all three adsorb to components of the lipopolysaccharide core (Wright *et al.,* 1980). In contrast, resistance to these phage is independent of resistance to phage T5, which adsorbs to a proteinaceous receptor.

There is one serious limitation to the standard procedure of isolating phage-resistant mutants by plating bacteria in the presence of excess phage. Such a procedure detects only those mutants that are completely resistant to the selecting phage, whereas bacterial mutants that are partially resistant (i.e., with an adsorption rate parameter that is reduced, but still greater than zero) are killed by the excess phage. It would seem logical to reduce the concentration of phage on the selective plates in order to isolate partially resistant mutants (as is done with antibiotics); but phage (unlike antibiotics) replicate on the sensitive bacteria, thereby defeating this approach. Thus, the standard procedure focuses on mutations of extreme phenotypic effect, and misses mutations that are more subtle but also of ecological and evolutionary significance. There have been a few studies of mutations conferring partial resistance. Lenski (1984a) demonstrated that mutations in *E. coli* B that confer complete resistance to phage T4 also confer partial resistance to phage T2, as indicated by a reduced adsorption rate. Some bacteria produce polysaccharide capsules that render them partially resistant to certain phage (Wilkinson, 1958; Paynter and Bungay, 1970); the expression of this trait depends on both the bacterial genotype and the culture medium.

2.2.2. Host-Range Mutations in Phage

Just as one can select mutant bacteria that are resistant to adsorption by a particular phage, it is also often possible to select mutant phage that are able to infect these resistant bacteria. Such phage are termed host-range mutants. The rate of host-range mutation can be estimated by procedures analogous to those used for estimating rates of bacterial mutation to resistance, except that host-range mutants are detected by plaque formation on lawns of resistant bacteria (Luria, 1945; Hershey, 1946).

Host-range mutants may differ from wild-type phage in two ways: (1) an altered specificity, whereby the configuration of the adsorption organelle of the phage is modified so as to permit the phage to bind to the surface of resistant bacteria; or (2) a reduced selectivity, whereby the threshold for conformational changes that cause irreversible binding is lowered so as to permit the genetic material of the phage to be released more readily into resistant bacteria. In support of the first mechanism, it is possible to substitute the tail fibers of T4 with those of T2 and thereby give resultant T4 the host range of T2 (Wright *et al.*, 1980). In support of the second mechanism, Crawford and Goldberg (1977) isolated host-range mutants of phage T4 that differed not in their tail fibers, but in the baseplate, whose expansion permits subsequent penetration of the phage's genetic material into the host.

More generally, it is noteworthy that most host-range phage mutants are capable of growth not only on bacteria resitance to wild-type phage, but also on wild-type bacteria (Luria, 1945; Chao *et al.*, 1977), and that various host-range mutants can usually be ordered in terms of increasing inclusivity (Hofnung *et al.*, 1976; Manning and Reeves, 1978; Schwartz, 1980). That is, most host-range mutants can be said to have an extended (not just a modified) host range. This seems more consistent with the reduced selectivity mechanism than with the altered specificity mechanism. Added support comes from the observation that many host-range mutants are relatively unstable, spontaneously ejecting their genetic material (and hence losing their infectivity) even in the absence of suitable hosts (Schwartz, 1980; Lenski and Levin, 1985). Thus, it appears that most host-range mutations generate phage with reduced selectivity, binding reversibly to the same basic receptor moiety, but "trigger happy" with respect to the irreversible events in the phage adsorption process (Schwartz, 1980).

As noted, many phage-resistant bacterial mutants have altogether lost particular receptor sites, not just changed their configuration, exposure, or density. The synthesis or incorporation of the receptor moiety may be blocked by any number of deletions, insertions, or point mutations; in short, even mutations that result in the loss of an existing gene function suffice to produce resistant bacteria. In contrast, very specific changes in the functions of one or more phage components would be necessary to generate host-range mutants capable of adsorbing to entirely different receptor sites on such resistant bacteria. Thus, "There can exist two broad classes of phage-resistant bacterial mutants: those for which one can select host-range phage mutants and those for which one cannot select host-range phage mutants" (Lenski, 1984b).

The most elegant demonstration of this fundamental asymmetry comes from the work of Hofnung *et al.* (1976) on phage Lambda and *E. coli* K12. These researchers generated a large set of resistant bacteria using a mutagen that causes single base substitutions. Corresponding host-range phage mutants could be isolated for only a fraction of the resistant clones. Bacterial clones were then tested for the responsiveness of their resistance mutations to nonsense suppressors, genes borne by certain laboratory vectors that allow the bacterial cell to translate a nonsense (i.e., stop) codon into a specific amino acid (Lewin, 1974). Those bacterial clones for which Hofnung and co-workers could isolate corresponding host-range mutants did not respond to nonsense repression, indicating that resistance was the result of missense mutations. In contrast, those bacterial mutants for which they could not isolate corresponding host-range mutants were responsive to nonsense repression, indicating that resistance was the result of nonsense mutations. Missense mutations yield an altered amino acid sequence in the proteinaceous phage receptor, whereas nonsense mutations yield a stop codon and consequently a complete loss of functional receptor sites. Host-range mutants exist only to counter the former. These results are summarized schematically in Fig. 1.

2.2.3. Restriction and Other Immunities

Resistance to adsorption is but one kind of bacterial defense against phage infection. A second kind of defense can be termed immunity, and differs from most resistance in several important ways: (1) immunity is manifest intracellularly, rather than on the cell surface; (2) immunity results in the death of the phage, rather than its rejection by the cell; (3) immunity results from the action of a specific gene product, rather than the inactivation of a gene coding for the receptor moiety; and (4) immu-

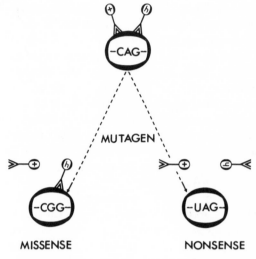

Figure 1. Schematic representation of the results of Hofnung *et al.* (1976). The plus symbol indicates a wild-type phage, while h indicates a host-range mutant. The three hypothetical codons correspond to a bacterial gene encoding a phage receptor, and are wild-type, missense mutant, and nonsense mutant, respectively.

nity is often conferred by the acquisition of an extrachromosomal element, rather than by a chromosomal mutation.

Restriction–modification systems represent one important class of bacterial immunity (Arber and Linn, 1969; Meselson *et al.*, 1972). Through the action of restriction endonucleases, phage DNA may be recognized as foreign and destroyed by the cleaving action of these enzymes. Corresponding modification enzymes protect the bacterial DNA by methylating nucleotide base sequences vulnerable to the restriction enzymes. However, there is some small probability (e.g., 10^{-4}) that the DNA of an infecting phage is accidentally modified. If so, all of the progeny phage are also modified, and after their lytic release they can infect that bacterial strain with full efficiency (Luria, 1953). These progeny are not protected, however, against restriction enzymes borne by other bacterial strains that differ in their specificity.

Restriction and modification enzymes may be encoded by chromosomal, prophage, or plasmid genes. It seems reasonable to postulate that these enzymes evolved as a defense against phage infection (Levin and Lenski, 1985), although they may also function in site-specific repair and recombination (Meselson *et al.*, 1972). Whatever their original function, restriction blocks the vegetative growth of phage, thereby permitting survival of a bacterium. Not surprisingly, many phage possess mechanisms that permit them to avoid the restriction immunity of bacterial hosts (Kruger and Bickle, 1983). For example, DNA of the T-even coliphage contains the base hydromethylcytosine in place of the normal cytosine, making it resistant to many restriction enzymes. This unusual base is also usually glucosylated, which protects it against the action of other bacterial enzymes. Phage T3 and T7, in contrast, avoid restriction and modification by the early expression of gene products that actively interfere with the host enzymes.

A second class of immunity is that of lysogenic bacteria to reinfection by temperate phage of the same type as the prophage. This immunity is also related to the specific action of a protein, the repressor, which binds to sites on the phage genome and thereby prevents transcription of those genes whose products lead to the lytic destruction of the bacterium (Ptashne *et al.*, 1980).

A third class of immunity derives from a variety of extrachromosomal genetic elements that result in abortive infections by certain phage (Duckworth *et al.*, 1981). For example, the prophage Lambda interferes with the vegetative growth of certain genotypes of the T-even phage, while carriage of an F plasmid results in nonproductive infections by phage T7. In some cases, such immunities may be of adaptive value to the bacteria (and hence also to semiautonomous elements that impart these effects), whereas in other cases the abortive phage infections nevertheless result in death for the infected bacteria. For example, Stone *et al.* (1983) produced a recombinant plasmid containing a portion of the T7 genome that rendered infections by whole T7 viruses nonproductive. This effect apparently derived from the premature expression of a gene that resulted in the lysis of bacteria prior to encapsulation of the replicated phage genomes.

3. Mathematical Models of the Interaction

There are a number of references dealing with various aspects of population modeling. Levins (1966) presents an overview of the strengths and weaknesses of three strategies (generality, realism, precision) in model building. May (1974) provides an extensive mathematical treatment of key issues in species interactions and community structure. May and Anderson (1983) discuss general epidemiological and population genetic models of host–parasite coevolution, while Levin and Lenski (1983) informally treat the coevolution of bacteria and their parasites (including plasmids and temperate and virulent phage). Williams (1972, 1980) and Bull and Slater (1982) present more general discussions of the modeling of microbial populations and interactions. Kubitschek (1970) summarizes some basic chemostat theory.

3.1. Variations on Lotka and Volterra

The most familiar mathematical model used to describe the dynamic relationship between predators and their prey (or parasites and their hosts) is that derived independently by Lotka (1925) and Volterra (1926). According to this model, the dynamics of predator and prey can be described by just four parameters: (1) the rate of growth of the prey population, which is assumed to be constant; (2) the rate of loss in the prey population, which is assumed to be an increasing linear function of the density of the predator population; (3) the rate of growth of the predator population, which is assumed to be an increasing linear function of the density of the prey population; and (4) the rate of loss in the predator population, which is assumed to be constant.

Although it is often criticized for its unrealistic assumptions, the Lotka–Volterra model provides a useful starting point for the development of a more realistic model of the interaction between bacteria and virulent phage. Models developed by Campbell (1961) and Levin et al. (1977) differ from the Lotka–Volterra model in two important ways. First, they include some form of density limitation acting on the prey population in the absence of the predator. Second, they include a time lag between the act of predation and the resulting increase in the predator population. A simple model which incorporates these additional factors is presented below.

3.1.1. Dynamic Model of Bacteria and Virulent Phage

Consider an open habitat (like a chemostat) that contains a population of virulent phage, a population of sensitive bacteria, and a potentially limiting bacterial resource. The habitat is liquid and thoroughly mixed, such that phage, bacteria, and resources encounter one another at random. The resource has a concentration C_0 (μg/ml) as it flows into the habitat at a rate ω (turnovers/hr). Uninfected bacteria, infected bacteria, free phage, and unutilized resource are washed out of the habitat at this same rate.

Uninfected bacteria multiply via binary fission at a per capita rate that is a hyperbolic function of the resource concentration in the habitat. The maximum specific growth rate is ψ (hr^{-1}), and K (μg/ml) is the resource concentration at which the bacteria grow at half this maximum rate. Each replication of a bacterium uses up ϵ (μg) of the resource.

Phage encounter and irreversibly adsorb to uninfected bacteria at a per capita rate that is a linear function of the bacterial density. The adsorption constant δ (ml/hr) corresponds to the "search and attack" efficiency of the phage. Each infection is lethal to a bacterium, and each yields β phage progeny after a latent period of τ (hr).

The following differential equations relate the concentrations of resource (C), uninfected bacteria (S), infected bacteria (I), and free phage (P):

$$dC/dt = (C_0 - C)\omega - \epsilon S\psi C/(K + C) \tag{1}$$
$$dS/dt = S\psi C/(K + C) - \delta SP - \omega S \tag{2}$$
$$dI/dt = \delta SP - e^{-\tau\omega}\delta S'P' - \omega I \tag{3}$$
$$dP/dt = \beta e^{-\tau\omega}\delta S'P' - \delta SP - \omega P \tag{4}$$

S' and P' are the concentrations of uninfected bacteria and free phage, respectively, at time $t - \tau$, and $e^{-\tau\omega}$ is the fraction of bacteria infected at time $t - \tau$ that has not washed out of the habitat before lysing.

We can solve for the equilibrium population densities by setting these differential equations equal to zero and performing some relatively simple algebraic manipulations. For the sensitive bacteria (and for the resource), there are actually two distinct equilibria, corresponding to the presence and absence of the virulent phage population. That which obtains in the absence of phage can be termed the resource-limited equilibrium, and is indicated by $\tilde{S}$:

$$\tilde{S} = [C_0 - \omega K/(\psi - \omega)]/\epsilon \tag{5}$$

That which obtains in the presence of phage can be termed the phage-limited equilibrium, denoted by $\hat{S}$:

$$\hat{S} = \omega/[\delta(\beta e^{-\tau\omega} - 1)] \tag{6}$$

Consider the following set of biologically plausible parameters: $\psi = 0.7$ hr^{-1}, $K = 5$ μg/ml, $\epsilon = 2 \times 10^{-6}$ μg, $C_0 = 100$ μg/ml, $\omega = 0.2$ hr^{-1}, $\delta = 1 \times 10^{-7}$ ml/hr, $\beta = 100$, and $\tau = 0.5$ hr. The phage-limited equilibrium density of sensitive bacteria is 2.2×10^4 ml^{-1}, or more than three orders of magnitude below the resource-limited equilibrium of 4.9×10^7 ml^{-1}. Note that the phage-limited equilibrium density of bacteria is in fact independent of the resource concentration, provided only that C_0 is sufficient to support bacterial growth in excess of washout.

The equilibrium density of the phage $\hat{P}$ does, however, depend on the resource concentration:

$$\hat{P} = [\psi\hat{C}/(K + \hat{C}) - \omega]/\delta \tag{7a}$$

$\hat{C}$ is the equilibrium concentration of resources when the sensitive bacteria are phage-limited. If $\hat{S}$ is much less that $\tilde{S}$, then phage-limited bacteria are able to utilize only a small fraction of the incoming resources, and $\hat{C}$ is very nearly equal to C_0. If, in addition, C_0 is much greater than K, then the phage-limited bacterial population is effectively growing exponentially at the rate ψ, and Eq. (7a) simplifies:

$$\hat{P} \approx (\psi - \omega)/\delta \qquad (7b)$$

These conditions hold for the parameters given above, and yield an equilibrium density for the virulent phage of 5×10^6 ml^{-1}, or roughly two orders of magnitude greater than the equilibrium density for their sensitive bacterial hosts.

Consider the same set of parameters, except now let $\delta = 1 \times 10^{-11}$ ml/hr. This much reduced adsorption rate yields a phage-limited equilibrium density for bacteria of 2.2×10^8 ml^{-1}, which is greater than the resource-limited equilibrium. However, inserting the resource-limited bacterial density into Eq. (4) yields a negative growth rate for the phage population, indicating that these phage cannot become established even when their bacterial hosts are at maximum density. In such cases, Eqs. (6) and (7) are invalid, as there does not exist a phage-limited state.

The model specified by Eqs. (1)–(4) corresponds closely to the basic description of lytic infections presented in Section 2.1. Of course, it contains many assumptions, two of which are as follows: (1) all of the parameters remain constant through time; and (2) there is no variation within the populations. In the next two sections, we will consider the effects of violating these assumptions.

3.1.2. Stability and Complexity

The Lotka–Volterra model is neutrally stable. That is, both predator and prey populations exhibit oscillations of constant frequency and amplitude, with the predator lagging one-quarter phase behind the prey. The phage–bacteria model just presented differs structurally from the Lotka–Volterra model in two respects. First, the rate of growth of the bacteria is a function of a potentially limiting resource. Second, there is a time lag between phage infection and multiplication. The former tends to stabilize the interaction, whereas the latter is destabilizing. The net effect of these opposing forces depends on the specific parameters used in the equations; the model can exhibit stable equilibria, stable oscillations, or oscillations leading to extinction (Levin et al., 1977).

The relative magnitudes of the phage-limited and resource-limited equilibrium densities of bacteria are especially important. If the difference is slight, then the extent to which bacteria overshoot the phage-limited equilibrium is reduced as resource limitation becomes a significant factor, and the interaction will be stabilized (Fig. 2a). On the other hand, if the difference is great, then resource limitation has little effect, and the destabilization that results from the time lag will predominate (Fig. 2b).

It is also of interest to ask how variation in environmental parameters influences the interaction. For example, consider virulent phage and bacteria interacting in a "seasonal" habitat, one that is closed except at occasional intervals when resources are renewed (Stewart and Levin, 1973). In Fig. 2c, the dynamics of virulent phage and bacteria is modeled for the same set of parameters as in Fig. 2a, except that the community is cultured serially, not continuously. That is, instead of a flow rate of 0.2 hr^{-1}, there is a transfer of 1% of the community into a fresh culture every 24 hr. While the interaction in chemostat culture is quite stable, it is not at all stable in serial culture. The virulent phage drive the sensitive bacteria to extinction even before the first transfer; extinction of the phage

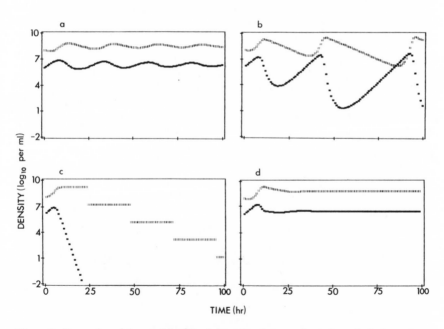

Figure 2. Dynamics of the model of the interaction between bacteria and virulent phage. Lighter symbols indicate free phage; darker symbols indicate uninfected bacteria. Parameter values are as in Section 3.1, except as noted. (a) $\delta = 1 \times 10^{-9}$ ml/hr, $C_0 = 25$ μg/ml. (b) $\delta = 1 \times 10^{-9}$ ml/hr, $C_0 = 100$ μg/ml. (c) As for (a), except in serial culture. (d) As for (b), except with nongenetic heterogeneity in bacterial vulnerability to phage infection.

results from subsequent daily dilutions. This instability occurs in serial transfer because continuous phage mortality due to washout is lacking.

I will present evidence later that many interactions between virulent phage and bacteria are more stable than anticipated from theory. One hypothesis that could account for this stability is the existence of refuges, either spatial or physiological, for sensitive bacteria. The simplest way of modeling a refuge is to allow two states for the bacteria, vulnerable and protected. The probabilities of a daughter cell being in one state or the other are *independent* of the state of the bacterium that is dividing; hence, the variation is not heritable. (The protected state is *not* equivalent to resistance, which is heritable. Resistance will be treated in Section 3.2.) In Fig. 2d, I have graphed the dynamics in chemostat culture using the same parameters as in Fig. 2b, except that each daughter cell now has one chance in ten of being protected from the phage.

It may be surprising that this variation, while greatly stabilizing the interaction, has little effect on the equilibrium densities of bacteria and phage. This can be understood, however, by recognizing that the pro-tected population is *not* self-sustaining. The maximum growth rate for the protected population is 0.7 hr^{-1}, but 90% of the bacteria that are pro-duced each generation are vulnerable to the phage. The remaining 10% are insufficient to offset washout at the rate of 0.2 hr^{-1}. Thus, the pro-tected population is maintained only as a "by-product" of the vulnerable population. The situation could be changed dramatically, however, if the probabilities that any given daughter cell is vulnerable or protected were reversed. No change is implied in the heritability (or rather lack thereof) for the trait; the probabilities are still independent of the state of the gen-erating bacteria. But with 90% of the bacteria produced each generation protected from the phage, the total bacterial population would rapidly approach resource limitation. Nonetheless, the virulent phage population could persist by exploiting the vulnerable population, which in this case would become a non-self-sustaining "by-product" of the protected population.

Smith (1972) provides an excellent discussion of the role of hetero-geneity in stabilizing populations, while Alexander (1981) reviews a vari-ety of factors that promote the coexistence of microbial parasites and hosts.

3.2. Evolutionary Considerations

The analysis of adaptive changes in the bacteria and phage popula-tions requires two distinct considerations. First, the model must indicate the rate at which mutants (or recombinants) appear in the community. Thus far, all processes in the model have been assumed to be deterministic. This is reasonable when rates are large relative to the inverse of population sizes, as is likely for most ecological processes. It may not be reasonable, however, for relatively rare events associated with genetic changes. Not only may the appearance of a new genotype be stochastic, but so may be its fate. That is, even if a new genotype has a net selective advantage, there is a certain probability that it will be lost (e.g., due to washout) before replicating. In this chapter, I will not specifically model the origin of new genotypes. Lenski and Levin (1985) discuss the stochastic appearance and fate of mutant phage-resistant bacteria and host-range phage. Levin (1981) models the appearance of recombinant genotypes in bacterial populations.

Second, the model must incorporate the subsequent growth of the new mutant (or recombinant) population and its interaction with other populations. In particular, it is of interest to ask whether the new population can become established, and if so, what are its equilibrium density and its effects on the equilibrium densities of the other populations.

We can easily write a differential equation for the dynamics of a resistant population, using subscript R to differentiate growth parameters from those for sensitive bacteria in Eq. (2). [It is assumed here that the resistance is absolute, although partial resistance can also be modeled (Levin *et al.*, 1977).] We have

$$dR/dt = R\psi_R C/(K_R + C) - \omega R \qquad (8)$$

Let us first consider whether a resistant population can invade a chemostat at phage-limited equilibrium. Recall that the concentration of resources in a phage-limited chemostat $\hat{C}$ is greater than the concentration in a resource-limited chemostat. If the growth parameters are identical for sensitive and resistant bacteria, then resistant bacteria can clearly invade as they have the same rate of replication, but fewer losses. In fact, they can increase even if they have a somewhat lower rate of growth, provided only that their rate of growth at resource concentration $\hat{C}$ exceeds the rate of flow through the habitat. In general, the greater the difference is between the phage-limited and resource-limited equilibrium densities of sensitive bacteria, the broader the conditions are for establishment of a resistant population. Resistant bacteria fail to invade only if they are so much less efficient at extracting resources and growing that they cannot offset losses due to washout.

If the resistant populations can invade, it increases until it becomes resource-limited, at which point its growth is exactly offset by losses to washout. If the growth parameters of the resistant bacteria are identical to those of the sensitive bacteria, then its resource-limited equilibrium is identical to that which occurs for the sensitive bacteria in the absence of phage. In this case, the sensitive population has no growth advantage, but sustains additional losses due to phage infection; it is driven to extinction as a consequence. With the extinction of the sensitive bacteria, the virulent phage population is also driven to extinction, unless host-range mutants arise that can infect the resistant bacteria.

On the other hand, if the resistant bacteria have a lower growth rate than the sensitive bacteria, then their resource-limited equilibrium is also lower. More important, the equilibrium concentration of resource that sustains the growth of the resistant population at a rate just sufficient to offset washout must allow a higher growth rate for the sensitive bacteria. Hence, sensitive bacteria can persist subsequent to the attainment of resource limitation by phage-resistant bacteria. In fact, the evolution of resistant bacteria has no effect at all on the equilibrium density of the phage-limited sensitive population; from Eq. (6), we see that this equilibrium is independent of resource concentration. In contrast, the equilibrium density of the virulent phage population is reduced by the evolution of a resistant bacterial population; from Eq. (7a), we see that this equilibrium is proportional to the growth rate of the sensitive population, and hence depends on the resource concentration. In effect, the resistant bacteria sap some (but not all) of the excess growth (relative to washout) of the sensitive bacteria, which supports the virulent phage.

Wild-type phage can thus persist subsequent to the evolution of resistant bacteria, provided that their sensitive hosts have a growth rate advantage when resources limit the resistant population. Nonetheless, there is strong selection for host-range phage mutants subsequent to the evolution of resistant bacteria; a resource-limited bacterial population provides a superabundance of potential hosts. I will not present an equation for the dynamics of a host-range phage population, as notation becomes further complicated and the equations for bacterial dynamics would have to be revised. However, the conditions for coexistence of wild-type and host-range phage populations are similar to those for sensitive and resistant bacteria. If sensitive bacteria have a growth rate advantage over resistant bacteria, *and if* host-range phage mutants exploit the sensitive bacteria less efficiently than wild-type phage, then the two phage genotypes and the two bacterial genotypes can coexist.

4. Laboratory Communities

From the mathematical model of the interaction between bacteria and virulent phage, two fundamental predictions should be emphasized. (1) Virulent phage can limit sensitive bacteria to a density below that set by resources. However, the stability of the interaction may depend on the availability of some refuge for sensitive bacteria, because in the absence of a refuge the interaction may produce oscillations leading to extinction. (2) Mutations conferring phage resistance in the bacterial population or extended host range in the phage population are generally favored. However, if phage-resistant bacteria are at a competitive disadvantage for resources, then sensitive bacteria can persist and thereby maintain virulent phage, even if no host-range phage mutants evolve.

4.1. Ecological Dynamics

It is possible to demonstrate an effect of virulent phage on bacteria in continuous culture by comparing bacterial densities either before and after the addition of virulent phage, or before and after the evolution of resistant bacteria. In principle, the former appraoch is preferable, since resistant bacteria may differ not only in their sensitivity to phage, but also in the efficiency with which they exploit their resources. In practice, however, the effects of virulent phage on bacterial densities in continuous culture are usually so great as to make this complication insignificant.

Data from a number of studies on the dynamics of bacteria and virulent phage in chemostat culture are summarized in Table I. The table includes "order-of-magnitude" approximations for (1) resource-limited equilibrium densities for bacteria $\tilde{S}$; (2) phage-limited equilibrium densities for bacteria $\hat{S}$; and (3) equilibrium densities for virulent phage prior to the evolution of resistant bacteria $\hat{P}$. With one exception, *these studies demonstrate a profound effect of virulent phage on the density of sensitive bacteria in chemostat culture.* Phage-limited densities were two to four orders of magnitude below resource-limited densities obtained in the

Table I. Summary of Chemostat Experiments on Virulent Phage and Bacteria[a]

Reference[b]	Phage	$\tilde{S}$	$\hat{S}$	$\hat{P}$
Paynter and Bungay (1969, Fig. 3)	T2	9	7	9
Horne (1970, Fig. 1A)	T3	8	4	6
Levin et al. (1977, Fig. 5A)	T2	8	4	6
Chao et al. (1977, Fig. 3)	T7	8	4	6
Lenski (1984a, Fig. 1)	T2	8	4	6
Lerner (1984, Fig. 1)	MS2	9	9	8
Lenski and Levin (1985, Fig. 1)	T4	8	4	6

[a]$\tilde{S}$ is the equilibrium density of resource-limited bacteria, $\hat{S}$ is the equilibrium density of phage-limited bacteria, and $\hat{P}$ is the equilibrium density of phage. All densities are expressed as $\log_{10}$ per ml.
[b]Approximate densities were obtained from the figure that is indicated for each reference. These values are necessarily rough because of the often pronounced variation even between successive samples. The various studies differed somewhat in culture conditions. Most notably, Paynter and Bungay (1969) used broth at 30°C, whereas all others used minimal media at 37°C. Rates of flow through the chemostats ranged from 0.04 to 0.3 turnover/hr.

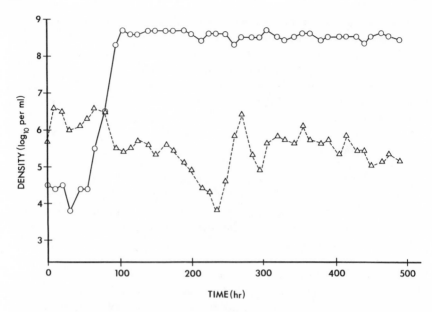

Figure 3. Dynamics of the interaction between *E. coli* B and virulent phage T4 in chemostat culture. Solid line gives bacterial density; dashed line gives phage density. [From Lenski and Levin (1985, Fig. 1), with permission of *The American Naturalist,* © 1985 by the University of Chicago.]

same medium either prior to the addition of virulent phage or subsequent to the evolution of resistant bacteria. This difference can be seen in Fig. 3.

In three of these studies, the authors obtained independent estimates of adsorption rate δ, burst size β, latent period τ, maximum bacterial growth rate ψ, and flow rate ω. By incorporating these parameters into Eqs. (6) and (7b), it is possible to estimate the equilibrium densities for phage-limited sensitive bacteria and for virulent phage. The parameter estimates and corresponding predicted equilibria are given in Table II. Equilibria predicted by Levin *et al.* (1977) working with *E. coli* B and phage T2, and by Lenski and Levin (1985) using B and T4, are within an order of magnitude of those observed (Table I). This degree of accuracy lends support to the general validity of the model, especially since the range of plausible equilibrium densities covers many orders of magnitude.

Very different results were observed by Lerner (1984) for the male-specific phage MS2 and a plasmid-bearing strain of *E. coli* K12 that is derepressed (i.e., fully expressed) for conjugative pili synthesis. The parameters estimated for this interaction yield a predicted phage-limited equilibrium of less than $10^7 \, \mathrm{ml}^{-1}$ (Table II), whereas the observed density of bacteria was about 10^9 ml^{-2}, indistinguishable from resource limitation (Table I). Moreover, the observed phage density was about three orders of magnitude below the anticipated equilibrium. These results indicate that this phage is much less effective at exploiting the bacteria than predicted from the model. But it is clear that phage were "eating" something in the chemostats; they persisted despite washout, and mutants resistant to the phage increased in frequency in the bacterial population.

It is likely that only a fraction of the derepressed bacteria were actually vulnerable to MS2 at any given time, a scenario that would be consistent with the very low adsorption rate. Although this hypothesis was not explicitly tested, it is supported by one difference between male-specific and most other phage. Whereas receptor sites for most phage are present at high multiplicities per bacterium (e.g., Schwartz, 1976), only one or a few conjugative pili are present on an individual cell. In fact, Brinton and Beer (1967) report that typically 50–90% of the bacteria in a fully expressed culture are actually free of any pili; adsorption by male-specific phage is necessarily restricted to those individuals that happen to have pili. These considerations, in conjunction with Lerner's (1984) chemostat observations, suggest a very high degree of nongenetic heterogeneity among individual bacteria in their vulnerability to male-specific phage. As discussed in Section 3.1.2, this variation could provide a refuge, and thereby cause a discrepancy between observed densities and those predicted by a model in which it is implicitly assumed that all individuals are identical.

According to the mathematical model, the stabilizing effect of bacterial resources on the interaction between sensitive bacteria and virulent phage is apparent only when bacteria are near their resource-limited equi-

Table II. Estimates of Parameters and Predicted Equilibrium Densities for Virulent Phage and Sensitive Bacteria[a]

Estimate	Levin et al. (1977) T2	Lerner (1984) MS2	Lenski and Levin (1985) T4
ψ (hr^{-1})	0.7	0.9	0.7
ω (hr^{-1})	0.1	0.2	0.3
δ (ml/hr)	6×10^{-8}	1×10^{-11}	3×10^{-7}
β	100	10,000	80
τ (hr)	0.5	0.8	0.6
$\hat{S}$	4.2	6.4	4.2
$\hat{P}$	7.0	10.9	6.1

[a]Predicted equilibria may be compared with observed densities presented in Table I. Symbols are defined in the text. Densities are expressed as log_{10} per ml.

librium. But the summary presented in Table I indicates that phage-limited bacterial densities are generally orders of magnitude below resource limitation, suggesting that the dynamics of the interaction could often be unstable. A perusal of the figures cited in Table I indicates that the interactions between virulent phage and sensitive bacteria are less stable than the interactions between bacteria and their resources. Rather than try to quantify this claim, I will let Fig. 3 be illustrative of this general tendency. Note in particular how much more variable are the phage densities than the bacterial densities, especially after the bacteria have attained resource limitation by evolving resistance.

Despite the apparent instability of the interaction between bacteria and virulent phage, reports of extinctions are few. Levin *et al.* (1977) observed that the model predicted oscillations of increasing amplitude leading to eventual extinction when the parameters were in the range estimated for *E. coli* B and virulent phage T2. In fact, oscillations tended to become less, not more, pronounced over the course of their experiments, and no extinctions were observed. Horne (1970) reports the coexistence of phage T3 and *E. coli* B for 80 weeks, and the coexistence of T4 and B for 52 weeks. Of the studies cited in Table I, only Lerner (1984) and Lenski and Levin (1985) report extinctions. Lerner (1984) observed extinctions of MS2 on plasmid-bearing strains of *E. coli* K12 that were *repressed* for conjugative pili synthesis. These extinctions, however, are perhaps better described as failures of the phage to become established. Phage inoculated into chemostats containing the repressed bacteria generally disappeared at a rate equal to the rate of washout. Lenski and Levin (1985) reported three extinctions, of both bacteria and phage, in ten replicate chemostats containing *E. coli* B and T4. Lenski and Levin (1985) also observed extinctions of virulent phage T5 in chemostats with *E. coli* B. The T5 extinctions, however, were subsequent to the evolution of resistant bacteria, and will be discussed later.

There is no continuous source of mortality to a phage population in serial culture comparable to washout in chemostat culture. Hence, the possibilities of stable coexistence would seem more limited, as virulent phage may reach high densities for a sufficient length of time to encounter and destroy all sensitive bacteria, as shown in Fig. 2c. It is certainly the conventional wisdom that in preparing phage lysates, sensitive bacteria are driven extinct as the culture is cleared, even though the culture may eventually become turbid with resistant bacteria. R. E. Lenski, B. R. Levin, and R. V. Evans (unpublished data) have examined the dynamics of virulent phage and bacteria in two serial transfer habitats. One of the habitats consisted of 10 ml of liquid broth, continuously agitated, from which 0.1 ml (or 1%) was removed daily and placed into 9.9 ml of fresh broth. The other habitat consisted of phage-infected colonies on agar plates, small fractions (approximately 3%) of which were removed each day and transferred to fresh agar plates. Densities of virulent phage and bacteria in liquid culture were assayed by standard dilution and plating techniques. In surface culture, the persistence of bacteria was documented simply by noting the formation of a visible colony subsequent to transfer; the persistence of phage was documented by transferring a portion of the colony to a lawn of sensitive bacteria and observing lysis if phage were present. In each habitat, we examined the interactions of an *E. coli* K12 host and six different virulent phage. For each phage, we ran three replicates in the liquid habitat and 50 replicates in the surface habitat. (The surface replicates were extremely simple, as all 50 communities could be maintained on one plate.)

Results in the surface habitat are best summarized by noting that the six virulent phage used in these experiments fell into two groups producing qualitatively different outcomes. Populations of phage T5 and T7 went extinct at very high rates, more than 90% of T5 and almost 40% of T7 after ten transfers. In contrast, *none* of the populations of T2, T4, T6, or a virulent mutant of Lambda were extinct after 12–20 transfers. The latter phage did not even approach extinction, but maintained high densities from transfer to transfer, indicating the persistence of bacteria that they could exploit. It is probably significant that these two groups also differ strikingly in the appearance of plaques and of infected colonies. Both T5 and T7 produce very large plaques, and infected colonies are translucent. In contrast, the other four phage produce small plaques, and infected colonies remain opaque but have "nibbled" outer edges. Thus, it appears that persistence of phage T2, T4, and T6 and virulent Lambda was due, at least in part, to the existence of spatial refuges for sensitive bacteria in the physically structured surface habitat.

In contrast, there would seem to be no opportunity for spatial refuges in liquid serial culture; vessels were replaced daily and only liquid was transferred. Yet for many of the phage, stable communities were maintained even in this habitat. In fact, only phage T5 behaved according to expectation. In all three replicates, T5 rapidly attained high densities; with subsequent dilutions, phage went extinct, although resistant bacteria flourished. A similar extinction was observed for one virulent Lambda population, but in two other replicates, phage persisted at high densities for 25 transfers, as seen in Fig. 4. Populations of phage T2, T4, T6. ¬nd T7 were also generally stable in liquid serial culture.

Although spatial refuges are unlikely to be important in liquid serial culture, several other factors could stabilize these interactions. One possibility is that some genetically sensitive cells are physiologically insensitive. This possibility was already raised in conjunction with the surprising results obtained in continuous culture with the male-specific

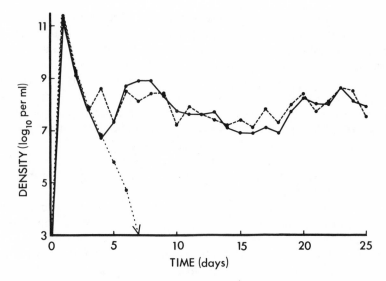

Figure 4. Dynamics of the interaction between *E. coli* K12 and a virulent mutant of phage Lambda in liquid serial culture. Separate lines correspond to three replicate experiments. Only phage population densities are shown; these were obtained just prior to daily transfers [R. E. Lenski, P. R. Levin, and R. V. Evans (unpublished data)].

phage MS2. Physiological protection against phage adsorption, though less dramatic, has also been demonstrated for other phage. Recall, for example, that phage Lambda adsorbs to receptors involved in the uptake of maltose, and that there can be considerable variation among bacteria in the number of these receptors, especially when bacteria are grown on other carbon sources. Physiological refuges may also arise as the consequence of starvation of bacteria (Delbrück, 1940a), depletion of a factor in the medium required for phage adsorption (Luria and Steiner, 1954), or bacterial clumping (Paynter and Bungay, 1970). Physiological refuges may even arise as a consequence of the phage–bacteria interaction; endolysin released by T7-infected bacteria can strip T7 receptors from the surface of surviving sensitive bacteria (Li *et al.*, 1961).

A second possibility is that genetically resistant bacteria either renew or protect the genetically sensitive population. Renewal could come about through back-mutation, although resistance may often arise by deletion and other irreversible events. Protection could occur if phage bind reversibly to (but do not infect) resistant bacteria. This would be equivalent to "hiding" a few sensitive bacteria among the many resistant cells. A test of this possibility was carried out by Stent and Wollman (1972) in their original demonstration of the two-step nature of phage adsorption (see Section 2.1.2). They found no effect of a high density of resistant bacteria on the rate of adsorption of phage T4 to sensitive cells.

A final possibility is that phage persist by continuously evolving host-range mutants (Rodin and Ratner, 1983). However, the experiments presented above do not support this hypothesis. For example, although bacteria resistant to Lambda evolved, no host-range mutants arose. [Host-range mutants of Lambda can be found that infect certain classes of resistant bacteria, but not all (Hofnung *et al.*, 1976).] In Section 4.2, I discuss more generally the limits to bacterial resistance and phage host range.

4.2. Evolutionary Change

A mutation that confers resistance to a virulent phage can increase the density of bacteria in chemostat culture by several orders of magnitude, as seen in the comparison of phage-limited and resource-limited bacterial densities (Table I). Accompanying this dramatic increase in bacterial density is a dramatic decrease in the availability of unused resource in the habitat. For example, Chao *et al.* (1977) found that the concentration of glucose in a phage-limited chemostat was indistinguishable from the 100 μg/ml concentration in the reservoir, while the concentration of free glucose in a resource-limited chemostat was only about 5 μg/ml. By the same token, a host-range mutation in a virulent phage population can cause a resource-limited chemostat to become phage-limited.

In all of the studies cited in Table I, resistant bacterial mutants were observed in at least some of the longer running experiments. It may take only tens of hours for *E. coli* B that are resistant to phage T4 to arise (Horne, 1970) or several hundred hours for the appearance of B mutants resistant to T2 (Levin *et al.*, 1977). The time required for this evolution depends on several investigator-controlled variables, as well as certain innate characteristics of the particular interaction. For example, Horne (1970) reports resource-limited communities of T4-resistant bacteria after only about 20–40 hr, whereas Lenski and Levin (1985) did not observe this state until about 80–140 hr (see Fig. 3). Two differences in experimental setup probably account for this difference. First, Horne (1970) began his experiments by adding the phage to a resource-limited population of sensitive bacteria, whereas Lenski and Levin (1985) began their experiments with bacteria at a density near their phage-limited equilibrium, several orders of magnitude lower. In the former case, resistant mutants probably were present at the start of the experiment (quite possibly tens or hundreds/ml), whereas they had to arise *de novo* in the latter case. Second, the flow rate in Horne's (1970) chemostats was only about

0.04 hr^{-1}, whereas it was about 0.3 hr^{-1} in the chemostats of Lenski and Levin (1985). A resistant mutant experiencing mortality due only to washout (and not to phage infection) requires only 1.06 hr to double at the lower flow rate, but 1.76 hr at the higher rate, assuming an intrinsic doubling time of 1 hr.

The time required for the evolution of resistant bacteria also depends, of course, on biological features of the interacting bacteria and phage. Resistance to most phage arises via a single genetic event; for example, the mutation conferring resistance to phage T4 by *E. coli* B occurs at a frequency of about 10^{-7} per cell generation. Lenski and Levin (1985) not only found reasonable agreement between observed and predicted equilibrium densities for B and T4, but also between observed and predicted times to the attainment of a phage-resistant, resource-limited bacterial population.

Resistance to virulent phage T2 is much more difficult to obtain. However, it was noted by Lenski (1984a) that T2 resistance by *E. coli* B evolves in chemostats more rapidly than anticipated with the rate of mutation estimated by the fluctuation test (see Section 2.2.1). This discrepancy occurs because only selection in chemostats can enrich mutants that are just partially resistant to phage. Lenski (1984a) demonstrated that resistance to T2 can evolve not only by a single rare mutation, but also by a pair of common mutations, one of which is the same mutation conferring resistance to phage T4. These T4-resistant intermediates adsorb T2 at only about half the rate of T4-sensitive bacteria, and mutate to complete T2 resistance at a rate about two orders of magnitude higher.

Although resistant bacteria appeared in at least some of the longer running experiments in all of these studies, this was not the case for host-range phage mutants. Host-range mutants were observed by Chao *et al.* (1977) for phage T7 and by Lenski and Levin (1985) for phage T2 and T7, but Lenski and Levin (1985) failed to detect host-range mutants of T4 or T5. Host-range phage mutants probably also appeared in the study by Horne (1970) with T3, judging by the time required for resistant bacteria to reach resource-limited density, but they were not specifically mentioned. Where host-range phage mutants evolved, higher order resistant bacteria also evolved, and *the most highly evolved communities were dominated by resource-limited bacterial genotypes that were resistant to all co-occurrring phage genotypes* (Chao *et al.*, 1977; Lenski and Levin, 1985). Despite this, only phage T5 was observed to go extinct with the evolution of resistant bacteria (Lenski and Levin, 1985). This general result is again illustrated in Fig. 3. Note that although the bacteria evolved resistance and there were no corresponding host-range mutations, the phage persisted indefinitely.

In fact, virulent phage may persist subsequent to the evolution of resistant bacteria, without extending their host range, by continuing to exploit sensitive bacteria that themselves persist due to superiority in competition for resources. However, these basic chemostat experiments do not exclude the possibility that phage are actually capable of growing on the "final" resistant bacterial genotype, but with such meager infectiousness that plaques are not observed. Verification of the former hypothesis thus requires an independent demonstration of one or more of the following: (1) in the presence of phage, sensitive bacteria persist subsequent to the attainment of the resource-limited equilibrium by resistant bacteria; (2) in the absence of phage, sensitive bacteria have a growth rate advantage over resistant bacteria at the resource-limited equilibrium; and (3) no phage genotype has the capacity to reproduce when only resistant bacteria are present.

Although the first of these tests is the most direct, it is also the most difficult. This is because sensitive bacteria are anticipated to be a small minority that cannot be directly selected. Recall from theoretical considerations presented in Section 3 that the phage-limited equilibrium density of sensitive bacteria is independent of the concentration of primary resource, and consequently is not affected by the presence of competitively inferior, but resource-limited, resistant bacteria. This phage-limited equilibrium is typically orders of magnitude lower than the equilibrium set by resources (Table I). And while it is possible to detect a minority of resistant cells by plating in the presence of phage, there is no analogous procedure that allows detection of a sensitive minority. Thus, explicitly demonstrating the persistence of sensitive cells subsequent to the evolution of the resistant population is like looking for a "needle in a haystack." A much cleverer approach was used by Chao *et al.* (1977). They initiated chemostat cultures with virulent phage T7 and both sensitive and resistant *E. coli* B strains, the sensitive carrying a selectable fermentation marker. A minority population of sensitive bacteria was thereby directly shown to persist even in the presence of phage.

Chao *et al.* (1977) also allowed the sensitive and resistant strains to compete in the absence of phage T7. As hypothesized, the sensitive bacteria prevailed, a result shown to be independent of the fermentation marker. Lenski and Levin (1985) found T2-, T4-, and T7-resistant strains to be at a competitive disadvantage with regard to the unselected *E. coli* B strain. The T4-resistant strain declined at a rate of 0.16 hr^{-1} in a chemostat with a flow rate of 0.3 hr^{-1}, indicating a selective disadvantage under these conditions of about 50%. Paynter and Bungay (1969) did not explicitly allow sensitive and resistant bacteria to compete in their chemostats, although they did note that many of the T2-resistant mutants had lower exponential growth rates and longer lag times when introduced into fresh medium.

In contrast, Lenski and Levin (1985) found no measurable competitive disadvantage associated with resistance to phage T5. In fact, T5
resistance has been widely used as a selectively neutral marker in other
studies on microbial competition and evolution (Dykhuizen and Hartl,
1983). Recall that only phage T5 was observed to go extinct subsequent
to the evolution of resistant bacteria, an outcome entirely consistent with
the failure to observe any competitive disadvantage.

Lenski and Levin (1985) also demonstrated that none of the phage
T2, T4, T5, and T7 could be supported by corresponding "final" resistant
bacterial populations. Furthermore, the maximum rate of host-range
mutation consistent with the failure to observe phage multiplication
could be back-calculated from the initial size of the phage population.
Using this approach, Lenski and Levin (1985) concluded that further
host-range extensions, if they exist at all, must occur at rates on the order
of 10^{-11} per virus replication or less. Thus, these experiments lend further
support to the notion of a fundamental asymmetry in the coevolutionary
relationship between bacteria and virulent phage. There exist bacterial
mutations conferring resistance to phage for which there are no (or only
extremely rare) corresponding host-range mutations. Virulent phage may
persist, however, because bacterial resistance often engenders a reduction
in competitive ability.

Whereas these studies indicate that a tradeoff between phage resistance and competitive ability occurs frequently in bacteria, it is not
known how this tradeoff affects the subsequent course of bacterial evolution. In the absence of phage, will resistant bacteria revert to sensitivity? This seems unlikely to be the general case, since many spontaneous
mutations conferring resistance could be the result of deletions and other
irreversible events. If not, will the resistant bacteria remain competitively
inferior through evolutionary time, or will they find alternative solutions
to metabolic limitations?

Just as phage-resistant mutants are often at a competitive disadvantage with regard to wild-type bacteria, so may host-range mutants be at a
disadvantage compared to wild-type phage when competing for sensitive
hosts. Chao et al. (1977) allowed wild-type and host-range phage T7
mutants to compete, and found that the wild-type phage did, in fact, have
an advantage in culture with sensitive hosts. Subsequent to the appearance of resistant bacteria, the host-range phage mutants increased, as
expected. Lenski and Levin (1985) found that a T7 host-range mutant
and one of two T2 host-range mutants survived more poorly than their
wild-type counterparts in the presence of bacteria resistant to both. These
disadvantages may result from the reduced selectivity of host-range
mutants (see Section 2.2.2), which increases the likelihood of errors in
adsorption and penetration.

One consequence of such tradeoffs, both in bacteria and phage, is to increase the genetic diversity that can be maintained in these communities, as discussed in Section 3.2. Chao *et al.* (1977) found that at least three bacterial genotypes (wild-type, first-order resistant, and second-order resistant) and two phage genotypes (wild-type and host-range mutants) could coexist stably in a chemostat with glucose as the limiting nutrient.

Malmberg (1977) investigated evolutionary changes in the virulent phage T4 due not just to mutation, but to recombination as well. Malmberg (1977) was able to manipulate the frequency of recombination in the phage population by controlling the multiplicity of phage infection; recall that recombination occurs when multiple phage infect a single bacterium. The phage population size was kept constant over all levels of recombination, and phage were restricted to a single round of adsorption and lytic replication per experimental generation. No evolution took place in the bacteria, because the same strain was cultured anew for each phage generation. The fitness of phage in any experimental treatment was defined as the number of progeny per parental phage per generation. The degree to which epistasis contributed to any increase in fitness was determined by crossing an evolved phage line with a phage line bearing a number of markers and examining the contribution of various regions of the phage genome to the observed change in fitness. Malmberg (1977) found that increases in the fitness of the phage were more rapid at the higher frequency of recombination, and that epistatic gene interactions were more important at the lower frequency of recombination. Malmberg (1977) also partitioned the fitness changes into several components of the phage life cycle. Burst size contributed much more than adsorption rate, a result not surprising given the imposition of a life cycle consisting of discrete (not overlapping) generations. Because bacteria did not coevolve in this experiment, the role of phage recombination in host-range shifts was not investigated.

5. Natural Communities

Can the ecological and evolutionary generalizations obtained from laboratory communities be extended to nature? Such extrapolation is difficult because natural communities contain so many additional complexities that have not been considered in the laboratory. Nonetheless, I will attempt to pursue one generalization that gives rise to reasonably specific predictions. That is, the apparent fundamental asymmetry in the coevolutionary potential of bacteria and virulent phage implies that most natural communities of bacteria (or at least coliform bacteria, on which the laboratory studies have focused) should consist of dominant clones resistant to all co-occurring phage. Virulent phage may be maintained by the existence of minority populations of sensitive bacteria that are more efficient in extracting resources than resistant bacteria. But if a virulent phage whose host range includes a dominant bacterial clone evolves or invades the community, then that clone will either evolve resistance or be replaced by another clone already resistant to that phage. Thus, resource-limited communities predominate, whereas communities in which the dominant bacterial clones are sensitive to co-occurring phage are transient aberrations.

5.1. Patterns of Abundance

An ideal data set for examining the role of phage in limiting bacteria in a natural community would include: (1) the host range of all phage determined for all bacteria; (2) all phage classified as temperate or virulent; (3) all bacteria insensitive to any phage classified as envelope resistance, restriction immunity, or the like; and (4) abundances of all bacteria and phage enumerated *in situ* over several sampling dates, with any changes in phage host range or bacterial sensitivity determined to be either an invasion of a new population or a genetic change in an existing population. I know of no study that contains anywhere near all-of this information, but there are bits and pieces that provide some insights into the structure of certain phage–bacteria communities outside the laboratory.

Scarpino (1978) has reviewed studies in which densities of coliform bacteria and coliphage were estimated in sewage. Ratios of coliphage to coliforms are generally much less than 1. This is similar to most laboratory communities in which the dominant bacterial clone is resistant to co-occurring virulent phage; it is quite different from most laboratory communities in which the dominant bacterial clone is sensitive to co-occurring virulent phage (Table I). While this is not proof, it does suggest that the dominant coliform populations in sewage communities are not phage-limited, consistent with the prediction based on the coevolutionary asymmetry.

There is a striking difference in the relative abundance of temperate and virulent phage in sewage and feces. Whereas virulent forms predominate in sewage, temperate forms predominate in feces (Table III). There are two hypotheses that could account for the difference in phage composition between these environments. According to one, sewer and gut communities differ not in the dynamics *per se* of bacteria and phage populations, but rather in the rate at which these communities are invaded by new bacteria and phage populations. A gut is a relatively closed hab-

Table III. Comparison of the Relative Abundance of Virulent and Temperate Phage in Sewage and Feces[a]

	Number of isolates	Percent temperate
Feces		
Furuse et al. (1983)		
Healthy humans	150	90
Human patients	101	77
Dhillon et al. (1976)		
Humans	(17)[b]	(82)[b]
Nonhumans	(29)[b]	(45)[b]
E. S. Anderson (1957)		
Humans	6	83
Sewage		
Dhillon et al. (1970)	77	1
E. S. Anderson (1957)	6	0

[a]Number of isolates refers to the number of single-plaque isolates characterized as virulent or temperate, and percent temperate is the percent of these isolates found to be temperate. E. S. Anderson (1957) isolated phage on *Salmonella;* all others used *E. coli.*
[b]Data not specified by single-plaque isolates; values indicate percent of samples in which phage were detected that included some or all temperate phage.

itat, at least by comparison to a sewer. Invasion by virulent phage may drive sensitive bacteria extinct, thereby causing their own subsequent extinction. Invasion by temperate phage may result in the formation of lysogens, which are immune to reinfection, ensuring the persistence of both bacteria and phage. In sewers, there would be a continual supply of invading bacteria and phage, with "blooms" in virulent phage density commonplace, and extinctions obscured by the high rate of turnover. In the gut, by contrast, virulent phage blooms would be much less frequent, and extinctions more apparent. Sewage would thus contain a higher proportion of virulent phage than feces. While this is consistent with the hypothesis that "Lysogeny provides a solution to the problem of exhaustion of the host supply" (Echols, 1972), it assumes that virulent phage usually drive their sensitive bacterial hosts extinct. But we saw in Section 4.1 that virulent phage and bacteria usually coexist in chemostat and even serial culture. If extinctions are atypical in these simple, relatively homogeneous habitats, it seems difficult to support a hypothesis that relies on high rates of extinction in much more complex and heterogeneous habitats such as the gut.

Alternatively, the contrast in phage composition between sewage and feces could reflect more fundamental differences in phage infectivity. Perhaps the intestinal environment is simply less able to support lytic replication of phage. This might be related to the biochemistry of the medium, to the nutritional status of the bacterial hosts, or to the structure of the physical environment. Phage rely on diffusion by random Brownian motion to encounter bacteria, and diffusion is limited by high viscosity. Any colloidal matter that bound phage would further limit adsorption. Roper and Marshall (1974) have demonstrated that sensitive bacteria are protected from phage in the presence of organic sediment, because phage and/or bacteria are adsorbed to the sediment and prevented from direct contact with one another. Thus, it is possible that many phage simply cannot be supported by lytic replication in the gut because the rate at which they encounter suitable hosts is too low. Whatever the explanation, it is clear that sewers and guts are very different in the phage–bacteria interactions that they support, and that more study of the causes of these differences is required.

Orpin and Munn (1974) observed the invasion of a bacteriophage active against a large bacterium (designated EO2) that is characteristic of the rumen of sheep. The density of EO2 in the rumen of a sheep dropped from about 10^7 ml^{-1} to zero within 10 days of the phage invasion, whereas no such decline occurred in a sheep whose rumen was not invaded by this phage. About 1 month subsequent to the disappearance of EO2, it reappeared, although it did not attain the high density previously observed. The authors did not detect any significant replacement of the EO2 population by competitors. It may be significant, however, that the sheep used in this experiment had been "cleansed" of their normal rumen biota and then reinoculated with defined bacterial populations. In particular, a closely related bacterium, EO1, apparently resistant to the phage, was not present in the experimental animals.

Emslie-Smith (1961) identified a clone of *E. coli* that was dominant in the feces of a human subject for 8 months. A mixture of phage active against this clone was administered orally, resulting in the disappearance of this clone and its replacement by another clone, presumably resistant to the phage, that remained dominant for some time. There is no mention of whether any of the phage were able to replicate and establish self-sustaining populations in the gut.

Do there exist in nature resistant bacterial mutants for which there are no corresponding host-range phage mutants? Perhaps the best data on sensitivity and host-range comes from the work of Reanney and co-workers on *Bacillus* and soil phage isolated from the same local environments (Reanney, 1976; Tan and Reanney, 1976). Many of the phage types had broad host ranges, and most of the bacterial strains were sensitive to one or more co-occurring phage types. However, Tan and Reanney (1976) indicate that a large fraction of the bacteria belonged to one strain that was resistant to all of the phage types that were characterized from the same local environment. Moreover, free phage were difficult to isolate without enrichment, suggesting that many of the phage were temperate (or pseudolysogenic).

5.2. Phage Therapy Revisited

Whereas bacteriophage have been tremendously important in basic scientific research, their practical significance is relatively minor and often indirect. Phage may be useful, as agents for determining the clonal identity of bacterial pathogens (Milch, 1978) and as indicators of contamination (Scarpino, 1978). Phage may be harmful, as vectors for genes that confer antibiotic resistance or toxin production to pathogenic bacteria (Williams Smith, 1972; Freeman, 1951) and as contaminants of processes that rely on bacterial metabolism, such as cheese-making (Whitehead, 1953). But there was a time when phage were perceived as a "magic bullet" against bacterial disease.

Bacteriophage were discovered in 1915 by Twort and in 1917 by d'Herelle (Duckworth, 1976). Almost immediately, they were seized upon as means for controlling bacteria, stimulated in part by d'Herelle's (1922) claim that bacteriophage form the natural basis of immunity to infectious disease. According to d'Herelle (Stent, 1963), "Pathogenesis and pathology of dysentery are dominated by two opposing factors: the dysentery bacillus as pathogenic agent and the filtrable bacteriophage as agent of immunity." Within a few years, hundreds of papers appeared in medical journals from around the world concerning phage and their medical significance (Peitzman, 1969); phage were isolated that could infect the bacteria responsible for cholera, diphtheria, gonorrhea, plague, and other dreaded diseases (Stent, 1963). In 1925, Sinclair Lewis published a popular novel, *Arrowsmith,* in which a young doctor by that name undertakes the implementation of phage therapy. By the 1930s, at least three major pharmaceutical companies were marketing phage preparations in the United States, as was an international firm founded by d'Herelle (Peitzman, 1969). Despite this promising start, the outlook began to change dramatically. Perhaps most importantly, there was little convincing evidence that phage therapy produced any demonstrable benefit, as determined by reviews commissioned by the American Medical Association (Peitzman, 1969). [The use of phage in the control of bacteria that are agricultural pests also appears to have been largely unsuccessful (Vidaver, 1976).] At best, benefits were inconsistent, and scientific claims lacked proper controls. Peitzman (1969) concludes that "sulfonamides and antibiotics served as the final forces in the demise of bacteriophage therapy."

Although poor science and the advent of antibiotics certainly contributed to the historical failure of phage therapy, they do not provide a biological explanation for this failure. In fact, phage would seem to have two advantages over antibiotics for the control of bacterial populations. First, phage are self-replicating entities, hence making their continual application unnecessary. Second, phage are evolving entities, hence counteracting the evolution of resistance by bacteria. However, a number of other factors could offset these advantages.

We have seen that temperate phage are much more common than virulent phage in feces (but not sewage), apparently because either the survival or the adsorption of free phage is inhibited *in vivo.* Although the factors responsible for this inhibition are not clear, they could have hindered certain therapeutic applications of phage. More generally, it is appropriate to recognize a fundamental difference between chemical and biological methods of control. Whereas both chemical and biological agents must adversely affect the target population, only biological agents must be adapted to the environment where they are to be released. It is not enough simply to find a "natural enemy" of the target organism. This point is well-illustrated by efforts to introduce parasites and predators of insect pests. DeBach (1971) reports that only about one-quarter of introduced parasites and predators have become established, and of these only about one-sixth (or 4% of the total) have exerted significant control over the target population. [Despite these low percentages, the benefit-to-cost ratios of biological control programs have been substantially higher than for chemical pesticides (DeBach, 1974).] We may similarly expect that only a small fraction of phage that can infect a bacterium *in vitro* would exert significant control *in vivo.*

We saw in Sections 2.2 and 4.2 that the coevolutionary potential of phage appears often to be less than that of their bacterial hosts; that is, there exist bacterial mutants resistant to phage for which there do not exist corresponding host-range phage mutants. This asymmetry could also have limited the efficacy of phage therapy. Furthermore, it is important to realize that most of the efforts at phage therapy were conducted *before* Lwoff (1953) had clearly established the phenomenon of lysogeny. Thus, it is quite possible that many of these efforts utilized temperate phage, and thereby generated lysogenic bacteria immune to reinfection.

In spite of the past history of failure, there has been some resurgence of interest in phage therapy in the last few years. This is due, at least in part, to the increasing prevalence of bacteria resistant to antibiotics, which has resulted primarily from the spread of plasmids (E. S. Anderson, 1968; Falkow, 1975). Bacteria responsible for nosocomial (hospital-acquired) infections are especially likely to be resistant to antibiotic therapy, and they plague patients whose immune systems are depressed; Shera (1970), for example, has explored the use of phage in combatting infections in burn patients.

Whether these newer (and more limited) applications can be successful remains to be seen. Success will almost certainly require the careful screening of possible phage agents, and not simply the isolation of any phage that happens to infect the target bacterium. (Advances in genetic engineering raise the possibility that appropriate phage might be constructed, rather than isolated *per se*.) In contrast to choosing an antibiotic, where a broad spectrum of sensitive bacteria is sought, the choice of an appropriate phage is likely to be specific to the target bacterium. One especially promising step in this direction has been taken by Williams Smith and Huggins (1982), who used phage successfully to treat experimental intramuscular and intracerebral infections in mice caused by a pathogenic strain of *E. coli*. They precluded the rise of resistant bacterial mutants by their very clever choice of phage. Pathogenicity of their target bacterium had been demonstrated previously to depend on the presence of the surface antigen K1 (Williams Smith and Huggins, 1980). Phage used to treat the infections were chosen because they adsorbed specifically to this antigen; bacterial mutants resistant to the phage lost the antigen, and hence their pathogenicity. Other phage that did not adsorb to this antigen were much less effective in treating these infections. In essence, Williams Smith and Huggins (1982) chose the phage such that there was a significant tradeoff between the resistance of the bacteria to the phage and the pathogenicity of the bacteria.

6. Summary

Bacteria and phage provide a powerful system for examining the dynamics of interacting populations, due in part to the strong conceptual and methodological framework provided by research in microbial genetics and molecular biology. The existence of phage-resistant bacterial mutants and extended host-range phage mutants is compelling evidence for their reciprocal adaptation, whereas the genetic and molecular characterizations of these mutants indicate significant constraints on their coevolution. In particular, there may exist phage-resistant bacteria for which there are not corresponding host-range phage mutants; and phage-resistant bacteria may have more limited metabolic capabilities than phage-sensitive bacteria.

The basic question that I have attempted to address is: How effective are phage infections in limiting the abundance of bacteria? I have focused on virulent phage because lytic infections are simpler and more adverse to bacteria than are lysogenic infections, which characterize temperate phage. However, the existence of temperate phage raises a number of other interesting questions that I have not addressed, concerning their own evolution and their role in bacterial evolution.

It is possible to modify the familiar Lotka–Volterra equations to provide a more realistic model of the interaction between bacteria and virulent phage. In particular, it is critical that the relation between bacteria and their own resources be incorporated. According to the model, virulent phage can coexist with sensitive bacteria in continuous culture, and may hold the density of sensitive bacteria well below that allowed by their resources. The model also predicts that phage-sensitive and phage-resistant bacteria can coexist in the presence of virulent phage, provided that the sensitive bacteria can outcompete the resistant bacteria in the absence of phage.

Bacteria and virulent phage populations generally coexist in laboratory communities. Phage-limited bacteria typically occur at densities several orders of magnitude below that set by their resources; phage outnumber sensitive bacteria often by two orders of magnitude. Equilibrium population densities can usually be predicted with reasonable accuracy by independently estimating the parameters of the model, but the dynamics appears to be generally more stable than anticipated from theory. The factors responsible for discrepancies between theory and experiment have not yet been determined, although nongenetic variation in the vulnerability of individual bacteria to phage infection could be important.

Evolution in coexisting bacteria and virulent phage populations can be very rapid and may have profound effects; for example, the appearance of a bacterial genotype that is resistant to all phage genotypes can increase the total bacterial density by several orders of magnitude. The study of coevolving laboratory populations corroborates the existence of constraints on phage and bacteria, and demonstrates that these constraints are important in structuring communities. Phage-resistant bacterial mutants usually appear, including some for which no corresponding host-range phage mutants appear. These resistant bacteria are often at a pronounced competitive disadvantage relative to sensitive bacteria, permitting their coexistence and the maintenance of virulent phage. The resulting community can be described as resource-limited, but contains a minority population of phage-limited sensitive bacteria. Total bacteria typically outnumber phage by an order of magnitude or more in such communities.

Determining whether these constraints are important in structuring natural communities is much more difficult. If the significant features of phage–bacteria interactions in nature depend on particular characteristics of their environments (intestine, sewer, soil, and so on), then broad generalizations may be impossible. For example, virulent phage are less common than temperate phage in feces, whereas in sewage the opposite is true. This suggests that the survival of free phage or their adsorption to suitable bacteria is inhibited in the gut.

More generally, it does appear that many bacteria are resistant or otherwise insensitive to co-occurring phage in nature. Lytic phage infections do not obviously limit most bacterial communities to densities well below that set by their resources, although component populations may be limited. If further research supports this interpretation, then the evolutionary constraints documented in the laboratory may also be significant in many natural communities. Much more research is necessary, however, especially concerning: (1) details of the dynamics and trophic relationships of co-occurring phage and bacteria in nature; (2) effects of physically structured habitats on the interactions of bacteria and phage; and (3) consequences of recombination for the coevolution of bacteria and phage in genetically diverse communities.

Phage were advanced as agents for the control of bacteria soon after their discovery, but "phage therapy" has generally been regarded as a failure. Historical factors, including especially the advent of antibiotics, contributed to this failure. However, the biological factors responsible for the failure are not clear; their self-replicating and coevolving nature would seem to give phage certain advantages over antibiotics. The increasingly widespread resistance of bacteria to antibiotics has prompted renewed interest in the use of phage for their control.

ACKNOWLEDGMENTS. I wish to thank Nelson Hairston for impressing on me the importance of experimental approaches to the science of ecology; Janis Antonovics for convincing me of the inseparability of ecological and evolutionary phenomena; Bruce Levin for introducing me to phage and bacteria, and for raising many of the interesting questions in this chapter; and Madeleine Lenski for her continuing patience and diversions.

References

Adams, M. H., 1959, *Bacteriophages,* Interscience, New York.

Alexander, M., 1981, Why microbial predators and parasites do not eliminate their prey and hosts, *Annu. Rev. Microbiol.* **35**:113.

Anderson, E. H., 1946, Growth requirements of virus-resistant mutants of *Escherichia coli* strain "B," *Proc. Natl. Acad. Sci. USA* **32**:120.

Anderson, E. S., 1957, The relations of bacteriophages to bacterial ecology, *Symp. Soc. Gen. Microbiol.* **7**:189.

Anderson, E. S. 1968, The ecology of transferable drug resistance in the Enterobacteria, *Annu. Rev. Microbiol.* **22**:131.

Anderson, T. F., 1948, The growth of T2 virus on ultraviolet-killed host cells, *J. Bacteriol.* **56**:403.

Arber, W., and Linn, S., 1969, DNA modification and restriction, *Annu. Rev. Biochem.* **38**:467.

Archibald, A. R., 1980, Phage receptors in Gram-positive bacteria, in: *Virus Receptors,* Part 1, *Bacterial Viruses* (L. L. Randall and L. Philipson, eds.), pp. 5–26, Chapman and Hall, London.

Bachmann, B. J., and Low, K. B., 1980, Linkage map of *Escherichia coli* K-12, edition 6, *Microbiol. Rev.* **44**:1.

Barksdale, L., and Arden, S. B., 1974, Persisting bacteriophage infections, lysogeny, and phage conversions, *Annu. Rev. Microbiol.* **28**:265.

Botstein, D., 1980, A theory of modular evolution for bacteriophages, *Ann. N.Y. Acad. Sci.* **354**:484.

Bradley, D. E., 1967, Ultrastructure of bacteriophages and bacteriocins, *Bacteriol. Rev.* **31**:230.

Braun, V., and Hantke, K., 1977, Bacterial receptors for phages and colicins as constituents of specific transport systems, in: *Microbial Interactions* (J. L. Reissig, ed.), pp. 99–137, Chapman and Hall, London.

Brinton, C. C., and Beer, H., 1967, The interaction of male-specific bacteriophages with F pili, in: *The Molecular Biology of Viruses* (J. S. Colter and W. Paranchych, eds.), pp. 251–289, Academic Press, New York.

Bronson, M. J., and Levine, M., 1971, Virulent mutants of bacteriophage P22. I. Isolation and genetic analysis, *J. Virol.* **7**:559.

Bull, A. T., and Slater, J. H., 1982, Microbial interactions and community structure, in: *Microbial Interactions and Communities,* Volume 1 (A. T. Bull and J. H. Slater, eds.), pp. 13–44, Academic Press, London.

Campbell, A. M., 1961, Conditions for the existence of bacteriophage, *Evolution* **15**:153.

Campbell, A., and Botstein, D., 1983, Evolution of the lambdoid phages, in: *Lambda II* (R. W. Hendrix, J. W. Roberts, F. W. Stahl, and R. A. Weisberg, eds.), pp. 365–380, Cold Spring Harbor Laboratory, Cold Spring Harbor, New York.

Chao, L., Levin, B. R., and Stewart, F. M., 1977, A complex community in a simple habitat: An experimental study with bacteria and phage, *Ecology* **58**:369.

Crawford, J. T., and Goldberg, E. B., 1977, The effect of baseplate mutations on the require-ment for tail-fiber binding for irreversible adsorption of bacteriophage T4, *J. Mol. Biol.* **111**:305.

DeBach, P., 1971, The use of imported natural enemies in insect pest management ecology, *Proc. Tall Timbers Conf.* **3**:211.

DeBach, P., 1974, *Biological Control by Natural Enemies,* Cambridge University Press, Cambridge.

Delbruck, M., 1940a, Adsorption of bacteriophages under various physiological conditions of the host, *J. Gen. Physiol.* **23**:631.

Delbruck, M., 1940b, The growth of bacteriophage and lysis of the host, *J. Gen. Physiol.* **23**:643.

Delbruck, M., and Luria, S. E., 1942, Interference between bacterial viruses. I. Interference between two bacterial viruses acting upon the same host, and the mechanism of virus growth, *Arch. Biochem.* **1**:111.

Demerec, M., and Fano, U., 1945, Bacteriophage-resistant mutants in *Escherichia coli, Genetics* **30**:119.

d'Herelle, F., 1922, *The Bacteriophage: Its Role in Immunity,* Williams and Wilkins, Baltimore.

Dhillon, T. S., Chan, Y. S., Sun, S. M., and Chau, W. S., 1970, Distribution of coliphages in Hong Kong sewage, *Appl. Microbiol.* **20**:187.

Dhillon, T. S., Dhillon, E. K. S., Chau, H. C., Li, W. K., and Tsang, A. H. C., 1976, Studies on bacteriophage distribution: Virulent and temperate bacteriophage content of mam-malian feces, *Appl. Environ. Microbiol.* **32**:68.

Doermann, A. H., 1948, Lysis and lysis inhibition with *Escherichia coli* bacteriophage, *J. Bacteriol.* **55**:257.

Duckworth, D. H., 1970, Biological activity of bacteriophage ghosts and "takeover" of host functions by bacteriophage, *Bacteriol. Rev.* **34**:344.

Duckworth, D. H., 1976, Who discovered bacteriophage?, *Bacteriol. Rev.* **40**:793.

Duckworth, D. H., Glenn, J., and McCorquodale, D. J., 1981, Inhibition of bacteriophage replication by extrachromosomal elements, *Microbiol. Rev.* **45**:52.

Dunn, G. B., and Duckworth, D. H., 1977, Inactivation of receptors for bacteriophage T5 during infection of *Escherichia coli* B, *J. Virol.* **24**:419.

Dykhuizen, D. E., and Hartl, D. L., 1983, Selection in chemostats, *Microbiol. Rev.* **47**:150.

Echols, H., 1972, Developmental pathways for the temperate phage: Lysis vs. lysogeny, *Annu. Rev. Genet.* **6**:157.

Ellis, E. L., and Delbruck, M., 1939, The growth of bacteriophage, *J. Gen. Physiol.* **22**:365.

Emslie-Smith, A. H., 1961, Observations on the secular succession of types of *Escherichia coli* and related organisms in the faecal flora of an adult human subject, *J. Appl. Bacteriol.* **24**:vii.

Falkow, S., 1975, *Infectious Multiple Drug Resistance,* Pion, London.

Fraser, D. K., 1957, Host range mutants and semitemperate mutants of bacteriophage T3, *Virology* **3**:527.

Freeman, V. J., 1951, Studies on the virulence of bacteriophage infected strains of *Corynebacterium diphtheriae, J. Bacteriol.* **61**:675.

Furuse, K., Osawa, S., Kawashiro, J., Tanaka, R., Ozawa, A., Sawamura, S., Yanagawa, Y., Nagao, T., and Watanabe, I., 1983, Bacteriophage distribution in human faeces: Continuous survey of healthy subjects and patients with internal and leukaemic diseases, *J. Gen. Virol.* **64**:2039.

Goldberg, E., 1980, Bacteriophage nucleic acid penetration, in: *Virus Receptors,* Part 1, *Bacterial Viruses* (L. L. Randall and L. Philipson, eds.), pp. 115–141, Chapman and Hall, London.

Hendrix, R. W., Roberts, J. W., Stahl, F. W., and Weisberg, R. A. (eds.), 1983, *Lambda II,* Cold Spring Harbor Laboratory, Cold Spring Harbor, New York.

Hershey, A. D., 1946, Mutation of bacteriophage with respect to type of plaque, *Genetics* **31**:620.

Hershey, A. D., and Rotman, R., 1949, Genetic recombination between host range and plaque-type mutants of bacteriophage in single bacterial cultures, *Genetics* **34**:44.

Hofnung, M., Jezierska, A., and Braun-Breton, C., 1976, *lamB* mutations in *E. coli* K12: Growth of Lambda host range mutants and effect of nonsense suppressors, *Mol. Gen. Genet.* **145**:207.

Horne, M. T., 1970, Coevolution on *Escherichia coli* and bacteriophages in chemostat culture, *Science* **168**:992.

Howes, W. V., 1965, Effect of glucose on the capacity of *Escherichia coli* to be infected by a virulent Lambda bacteriophage, *J. Bacteriol.* **90**:1188.

Koerner, J. F., and Snustad, D. P., 1979, Shutoff of host macromolecular synthesis after T-even bacteriophage infection, *Microbiol. Rev.* **43**:199.

Krueger, A. P., 1931, The sorption of bacteriophage by living and dead susceptible bacteria. I. Equilibrium conditions, *J. Gen. Physiol.* **14**:493.

Kruger, D. H., and Bickle, T. A., 1983, Bacteriophage survival: Multiple mechanisms for avoiding the deoxyribonucleic acid restriction systems of their hosts, *Microbiol. Rev.* **47**:345.

Kruger, D. H., and Schroeder, C., 1981, Bacteriophage T3 and bacteriophage T7 virus–host cell interactions, *Microbiol. Rev.* **45**:9.

Kubitschek, H. E., 1970, *Introduction to Research with Continuous Cultures,* Prentice-Hall, Englewood Cliffs, New Jersey.

Lenski, R. E., 1984a, Two-step resistance by *Escherichia coli* B to bacteriophage T2, *Genetics* **107**:1.

Lenski, R. E., 1984b, Coevolution of bacteria and phage: Are there endless cycles of bacterial defenses and phage counterdefenses?, *J. Theor. Biol.* **108**:319.

Lenski, R. E., and Levin, B. R., 1985, Constraints on the coevolution of bacteria and virulent phage: A model, some experiments, and predictions for natural communities, *Am. Nat.* **125**:585.

Lerner, F., 1984, Population biology of male-specific bacteriophage, PhD. Dissertation, University of Massachusetts, Amherst.

Levin, B. R., 1981, Periodic selection, infectious gene exchange and the genetic structure of *E. coli* populations, *Genetics* **99**:1.

Levin, B. R., and Lenski, R. E., 1983, Coevolution in bacteria and their viruses and plasmids, in: *Coevolution* (D. J. Futuyma and M. Slatkin, eds.), pp. 99–127, Sinauer, Sunderland, Massachusetts.

Levin, B. R., and Lenski, R. E., 1985, Bacteria and phage: A model system for the study of the ecology and co-evolution of hosts and parasites, in: *Ecology and Genetics of Host–Parasite Interactions* (D. Rollinson and R. M. Anderson, eds.), pp. 227–242, Academic Press, London.

Levin, B. R., Stewart, F. M., and Chao, L., 1977, Resource-limited growth, competition, and predation: A model and experimental studies with bacteria and bacteriophage, *Am. Nat.* **111**:3.

Levins, R., 1966, The strategy of model building in population biology, *Am. Sci.* **54**:421.

Lewin, B., 1974, *Gene Expression,* Volume 1, *Bacterial Genomes,* Wiley, London.

Li, K., Barksdale, L., and Garmise, L., 1961, Phenotypic alterations associated with the bacteriophage carrier state of *Shigella dysenteriae, J. Gen. Microbiol.* **24**:355.

Lotka, A. J., 1925, *Elements of Physical Biology,* Williams and Wilkins, Baltimore.

Luria, S. E., 1945, Mutations of bacterial viruses affecting their host-range, *Genetics* **30**:84.

Luria, S. E., 1953, Host-induced modifications of viruses, *Cold Spring Harbor Symp. Quant. Biol.* **18**:237.

Luria, S. E., and Delbruck, M., 1943, Mutations of bacteria from virus sensitivity to virus resistance, *Genetics* **28**:491.

Luria, S. E., and Steiner, D. L., 1954, The role of calcium in the penetration of bacteriophage T5 into its host, *J. Bacteriol.* **67**:635.

Luria, S. E., Darnell, J. E., Jr., Baltimore, D., and Campbell, A., 1978, *General Virology,* Wiley, New York.

Lwoff, A., 1953, Lysogeny, *Bacteriol. Rev.* **17**:269.

Malmberg, R. L., 1977, The evolution of epistasis and the advantage of recombination in populations of bacteriophage T4, *Genetics* **86**:607.

Manning, P. A., and Reeves, P., 1978, Outer membrane proteins of *Escherichia coli* K-12: Isolation of a common receptor protein for bacteriophage T6 and colicin K, *Mol. Gen. Genet.* **158**:279.

Marvin, D. A., and Hohn, B., 1969, Filamentous bacterial viruses, *Bacteriol. Rev.* **33**:172.

Mathews, C. K., Kutter, E. M., Mosig, G., and Berget, P. B. (eds.), 1983, *Bacteriophage T4,* American Society for Microbiology, Washington, D.C.

May, R. M., 1974, *Stability and Complexity in Model Ecosystems,* Princeton University Press, Princeton, New Jersey.

May, R. M., and Anderson, R. M., 1983, Parasite–host coevolution, in: *Coevolution* (D. J. Futuyma and M. Slatkin, eds.), pp. 186–206, Sinauer, Sunderland, Massachusetts.

Meselson, M., Yuan, R., and Heywood, J., 1972, Restriction and modification of DNA, *Annu. Rev. Biochem.* **41**:447.

Milch, H., 1978, Phage typing of *Escherichia coli,* in: *Methods in Microbiology,* Volume 11 (T. Bergan and J. R. Norris, eds.), pp. 88–155, Academic Press, London.

Morona, R., and Henning, U., 1984, Host range mutants of bacteriophage Ox2 can use two different outer membrane proteins of *Escherichia coli* K-12 as receptors, *J. Bacteriol.* **159**:579.

Morona, R., and Henning, U., 1986. New locus (*ttr*) in *Escherichia coli* K-12 affecting sensitivity to bacteriophage T2 and growth on oleate as the sole carbon source, *J. Bacteriol.* **168**:534.

Orpin, C. G., and Munn, E. A., 1974, The occurrence of bacteriophage in the rumen and their influence on rumen bacterial populations, *Experientia* **30**:1018.

Paynter, M. J. B., and Bungay, H. R., III, 1969, Dynamics of coliphage infections, in: *Fermentation Advances* (D. Perlman, ed.), pp. 323–335, Academic Press, New York.

Paynter, M. J. B., and Bungay, H. R., III, 1970, Capsular protection against virulent coliphage infection, *Biotechnol. Bioeng.* **12**:341.

Peitzman, S. J., 1969, Felix d'Herelle and bacteriophage therapy, *Trans. Stud. Coll. Physicians Phila.* **37**:115.

Ptashne, M., Jeffrey, A., Johnson, A. D., Maurer, R., Meyer, B. J., Pabo, C. O., Roberts, T. M., and Sauer, R. T., 1980, How the Lambda repressor and Cro work, *Cell* **19**:1.

Reanney, D., 1976, Extrachromosomal elements as possible agents of adaptation and development, *Bacteriol. Rev.* **40**:552.

Reanney, D. C., and Ackermann, H. W., 1982, Comparative biology and evolution of bacteriophages, *Adv. Virol. Res.* **27**:205.

Reanney, D. C., Gowland, P. C., and Slater, J. H., 1963, Genetic interactions among microbial communities, *Symp. Soc. Gen. Microbiol* **34**:396.

Rodin, S. N., and Ratner, V. A., 1983, Some theoretical aspects of protein coevolution in the ecosystem "phage–bacteria." I. The problem. II. The deterministic model of microevolution, *J. Theor. Biol.* **100**:185.

Roper, M. M., and Marshall, K. C., 1974, Modification of the interaction between *Escherichia coli* and bacteriophage in saline sediment, *Microb. Ecol.* **1**:1.

Ryter, A., Shuman, H., and Schwartz, M., 1975, Integration of the receptor for bacteriophage Lambda in the outer membrane of *Escherichia coli:* Coupling with cell division, *J. Bacteriol.* **122**:295.

Scarpino, P. V., 1978, Bacteriophage indicators, in: *Indicators of Viruses in Water and Food* (G. Berg, ed.), pp. 201–227, Ann Arbor Science, Ann Arbor, Michigan.

Schlesinger, M., 1932, Ueber die Bindung des bakteriophagen an homologe Bakterien. II. Quantitative Untersuchungen ueber die Bindungsgeschwindigkeit und die Saettigung. [English translation in G. S. Stent (ed.), *Papers on Bacterial Viruses,* Little and Brown, Boston (1960).]

Schwartz, M., 1976, The adsorption of coliphage Lambda to its host: Effect of variations in the surface density of receptor and in phage-receptor affinity, *J. Mol. Biol.* **103**:521.

Schwartz, M., 1980, Interaction of phages with their receptor proteins, in: *Virus Receptors,* Part 1, *Bacterial Viruses* (L. L. Randall and L. Philipson, eds.), pp. 59–94, Chapman and Hall, London.

Shera, G., 1970, Phage treatment of severe burns, *Br. Med. J.* **1**:568.

Smith, F. E., 1972, Spatial heterogeneity, stability, and diversity in ecosystems, in: *Growth by Intussusception* (E. S. Deevey, ed.), pp. 307–335, Connecticut Academy of Arts and Sciences, New Haven, Connecticut.

Stent, G. S., 1963, *Molecular Biology of Bacterial Viruses,* Freeman, San Francisco.

Stent, G. S., and Wollman, E. L., 1952, On the two-step nature of bacteriophage adsorption, *Biochim. Biophys. Acta* **8**:260.

Stewart, F. M., and Levin, B. R., 1973, Partitioning of resources and the outcome of interspecific competition: A model and some general considerations, *Am. Nat.* **107**:171.

Stewart, F. M., and Levin, B. R., 1984, The population biology of bacterial viruses: Why be temperate, *Theor. Popul. Biol.* **26**:93.

Stone, J. C., Smith, R. D., and Miller, R. C., Jr., 1983, A recombinant DNA plasmid which inhibits bacteriophage T7 reproduction in *Escherichia coli, J. Gen. Virol.* **64**:1615.

Szmelcman, S., and Hofnung, M., 1975, Maltose transport in *Escherichia coli* K-12: Involvement of the bacteriophage Lambda receptor, *J. Bacteriol.* **124**:112.

Tan, J. S. H., and Reanney, D. C., 1976, Interactions between bacteriophages and bacteria in soil, *Soil Biol. Biochem.* **8**:145.

Vidaver, A. K., 1976, Prospects for control of phytopathogenic bacteria by bacteriophages and bacteriocins, *Annu. Rev. Phytopathol.* **14**:451.

Volterra, V., 1926, Fluctuations in the abundance of a species considered mathematically, *Nature* **118**:558.

Whitehead, H. R., 1953, Bacteriophage in cheese manufacture, *Bacteriol. Rev.* **17**:109.

Wilkinson, J. F., 1958, The extracellular polysaccharides of bacteria, *Bacteriol. Rev.* **22**:46.

Williams, F. M., 1972, Mathemetics of microbial populations, with emphasis on open systems, in: *Growth by Intussusception* (E. S. Deevey, ed.), pp. 395–426, Connecticut Academy of Arts and Sciences, New Haven, Connecticut.

Williams, F. M., 1980, On understanding predator–prey interactions, in: *Contemporary Microbial Ecology* (D. C. Ellwood, J. N. Hedger, M. J. Latham, J. M. Lynch, and J. H. Slater, eds.), pp. 349–375, Academic Press, London.

Williams Smith, H., 1972, Ampicillin resistance in *Escherichia coli* by phage infection, *Nature* **238**:205.

Williams Smith, H., and Huggins, M. B., 1980, The association of the O18, K1 and H7 antigens and the ColV plasmid of a strain of *Escherichia coli* with its virulence and immungenicity, *J. Gen. Microbiol.* **121**:387.

Williams Smith, H., and Huggins, M. B., 1982, Successful treatment of experimental *Escherichia coli* infections in mice using phage: Its general superiority over antibiotics, *J. Gen. Microbiol.* **128**:307.

Wright, A., McConnell, M., and Kanegasaki, S., 1980, Lipopolysaccharide as a bacteriophage receptor, in: *Virus Receptors,* Part 1, *Bacterial Viruses* (L. L. Randall and L. Philipson, eds.), pp. 27–57, Chapman and Hall, London.

2

Methods for Detecting Genetically Engineered Microorganisms in the Environment

SIMON FORD and BETTY H. OLSON

1. Introduction

The ability to monitor accurately genetically engineered microorganisms (GEMs) is critical in determining their potential impact on a given environment. Many articles have delineated the crucial issues surrounding the release of GEMS (Rissler, 1984; Alexander, 1986; Strauss *et al.,* 1985; Gillett *et al.,* 1986; Lenski, 1987). Five major concerns are repeatedly referred to by various authors: (1) incorporation of the novel gene or genes into natural microorganisms, (2) ability of the novel organism to survive in the environment, (3) ability of the novel organism to multiply in the environment, (4) interaction of the novel organism with biological systems that could be injurious to other organisms, and (5) the assessment of harm caused by the organism. The ability to address the first four of these concerns is dependent upon the development of appropriate methodologies. The types of ecological questions that are dependent on adequate methodological techniques for answers are shown in Table I. These questions must be answered before the release of GEMs can become as routine a practice as is now the case for the use of licensed pesticides.

Until recently, ecological studies have been almost completely confined to measuring the occurrence or activity of observable phenotypes.

SIMON FORD and BETTY H. OLSON • Program in Social Ecology, University of California, Irvine, California 91717.

Table I.   General Questions on the Biotic Effects of Deliberately Released
Genetically Engineered Microorganisms That Can Be Answered through
Monitoring Techniques[a]

1. How can the GEMs be specifically identified and detected?
2. Can the population size be determined at sufficiently low levels?
3. Can survival of GEMs and persistence of R-DNA be monitored accurately in the
   introduced environment (or in an environment to which they may be transported)?
4. How can R-DNA transport be detected?
5. What are the levels of nonengineered host organisms in the environment and what is
   their effect on monitoring the novel organism?
6. Do GEMs have a selective advantage?
7. Are the GEMs genetically as stable as unaltered genotypes in the environment?
8. What is the quantitative population change over time in the environment in which
   the microorganism was released or to which it was transported?
9. How do environmental conditions affect the multiplication and persistence of GEMs?
10. Can environmental growth requirements be used to contain GEMs?
11. What species or biological systems are exposed to the novel mircoorganisms?
12. Are existing exposure assessment methodologies for microorganisms satisfactory for
    GEMs?
13. Can GEMs be subjected to a sufficient diversity of environments such that all "new"
    genes will be expressed?
14. Do GEMs have unique dose–response relationships or thresholds of activity that
    cannot be predicted from knowledge of unaltered parent organisms?
15. What level of characterization of gene segments is sufficient to predict associated
    adverse effects?

[a]Adapted from Rissler (1984).

The advent of advanced techniques in molecular biology enabled the microbial ecologist to examine the genotypic response of microorganisms, primarily bacteria, to a variety of environmental conditions. The importance of genotypic studies has been recognized by a broad spectrum of scientists because of the release of GEMs and the anticipated widespread use of bioengineered organisms in both terrestrial and aquatic systems. Some of the intended uses of such organisms include biomass conversion, pest management, agricultural production, pollution control, mining and mineral recovery, and microbial enhancement of oil recovery (Saunders and Saunders, 1987).

Drawing from traditional ecological methodologies and molecular techniques, the occurrence and distribution of a genetically engineered microorganism may be monitored at the level of the phenotype or the genotype. Definitions of these terms drawn from the glossary of a recent report (Gillett *et al.*, 1985) helps clarify the difference in emphasis achieved by monitoring at either of these levels. Phenotypic approaches measure "the perceptible properties or other characteristics of an organism resulting from the interaction of its genetic constitution and the environment." In contrast, genotypic approaches directly detect "the genetic constitution of an organism, as distinct from its physical and functional appearance."

Little information is available on how the phenotype of GEMs may be altered by physical and chemical factors in the environment, but there is evidence that microorganisms can lose the ability to express a carried trait. Loss of expression could have dramatic consequences on phenotypic monitoring results and points to the importance of assessing the occurrence of the trait of interest directly by a genotypic method. As more information is generated regarding the stability of engineered genes in the environment, it may well be possible to select a monitoring strategy that only includes the assessment of the phenotypic characteristic of the trait. Until the time such an approach is feasible, it would appear advisable to monitor the fate and survival of GEMs at both the phenotypic and genotypic levels. Methods should be capable of detecting and discriminating between both the original GEM and the recombinant DNA (R-DNA) even when the R-DNA has been transferred into naturally occurring bacteria in the community where it may not be expressed. It is not the purpose of this review to focus on the ways that R-DNA or GEMs can be distributed among bacterial populations, as this topic has been reviewed by others (Stotzky and Babich, 1984, 1986; Barnthouse and Palumbo, 1986). However, it is important to discuss briefly the implications that mechanisms of genetic spread have on the methods that should be used to determine the presence of GEMs and the R-DNA they contain.

The importance of conjugation, transduction, and transformation has been discussed in several reviews (Reanny, 1976, 1977; Datta, 1984). Less frequently discussed, but perhaps of equal importance, is the role of transposable genetic elements, such as transposons and insertion sequences, in the translocation of DNA within a host cell (Levin, 1986). The illegitimate recombination events mediated by this class of elements may result in an increase in the probability of transfer of R-DNA sequences to a variety of host organisms. The importance of this phenomenon is readily demonstrated in the case of epidemiological studies on the spread of antibiotic resistance genes among bacteria of clinical significance (Datta and Hughes, 1983; Hughes and Datta, 1983). The major mode of acquisition of antibiotic resistance by bacteria appears to have been via the insertion of antibiotic resistance determinants into plasmids. This observation stresses the importance of characterizing the transposition status of genetic elements present in recombinant molecules when assessing the presence of GEMs in the environment.

In this review we describe methods that can be used in the detection of GEMs in natural environments. These methods are equally applicable to the general study of genotypic adaptation of microbial communities *in situ.* Existing technologies for analysis of the phenotype expressed by a trait of interest or an introduced marker and methods for identifying the occurrence of specific sequences of DNA will be compared. This review is divided into three major subjects: Section 2 deals with the criteria that must be considered in the development of methods of detection, Section 3 introduces phenotypic methods of analysis, and Sections 4 and 5 discuss aspects of genotypic methods of analysis.

2. Criteria for Methods of Detection

The ability to determine whether a particular microorganism is present in an environment is a difficult task. Factors that affect this ability are controlled by population characteristics such as density, distribution, and nutritional, temperature, oxygen, and pH requirements, as well as physical constraints such as substrate uniformity and adsorption characteristics. If the organism is introduced, as in the case of GEMs, the ability to detect this organism is likely to depend on (1) the initial inoculum, (2) the effectiveness of dispersion, and (3) the ability of the organism to establish itself or its genetic trait in the effected environment. In order to select appropriate methodologies an investigator has to determine the primary goal of the investigation. The two main areas of concern in monitoring GEMs are generally the occurrence of the GEM or R-DNA and the fate of the GEM or R-DNA within the environment. The selection of appropriate techniques is dependent on the goals of a given investigation. Regardless of which technique is employed, it should conform to certain stringent criteria. Factors that are important in assessing the usefulness of genotypic methods for the detection of GEMs are shown in Table II. The criteria of sensitivity, specificity, reproducibility, and practicality of a technique are equally important in its successful use in monitoring any foreign substance in the environment, be it toxic chemicals or introduced organisms.

Besides the important considerations of the methodology, there are general considerations that must be addressed. The validity of studies on the incidence of GEMs in natural environments depends not only on the selection of an appropriate detection methodology, but is also highly dependent upon the sampling strategy (Glaser *et al.,* 1986). The location for testing may be aquatic or terrestrial, carrying with it the particular constraints associated with that site. Determining how to collect representative samples is of importance in ensuring that the material obtained and thus the results of the analysis are indeed meaningful (APHA, 1985; Page, 1985; Tsernoglon and Anthony, 1971). The significance of repre-

Table II. Critical Factors in the Accuracy of Genotypic Methods of Analysis

	Sensitivity	Specificity	Reproducibility	Practicality
Direct plating method	1. Number of organisms that can be plated (up to 5×10^5 per plate) 2. Specific activity of probe 3. Choice of plating media	1. Uniqueness and length of probe sequence 2. Stringency of hybridization 3. Inability to control adequately for false negatives	Excellent in aquatic habitats	1. Easy to process large number of samples, providing no other information required 2. Technically undemanding 3. Hazards associated with isotopes 4. High cost of reagents
Colony hybridization method	1. Number of organisms that can be analyzed 2. Choice of plating media	1. Uniqueness and length of probe sequence 2. Stringency of hybridization 3. Excellent controls	Excellent in soil and aquatic sediment habitats	1. Impractical to process large number of samples 2. Technically undemanding 3. Hazards associated with isotopes 4. High cost of reagents
Direct extraction method	1. Size of sample that can be processed 2. Efficiency of DNA extraction procedure 3. Specific activity of probe	1. Uniqueness and length of probe 2. Stringency of hybridization 3. Representativeness of DNA extraction procedure	Untested	1. Technique in development stage 2. Expensive equipment

sentativeness in the experimental protocol must be adequately addressed in order to assess accurately the presence or absence of genetically engineered organisms (Bordner and Winter, 1978; Litchfield *et al.*, 1975; Pramer and Bartha, 1972). In terrestrial environments, soils can be highly heterogeneous in composition and therefore it may be required to collect a large number of samples to ensure adequate representation (Jensen, 1968; Heal, 1970). In contrast to soil environments, the aquatic environment is usually more uniform in its physical composition. However, the numbers of organisms present may be low, especially in oligotrophic waters. Therefore, other considerations regarding sampling techniques must be considered, such as sample concentration. In order to assure that the chosen sampling procedure is adequate, the reader is referred to references on sampling techniques (APHA, 1985; Page, 1985).

The efficiency of recovery of organisms from an environmental substratum must also be addressed. The reliability of the recovery method may be of more significance than the efficiency of recovery as long as the efficiency meets the minimum level of detection required. A procedure that constantly yields a 30% recovery may be more valuable than one in which recovery yields vary from 20 to 80% (Bohlool and Schmidt, 1980). The type of host selected for the R-DNA will influence primarily the isolation procedures. In this review we focus on bacteria, because the first releases of GEMs are likely to involve this group of organisms. Certainly many other microorganisms, such as viruses, protozoa, and fungi, will be utilized as genetically engineered systems are more widely employed.

A frequent problem with environmental studies is inadequate description of the physical and chemical variables of the environment being studied. To date there is not an adequate understanding of how many of these variables simultaneously affect microorganisms, although a number of studies have delineated the impact of a specific variable on microbial survival, growth, or adaptation to an environment (Atlas and Bartha, 1981). Adequate definition of physical and chemical variables will be useful in developing predictive models on dieoff times and rates.

2.1. Sensitivity

The sensitivity of the method being used will establish the level of detection. Analyses that contain a plating step require consideration of two types of sensitivity: (1) the sensitivity of the isolation procedure, and (2) the sensitivity of the method for detecting the GEM or R-DNA. If plating techniques are utilized in the isolation procedure, a reduction in the recovery of a number of organisms present in a sample must be expected. For example, it is estimated that plating techniques for soil samples recover maximally 1% of a soil bacterial community and may be as low as 0.01%, depending on the medium used. A similar situation exists in the aquatic environment; a comparison of direct counts and viable counts from potable water samples indicated 2–3 log lower counts on solid medium (McCoy and Olson, 1985). The low bacterial recovery achieved in plating soil or water samples may be attributed to such limitations as the inability of the nutrient composition of the plating medium to recover all microorganisms present in a sample. Direct count techniques, on the other hand, give higher numbers of microorganisms, but do not distinguish viable from nonviable cells.

In the case of the detection, the important factor is the number of cells required to produce an adequate signal, such as the formation of a colony on selective media or the detection of a cell in a microscopic field. When labeled genetic probes are used, the strength, of the signal will also vary, depending on whether radioisotopes or non-autoradiographic labels such as biotin are utilized.

2.2. Specificity

The specificity of a method is concerned with the relative assurance that the trait of interest is being measured. Two types of errors can occur: (1) conditions may be so highly restrictive that positive organisms are missed, yielding false negatives, or (2) lack of specificity of the detection method may result in false positives. The first error is failure to detect the GEM or R-DNA when it is present in the environment. The second error concerns a result that indicates the presence of the microorganism or the R-DNA when it is absent. False negatives are unlikely to occur with genetic probes. However, the potential for false positives is especially important, as the natural environment contains a number of organisms that may be similar to the released GEM. It is therefore necessary to use a method that is capable of distinguishing between normal inhabitants of the environment and the novel organism. Both types of errors can be controlled and a method should be sufficiently tested in these respects before it is considered for use in the environment. The exact considerations necessary to prevent these types of errors are specific to the methodologies selected and are discussed in more detail with each technique described in the later sections of this chapter.

2.3. Reliability

Any procedure selected for use must be reproducible. The reliability of a system therefore depends in part on the stability of the system being used. For example, the trait being measured must be stable and the method for measurement must also be stable. Two main considerations are then important. First is the effect the environment will have on the trait. It is possible that an introduced trait may not be stable and that part of the genetic material may be deleted. Thus, monitoring methods should be designed to detect these types of modifications. If indeed the trait is no longer expressed, then the phenotype will change. Loss of phenotypic expression may be associated with only partial deletion of sequences of R-DNA. Under these circumstances, R-DNA may still be detected by genotypic analysis. This again points to the importance of using methodologies that allow assessment of both the phenotype and the genotype.

It is also important that the methodology chosen produces consistent results regardless of the type of environment in which it is applied. It must be determined that a test that may be performed successfully in the aquatic environment will also have the same level of reliability or reproducibility in the soil environment. Factors that are commonly important under these circumstances include levels of detection, interference problems, and recovery of microorganisms from the sample. In the following sections we discuss some of the problems associated with investigation of samples from soil, sediment, and aquatic environments.

2.4. Practicality

The practicality of the method selected will be based on cost and ease of use, provided that the techniques are sensitive, specific, and reliable for the needs of the designated experiment. Usually there are compromises involved in the cost of the procedure, simplicity, specialized equipment necessary, or personnel. Monitoring methods are likely to be used, under a variety of environmental conditions, by different personnel. This requires that the method be carefully reviewed as to its limitations prior to selection. As with any monitoring program, it is likely that a large number of samples will have to be processed, and therefore the method selected should facilitate this. Considerations regarding this aspect of monitoring for GEMs by genotypic methods are detailed in Table II.

3. Phenotypic Analysis

3.1. Markers for Detection of Novel Microorganisms in the Natural Environment

The use of marker genes to track GEMs introduced into the environment include chromogenic markers, resistance markers and metabolic markers (Glaser *et al.*, 1986). Chromogenic markers, which result in pigmented microorganisms on the appropriate indicator medium have the advantage that they produce an easily scorable phenotype. Resistance markers include heavy metal resistance genes and antibiotic resistance genes, and require the use of selective media. Metabolic markers, such as raw sugar or other carbon source genes may be utilized to detect GEMs. Such an approach requires the plating of the organism on supplemented minimal medium or an appropriate indicator medium.

The choice of the type of marker to be used in following the fate of a novel microorganism introduced into the natural environment is influenced by a number of factors, including the availability of suitable markers in the introduced organism, the incidence of occurrence of the marker in the natural community of a given environment, the practicality of assaying for the occurrence of that marker in different environments, and the stability of the marker. Some of these aspects may be anticipated in the design of the GEM. Tracking of the GEM may be facilitated by the inclusion of a gene cassette in the genetic complement of the microorganism. Such a casette would contain a gene that is not represented among the organisms normally occupying the environments into which the GEM is to be released (Sayler and Stacey, 1986). Alternatively, a gene that occurs naturally in the strain of interest may be used for this purpose. In both of these cases monitoring is achieved by detection of the appropriate phenotype and as such the ability to detect the novel organism is dependent on expression of this marker gene. Failure of the gene to function in a novel host, which may occur as a possible consequence of gene transfer, will result in false-negative results. This is most likely a common drawback of any detection method that is dependent solely on expression of a marker gene. Although the ability of a gene to express itself in certain hosts may be determined in the laboratory, it is impractical to test each marker gene in all possible bacterial hosts. Microcosm experiments may be performed prior to field studies to investigate the practicality of such an approach for a given experimental system.

3.2. Detection of Microorganisms by Bacteriophage Sensitivity

Phage specific for bacterial species have been widely used to identify pathogens of plants and animals (Cherry et al., 1954; Brown and Cherry, 1955; Billing and Garrett, 1980; Hirsch and Martin, 1983). This method of identification has the advantage of rapidity and results can usually be read within 24 hr of inoculation. This method has been successfully employed in the detection of *Pseudomonas syringae* pv. *tomato,* the causal agent of bacterial speck, a disease of tomatoes that results in spotted fruit (Cuppels, 1984). Phage characterization is particularly useful here because serological methods are not available and therefore conventional biochemical characterization is necessary to identify the pathogen, which is a highly labor-intensive task. By using a bacteriophage sensitivity test, Cuppels was able to discriminate between *P. syringae* pv. *tomato* and a variety of other bacterial pathogens of tomato.

The main limitation of this approach is the availability of a virulent phage specific to the given organism. This problem is particularly significant in the case of the application of this method for the detection of the sensitivity of GEMs. The GEM may differ from wild-type organisms by only one locus. Lenski (1984) has suggested the use of bacteriophage as a means of limiting the spread of GEMs in the environment, a method that is dependent on the isolation of a bacteriophage whose site of adsorption is a cell surface antigen unique to the GEM. The practicality of such a strategy for the detection of GEMs remains to be proven. However, a strategy that included the isolation of a bacteriophage specific for a pathogenic bacterial antigen has been successfully employed in treating bacterial infections of mice (Williams Smith and Huggins, 1980, 1982).

There are several potential drawbacks to bacteriophage sensitivity approaches to detection of GEMs. Perhaps most significant among these is the observation that mutation or low nutrient conditions can significantly alter outer membrane surface properties of bacteria (Ellwood and Tempest, 1972; Lugtenberg *et al.,* 1976; Brown and Williams, 1985). Importantly, these outer membrane surface changes have been shown to modify virus attachment (Farkas-Himsley, 1964), which has been shown to contribute to phage immunity (Chai, 1983).

3.3. Detection of Microorganisms by Differential Plating Techniques

One of the most common ways of identifying specific microorganisms in medical research and molecular biology is by the use of antibiotic resistance genes. In this case multiple antibiotic resistance genes may be included in the genetically engineered microorganism during construction. Subsequent tracking of the GEM is achieved by assaying for microorganisms with the characteristic pattern of drug resistance genes. Usually more than one marker is used, because of the likelihood that other organisms in an environmental sample may have resistance to the test antibiotic. In designing a monitoring procedure using antibiotic resistance it is essential to pretest the area in which the organism is going to be released. If widespread use is anticipated, samples from several environments in which it is likely to be introduced must be assayed for the frequency of antibiotic resistance in those areas.

Antibiotic resistance has the advantage that it is easy to assay and relatively stable under environmental conditions. It should be noted that antibiotic resistance patterns do change in the environment regardless of whether or not selective pressure is maintained. The trait is usually more stable under selective conditions, but also has a greater opportunity for dissemination under those conditions. Thus, if an antibiotic resistance gene is selected for, care must be taken to ensure that it has a low probability for spread. Graham and Istock (1979) designed recipients in an experiment such that three antibiotic resistance markers were used thereby insuring that the probability that a mutation could have accounted for the presence of all three traits simultaneously is extremely low (10^{-18}). Thus, the detection of the three markers have high assurance that the test strain was recovered.

A method originally intended to estimate the density of individual bacterial populations introduced into natural ecosystems (Danso et al., 1973) may be used to assess the survival of GEMs in environment samples. This method (Mallory et al., 1982) relies on using differences in antibiotic resistance patterns between the introduced GEM and natural bacterial populations. The authors suggest that this method can detect extremely small numbers, as low as 1 organism/ml, of antibiotic-resistant test organisms in model ecosystems. Samples from sewage were plated on agar media containing antibiotics in concentrations and combinations that eliminated the growth of almost all of the indigenous microbial population. This method of investigation is simple and sensitive and may serve as a rapid method to monitor the behavior of specific bacterial populations in natural environments.

There are several problems associated with using antibiotic resistance genes as markers in monitoring GEMs, apart from the fact that antibiotic resistance genes are generally encoded on plasmids and therefore may not display the degree of stability required of a marker for monitoring experiments. In addition, antibiotic resistance genes are frequently transposable or contain insertion sequences or other components of transposition systems. The inclusion of such elements in a GEM would contribute to genetic instability and may result in an increase in transfer of R-DNA.

There are problems associated with transfer of antibiotic resistance markers into the chromosome of GEMs. If the antibiotic resistance gene is already commonly found in the environment, then detection of the GEM will prove to be difficult due to false negatives from the indigenous population. Alternatively, if the antibiotic resistance gene in question is one that is not commonly found in the environment, then it is impractical to release it in large quantities into the environment, because by so doing the experimenter may promote resistance to an antibiotic that may be of current of future clinical importance. Similar arguments can be used in the case of constructing multiple resistance factors containing antibiotic resistance genes in combinations uncommon in nature; such an action would contribute to the already escalating incidence of multiple drug-resistant pathogens and opportunists.

3.4. Detection of Specific Bacterial Antigens by Fluorescent Antibody Detection Methods

Fluorescent antibody (FA) detection systems have been developed and used primarily for the identification and localization of components of bacterial and viral pathogens (Bohlool and Schmidt, 1980). Two advantages of this technique include the small quantity of fluorescent antibody necessary for reaction and the high specificity of the antibody. FA detection systems are dependent on the preparation of an active antiserum to the antigenic material of the organism of interest. Two types of fluorescent antibody systems are commonly used, direct and indirect. The direct system utilizes an antibody developed specifically to surface antigens of the microorganism. In contrast, the indirect system uses an antibody made to the antibody for the microorganism. This second antibody is labeled, while the antibody to the microorganism is unlabeled. The process is sometimes referred to as the "sandwich technique." Antibody to a microorganism is prepared by injection of the antigen of interest into rabbits and the indicator antigen by the injection of rabbit antiserum into, for example, a goat. Antiserum is labeled with a fluorescent dye, such as fluorescein isothiocyanate (FITC). Thus, the fluorescent antibody marker can be used to identify the presence of the original antigenic marker (direct) or the antibody to the original antigen (indirect). The advantage of the indirect method is that several antibodies can be tested for simultaneously. The specificity of the technique is dependent on the uniqueness of the antigenic material selected. Several reviews cover the development of antibodies and use of immunofluorescence in environmental samples (Garvey et al., 1977; Bohlool and Schmidt, 1980).

As most of the techniques have been prepared for use in the medical field, it is important to note that several factors make the application of this technique more difficult in the environment. The serology of many organisms in aquatic and terrestrial habitats is unknown and the ways that the physical, chemical, and biological diversity of an environment may affect serology is unclear. Thus, adequate control procedures are vital for the development of a successful procedure. A review published by Schmidt (1973) is useful in describing many of these considerations. Seven considerations in the use of fluorescent antibody techniques should be closely monitored. These include (1) technical precision, (2) specificity of the antibody, (3) nonspecific staining, (4) autofluorescence, (5) stability of the antigen under environmental conditions, (6) inability to distinguish between viable and nonviable cells, and (7) quantification.

FA procedures rely in part on the subjective judgment of the individual doing the microscopic observations. Thus, the training of technical assistants to operate microscopic analysis is critical, particularly when working with the heterogeneous and highly complex samples typical of monitoring GEMs. Appropriate controls are therefore necessary to ensure the accuracy of the microscopic observations.

In considering the question of specificy it is important to determine the exact goal of the use of the technique. In the case of studies involving GEMs the goal is most likely the detection of the novel organism through the presence of a natural antigen or an antigen specifically introduced into the novel organism. As the antibody reaction occurs with antigens on the surface of the cell, it is unlikely that a trait of interest in the GEM can be directly detected by this method.

Cross reactivity of the antibody is a continual concern. There is cross-reactivity among related oragnisms (Fliermans *et al.,* 1974), unrelated microorganisms (Joklik and Willett, 1976; Heidelberger and Elliot, 1966; Dudman and Heidelberger, 1969), and "universal acceptors" (Bohlool and Schmidt, 1980). Universal acceptors are usually fungal spores and may react with any FA stain.

Autofluorescence and nonspecific staining are troublesome when working with natural samples. Plant material can be particularly difficult in terms of autofluorescence, whereas soils and sediments exhibit nonspecific staining. Auto fluorescence can mask stained cells, thus reducing the sensitivity of the method; furthermore, nonspecific staining can result in false-positive results. Several techniques have been developed to reduce autofluorescence and nonspecific staining (Bohlool and Schmidt, 1968; Schmidt *et al.,* 1974; Bohlool and Brock, 1974).

Although only a few antigens have been tested for their stability in the environment, those that have appear to be relatively stable. Rhizobia, for example, have been extensively tested on different media, in different soils, and as bacteroids in nodules (Bohlool and Schmidt, 1980). The antigens associated with this genus appear to be highly stable. A 12-year study on the stability of rhizobial antigens in Australian soils indicated that colony morphology and antigen characteristics did not change over the time period, although antibiotic resistance was slightly altered (Diatloff, 1977). Unfortunately, the antigenic stability for certain genera that are likely to be released as GEMs, for example, *Pseudomonas,* are not well characterized from an environmental perspective. The conserved nature of antigens is demonstrated by the cross-reactivity among certain strains of a bacterial genera isolated on a worldwide basis, such as *Sulfolobus* (Bohlool, 1975).

A distinct limitation of FA procedures is their inability to distinguish between viable and nonviable cells. In the case of GEM monitoring experiments this may be a significant limitation. Intact R-DNA sequences contained in nonviable cells may be acquired by other organisms in the environment by transformation. A few studies have been made on dead cells and FA reactions in the soil environment. Studies have included *Rhizobium japonicum, Escherichia coli,* and *Nitrobacter.* The results indicated that recoverable nonviable bacterial isolates tended to disappear quickly, within 1–2 weeks, from the test soils (Bohlool and Schmidt, 1973; Schmidt, 1974; Rennie and Schmidt, 1977). Evidence based on the nucleic acid contents of the humic acid components of soil indicates that microbial DNA may survive in the soil after decomposition of the cell structure(Anderson, 1958, 1979). It is therefore important that studies be performed to investigate how long DNA sequences are biologically active and persist in different environments.

Quantification of cell numbers by FA techniques can be difficult in environmental samples. Most troublesome are soil samples or water samples with high levels of turbidity. To quantify bacterial populations associated with soil or sediments it is necessary to separate the bacteria from the substratum. It should be noted that a concentration step is also often required. Thus, the efficiencies of the extraction and concentration procedures are two additional variables that require proper controls.

It is important to consider the efficiency of recovery in different media. Vidor and Miller (1980) found that recoveries varied markedly, depending upon soil type. The most important factor in recovery is consistency. A method that is highly variable may be less desirable than a very stable one in which recovery efficiency is low. However, low recovery rates can become a problem when the density of the population of interest drops below 10^5 bacteria/g. It will be necessary to determine the efficiency of recovery of the method in any system under study. The approach suggested by Bohlool and Schmidt (1980) consists in adding the organism or interest in pure culture to the soil and comparing plating recoveries with fluorescent antibody detection methods.

4. Genotypic Analysis: The Use of Genetic Probes

4.1. The Use of Genetic Probes for the Detection of Specific DNA Sequences

The development of techniques utilizing hybridization with labeled genetic probes has resulted in many of the recent advances in molecular genetics and the biomedical sciences. Among the techniques based on this approach are Southern blot analysis (Southern, 1975), Northern analysis (Alwine *et al.,* 1979), *in situ* hybridization (Chandler and Yunis, 1978), colony hybridization (Grunstein and Hogness, 1975), and plaque hybridization (Jones and Murray, 1975; Benton and Davis, 1977). All of these techniques are based on the same underlying principle: the complementarity of related single-stranded DNA or RNA species. That is, two single-stranded sequences of opposite polarity may form a duplex or double-stranded molecule if they are identical or sufficiently similar in sequence. In this way a labeled probe sequence may be used to investigate the occurrence of a complementary target sequence.

A number of studies have utilized probe/target sequence combinations for identifying microorganisms in natural environments and foods (Table III). The choice of target sequence/probe combinations for detection of GEMs is dependent on the availability of unique DNA sequences within the GEM and the frequency of occurrence of cross-reactive sequences in the natural bacterial population in the site to which the organism is to be introduced.

For the purpose of initial field trials of GEMs it may be advisable to introduce marker sequences into the genotype of the GEM during construction in a gene cassette. Such sequences may be generated by *in vitro* synthesis or alternatively may be derived from DNA sequences not normally found in the environment being tested, for example, genes isolated from higher organisms. This approach would facilitate tracking of the GEM by allowing the investigator to control interference from cross-reactive sequences in the environment. An approach of this type may be the method of choice for tracking GEMs that differ from wild-type organisms only by deletion of genetic information. This type of approach to marking an organism enables the investigator to use a detection method that is not dependent on expression of the sequence. The use of a marker

Table III. Use of Genetic Probes in Environmental Samples

1. Use of three *Escherichia coli* enterotoxin gene probes (Moseley *et al.*, 1982)
2. Detection of virulent *Yersinia enterocolitica* in food (Hill *et al.*, 1983)
3. Detection of *Neisseria gonorrhoeae* in cases of urethritis (Totten *et al.*, 1983)
4. Identification of enterotoxigenic *E. coli* in children (Echeverria *et al.*, 1984a)
5. Prevalence of heat-stable II enterotoxigenic *E. coli* in pigs, water, and humans (Echeverria *et al.*, 1984b)
6. Identification of enterotoxigenic *E. coli* (Hill and Payne, 1984)
7. Detection of catabolic genotypes (Sayler *et al.*, 1985)
8. Detection of mercury resistance genes in natural bacterial populations (Barkay *et al.*, 1985)
9. Incidence of *Pseudomonas* strains in meat (Ursing, 1986)
10. Phenotypic and genotypic adaptation of aerobic heterotrophic sediment bacterial communities to mercury stress (Barkay and Olson, 1986)
11. Identification of 4-chlorobiphenyl degradative plasmids (Pettigrew and Sayler, 1986)

sequence which expressed in a wide range of hosts would have the advantage that it would allow the investigator to integrate phenotypic and genotypic methods of detection. The effectiveness of gene cassette methods are dependent on maintaining linkage of the target cassette to the R-DNA. The parameters that might optimize linkage of the marker sequence to the R-DNA sequence have not yet been defined and have to be determined experimentally, but the inclusion of different domains of the marker sequence flanking the R-DNA sequence may reduce the frequency of loss of linkage to the R-DNA sequence. Inclusion of multiple copies of the same marker sequence would not be recommended, as such sequences could act as sites for recombination and the resulting genetic instability could lead to excision of R-DNA sequences.

A variety of different methodologies are available to the experimental molecular biologist for preparing labeled probes. These methods differ in respect to the type of label used and the mechanism utilized for incorporating that label into the probe. Label may be either a radioisotope, such as ^{32}P, or a nonradioactive label, such as biotin. Either of these labels may be incorporated into double-stranded DNA probes by nick translation or by *in vitro* transcription for RNA probes.

Special considerations apply when using a genetic probe in the environment. For example in the case of probes prepared by nick translation there is the question of whether the probe should be prepared from the whole plasmid, i.e., probe sequence and vector, or from a purified restriction fragment. Most plasmids used for subcloning and maintenance of probe sequences have antibiotic resistance genes or origin of replication sequences that may cross-hybridize with related sequences present among naturally occurring bacterial populations. Events like these could result in an increased frequency of false-positive identifications, and therefore it is advisable to prepare nick-translated probes from purified restriction fragments (Barkay *et al.*, 1985). Cross-hybridization between the *Pseudomonas* degradative plasmids TOL and NAH has been documented by Sayler *et al.* (1985). A control may be included to ensure that the probe preparation is free from contamination by vector plasmid DNA sequences. A probe preparation free of vector DNA contamination will give a negative result upon hybridization with a test organism that contains the vector plasmid without inserted cloned DNA.

4.2. Nick-Translated Probes

Nick translation (Kelly *et al.*, 1970; Rigby *et al.*, 1977) is the most common method currently employed in molecular biology laboratories for the labeling of probes. The substrate for the nick-translation reaction is a double-stranded DNA molecule, which may be either an entire plasmid or a restriction fragment. In the latter case the fragment is prepared by restriction enzyme digestion and purified away from the vector fragments by preparative agarose gel electrophoresis (Maniatis et al., 1982).

A nick is introduced into the duplex plasmid or restriction fragment DNA molecule using DNaseI. This nick is translated along the probe DNA by the combined $5' > 3'$ exonuclease and $5' > 3'$ polymerase activity of Escherichia coli DNA polymerase I. The reaction is performed in a buffer containing all four deoxynucleoside triphosphates (dNTPs), one or more of which is labeled. As a consequence of the turnover reaction of nick translation, the dNTPs, including the labeled dNTP, are successively incorporated into the probe DNA. The probe DNA is then separated from the unincorporated dNTPs by, for example, sequential ethanol precipitation. The resulting labeled probe is a double-stranded DNA molecule, which is denatured by boiling prior to hybridization.

4.3. Transcription Probes

Use of asymmetric RNA probes results in a dramatic increase in hybridization efficiency relative to nick-translated DNA probes (Zinn et al., 1983; Cox et al., 1984; Church and Gilbert, 1984). Different transcription vector–in vitro transcription systems are available that can facilitate the production of large quantities of RNA probe of high specific activity (Kassavetis et al., 1982; Melton et al., 1984). The probe DNA sequence is inserted into a cloning site in the transcription vector. Adjacent to the cloning site is positioned a promoter sequence which displays a specificity for a given RNA polymerase. Commonly used promoter/polymerase systems are derived from bacteriophages SP6, T3, and T7 (Chamberlain et al., 1970; Butler and Chamberlain, 1982; Davanloo et al., 1984; Melton et al., 1984). Transcription vectors are available in which the polylinker is flanked by two different promoters, the polarity of which allows differential RNA synthesis using either strand of the plasmid as template.

Prior to transcription the vector plasmid is linearized by restriction enzyme cleavage within the cloning site, $3'$ to the probe sequence. This ensures precise termination of transcription. Transcription with the appropriate RNA polymerase takes place in the presence of all four NTPs, one or more of which may be labeled. Termination of the reaction and removal of the vector is achieved by treatment with RNase-free DNase and phenol extraction (Green et al., 1983). Unincorporated label is removed by sequential ethanol precipitation. The elimination of self-hybridization contributes to the enhanced efficiency of hybridization achieved with RNA probes. In addition, purification of the probe restriction fragment prior to labeling is not necessary.

4.4. Choice of Label for Genetic Probes

Most laboratories label probes by nick translation with ^{32}P-labeled nucleotides. Hybridization is visualized by autoradiography as shown in Fig. 1. Although protocols for hybridization using ^{32}P-labeled probes are well established, there are a number of drawbacks. First, radioactive isotopes require special precautions in handling, storage, and disposal, with attendant hazard and expense. Second, the short half-life of ^{32}P limits the shelf-life of labeled probes to about 2 weeks, necessitating frequent shipment of radioisotopes and preparation of probes.

A number of alternative methods have been published for the detection of hybridization by nonautoradiographic methods. Two labels are commonly used, biotin (Langer *et al.,* 1981; Leary *et al.,* 1983) and 2-acetylaminofluorene (AAF) (Landegent *et al.,* 1985). Either may be directly coupled to the nucleotide that will be incorporated into the probe or introduced via specific antibodies after hybridization. Biotin has been widely adopted as a nonradioactive label and may be incorporated into DNA fragment probes by nick translation with Biotin-11-dCTP as label or incorporated into RNA transcript probes by transcription with Biotin-11-UTP as label (Langer *et al.,* 1981). Hybridized probe DNA is detected by a soluble detection complex; for example, a streptavidin–horseradish peroxidase complex visualization method may be used for colony hybridizations (see Fig. 2). Special precautions have to be taken when probing colony lysates with biotin-labeled probes, as proteins present in the bacterium may give rise to nonspecific background. This problem may be resolved by removal of protein from the filter by protein blocking. Protein blocking may be achieved by either phenol/chloroform treatment or proteinase K treatment of the filters after lysis (Taub and Thompson, 1982; Langner and Girton, 1985).

One major drawback of nonautoradiographic labeling systems relative to autoradiographic methods is a reduction in sensitivity of hybridization detection. This reduction in efficiency may be compensated for by labeling probes by *in vitro* transcription.

4.5. Stringency

The term stringency is used to describe the conditions for washing hybrids that result from the annealing of complementary nucleic acids. Stringency relates to the melting temperature T_m of a hybrid. The conditions under which duplex hybrids melt are dependent on the washing temperature, salt concentration, sequence homology, and sequence length. Stringency is considered to be high at elevated temperatures and low salt concentrations. Low stringency occurs at reduced temperatures and high salt concentrations.

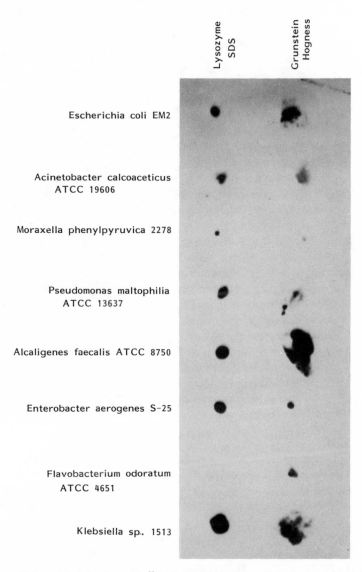

Figure 1. Colony hybridization with ^{32}P-labeled DNA probe. Colonies were lysed by the lysozyme–SDS method and the Grunstein and Hogness (1975) method. Filters were hybridized with a ribosomal gene probe labeled with ^{32}P by nick translation. Autoradiography was overnight at $-70°$C using Kodak XRP-5 film.

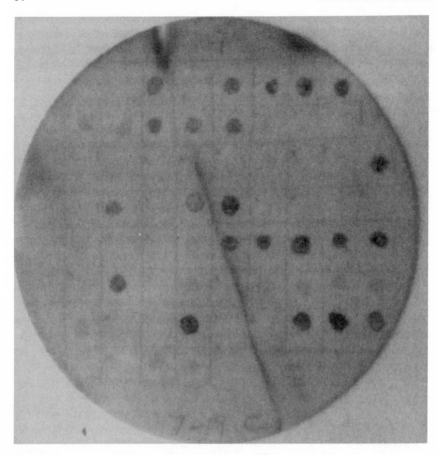

Figure 2. Colony hybridization with a biotinylated DNA probe. The filter, spotted with clinical isolates of *Neisseria gonorrhoeae*, was hybridized with a biotinylated probe derived from an *E. coli* beta lactamase plasmid. Detection was by the streptavidin–horseradish peroxidase complex method. Ampicillin-resistant strains were identified by brown color; ampicillin-sensitive strains were colorless or showed a weak response. [After H.-L. Yang, Enzo Biochem. Inc., by permission.]

The T_m of a DNA duplex decreases by 1.5°C for every 1% increase in the number of mismatched base pairs. Therefore, duplexes formed from sequences that are divergent will melt at lower temperatures than duplexes formed from more closely related sequences. In the case of hybridizations performed on environmental isolates this has great importance. For example, in studies on the distribution of mercury resistance genes among the natural bacterial populations of mercury-contaminated soils a probe, *mer*, was derived from the mercury resistance operon of the plasmid R100 (Davies and Rownd, 1972). This *mer* probe shares homology with the mercury resistance systems of a number of different Gram-negative soil microorganisms (Barkay *et al.,* 1985). In cases like this particular care has to be taken with regard to the selection of washing conditions. Stringency should be sufficiently high that it discriminates against other genetic systems that have only a slight degree of sequence homology with the mercury resistance system, yet low enough that it allows the identification of mercury resistance genes that may be only 75% homologous. The optimal conditions for washing hybrids may be determined empirically (Barkay *et al.,* 1985).

In the case of identifying the incidence of specific DNA sequences introduced into microorganisms by genetic engineering, the investigator would most likely be able to use as high a stringency as is possible. In such a case a genetic probe that is completely homologous to the sequence of interest will usually be available. Consequently, the occurrence of false positives due to cross-hybridization with other genes in the environment will be reduced. Maniatis *et al.* (1982) recommend the use of a combination of salt concentration and temperature that is 5°C below the T_m of the hybrid. This combination of conditions may be determined empirically.

In cases where the DNA sequence to be detected is a novel sequence not normally present in the bacterial community of that environment, higher stringency washes may be used. This will reduce the occurrence of false positives due to cross-hybridization with other genes in the environment. Conditions of stringency for a variety of buffer systems and hybridization supports have been described (Maniatis *et al.,* 1982).

5. Genotypic Analysis: Methods of Analysis

Methods for investigating the occurrence of specific DNA sequences in bacterial populations by DNA hybridization fall into two main categories, those that include plating of organisms isolated from a given location and those that are not dependent on plating of organisms. In each case it is important that methods yield results that accurately represent the different organisms present in the original sample. For methods based on plating techniques, such as colony hybridization to isolated microorganisms (Barkay and Olson, 1986) and direct hybridization to high-density bacterial platings (Sayler *et al.,* 1985; Pettigrew and Sayler, 1986), the choice of plating medium becomes an important consideration (see Section 2.1 for a discussion of the importance of plating medium).

Also of considerable importance in these methods is the efficiency of the technique used to extract DNA from the microorganisms. Grunstein and Hogness (1975) describe a high pH method for the lysis of *E. coli* colonies on nitrocellulose. This method has been used by Sayler *et al.* (1985) to analyze aquatic sediments. A lysozyme–SDS modification of this technique has been developed (Barkay, unpublished data), which results in lysis superior to the Grunstein and Hogness method for a number of environmentally important Gram-negative species, including *Enterobacter, Alcaligenes, Klebsiella, Pseudomonas, Acinetobacter,* and *Moraxella* (see Fig. 1). Methods used for the lysis of some Gram-positive organism, such as *Bacillus subtilis* in liquid culture (Niaudet and Ehrlich, 1979; Hofemeister *et al.,* 1983), differ little from the lysozyme–SDS modification of the Grunstein and Hogness method, although treatments are somewhat harsher. The efficiency of lysing environmentally significant Gram-positive organisms on nitrocellulose using the lysozyme–SDS modified approach awaits evaluation.

Criteria applied in evaluating a lysis procedure include reproducibility, ability of the method to lyse all organisms represented in the bacterial community with approximately equal efficiency, and ability to assess the efficiency of lysis. Among the approaches available for testing the efficiency of lysis are: (1) hybridization with probe derived from a bacterial ribosomal gene probe, such as plasmid pSPC-1 (Fox *et al.,* 1980; Mark *et al.,* 1983); such probes may be cross-reactive for a range of bacterial isolates, or (2) the use of dyes such as methylene green or methyl green, which specifically stain nucleic acid immobilized on filters (Grunstein, 1983). These two methods of lysis assessment may be used independently and can successfully discriminate between lysed and unlysed bacterial spots on both nitrocellulose and nylon-based supports (S. Ford and B. H. Olson, unpublished data).

In the case of methods that are not dependent on plating, the representativeness of the procedure used to extract the DNA is generally important in determining the validity of the approach. In this case it is more difficult to determine that the bacterial lysate is representative of the original bacterial community. One approach to determine the heterogeneity of the DNA is reassociation kinetics (Britten and Kohne, 1968; Torsvik, 1980; Sayler and Stacey, 1986).

5.1. Hybridization to Plated Bacteria

5.1.1. Direct Hybridization to High-Density Bacterial Plating

This method is currently state-of-the-art technology for high sensitivity of detection of genetic sequences of interest in aquatic sediments (Sayler *et al.,* 1985; Pettigrew and Sayler, 1986). However, the precision of

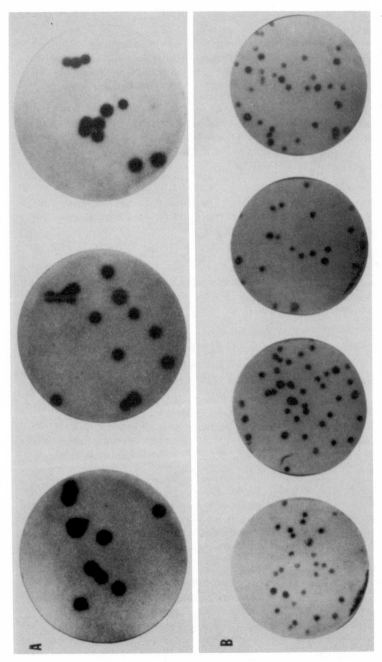

Figure 3. Direct hybridization to plated bacteria. Bacterial lawns were transferred to filters and treated as described by Sayler *et al.* (1985). Filters were hybridized with the *Pseudomonas putida* toluene catabolic (TOL) plasmid as probe. Probe was labeled by nick translation with ^{32}P. Plates contained *E. coli* and *P. putida* in varying ratios. (A) High density of plating; 5×10^5 CFU per plate (low frequency *P. putida*, high frequency *E. coli*: 2.3×10^{-5} *P. putida*). (B) Low density of plating; 77 CFU per plate (high frequency *P. putida*, low frequency *E. coli*: 4.6×10^{-1} *P. putida*).

this method for conditions encountered in terrestrial samples, which often show high variability in designated parameters within a few feet, is not well delineated.

This method is similar to plasmid screening at high colony density (Hanahan and Meselson, 1980). The method consists in spread plating of bacteria followed by transfer to a solid support, lysis (Grunstein and Hogness, 1975), and subsequent hybridization (see Fig. 3). The method may be used in a screening strategy similar to that employed in the identification of cosmid clones (Grosveld et al., 1981; Ish-Horowicz and Burke, 1981). Bacteria from environmental samples may be plated at a range of different cell densities. High-density plating (Sayler et al., 1985) facilitates the screening of up to 5×10^5 cells per plate, although it is not possible to recover directly pure cultures of organisms of interest. In addition, it is difficult to ensure that bacterial lysis is uniform in samples that may be highly heterogeneous. Low-density plating (Pettigrew and Sayler, 1986) may be used as a secondary screening stage. By analogy with the cosmid screening strategy, a mixed culture of bacteria isolated from a high-density bacterial lawn may be screened at low plating density. In this way a pure culture of a positively hybridizing organism may be obtained for further characterization. A tertiary round of low-density screening may be necessary to obtain a pure culture. Such a screening strategy allows for tremendous sensitivity, although an inherent limitation is the inability to ensure adequate lysis at high plating densities.

5.1.2. Colony Hybridization and Identification

This technique is based on a method developed for identifying recombinant DNA clones in *Escherichia coli* K12 by hybridization (Grunstein and Hogness, 1975). The technique has been modified in this laboratory for the analysis of Gram-negative microorganisms commonly found in the soil environment and aquatic sediments (Barkay and Olson 1986). The colony hybridization method consists in individually applying previously isolated bacteria to a support filter (nitrocellulose) and subsequently lysing the bacteria on the filter, followed by hybridization to the probe of interest. The isolates from the sample may also be subjected to a number of other tests, for example, identification and phenotypic characterization. Although this method limits the sensitivity of R-DNA detection in that a limited number of colonies are examined, it allows a more thorough examination of the community via identity of the host and the comparison of genotype and phenotypic expression. Modifications to this technique have assisted in handling environmental samples and facilitated accurate determination of lysis.

A modification of this method that avoids several problems that occur with colony hybridization is the application of pure broth cultures of isolated bacteria to the filter using a "dot-blot" type apparatus, which is designed to apply small volumes of liquid cell cultures to nitrocellulose by vacuum filtration. The application of bacteria in this way enables the accurate quantitation of the number of bacteria applied and adjustment may be made for organisms that grow slowly. Further, it alleviates the need for growth of the bacteria after application to the filter, avoiding the problems of smearing encountered when swarming bacteria are present in test samples. This method also makes the results amenable to densitometer analysis.

Samples applied using a "slot-blot" apparatus result in the deposition of the bacteria in the form of a bar. In this case the filter may be cut along the center of each sample bar. Each half of the filter may then be hybridized separately to different probes. In this way it is possible to probe the same filter simultaneously with two separate probes. In addition the filter may be stained after hybridization with a dye for detecting the efficiency of lysis (see introduction to Section 5). Known organisms should be applied to each filter. These should be appropriate to the probe(s) to be used and include positive and negative controls.

5.2. Direct Extraction

The preceding methods of analysis are based on plating techniques. Alternative approaches in which DNA is extracted directly from environmental samples have the advantage that they are not dependent on subculturing of organisms. Studies on the separation and purification of bacteria from soil illustrate the importance of direct extraction approaches for investigating the occurrence of GEMS and R-DNA. A number of methods are available for separating soil bacteria from other components of the soil. Physical properties of the soil significantly influence both the efficiency of recovery and purity of the bacterial fraction. Bakken (1985) found that, using a blending-centrifugation procedure, the percentage yield of bacterial cells was negatively correlated with the clay content of the soil, whereas the purity was positively correlated with it. Faegri et al. (1977) describe a technique for preparing a relatively pure soil bacterial fraction, free of other organisms. Estimation of the DNA content of soil bacteria from a sample containing 1.1×10^{10} bacteria/g dry weight gave values of 8.4 fg DNA/cell or 90 μg DNA/g soil dry weight (Torsvik and Goksoyr, 1978). The authors concluded that there is very good agreement between the DNA content of bacteria as determined by their method and the microscopic count of bacteria, whereas plate counts of bacteria from this same soil gave values 2000 times lower than the microscopic count. This strongly suggests that the recovery of organisms by plating methods is an underestimate and may not be representative of the original soil community.

Estimation of the nucleic acid contents of Soviet zonal soils indicates total DNA contents ranging between 40 and 300 μg DNA/g soil (Aseyeva and Samko, 1984). Nucleic acid content was found to vary, depending on the type of soil; soils of taiga forest and humid subtropical forest regions showed highest nucleic acid concentrations, whereas the smallest quantities of nucleic acids were found in desert-steppe regions. Nucleic acid content was found to be correlated with the concentration of organic matter. Quantitation of RNA demonstrated that RNA comprises between 33% and 42% of the total nucleic acid of these soils. An alternative method for the determination of RNA and DNA concentrations in soil is described by Nannipieri et al. (1986). They used colorimetric methods to determine the concentration of nucleic acids after removal of low-molecular weight ribose- and deoxyribose-containing compounds. Although reliable for soils with organic carbon contents below 2%, interference due to contaminating humic molecules is reported in richer soils.

 A method for preparing relatively pure DNA from soil samples without culturing of organisms is described by Torsvik (1980). In this method the bacterial fraction is prepared from homogenized soil suspensions by a blending-centrifugation method accompanied by washes with sodium hexaphosphate to remove as much humic material as possible. A modification of the method of Marmur (1961) is used to lyse the bacterial fraction and the resulting bacterial lysate purified on hydroxyapatite. DNA preparations resulting from this procedure were of an apparently high degree of purity. Molecular weight determinations based on analytical ultracentrifugation indicated that DNA was native, of relatively high molecular weight, but displayed great variation in size range [(2.3–10.1) × 10⁵ daltons]) (Torsvik, 1980). Agarose gel electrophoresis of DNA prepared by this method from soils confirms the heterogeneity of size of the DNA preparation (S. Ford and B. H. Olson, unpublished observation). It seems likely that with some modification, such as minimizing physical shearing and CsCl isopycnic centrifugation it may be possible to improve the recovery of high-molecular weight DNA. Yields of up to 1.5 mg DNA/90 g wet soil (37 g dry weight) are reported for this method (Torsvik, 1980). Thermal denaturation analysis of isolated DNA demonstrated that the soil bacterial DNA has a broad melting profile characteristic of highly heterogeneous DNA (Torsvik, 1980). On the basis of these data the author suggested that the isolated DNA contains the genetic information of nearly all the bacteria present in the soil in amounts that are proportional to their relative abundance in the soil.

This method or a modification of this method may be used for preparing DNA directly from soils and adapted for use on samples collected from other environments. This could prove to be a powerful tool for investigating the occurrence of R-DNA sequences. There are a variety of ways in which DNA prepared by this method may be analyzed in investigating the distribution of R-DNA sequences. First DNA may be applied directly to a nitrocellulose filter using a slot blot apparatus. The occurrence and relative abundance of the R-DNA sequence in the sample may be determined by hybridization with a labeled probe. This type of analysis will give information on the apparent number of copies of the DNA sequences capable of hybridizing with the probe in terms of the number of grams of soil from which the sample was prepared. It should also be noted that dot blot analysis, unlike fingerprinting (see below), is not limited by partially degraded DNA preparations.

DNA may be subjected to restriction pattern analysis or "fingerprinting," a method that compares DNA molecules on the basis of the positions of restriction sites. Fingerprinting has proved to be an extremely useful epidemiological tool for tracing the transmission of plasmids from one bacterial strain to another and for tracing the sources of epidemic plasmids (Jameison *et al.,* 1979; Portnoy and Falkow, 1981; Schaberg *et al.,* 1981; O'Brien *et al.,* 1982; Farrar, 1983). Common or conserved DNA sequences have similar restriction patterns due to conservation of restriction sites, whereas unlike sequences differ in their restriction patterns. Therefore the occurrence of an R-DNA sequence may be determined by identification of specific restriction patterns. The DNA is cleaved with restriction enzymes, separated by agarose gel electrophoresis, and transferred to nitrocellulose by the method of Southern (1975). The nitrocellulose filter may then be hybridized with the probe of interest. Choice of the restriction enzyme for cleavage of the DNA is important and information derived from the nucleotide sequence or restriction map of the sequence of interest can be utilized in this process. In order to quantitate the incidence of an R-DNA sequence, a restriction enzyme that cuts infrequently, but at least twice, within the R-DNA sequence may be used that results in a restriction fragment characteristic of the R-DNA sequence. Analysis of a DNA sample with more than one restriction enzyme will facilitate accurate identification of the occurrence of the sequence. More qualitative data about the R-DNA sequence may be gained by using a restriction enzyme that does not cut within the R-DNA sequence or cuts the R-DNA sequence only once. Cleavage with this enzyme will enable the experimenter to determine the sizes of the restriction fragments that flank the R-DNA. Analysis of this type would have great value in cases where translocation or transfer of R-DNA sequences had taken place. In effect this approach is a measure of the position of the R-DNA within the genome. A limitation of fingerprinting as a method of analysis for DNA prepared by the method of Torsvik is that a DNA preparation of high molecular weight (10^7 dalton minimum average size) is required. If a degraded preparation is used, extensive background will be apparent on hybridization and precise determination of the restriction pattern will be difficult.

In the case of monitoring plasmid-borne traits, pulsed-field electrophoresis may be employed to determine sizes of the plasmids involved. This gel system (Carle and Olson, 1984; Carle *et al.*, 1986; Schwartz and Cantor, 1984) separates large DNA molecules (up to 2000 kb) and is thus highly suitable for resolving environmental plasmids, which are often too large to resolve by conventional agarose gel electrophoresis. Undigested DNA samples separated by pulsed-field electrophoresis may also be analyzed by Southern blot transfer (Southern, 1975), which will allow for the identification of the replicon on which R-DNA sequences are carried.

6. Summary

We have outlined some techniques for monitoring the occurrence of a GEM. These methods assay a phenotype exhibited by the organism or monitor directly the occurrence of specific DNA sequences in that organism. Practical application of this technology in the monitoring of field trials of GEMs will require the elaboration of detailed strategies for the investigation of widely differing organisms under a variety of different environmental conditions.

Appropriate combinations of techniques may be used in a general monitoring strategy that could be applied to a variety of free release field trials. The development of such a strategy is currently dependent upon carefully controlled comparative evaluations of methods of detection. Evaluations using microcosm studies and field trials must include testing under different physical and chemical conditions and take into account sampling approach, seasonal aspects, and other specific environmental conditions.

In deploying the available techniques in an investigative strategy it is useful to consider such techniques in quantitative and qualitative terms. For the initial test in such a strategy accurate quantitation at maximum sensitivity should be the goal. Thus, in cases where the use of a probe is practical, a determination may be made of the incidence of the R-DNA sequence in the sampling site and the need for further investigation can be assessed. In this way it may be found, for example, that transport of a GEM had occurred from the introduced environment to a different environment. The type of approach that might serve best for this initial determination may be dot-blot analysis and hybridization to DNA directly extracted from an environmental sample by a modification of the method of Torsvik. If this initial test yields positive identifications of R-DNA sequences, a combination of other approaches, tailored to the specific experimental circumstances, may be employed to give more qualitative information on the events that contributed to the occurrence of the R-DNA at that particular location. In this way it should be possible to produce a flexible strategy for evaluating the spread of GEMS and R-DNA in a variety of different environmental circumstances.

References

Alexander, M., 1986, Survival and growth of bacteria, *Environ. Manage.* **10**:464–469.

Alwine, J. C., Kemp, D. J., Parker, B. A., Reiser, J., Renart, J., Stark, G. R., and Wahl, G. M., 1979, Detection of specific RNAs or specific fragments of DNA by fractionation in gels and transfer to diazobenzylmethyl paper, *Meth. Enzymol.* **68**:220–242.

Anderson, G., 1958, Identification of derivatives of deoxyribonucleic acid in humic acid, *Soil Sci.* **86**:169–174.

Anderson, G., 1979, Bacterial DNA in soil, *Soil Biol. Biochem.* **11**:213.

APHA (American Public Health Association), 1985, *Standard Methods for the Examination of Water and Wastewater*, 16th ed., American Public Health Association.

Aseyeva, I. V., and Samko, O. T., 1984, Nucleic acids and biomass of microcosms in zonal soils, *Sov. Soil Sci.* **24**:84–90.

Atlas, R. M., and Bartha, R., 1981, *Microbial Ecology, Fundamentals and Applications,* Addison-Wesley, Reading, Massachusetts.

Bakken, L. R., 1985, Separation and purification of bacteria from soil, *Appl. Environ. Microbiol.* **49**:1482–1487.

Barkay, T., and Olson, B. H., 1986, Phenotypic and genotypic adaptation of aerobic heterotrophic sediment bacterial communities to mercury stress, *Appl. Environ. Microbiol.* **52**:403–406.

Barkay, T., Fouts, D. L., and Olson, B. H., 1985, Preparation of a DNA gene probe for detection of mercury resistant genes in Gram-negative bacterial communities, *Appl. Environ. Microbiol.* **49**:686–692.

Barnthouse, L. W., and Palumbo, A. V., 1986; Assessing the transport and fate of bioengineered microorganisms in the environment, In: *Biotechnology Risk Assessment, Issues and Methods for Environmental Introductions,* Fiksel, J., and Covello, V. T., eds., Pergamon Press, NY.

Benton, W. D., and Davis, R. W., 1977; Screening lambda gt recombinant clones by hybridization to single plaques in situ, *Science* **196**:180–182.

Billing, E., and Garrett, C. M. E., 1980, Phages in the identification of plant pathogenic bacteria, in *Microbial Classification and Identification of Plant Pathogenic Bacteria* (M. Goodfellow and G. Board, eds.), pp. 319–338, Academic Press, Toronto.

Bohlool, B. B., 1975, Occurrence of *Sulfolobus acidocaldarius,* an extremely thermophilic bacterium, in New Zealand hot springs: Isolation and immunofluorescence characterization, *Arch. Microbiol.* **106**:171–174.

Bohlool, B. B., and Brock, T. D., 1974, Immunofluorescence characterization, *Arch. Microbiol.* **106**:171–174.

Bohlool, B. B., and Schmidt, E. L., 1968, Nonspecific staining: Its control in immunofluorescence examination of soil, *Science* **162**:1012–1014.

Bohlool, B. B., and Schmidt, E. L., 1973, Persistence and competition aspects of *Rhizobium japonicum* in soil, *Soil Sci.* **110**:229–236.

Bohlool, B. B., and Schmidt, E. L., 1980, The immunofluorescence approach in microbial ecology, in *Advances in Microbial Ecology,* Vol. 4 (M. Alexander, ed.), pp. 203–236, Plenum Press, New York.

Bordner, R., and Winter, J., 1978, *Microbiological Methods for Monitoring the Environment, I. Water and Wastes,* Environmental Protection Agency, Cincinnati, Ohio.

Britten, R. J., and Kohne, D. E., 1968, Repeated sequences in DNA, *Science* **161**:529–540.

Brown, E. R., and Cherry, W. B., 1955, Specific identification of *Bacillus anthracis* by means of a varient bacteriophage, *J. Infect. Dis.* **96**:34–39.

Brown, M. R. W., and Williams, P., 1985, The influence of environment on envelope properties affecting survival of bacteria in infections, *Annu. Rev. Microbiol.* **39**:527–556.

Butler, E. T., and Chamberlin, M. J., 1982, Bacteriophage SP-6 RNA polymerase I. Isolation and characterization of the enzyme, *J. Biol. Chem.* **257**:5772–5778.

Carle, G. F., and Olson, M. V., 1986, Elctrophoretic separations of large DNA molecules by periodic inversion of the electric field, *Science* **232**:65–68.

Chai, T.-J., 1983, Characteristics of *E. coli* grown in bay water as compared with rich medium, *Appl. Environ. Microbiol.* **45**:1316–1323.

Chamberlin, M., McGrath, J., and Waskell, L., 1970, New RNA polymerase from *E. coli* infected with bacteriophage T7, *Nature* **228**:227–231.

Chandler, M. E., and Yunis, J. J., 1978, A high resolution *in situ* hybridization technique for the direct visualization of labeled G-banded early metaphase and prophase chromosomes, *Cytogenet. Cell Genet.* **22**:352–356.

Cherry, W. B., Davies, B. R., Edwards, P. R., and Hogan, R. B., 1954, A simple procedure for the identification of the genus *Salmonella* by means of a specific bacteriophage, *J. Lab. Clin. Med.* **44**:51–55.

Church, G. M., and Gilbert, W., 1984, Genomic sequencing, *Proc. Natl. Acad. Sci. USA* **81**:1991–1995.

Cox, K. H., DeLeon, D. V., Angerer, L. S., and Angerer, R. S., 1984, Detection of mRNAs in sea urchin embryos by *in situ* hybridization using asymmetric RNA probes, *Dev. Biol.* **101**:485–502.

Cuppels, D. A., 1984, The use of pathovar-indicative bacterophages for rapidly detecting *Pseudomonas syringae* pv. *tomato* in tomato leaf and fruit lesions, *Phytopathology* **74**:891–894.

Danso, S. K. A., Habte, M., and Alexander, M., 1973, Estimating the density of individual bacterial populations introduced into natural ecosytems, *Can. J. Microbiol.* **19**:1450–1451.

Datta, N., 1984, Bacterial resistance to antibiotics, in: *Origins and Development of Adaptation,* pp. 204–218, Pitman, London.

Datta, N., and Hughes, V. M., 1983, Plasmids of the same *Inc* groups in enterobacteria before and after the medical use of antibiotics, *Nature* **306**:616–617.

Davanloo, P., Rosenberg, A. H., Dunn, J. J., and Studier, F. W., 1984, Cloning and expression of the gene for bacteriophage T7 RNA polymerase, *Proc. Natl. Acad. Sci. USA* **81**:2035–2039.

Davies, J. E., and Rownd, R., 1972, Transmissible multiple drug resistance in Enterobacteraiceae, *Science* **176**:758–768.

Diatloff, A., 1977, Ecological studies of root-nodule bacteria introduced into field environments. 6. Antigenic stability in *Lotononis* rhizobia over a 12 year period, *Soil Biol. Biochem.* **9**:85–88.

Dudman, W. F., and Heidelberger, M., 1969, Immunochemistry of newly found substituents of polysaccharides of *Rhizobium* species, *Science* **164**:954–955.

Echeverria, P., Seriwatana, J., Leksomboon, U., Tirapat, C., Chaicumpa, W., and Rowe, B., 1984a, Identification by DNA hybridization of enterotoxigenic *Escherichia coli* in homes of children with diarrhea, *Lancet* **1**:63–66.

Echeverria, P., Seriwatana, J., Patamaroj, U., Moseley, S. L., MacFarland, A., Chityothin, O., and Chaicumpa, W., 1984b, Prevalence of heat-stable enterotoxigenic *Escherichia coli* in pigs, water, and people at farms in Thailand as determined by DNA hybridization, *J. Clin. Microbiol.* **19**:489–491.

Ellwood, D. C., and Tempest, D. W., 1972, Effects of environment on bacterial wall contents and composition, *Adv. Microbiol. Phys.* **7**:83–117.

Faegri, A., Torsvik, V. L., and Goksoyr, J., 1977, Bacterial and fungal activities in soil: Separation of bacteria and fungi by a rapid fractionated centrifugation technique, *Soil Biol. Biochem.* **9**:105–112.

Farkas-Himsley, D., 1964, Killing of chlorine-resistant bacteria by chlorine-bromine solutions, *Appl. Microbiol.* **12**:1–6.

Farrar, W. E., 1983, Molecular analysis of plasmids in epidemiological investigation, *J. Infect. Dis.* **148**:1–5.

Fliermans, C. B., Bohlool, B. B., and Schmidt, E. L., 1974, Autecological study of the chemoautotroph *Nitrobacter* by immunofluorecence, *Appl. Microbiol.* **27**:124–129.

Fox, G. E., Stackebrandt, E., Hespell, R. B., Gibson, J., Maniloff, J., Dyer, T. A., Wolfe, R. S., Balch, W. E., Tanner, R. S., Magrum, L. J., Zablen, L. B., Blakemore, R., Gupta, R., Bonen, L., Lewis, B. J., Stahl, D. A., Luehrsen, K. R., Chen, K. N., and Woese, C. R., 1980, The phylogeny of prokaryotes, *Science* **209**:457–463.

Garvey, J. S., Cremer, N. F., and Susdorf, D. H., 1977, *Methods in Immunology*, 3rd ed., Benjamin, Reading, Massachusetts.

Gillett, J. W., Stern, A. M., Levin, S. A., Harwell, M. A., Alexander, M., and Andow, D. A., 1986, Potential impacts of environmental release of biotechnology products: Assessment, regulation, and research needs, *Environ Manage* **10**:433–563.

Glaser, D., Keith, T., Riley, P., Chambers, A., Manning, J., Hattingh, S. and Evans, R., 1986, *Monitoring techniques for genetically engineered microorganisms in biotechnology and the environment, Research Needs.* Omenn, A. S. and Teich, A. H., eds.; Noyes Data Corporation, New Jersey.

Graham, J. B., and Istock, C. A., 1979, Gene exchange and natural selection cause *Bacillus subtilis* to evolve in soil culture, *Science* **204**:637–638.

Gray, T. R. G., 1973, The use of the fluorescent-antibody technique to study the ecology of *Bacillus subtilis* in soil, *Bull. Ecol. Res. Commun. (Stockholm)* **17**:119–122.

Green, M. R., Maniatis, T., and Melton, D. A., 1983, Human betaglobin pre-mRNA synthesized *in vitro* is accurately spliced in *Xenopus* oocyte nuclei, *Cell* **32**:681–694.

Grosveld, F. G., Dahl, H.-H. M., de Boer, E., and Flavell, R. A., 1981, Isolation of beta-globin related genes from a human cosmid library, *Gene* **13**:227–237.

Grunstein, M., 1983, *Methods for Screening, Colony Hybridization, Plaque Hybridization,* Schleicher and Schuell, Keene, New Hampshire.

Grunstein, M., and Hogness, D. S., 1975, Colony hybridization: A method for the isolation of cloned DNAs that contain a specific gene, *Proc. Natl. Acad. Sci. USA* **72**(10):3961–3965.

Hanahan, D., and Meselson, M., 1980, Plasmid screening at high colony density, *Gene* **10**:63–67.

Heal, O. W., 1970, Methods of study of soil protozoa, in: *Methods of Study in Soil Ecology* (J. Phillipson, ed.), pp. 119–126, UNESCO, Paris.

Heidelberger, M., and Elliot, S., 1966, Cross-reactions of *Streptococcus* (Group N teichoic acid in antipneumonococcal horse sera of type VI, XIV, XVI, XXVII, *J. Bacteriol.* **92**:281–283.

Hill, W. E., and Payne, W. L., 1984, Genetic methods for the detection of microbial pathogens. Identification of enterotoxigenic *Escherichia coli* by DNA colony hybridization: Collaborative study, *J. Assoc. Off. Anal. Chem.* **67**:801–807.

Hill, W. E., Payne, W. L., and Aulisio, C. C. G., 1983, Detection and enumeration of virulent *Yersinia enterocolitica* in food by DNA colony hybridization, *Appl. Environ. Microbiol.* **46**:636–641.

Hirsch, D., and Martin, L. D., 1983, Rapid detection of *Salmonella* spp. by using Felix-01 bacteriophage and high-performance liquid chromatography, *Appl. Environ. Microbiol.* **45**:260–264.

Hofemeister, J., Israeli-Reches, M., and Dubnau, D., 1983, Integration of plasmid pE194 at multiple sites on the *Bacillus subtilis* chromosome, *Mol. Gen. Genet.* **189**:58–68.

Hughes, V. M., and Datta, N., 1983, Conjugative plasmids in bacteria of the preantibiotic era, *Nature,* **302**:725–726.

Ish-Horowicz, D., and Burke, J. F., 1981, Rapid and efficient cosmid cloning, *Nucleic. Acid Res.* **13**:2989–2998.

Jamieson, A. F., Bremner, D. A., Bergquist, P. L., and Lane, H. E. D., 1979, Characterization of plasmids from antibiotic-resistant *Shigella* isolates by agarose gel electrophoresis, *J. Gen. Microbiol.* **113**:73–81.

Jensen, V., 1968, The plate count technique, in: *The Ecology of Soil Bacteria* (T. R. G. Gray and D. Parkinson, eds.), pp. 158–170, University of Toronto Press, Toronto.

Joklik, W. K., and Willett, H. P., 1976, *Zinsser's Microbiology,* 16th ed., Appleton-Century-Crofts, New York.

Jones, K., and Murray, K., 1975, A procedure for detection of heterologous DNA sequences in lambdoid phage by *in situ* hybridization, *J. Mol. Biol.* **51**:393–409.

Kassavetis, G. A., Butler, E. T., Roulland, D., and Chamberlin, M. J., 1982, Bacteriophage SP6-specific RNA polymerase, *J. Biol. Chem.* **257**:5779–5788.

Kelly, R. B., Cozzarelli, N. R., Deutscher, M. P., Lehman, I. R., and Kornberg, A., 1970, Enzymatic synthesis of deoxyribonucleic acid, *J. Biol. Chem.* **245**:39.

Landegent, J. E., Jansen in de Wal, N., van Ommen, G.-J., Baas, F., de Vijlder, J. J. M., van Duijn, P., and van der Ploeg, M., 1985, Chromosomal localization of a unique gene by non-autoradiographic *in situ* hybridization, *Nature* **317**:175–177.

Langer, P. R., Waldrop, A. A., and Ward, D. C., 1981, Enzymatic synthesis of biotin-labeled polynucleotides: Novel nucleic acid affinity probes, *Proc. Natl. Acad. Sci. USA* **78**:6633–6637.

Langner, K., and Girton, J., 1985, DNA detection spot plaque hybridization, *Focus (BRL)* **7**(2):11.

Leary, J. J., Brigati, D. J., and Ward, D. C., 1983, Rapid and sensitive colorimetric method for visualizing biotin-labeled DNA probes hybridized to DNA or RNA immobilized on nitrocellulose: Bio-blots, *Proc. Natl. Acad. Sci. USA* **80**:4045–4049.

Lenski, R. E., 1984, Releasing "ice-minus" bacteria, *Nature* **307**:8.

Lenski, R. E., 1987, The infectious spread of engineered genes, In: *Application of biotechnology: Environmental and policy issues.* In press.

Levin, B. R., 1986, The maintenance of plasmids and transposons in natural populations of bacteria, In: *Banbury Report 24, Antibiotic Resistance Genes: Ecology, Transfer and Expression,* Cold Spring Habor Laboratory, NY, 57–70.

Litchfield, C. D., Raker, J. B., Zindulis, J., Esysnsbe, R. T., and Stein, D. J., 1975, Optimization of procedures for the recovery of heterotrophic bacteria from marine sediments, *Microb. Ecol.* **1**:219–233.

Lugtenberg, B., Peters, R., Bernheimen, H., Bernheimen, W., and Berendsen, W., 1976, Influence of cultural conditions and mutations on the composition of the outer membrane properties of *E. coli, Mol. Gen. Genet.* **147**:251–262.

Mallory, L. M., Sinclair, J. L., Liang, L. N., and Alexander, M., 1982, A simple and sensitive method for assessing the survival in environmental samples of species used in recombinant DNA research, *Recomb. DNA Tech. Bull.* **5**:5–6.

Maniatis, T., Fritsch, E. F., and Sambrook, J., 1982, *Molecular Cloning, A Laboratory Manual,* Cold Spring Harbor Laboratory, Cold Spring Harbor, New York.

Mark, L. G., Sigmund, C. D., and Morgan, E. A., 1983, Spectinomycin resistance due to a mutation in an rRNA operon of *Escherichia coli, J. Bacteriol.* **155**(3): 989–994.

Marmur, J., 1961, A procedure for the isolation of deoxyribonucleic acid from microorganisms, *J. Mol. Biol.* **3**:208–218.

McCoy, W., and Olson, B. H., 1985, Fluorometric determination of the DNA concentration in municipal drinking water, *Appl. Environ. Microbiol.* **49**:811–817.

Melton, D. A., Krieg, P. A., Rebagliati, M. R., Maniatis, T., Zinn, K., and Green, M. R., 1984, Efficient *in vitro* synthesis of biologically active RNA and RNA hybridization probes containing a bacteriophage SP6 promoter, *Nucleic Acids Res.* **12**:7035–7056.

Moseley, S. L., Echeverria, P., Seriwatana, J., Tirapat, C., Chaicumpa, W., Sakuldaipaera, T., and Falkow, S., 1982, Identification of enterotoxigenic *Escherichia coli* by colony hybridization using three enterotoxin gene probes, *J. Infect. Dis.* **145**:863–869.

Nannipieri, P., Ciardi, C., Badalucco, L., and Casella, S., 1986, A method to determine soil DNA and RNA, *Soil Biol. Biochem.* **18**:275–281.

Niaudet, B., and Ehrlich, D. S., 1979, *In vitro* genetic labeling of *Bacillus subtilis* cryptic plasmid pHV400, *Plasmid* **2**:48–58.

O'Brien, T. F., Hopkins, J. D., and Gilleece, E. S., 1982, Molecular epidemiology of antibiotic resistance in *Salmonella* from animals and human beings in the United States, *N. Engl. J. Med.* **307**:1–6.

Page, A. L., 1985, *Methods of Soil Analysis,* Part 2: *Chemical and Microbiological Properties,* American Society of Agronomy, Madison, Wisconsin.

Pettigrew, C. A., and Sayler, G. S., 1986, The use of DNA:DNA colony hybridization in the rapid isolation of 4-chlorobiphenyl degradative bacterial phenotypes, *J. Microbiol. Meth.* **5**:205–213.

Portnoy, D. A., and Flakow, S., 1981, Virulence-associated plasmids from *Yersinia enterocolitica* and *Yersinia pestis, J. Bacteriol.* **148**:877–883.

Pramer, D., and Bartha, R., 1972, Preparation and processing of soil samples for biodegradation studies, *Environ. Lett.* **2**:217–224.

Reanney, D., 1976, Extrachromosomal elements as possible agents of adaptation and development, *Bacteriol. Rev.* **40**:552–590.

Reanney, D., 1977, Gene transfer as a mechanism of microbial evolution, *Bioscience* **27**:340–344.

Rennie, R. J., and Schmidt, E. L., 1977, Immunofluorescence studies of *Nitrobacter* populations in soils, *Can. J. Microbiol.* **23**:1011–1017.

Rigby, P. W. J., Dieckmann, M., Rhodes, C., and Berg, P., 1977, Labeling deoxyribonucleic acid to high specific activity *in vitro* by nick-translation with DNA polymerase I, *J. Mol. Biol.* **113**:237–251.

Rissler, J. F., 1984, Research needs for biotic environmental effects of genetically-engineered microorganisms, *Recomb. DNA Tech. Bull.* **7**:20–30.

Saunders, V. A., and Saunders, J. R., 1987, *Environmental biotechnology in microbial genetics applied to biotechnology,* Macmillan, New York, 384–406.

Sayler, G. S., Shields, M. S., Tedford, E. T., Breen, A., Hooper, S. W., Sirotkin, K. M., and

Davis, J. W., 1985, Application of DNA–DNA colony hybridization to the detection of catabolic genotypes in environmental samples. *Appl. Environ. Microbiol.* **49**(5):1295–1303.

Sayler, G., and Stacey, G., 1986, Methods for evaluation of microorganism properties, In: *Biotechnology Risk Assessment Issues and Methods for Environmental Introductions* (Fiskel, J., and Corello, V. T., eds.), Pergamon Press, NY.

Schaberg, D. R., Tomkins, L. S., and Falkow, S., 1981, Use of agarose gel electrophoresis of plasmid deoxyridonucleic acid to fingerprint Gram-negative bacilli, *J. Clin. Microbiol.* **13**:1105–1108.

Schmidt, E. L., 1973, Fluorescent antibody technique for the study of microbial ecology, *Bull. Ecol. Res. Commun. (Stockholm)* **17**:67–76.

Schmidt, E. L., 1974, Quantitative autecological study of microorganisms in soil by immunofluorescence. *Soil Sci.* **118**:141–149.

Schmidt, E. L., Biesbrock, J. A., Bohlool, B. B., and Marx, D. H., 1974, Study of *Mycorrhizae* by means of fluoroscent-antibody technique, *J. Gen. Microbiol.* **20**:137–139.

Schwartz, D. C., and Cantor, C. R., 1984, Separation of yeast chromosome-sized DNAs by pulsed field gradient gel electrophoresis, *Cell* **37**:67–75.

Shapton, D. A., and Board, R. G., 1971, *Isolation of Anaerobes* (Society for Applied Bacteriology Technical Series No. 5), Academic Press, London.

Southern E., 1975, Detection of specific sequences among DNA fragments separated by gel electrophoresis, *J. Mol. Biol.* **98**:503–517.

Stotzky, G., and Babich, H., 1984, Fate of genetically-engineered microbes in natural environments, *Recomb. DNA Tech. Bull.* **7**:163–188.

Stotzky, G., and Babich, H., 1986, Survival of and genetic transfer by genetically engineered bacteria in natural environments, *Adv. Appl. Microbiol.* **31**:93–165.

Strauss, H., Hattis, D., Page, G. S., Harrison, K., Vogel, S. R., and Caldart, C. C., 1985, Direct release of genetically-engineered microorganisms: A preliminary framework for risk evaluation under TSCA, Center for Technology, Policy and Industrial Development, Massachusetts Institute of Technology, Cambridge, Massachusetts.

Taub, F., and Thompson, E. B., 1982, An improved method for preparing large arrays of bacterial colonies containing plasmids for hybridization: *In situ* purification and stable binding of DNA on paper filters, *Anal. Biochem.* **126**:222–230.

Torsvik, V. L., 1980, Isolation of bacterial DNA from soil, *Soil Biol. Biochem.* **12**:15–21.

Torsvik, V. L., and Goksoyr, J., 1978, Determination of bacterial DNA in soil, *Soil Biol. Biochem.* **10**:7–12.

Totten, P. A., Holmes, K. K., Handesfield, J. S., Knapp, P. L., Perine, P. L., and Falkow, S., 1983; DNA hybridization technique for the detection of *Neisseria gonorrhoeae* in men with urethritis, *J. Infect. Dis.* **148**:462–471.

Trevors, J. T., Barkay, T., and Bourquin, A. W., 1987, Gene transfer among bacteria in soil and aquatic environments: A review, *Can. J. Microbiology* **33**:191–198.

Tsernoglou, D., and Anthony, E. H., 1971, Particle size, water-stable aggregates, and bacterial populations in lake sediments, *Can. J. Microbiol.* **17**:217–227.

Ursing, J., 1986, Similarities of genome deoxyribonucleic acids of *Pseudomonas* strains isolated from meat, *Curr. Microbiol.* **13**:7–10.

Vidor, C., and Miller, R. H., 1980, Relative saprophyte competence of *Rhizobium japonicum* strains in soils as determined by quantitative fluorescent antibody technique, *Soil Biol. Biochem.* **12**:483–487.

Williams Smith, H., and Huggins, M. B., 1980, The association of the O18, K1 and K7 antigens and the ColV plasmid of a strain of *E. coli* with its virulence and immunogenicity, *J. Gen. Microbiol.* **121**:387–400.

Williams Smith, H., and Huggins, M. B., 1982, Successful treatment of experimental *Escherichia coli* infection in mice using phage: Its general superiority over antibiotics, *J. Gen. Microbiol.* **128**:307.

Zinn, K., DiMaio, D., and Maniatis, T., 1983, Identification of two distinct regulatory regions adjacent to the human beta-interferon gene, *Cell* **34**:865–879.

3

Ecology and Recognition in the Nematode–Nematophagous Fungus System

BIRGIT NORDBRING-HERTZ

1. Introduction

Nematophagous fungi have continued to attract attention since their function as predators of nematodes was first recognized at the end of the last century (Zopf, 1888). This fascination is in large part due to the dramatic capturing of nematodes, which can easily be demonstrated when a petri dish sprinkled with soil of appropriate origin is observed under the light microscope. The remarkable morphological adaptations among these fungi are further reasons for continuing interest. The role of these fungi in natural or applied biological control of plant parasitic nematodes has attracted more recent attention. The lack of promising results in early studies within this field, however, has demanded a better understanding of the ecology of these fungi (Cooke, 1968). The continued need for alternative control methods has aroused an interest among current researchers in this field who hope to correlate laboratory results with field investigations for a better understanding of the function of these fungi in soil.

Nematophagous fungi comprise different kinds of fungi, which can be characterized largely as *predators* or *endoparasites of vermiform nematodes* and *parasites of cysts or eggs*. The predatory fungi are those that

BIRGIT NORDBRING-HERTZ • Department of Microbial Ecology, University of Lund, S-223 62 Lund, Sweden.

form trapping structures of different types, and the endoparasites are those that infect nematodes with the aid of their spores and exist mainly within the nematode. The cyst-nematode parasites, finally, infect mainly or exclusively the females, eggs, or larvae of cyst-nematodes. Fungi in this last group have attracted much interest during recent years and have been reviewed several times (Kerry, 1980, 1984; Mankau, 1980, 1981), and high expectations have been attached to their use as biological control agents.

For general information on the biology of these fungi the reader is referred to several recent reviews. Barron's book (1977) is an excellent source of information on all aspects of biology of the fungi, complemented by his later reviews (Barron, 1981, 1982). Different aspects of the biology and ecology of the predatory and endoparasitic fungi have further been discussed in two review articles (Lysek and Nordbring-Hertz, 1983; Nordbring-Hertz and Jansson, 1984) and in reviews by Mankau (1980) and Kerry (1984). Specific attention was paid to a recognition mechanism based on a lectin–carbohydrate interaction in relation to the mycelial development of nematophagous fungi (Nordbring-Hertz, 1984), and the role of fungal lectins in interactions between fungi and higher organisms was discussed by Nordbring-Hertz and Chet (1986). Zuckerman (1983) and Zuckerman and Jansson (1984) evaluated the role of host recognition in nematode–plant and in nematode–fungus interactions as a basis of a new approach for biological control of plant-parasitic nematodes.

The aim of this chapter is mainly to discuss results from studies of the predatory and endoparasitic fungi, which preferably attack vermiform nematodes. Special emphasis is placed on (1) present knowledge on the ecology of these organisms, and (2) basic studies on prey recognition that have been carried out in several laboratories during recent years.

2. Morphological Adaptations

Morphogenesis plays a crucial role in the function of all types of nematophagous fungi. In the predatory fungi, the formation of trapping organs is a prerequisite for the predaceous habit of the fungi (Fig. 1). Also, endoparasites infect nematodes with the aid of spores, which have undergone considerable morphological adaptations (Fig. 2) (Barron, 1977). Some predatory fungi may function as mycoparasites as well, forming hyphal coils around host fungi (Fig. 3) (Tzean and Estey, 1978; Persson *et al.*, 1985). In parasites of the cyst-nematodes a specific morphological change is not so obvious and several of these fungi seem to enter their hosts using hyphae (C. Dackman, personal communication). However,

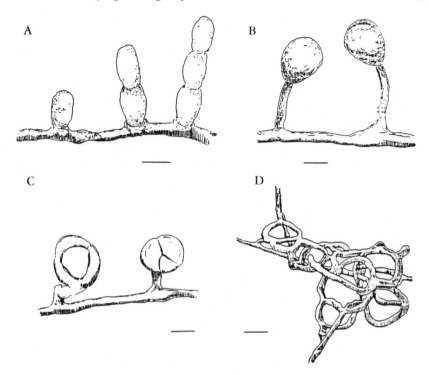

Figure 1. Nematode-trapping structures of some predatory fungi. (A) Adhesive branches of *Monacrosporium cionapagum;* bar: 5 μm. (B) Adhesive knobs of *Dactylaria candida;* bar: 2 μm. (C) Constricting rings of *Arthrobotrys dactyloides;* bar: 10 μm. (D) Adhesive network of *Arthrobotrys oligospora;* bar: 10 μm. [Drawings by Per Helin.]

appressoria-like hyphal swellings have been occasionally reported (e.g., Nigh *et al.,* 1980) and the development of zoospores is essential for some of these parasites (Kerry, 1980). Finally, chlamydospores are common both within cyst parasites and predatory and endoparasitic fungi, again showing that specialized morphological forms play a role in the survival of these fungi (Barron, 1977; Kerry, 1984).

In the classical work on these fungi the different types of morphological adaptations are described in detail and their function demonstrated [cf. reviews by Drechsler (1941) and Duddington (1962)]. Some examples of structures responsible for infection of nematodes by predatory and endoparasitic fungi are shown in Figs. 1 and 2, respectively. Spore morphology and type of trapping structure are helpful keys to the identity of the fungi (Cooke and Godfrey, 1964).

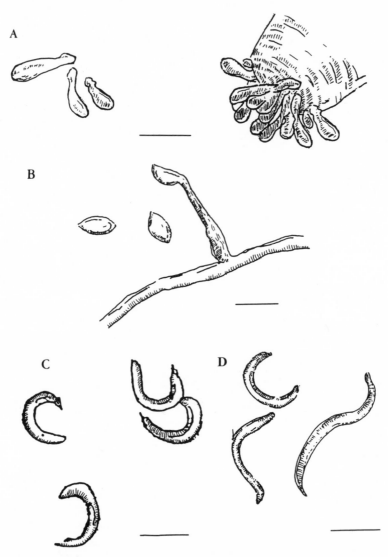

Figure 2. Nematode-infecting conidia of some endoparasitic fungi. (A) Left: adhesive conidia of *Meria coniospora;* right: conidia infecting nematode; bar: 5 μm. (B) Left: conidia of *Hirsutella* sp.; right: conidia on flask-shaped phialide; bar: 10 μm. (C) Conidia of *Harposporium anguillulae;* bar: 10 μm. (D) Conidia of *Harposporium helicoides;* bar: 10 μm. [Drawings by Per Helin.]

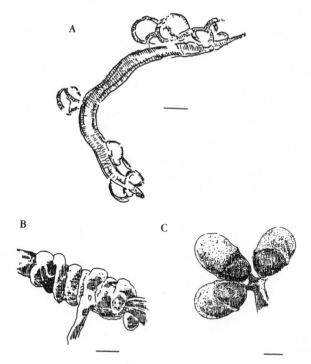

Figure 3. Typical structures of *Arthrobotrys oligospora*. (A) Nematode captured in adhesive network; bar: 20 μm. (B) Hyphal coils of *A. oligospora* around *Rhizoctonia* hypha; bar: 5 μm. (C) Pear-shaped, two-celled conidia; bar: 5 μm. [Drawings by Per Helin.]

2.1. Influence of Biotic and Abiotic Factors on Morphogenesis

In predatory fungi the transition from a saprophytic phase of growth to a parasitic phase, when traps are formed, is influenced by both biotic and abiotic factors. Fungal mycelium, in the presence of nematodes, is induced to form the structures that eventually capture the nematodes. Pramer and Kuyama (1963) ascribed this effect to nemin, a substance liberated by enzymatic hydrolysis of protein and present in nematodes. Later, small peptides with a high proportion of nonpolar and aromatic amino acids were shown to induce heavy trap formation in *Arthrobotrys oligospora* (Nordbring-Hertz, 1973), and such peptides were present in exudates and homogenates of nematodes (Nordbring-Hertz, 1977). Similar compounds produced by biological degradation might assist in bringing about morphogenesis in nature. Volatiles, excreted from the soil microfauna, might also effect morphogenesis, either directly or by changing the environment at the nematode–hyphal interface (Nordbring-Hertz and Odham, 1980).

It is very likely that nematode degradation products are responsible for the heavy trap formation that follows consumption of nematodes. However, experiments with single living nematodes point in addition to a physical action. An active, solitary nematode can induce trap formation in *A. oligospora* in a very short time (~2 hr), provided that the fungus is grown on a medium low in nutrients (Nordbring-Hertz, 1977). The time required for trap formation to start is independent of the number of nematodes added. This is taken as an indication that *initiation* of trap formation is of a physical nature rather than a chemical one. This view was substantiated by the application of an electrical potential to fungal colonies which induced highly synchronized trap formation (B. Nordbring-Hertz, unpublished results) similar to that observed after the addition of living nematodes (Nordbring-Hertz, 1977). It has been suggested that initiation of trap formation could be due to changes in membrane potential as a result of nematode movement, which leads to transport of solutes across the plasma membrane. This view is supported by the observation that changes in the ionic environment of the fungi influence morphogenesis (Lysek and Nordbring-Hertz, 1981, 1983, and unpublished observations).

It is obvious that these fungi function as antagonists of nematodes only if the morphogenic process can proceed undisturbed. Morphogenesis in fungi in general seems to be more easily affected by inhibitory concentrations of various chemicals and volatiles than is vegetative growth. Different results were obtained in nematophagous fungi in this respect. The effect of volatile compounds from nematodes on different fungi varied greatly; e.g., increased levels of CO_2 (5–10%) completely inhibited morphogenesis in *A. oligospora*, whereas in one strain of *A. conoides* high levels of CO_2 induced trap formation (Nordbring-Hertz and Odham, 1980, and references therein).

The only study on the effect of heavy metals on growth and trap formation (Rosenzweig and Pramer, 1980) shows that the response to Cd, Pb, and Zn varied among the seven species investigated. In most cases, inhibition of growth (on corn meal agar) was directly correlated with a decreased capacity for nematode-induced trap formation. In some cases, however, trap formation was either more or less inhibited than was growth. Cadmium exhibited the greatest toxicity. On the other hand, production of collagenase, an enzyme implicated in the penetration of the nematode cuticle, was less affected than was growth and trap formation (Schenck *et al.,* 1980; Rosenzweig and Pramer, 1980). In another study, the effect of cadmium on growth (on dilute corn meal agar) and trap formation of *A. oligospora, A. superba,* and *Dactylaria candida* was a pronounced inhibition of trap formation at concentrations where vegetative growth was unaffected (M. Grip and B. Nordbring-Hertz, unpublished results).

Such differing results suggest that there is an urgent need for critical studies on the effect of environmental factors on growth and development in nematophagous fungi and the ultimate effects of these factors on the soil nematode population.

2.2. Morphology and Saprophytic/Predaceous Ability

The tendency to form trapping devices varies greatly among species of predatory fungi. Fungi that form adhesive network traps are fairly good saprophytes, but can be induced to shift to a predaceous phase by environmental factors. Nematode-trappers with adhesive branches, knobs, or constricting and nonconstricting rings are usually more predaceous than the net-formers. Finally, many endoparasites are apparently obligate parasites, although some can be grown on laboratory media. There is good agreement between laboratory studies of saprophytic/predaceous ability and field experiences: Cooke (1963) found that fungi with adhesive network traps were good competitors in soil, but were unable to decrease nematode numbers, whereas the more predaceous branch- or knob-formers decreased nematode numbers significantly in soil. In a soil microcosm study using *A. oligospora* (adhesive networks), *D. candida* (adhesive knobs), and *Meria coniospora* (adhesive conidia), again the net-former was the least effective in reducing nematode numbers (predacity) (Jansson, 1982b).

Estimations of nematophagous fungi in soil partly confirm these laboratory studies (see Section 3).

3. Nematophagous Fungi in Soil

Nematophagous fungi, often studied in the laboratory because of their intriguing life cycles, have failed to unmask their secrets in the natural environment where they are found. We know that they are present in most types of soil in all parts of the world, especially in soil rich in decomposing organic matter, but we do not know what role they play in regulating the natural nematode population. Cooke (1968), in his critical and foresighted review, stressed the urgent need for research on the ecology of these fungi before any application of biological control could be considered. He based this opinion on the development of biological control methods in his own laboratory soil experiments in the early 1960s and on field biocontrol experiments by others that mostly had failed to bring about significant biological control of the parasitic nematode population (Cooke, 1968). It is clear that estimation of the importance of nematophagous fungi has faced great technical problems because special techniques are required for their recovery and isolation. Furthermore, some of these organisms are obligate parasites and do not grow *in vitro*, whereas others are fairly good saprophytes. The recent development toward an appreciation of the importance of the endoparasites and the cyst-nematode parasites and some members of the predatory group as biological control agents (Kerry, 1984; Stirling *et al.*, 1979) has revealed an increased need for methods not only to detect and describe the organisms, but also to quantify nematode-destroying fungi.

3.1. Techniques to Recover and Isolate Nematophagous Fungi from Soil

Barron (1982) described in detail methods used to isolate and maintain both fungi and nematodes and to estimate the activity of nematode-destroying fungi in soil. These methods are usually based on the sprinkled plate technique for predators and differential centrifugation or Baermann funnel techniques for endoparasites. Nematodes, either those present in the soil sample or those added to the detection plates, are necessary in these techniques both to detect the fungi and to maintain and purify them. A simplified scheme of the methods most often used, compiled from Barron (1982), Gray (1984c), and Dackman *et al.* (1987), is given in Fig. 4 [for further references see Barron (1982)]. With the use of these techniques, nematophagous fungi have been isolated and identified from all types of soil in all parts of the world. Around 200 species have been identified.

3.2. Distribution and Quantification

Recently several investigators have attempted to evaluate the distribution of predatory and endoparasitic fungi in different environments. In a series of investigations, Gray applied common ecological sampling techniques and sprinkled plate and Baermann funnel techniques before the detection plates were carefully examined for predators and endoparasites (Gray, 1983, 1984a–c, 1985). He compared different types of soil, from the maritime antarctic (Gray and Lewis Smith, 1984) to different terrestrial habitats, such as deciduous woodland and agricultural and grassland soil (Gray, 1983, 1984b,c, 1985; Gray and Bailey, 1985) and activated sludge (Gray, 1984a). In all types of habitats endoparasites tended to be more abundant than predators and, among the predators, those with branches, knobs, and constricting rings (Fig. 1) were more abundant than network-formers. Among the endoparasites, those with

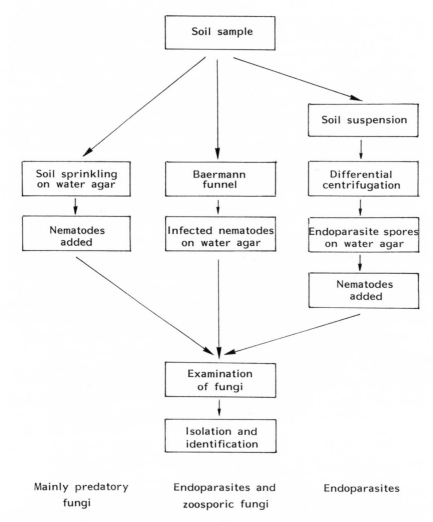

Figure 4. Techniques used to collect, extract, enumerate, and isolate nematophagous fungi from soil. [Compiled from Barron (1982), Gray (1984c), and Dackman *et al.* (1987).]

adhesive spores appeared to be more frequent than those with nonadhesive spores.

A major problem in such studies is to quantify and to estimate the activity of these fungi in soil. Barron (1982) rightly states that "it is difficult to measure accurately the numbers and kinds of nematode-destroying fungi in soil or organic debris. It is equally difficult to estimate the levels of activity of fungi or effects of various types of soil amendments or treatments upon these activities." Gray subjected his data to statistical analysis and stated that most investigations in the past provided an insufficient set of data for such analyses. On an average, many of the fungi isolated occurred in low frequencies and thus were not always suitable for the type of statistical analysis performed. However, these are the first investigations conducted to determine the distribution of nematophagous fungi on a larger scale (Gray, 1983) and to estimate the effect of soil moisture, organic matter, pH, and nematode density on this distribution (Gray, 1985).

All attempts to quantify nematophagous fungi are extremely time-consuming. This is true both for the isolation methods used by Gray and the following statistical treatment (Gray, 1984a–c, 1985) and for detection methods in combination with an MPN estimation (Eren and Pramer, 1965; Dackman et al., 1987). Eren and Pramer (1966, 1978) further applied an immunofluorescent staining technique for laboratory studies of growth and activity of Arthrobotrys conoides in soil and determined the influence of different soil treatments on hyphal length. Such quantification of the fungi gives valuable information on their occurrence and distribution, and it is also an estimation of the activity of the fungi, provided nematode numbers are determined concomitantly. This was also done in a number of investigations (Eren and Pramer, 1978; Dackman et al., 1987), but a clear correlation between nematode-destroying activity and hyphal length or biomass is not available. Again, the complexity of the system is due to the fact that only certain mycelial adaptations are responsible for the nematode-destroying activity, and this activity is not necessarily correlated with hyphal length (Dackman et al., 1986; Hayes and Blackburn, 1966; Jansson, 1982b).

3.3. Survival Strategies

Chlamydospores are survival structures responsible for the persistence of Hyphomycetes in soil. Although found in both predatory and endoparasitic fungi, they have seldom been detected in soil (Barron, 1977, 1981). In contrast, the thick-walled oospores of the female cyst-nematode parasite Nematophthora gynophila can survive at least 5 years in soil (Kerry, 1984), and the extraction of these structures from soil can be used quantitatively to estimate the level of this parasite (Crump and Kerry, 1981; Dackman and Nordbring-Hertz, 1985). Several of the predators have been shown to act as mycoparasites on other soil fungi (Tzean and Estey, 1978; Persson et al., 1985) by coiling around the host hyphae. Tzean and Estey (1978) observed that the coiling was induced by the presence of hyphae of only a limited number of fungi, only 3 out of 30. Persson et al. (1985) found a wider host range for A. oligospora; 6 out of 13 species of different taxonomic groups were attacked by *A. oligospora* by coiling. The interaction between *A. oligospora* and *Rhizoctonia solani* was studied in detail by light and electron microscopy (Fig. 3). The coils possessed a high metabolic activity as indicated by fluorescein diacetate (FDA) staining. On the ultrastructural level, the coils showed an abundance of membranous vesicles. After some time the coiled cells died, as indicated by vital staining; penetration of intact cells was not observed. The interaction therefore was interpreted in terms of competition for nutrients (Persson *et al.*, 1985). Although hyphal coiling by *A. oligospora* in natural soil has not yet been detected, the possibility of survival as a mycoparasite cannot be neglected.

3.4. Biocontrol

Ever since an interest in the study of this group of fungi arose, the question of their use in natural or applied biocontrol of parasitic nematodes has been the aim of many studies. There are several reviews describing these attempts (Mankau, 1980, 1981; Kerry, 1980, 1984; Stirling *et al.*, 1979) and it is not the purpose of this review to go into details of biocontrol. In recent years, however, the aspect of host recognition has been taken into account when designing biocontrol experiments (Zuckerman, 1983; Zuckerman and Jansson, 1984). Furthermore, the endoparasites, both those that attack free-living nematodes and those parasitic on eggs and females, are the organisms preferred as possible control agents today. Examples of the former are *Meria coniospora* (Jansson *et al.*, 1985b) and *Hirsutella rhossiliensis* (Jaffee and Zehr, 1985), of the latter, *Dactylaria oviparasitica* (Stirling *et al.*, 1979), *Nematophthora gynophila*, and *Verticillium* sp. (Kerry, 1984).

Meria coniospora was used in biocontrol experiments because of its high ability to attract nematodes (Jansson, 1982a), its known specificity of conidial adhesion to nematodes (Fig. 2) (Jansson *et al.*, 1984a; Jansson and Nordbring-Hertz, 1983, 1984), and its virulence in a soil microcosm (Jansson, 1982b). *Meria coniospora* significantly reduced root-knot nematode, *Meloidogyne* spp., galling on tomatoes in greenhouse pot trials (Jansson *et al.*, 1985b). The results were promising for further development of microplot trials. To overcome a possible fungistatic effect, nonparasitic nematodes infected with *M. coniospora* were used as inoculum, and it is speculated that this method enables the fungus to become established in the rhizosphere (Jansson *et al.*, 1985b).

The growth characteristics and mode of infection of *M. coniospora* is very similar to another endoparasitic fungus, *Hirsutella* sp., which was also suggested as a promising candidate for use in biological control of phytonematodes. This newly detected parasitic fungus seems to be a useful biological control agent, as indicated by its host specificity (Sturhan and Schneider, 1980; Dürschner, 1983) and the observation that nematode populations decline in habitats where the fungus is present at sufficiently high levels (Jaffee and Zehr, 1985).

Hirsutella heteroderae was first described as a nematode pathogen in Germany (Sturhan and Schneider, 1980). The organism infects its host with adhesive spores borne on flask-shaped phialides (Fig. 2). The fungus showed a certain selectivity in the choice of prey, and especially nematode larvae of the plant-parasitic genera *Heterodera, Globodera, Pratylenchus, Meloidogyne, Ditylenchus,* and *Aphelenchoides* were attacked. Investigations on *H. rhossiliensis,* attacking another plant-parasitic nematode, *Criconemella xenoplax* (Jaffee and Zehr, 1982, 1985), and probably synonymous with *H. heteroderae,* confirmed the ability of this genus to reduce nematode populations in peach orchard soil (Eayre *et al.,* 1983). These authors determined the relative saprophytic and parasitic abilities of this fungus in sterile and nonsterile soil and suggested that the fungus was a better parasite than saprophyte. It may be specialized for attacking nematodes (Jaffee and Zehr, 1985) and thus serve as a biological control agent (Eayre *et al.,* 1983). Parasitic and saprophytic abilities are often inversely related, so that efficient parasites are inefficient saprophytes and vice versa (Cooke, 1963; Jansson, 1982b; Jansson and Nordbring-Hertz, 1979).

4. Host Specificity

The question of host specificity in the nematophagous fungus–nematode system has been much discussed, especially in connection with recognition of prey (see Section 5). In general, most nematophagous fungi seem to attack mainly nematodes, although some species also capture other minor animals (Barron, 1977). Because of the trap size, capture might also be restricted by the size of the host animal.

In laboratory studies the predatory fungi do not seem to discriminate among different species of nematodes, as far as the mere attachment is concerned. Six nematode species (bacterial feeding, fungal feeding, and plant parasitic) were all captured within ½ hr when added to preformed traps of *A. oligospora* (Jansson and Nordbring-Hertz, 1980). Seven species of fungi captured nine different nematodes in another study under similar conditions (Rosenzweig *et al.,* 1985). However, Dowe (1966, 1972) found that there was a difference in infectivity and parasitic development when three plant-parasitic nematodes were captured by four different types of predatory fungi, in the order *Ditylenchus dipsaci* > *Pratylenchus penetrans* > *Heterodera schachtii* larvae. This might indicate that the nature of the cuticle has a pronounced influence on the further development of the infection process. This phenomenon was illustrated with the endoparasitic fungus *M. coniospora*, which infects its host by means of adhesive conidia. Adhesion of conidia did not always result in infection. Ten out of 17 nematode species completed the predatory process within 3–10 days (Jansson *et al.*, 1985a). These results indicate that there might be some host specificity in *M. coniospora* (Jansson and Nordbring-Hertz, 1983); Nordbring-Hertz and Jansson, 1984). Also, Dürschner (1983) found a restricted host range for *M. coniospora* and several other endoparasites. In contrast, some endoparasites have shown a rather wide host range, e.g., *Catenaria* species, which attack not only nematodes, but also other minute soil animals [for references see Barron (1977)]. Barron (1978) showed that 32 out of 40 endoparasitic species were capable of attacking the soil nematode *Rhabditis terricola*. Thus, although several laboratory studies point to a wide host range for most predatory and endoparasitic fungi, on the whole the endoparasites seem to possess a somewhat greater host specificity than the predatory fungi. This is especially true for those fungi that attack cyst-nematode eggs and females. The female parasite *Nematophthora gynophila* attacked several species of *Heterodera*, but not the potato cyst-nematode *Globodera rostochiensis* (Kerry and Crump, 1977; Kerry, 1980). The reason for this difference is not known.

The differing results reported here indicate that it is time to look more closely at the entire infection process in determining host specificity and to relate broad and narrow specificity with degree of virulence, in line with host–parasite relationships in other groups of organisms, as suggested by Barron (1977).

5. Recognition of Prey

The question of how nematophagous fungi recognize their prey has been dealt with from different points of view by mycologists, nematologists, and microbial ecologists. From the discussion above (Section 4), it is clear that there is no simple host specificity in any of the nematophagous species that would indicate a recognition on the molecular level between a fungus and its prey. Clarke and Knox (1978), reviewing cell recognition in plants, defined recognition in a rather broad sense as "the initial events of cell–cell communication which elicits a defined biochemical, physiological or morphological response." Applied to the nematode–nematophagous fungus system, it is suggested here that initial events would include both *chemoattraction* of nematodes to the vicinity of the predator and *adhesion* of the nematodes to surface structures of the fungi.

The types of communication involved are, in the first case, the release of chemoattractants and their chemoreception and, in the second case, a short-range interaction mediated by a lectin–carbohydrate binding and/ or other adhesion mechanism. To obtain insight into the mechanisms of these events, biological, biochemical, and ecological experiments have to be performed. Light and electron microscope observations give valuable information when studies on the biochemical and molecular levels are correlated with biological and ecological studies of the interaciton between two living organisms.

Such studies are now going on in several laboratories. At present we can only speculate on the role of our findings in the natural environment, but it is interesting that interference with nematode chemotactic behavior and adhesion to surfaces are now being considered as a possible basis for development of biological control methods (Zuckerman, 1983; Zuckerman and Jansson, 1984).

5.1. Chemoattraction of Nematodes

It is not easy to understand how nematodes, probably to a large extent moving randomly in soil, can be attracted by special stimuli under natural conditions. Carlile (1980), considering the role of motility, taxis, and tropism among microorganisms in the natural environment, suggested that active movement, guided by means of a sensory system, is complementary to passive movement over long distances and is important at the beginning and end of long passive movements. Nematodes are known to be attracted to various food sources, such as plants, fungi, and bacteria. It is more surprising that they are also attracted to living mycelium of their natural enemies (Field and Webster, 1977; Jansson and Nordbring-Hertz, 1979, 1980) and thus are lured into their death. It appears that some factor other than major nutrients might function to guide the nematode prey to its fungal predator. Chet and Mitchell (1976), discussing predator–prey relationships between microorganisms, suggested that the ability of a predaceous organism to be attracted to its prey would give this organism a strong selective advantage. In the nematophagous fungus system again the predaceous organism is supported, although the situation is reversed as the nematode prey is attracted to its predator.

5.1.1. Chemoattraction in Relation to Morphology and Saprophytic/ Predaceous Ability

There is a close connection between the type of infection structure and the saprophytic/predaceous ability of the fungus (see Section 2)

(Cooke, 1963; Jansson, 1982b). Both the morphology and, consequently, the saprophytic/predaceous ability strongly influence the attractiveness of the fungi. In a series of investigations, the behavior of different types of nematodes in relation to different types of nematophagous fungi was observed using an agar plate attraction assay (Jansson and Nordbring-Hertz, 1979, 1980; Jansson, 1982a). The most important results of these studies are: (1) About 75% of the nematophagous fungi attracted the bacterial-feeding nematode *P. redivivus.* (2) Fungi that were more predaceous showed a stronger attraction than the more saprophytic ones; that is, the endoparasitic species infecting the nematodes with their conidia were more efficient than the more saprophytic species with different kinds of trapping devices. (3) There were variations in the attractiveness of different fungi to different species of nematodes. These differing attracting abilities may reflect some possible specificity in the choice of prey among these fungi.

In a sterile soil microcosm experiment, a very good correlation was found between the ability to attract nematodes and the predacity, defined as the ability of the fungi to reduce nematode numbers (Jansson, 1982b; Nordbring-Hertz and Jansson, 1984). As the nematode-destroying activity is always dependent on the presence of infective structures, the type of morphological adaptations also influences the attractiveness of the fungi. In *A. oligospora,* presence of traps increased the attractiveness by a factor of two (Jansson, 1982b). Among the endoparasitic fungi, conidia that were adhesive were attractive, whereas those without an adhesive part were not (Jansson, 1982a).

All the studies of the behavior of nematodes in the vicinity of the fungi have been performed in the laboratory. Thus, there is need for caution when these results are translated to field conditions. However, the consistent pattern of strong attractiveness, in the more predaceous forms especially, should have some significance in the natural environment.

5.1.2. Chemoattraction in Relation to Adhesion Mechanism

The sensory organs of the nematodes which contain the receptors for chemoattraction, are situated mainly in the head region. When nematodes were infected by adhesive conidia these were frequently attached to the cephalic region (Jansson, 1982a; Jansson and Nordbring-Hertz, 1983). Nematodes with adhered conidia lost their ability to be attracted to various food sources (Jansson and Nordbring-Hertz, 1983). As a consequence of these results, it became possible to interfere with nematode chemotactic behavior (Zuckerman, 1983; Zuckerman and Jansson, 1984) by blocking or obliterating the receptor sites. The adhesion of conidia of *M. coniospora* to the nematode cuticle seems to be lectin-mediated in that a fungal carbohydrate-binding protein binds to nematode sialic acid (Jansson and Nordbring-Hertz, 1983, 1984) (see Section 5.2 and Table II). The characteristics of the relationship between attraction and adhesion in *M. coniospora*-nematode interaction are further discussed in Section 5.2.

5.2. Adhesion of Nematodes

The most important initial event in the entire infection process in the nematophagous fungus–nematode system is the firm anchoring of trapping devices or conidia to nematodes. The inevitable fate of the nematode is that its cuticle is penetrated and trophic hyphae developed within the body, digesting its contents. As in other host–parasite relationships, this sequence of events depends on a series of signal–response reactions, the details of which are not yet known. The most remarkable of these steps is certainly the rapid adhesion of a large nematode to a much smaller trapping structure.

It has been suggested that this sequence of events is initiated by a lectin–carbohydrate binding between complementary molecules on the trap and nematode surfaces (Nordbring-Hertz and Mattiasson, 1979). This finding has been supported by a number of studies, which will be described in some detail below (Table I; see Sections 5.2.2 and 5.2.3).

Table I. Steps of Evidence for a Trap Lectin in *Arthrobotrys oligospora*

1. Inhibition of capture of nematodes by *N*-acetyl-D-galactosamine (GalNAc)	Nordbring-Hertz and Mattiasson (1979), Premachandran and Pramer (1984)
2. Binding of red blood cells to traps	Nordbring-Hertz and Mattiasson (1979)
3. Demonstration of GalNAc on nematode surface	Nordbring-Hertz and Mattiasson (1979)
4. Inhibition of capture by trypsin and glutaraldehyde treatment of traps	Nordbring-Hertz *et al.* (1982)
5. Binding of ^{125}I-labeled fungal homogenate to GalNAc–Sepharose	Mattiasson *et al.* (1980)
6. Inhibition of binding of homogenate by trypsin treatment	Mattiasson *et al.* (1980)
7. Isolation and characterization of a carbohydrate-binding, Ca^{2+}-dependent protein (subunit molecular weight 20,000)	Borrebaeck *et al.* (1984), Premachandran and Pramer (1984)
8. Isolation of one major lectin-binding glycoprotein (molecular weight 65,000) from nematode cuticle	Borrebaeck *et al.* (1985)
9. Localization of the trap lectin on the surface of the trap by the immunogold technique	C. A. K. Borrebaeck, B. Mattiasson, M. Veenhuis, W. Harder, and B. Nordbring-Hertz, unpublished results

5.2.1. *Arthrobotrys oligospora as a Model Organism*

Arthrobotrys oligospora (Fig. 5) is probably the most investigated of all nematophagous fungi. There are several reasons for this: It was the first species reported to capture nematodes (Zopf, 1888) and it is widely spread in all types of habitats around the world. However, compared to other types of nematophagous fungi, it does not seem to be either the most predaceous or the most frequently recorded (Gray, 1983). The reason for this is probably its occurrence in either a saprophytic phase or a predaceous phase, the transition of the former to the latter depending on environmental conditions and presence of nematodes. A view of *A. oligospora* with hyphae, traps, conidiophores with conidia, and captured nematodes in different phases of digestion is shown in Fig. 5.

One advantage of *A. oligospora* as a model for studying recognition phenomena is that it can be grown as a saprophyte without traps or as a predator with traps. Thus, when *A. oligospora* (ATCC 24927) is grown on ordinary media, traps are rarely formed. On the other hand, heavy trap formation can be induced on media low in nutrients supplemented with trap-inducing peptides or amino acids both on solid media (Nordbring-Hertz, 1973) and in liquid cultures (Friman *et al.,* 1985). This technique enables us to compare normal mycelium with mycelium bearing traps in many aspects related to recognition, such as ability to attract nematodes (Jansson, 1982b), adhesion of nematodes and its inhibition (Nordbring-Hertz and Mattiasson, 1979; Nordbring-Hertz, 1984), and isolation of a carbohydrate-binding protein (Borrebaeck *et al.,* 1984).

5.2.2. *A Trap Lectin of Arthrobotrys oligospora*

Lectins are carbohydrate-binding proteins or glycoproteins of non-immune origin that agglutinate cells and/or precipitate glycoconjugates (Goldstein *et al.,* 1980). Most of the lectins isolated [about 100 (Lis and Sharon, 1984)] originate from higher plants. Lectins have also been detected in and isolated from mammalian tissues, invertebrates, algae, slime molds, and true fungi, as well as bacteria. Their ubiquitous occurrence indicates one or more general functions in nature. Although several possibilities have been suggested, final evidence for their function is lacking. As with plant lectins, microbial lectins have received an increasing interest in recent years (Mirelman, 1986). Fungal lectins or agglutinins have been isolated from various fruiting bodies and in some cases from mycelium of Hyphomycetes. The importance of fungal lectins in interactions with higher plants, nematodes, other fungi, and lichens was recently discussed (Nordbring-Hertz and Chet, 1986).

The possibility that a molecular recognition based on a lectin–car-

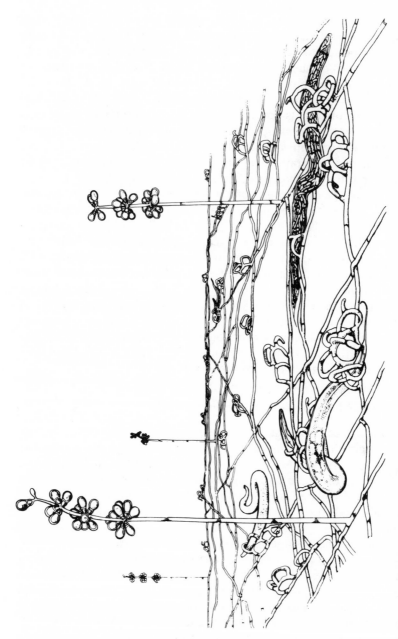

Figure 5. Predaceous activity of *Arthrobotrys oligospora.* Foreground left shows recently caught nematode, foreground right shows nematode about 24 hr after capture. Vertical conidiophores with clusters of two-celled conidia are scattered over the substrate. [From Barron (1977), with permission.]

bohydrate interaction plays an important role in the adhesion of nematodes to nematophagous fungi was first indicated in biological interaction studies of intact organisms (Nordbring-Hertz and Mattiasson, 1979). These studies were based on the assumption that if a lectin–carbohydrate interaction is effective in the adhesion of nematodes to surface structures of traps or conidia, capture should be inhibited when the lectin-bearing structure is treated with the appropriate carbohydrate or carbohydrate sequence. A suitable assay system for such inhibition studies of the predatory fungi was developed for *A. oligospora* using a dialysis membrane technique (Nordbring-Hertz and Mattiasson, 1979; Nordbring-Hertz, 1983; Nordbring-Hertz *et al.*, 1984) in which the trap-containing mycelium could be easily flooded with appropriate solutions and the effects tested after addition of a nematode suspension. In hapten inhibition experiments with about 20 simple carbohydrates, capture of nematodes *(P. redivivus)* was inhibited mainly by *N*-acetyl-D-galactosamine (GalNAc) (Nordbring-Hertz and Mattiasson, 1979), indicating the presence of a carbohydrate-binding protein, a lectin, on the traps of *A. oligospora* and this carbohydrate on the nematode surface. Residues of GalNAc were subsequently detected on the nematode surface by means of a lectin-peroxidase assay. Red blood cells carrying GalNAc residues accumulated around the traps and, after about 1 hr, lysed as an effect of the interaction with the trap. The protein nature of the binding molecule was further substantiated by treatment of the fungus with trypsin or glutaraldehyde, which completely inhibited capture (Nordbring-Hertz *et al.*, 1982). The specificity for GalNAc was never complete in these *in vivo* tests. Nevertheless, a GalNAc-substituted affinity gel could successfully be used to isolate the carbohydrate-binding protein by affinity chromatography (Borrebaeck *et al.*, 1984).

A major problem in characterizing this protein was to obtain enough trap-bearing mycelium. Initially, trap formation was induced either by nematodes (Nordbring-Hertz and Mattiasson, 1979) or by peptides (Mattiasson *et al.*, 1980) on cultures growing on dialysis membranes over solid media. Later, a liquid culture method was developed for heavy trap formation of *A. oligospora* (Friman *et al.*, 1985). The fungus was grown in modified separatory funnels in a very dilute medium (0.01% soya peptone) supplemented with a trap inducer, phenylalanylvaline, or its constituents phenylalanine and valine (0.005–0.01%). Trap formation started after 2 days, and after 6 days abundant traps had developed. To isolate the protein, mycelium with fully developed traps was surface radiolabeled with (^{125}I)-iodosulfanilic acid and then homogenized and subjected to two affinity chromatography steps (Borrebaeck *et al.*, 1984). In the first, a GalNAc-substituted gel was used to isolate any GalNAc-binding protein (Fig. 6). A peak of labeled protein could be eluted from the affinity

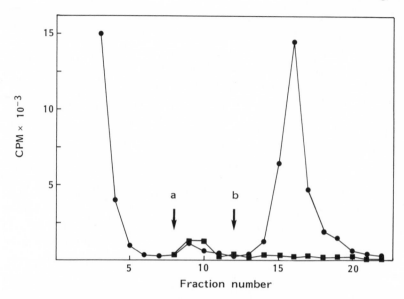

Figure 6. (●) Affinity chromatogram from a GalNAc-containing gel. Arrow *a* indicates a change in buffer from 0.1 M triethanolamine buffer (pH 7.2), containing 1 mM each of CaCl₂, MnCl₂, and ZnCl₂ to 10 mM potassium phosphate buffer (pH 7.2), containing 1 M NaCl. Arrow *b* indicates a change in buffer to glycine–HCl buffer (pH 3.0). (■) Radiolabeled fungal material containing no trap-bearing mycelium, run as a control. [From Borrebaeck *et al.* (1985), with permission.]

gel with 0.1 M glycine–HCl buffer. Only minor amounts of fungal molecules were adsorbed nonspecifically to the matrix. A homogenate of radiolabeled hyphae without traps gave no significant binding (Fig. 6) (Borrebaeck *et al.*, 1984, 1985). The metal ion dependence of the carbohydrate-binding protein was shown by using a Ca^{2+}-containing affinity gel (Fig. 7). Sodium dodecyl sulfate (SDS)–polyacrylamide gel electrophoresis and autoradiography showed that both the sample from the GalNAc affinity gel and that from the Ca^{2+} binding gel gave one major band at a molecular weight of ~20,000 (Fig. 8).

The conclusion is that a GalNAc-specific and Ca^{2+}-dependent protein was isolated, and we believe it is developmentally regulated, as it is present on the traps of *A. oligospora*. It has been suggested that this trap lectin initiates a series of signal–response reactions leading eventually to penetration of the nematode cuticle and digestion of the nematode (Nordbring-Hertz and Mattiasson, 1979; Borrebaeck *et al.*, 1984; Nordbring-Hertz and Jansson, 1984). The isolation of the lectin from trap-

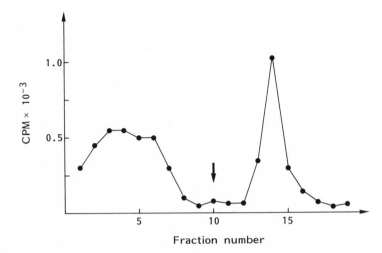

Figure 7. Metal chelate affinity chromatogram from a Ca^{2+}-containing gel. The sample was from the GalNAc affinity chromatography step described in the legend to Fig. 6. The arrow indicates the change of buffer from 0.05 M Tris–acetate buffer (pH 8.2) containing 0.5 M NaCl to the same buffer containing 10 mM EDTA. [From Borrebaeck *et al.* (1984), with permission.]

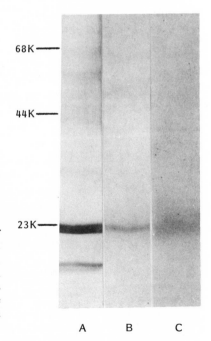

Figure 8. (A) Sodium dodecyl sulfate–polyacrylamide gel electrophoresis slab stained by the basic silver stain procedure. The sample was from the GalNAc affinity chromatography step described in the legend to Fig. 6. (B) Autoradiogram of the gel slab in lane A; 38 days of exposure. (C) Autoradiogram of polyacrylamide gel slab containing the eluted peak from the metal affinity chromatography (Fig. 7) as a sample; 60 days of exposure. Molecular weights are indicated on the left-hand side of the figure ($K = 10^3$). [From Borrebaeck *et al.* (1984), with permission.]

bearing cultures of *A. oligospora* is an important step toward an understanding of the biological function of this fungal lectin.

In a similar study with another strain of *A. oligospora,* the presence of a GalNAc-specific protein was confirmed (Premachandran and Pramer, 1984). The GalNAc inhibited capture at a capacity of 35–50% of the controls. To isolate the GalNAc-specific protein, the fungus was grown in corn meal broth for 2 weeks. The mycelial mats (apparently without traps) were then homogenized and centrifuged. The supernatant was analyzed by affinity chromatography, using an affinity gel with specificity for GalNAc-binding proteins. The sugar-specific protein was eluted with 0.3 M lactose. This protein was homogeneous and displayed a molecular weight of about 22,000 by SDS–polyacrylamide gel electrophoresis. Pretreatment of nematodes with the purified protein reduced entrapment, indicating a role for the sugar-binding protein in recognition and capture of prey by the fungus. However, the authors point out that it is doubtful that lectins can account fully for the entire adhesion process. This point is further discussed in Sections 5.2.3 and 5.2.5.

5.2.3. Proposed Trap Lectins of Other Predatory Fungi

The same hapten inhibition technique described in the preceding section has been used to detect trap lectins in other fungi with different types of traps (Table II). Rosenzweig and Ackroyd (1983) found that glucose/mannose inhibited capture of the nematode *P. silusiae* by *A. conoides* and fucose prevented capture by *Monacrosporium eudermatum,*

Table II. Binding Characteristics of Proposed Lectins of Nematophagous Fungi

Specificity	Fungus	Structure	Reference
N-Acetyl-D-galactosamine (GalNAc)	*Arthrobotrys oligospora*	Adhesive network (Fig. 1D)	Nordbring-Hertz and Mattiasson (1979), Borrebaeck *et al.* (1984, 1985), Premachandran and Pramer (1984)
D-Glucose/D-mannose	*Arthrobotrys conoides*	Adhesive network (Fig. 1D)	Rosenzweig and Ackroyd (1983, 1984)
L-Fucose	*Monacrosporium eudermatum*	Adhesive network (Fig. 1D)	Rosenzweig and Ackroyd (1983, 1984)
2-Deoxyglucose	*Monacrosporium rutgeriensis*	Adhesive network (Fig. 1D)	Rosenzweig and Ackroyd (1983, 1984)
2-Deoxyglucose	*Dactylaria candida*	Adhesive knob (Fig. 1B)	Nordbring-Hertz *et al.* (1982)
Sialic acid	*Meria coniospora*	Adhesive conidia (Fig. 2A, right)	Jansson and Nordbring-Hertz (1983, 1984)

whereas 2-deoxyglucose inhibited capture by *M. rutgeriensis*. Out of 18 carbohydrates tested, 2-deoxyglucose was the only sugar that inhibited capture by *Dactylaria candida,* whereas *M. cionopagum* was not inhibited by any of the sugars tested (Nordbring-Hertz *et al.,* 1982). Thus, different fungi seem to carry different carbohydrate-binding proteins with different carbohydrate specificities. Rosenzweig *et al.* (1985) raised the question of the role of trap lectins in the specificity of capture of nematodes. The fact that different fungi seem to carry lectins with different carbohydrate specificities, whereas the fungi show little selectivity in the choice of prey, seems to indicate either that lectins on the trap surface are not the basis of capture of prey or that the cuticle surface of nematodes carry several saccharides (Rosenzweig *et al.,* 1985). Questions of this type have prompted investigations into the surface structures of nematodes and surface structures of adhesive traps and conidia, respectively. These questions will be dealt with in Section 5.2.5.

5.2.4. Endoparasitic Fungi

The recent interest in the endoparasite *M. coniospora* as a possible biological control agent is based partly on its frequency in the natural environment (Section 3) and partly on the studies of nematode chemotaxis (Section 5.1). These studies in turn are based on the specific adhesion of conidia to sensory organs of nematodes. Hapten inhibition experiments show that this adhesion was specifically inhibited only by sialic acid, and this carbohydrate was demonstrated on the nematode surface with the aid of sialidase, indicating the presence of a carbohydrate-binding protein on the adhesive conidia (Jansson and Nordbring-Hertz, 1983). This result was supported by the finding that treatment of conidia with trypsin and glutaraldehyde inhibited adhesion (Jansson and Nordbring-Hertz, 1984). Since the adhesion of conidia is preferably located at the receptors in the cephalic region, subsequent studies were directed toward interference with both chemotaxis and adhesion. Furthermore, treatment of the nematode *P. redivivus* with sialidase not only inhibited adhesion of conidia, but also reduced attraction to fungi (Jansson and Nordbring-Hertz, 1984). Treatment with mannosidase and sialidase completely and with trypsin partially inhibited chemotactic response to a bacterial filtrate by both *Caenorhabditis elegans* and *P. redivivus,* whereas other enzymes were ineffective (Jansson *et al.,* 1984b). These results indicated the presence of sialic acid and mannose on these nematodes. Further evidence for the presence of these sugars on chemoreceptors of the nematodes is that treatment of nematodes with the lectins Con A and limulin (with specificities for mannose/glucose and sialic acid, respectively) brought about a partial inhibition of chemotaxis (Jansson and

Nordbring-Hertz, 1984; Jeyaprakash *et al.,* 1985). The evidence for a sialic acid-specific lectin on *Meria* conidia is still indirect and isolation of the lectin has not been achieved.

5.2.5. Surface Structures of Nematodes

The differential adhesion and infection of nematodes by *Meria* conidia (Jansson *et al.,* 1985a) point to differences in cuticle composition important to aspects of recognition. Investigations into surface structures of nematodes have been made from the points of view of the plant pathologist and the microbial ecologist. The sugar residues identified so far on different nematode cuticle surfaces are shown in Table III. Some of the sugars are identified on different types of nematodes and, in addi-

Table III. Carbohydrates Detected on Nematode Cuticles

Carbohydrate residues[a]	Nematode[b]	Method[c]	Reference
Gal/GalNAc	BF	Enzyme-lectin assay	Nordbring-Hertz and Mattiasson (1979)
	BF, PP, AP	Enzyme-lectin assay	Rosenzweig *et al.* (1985)
	PP	Galactose oxidase treatment, fluorescent and ferritin labeling	Spiegel *et al.* (1982, 1983)
	AP	FITC-lectins	Bone and Bottjer (1985)
	PP	TRITC-lectins	Forrest and Robertson (1986)
Glc/Man, GlcNAc	BF, PP	Con A-hemocyanin	McClure and Zuckerman (1982)
	BF	^{125}I-lectins	Zuckerman and Kahane (1983)
	BF, PP, AP	Enzyme-lectin assay	Rosenzweig *et al.* (1985)
	PP	FITC-lectins	Spiegel and Cohn (1982)
	AP	FITC-lectins	Bone and Bottjer (1985)
	BF	TRITC-lectins	Jansson *et al.* (1986)
	PP	TRITC-lectins	Forrest and Robertson (1986)
Sialic acid	BF	Indirect evidence	Jansson and Nordbring-Hertz (1983)
	PP	Periodate oxidation, fluorescent and ferritin labeling	Spiegel *et al.* (1982, 1983)

[a]Gal, Galactose; GalNAc, *N*-acetyl-D-galactosamine; Man, mannose.
[b]BF, Bacteria-feeding; PP, plant-parasitic; AP, animal-parasitic nematode.
[c]FITC, fluorescein isothiocyanate; TRITC, tetra methylrhodamine isothiocyanate.

tion, different parts of a nematode may carry different sugar residues. This situation, together with the fact that different nematophagous fungi carry lectins with different specificities, could explain the wide host range among nematophagous fungi.

To gain more information on the character of the receptor on the nematode cuticle of the fungal lectin of *A. oligospora,* nematodes were treated with the lectins from *Glycine max* (SBA; specific for GalNAc), *Conavalia ensiformis* (Con A; Glu/Man), *Dolichos biflorus* (DBA; GalNAc), *Phaseolus vulgaris* (PHA; GalNAc), *Triticum vulgare* (WGA; GlcNAc), and *Helix pomatia* (HPA; GalNAc). These lectin-treated nematodes were then added to trap-containing mycelia on dialysis membrane (Borrebaeck *et al.,* 1985). SBA showed about 50% inhibition of capture compared to the control. This indicated that SBA bound to the same surface receptor as the fungal lectin or to a site in close connection to that receptor. Based on this experience, an affinity gel containing immobilized SBA was used to isolate the apparent receptor molecules from the nematode cuticle. One major protein of molecular weight $\sim 65,000$ was particularly prominent on this affinity column. Treatment of the traps with this glycoprotein resulted in a somewhat greater ability to inhibit capture than was exhibited by the unbound fraction (Borrebaeck *et al.,* 1985).

5.2.6. Surface Properties of Adhesive Infectious Structures

Over the years, many light and electron microscope studies of nematophagous fungi have been published illustrating the morphology of the trapping structure, the outer surface of the trap, the development and fate of typical organelles, and the penetration and colonization of the nematode [e.g., in the predatory fungi, Heintz and Pramer (1972), Nordbring-Hertz (1972), Dowsett *et al.* (1977), Nordbring-Hertz and Stålhammar-Carlemalm (1978), Dowsett and Reid (1979), Wimble and Young (1983, 1984), and Veenhuis *et al.* (1984, 1985a,b); and in the endoparasitic fungi, Saikawa (1982) Saikawa *et al.* (1983) and Jansson *et al.* (1984a)]. Pertinent to the study of recognition is the role of the adhesive in the capture of nematodes. Premachandran and Pramer (1984) pointed out that it is doubtful that lectins can account fully for the remarkable tenacity of the mucilage produced by *A. oligospora* and other species. The fact that a nematode attached only at the surface of a trap stays captured in spite of its violent movement indicates the existence of a very efficient adhesive. An increased secretion of a fibrillar adhesive in the presence of prey has been reported by several authors (Dowsett and Reid, 1979; Nordbring-Hertz and Stålhammar-Carlemalm, 1978). Invariably in transmission electron microscopy, an electron-dense layer is seen between the trap and the nematode, and the fibrils of the adhesive become oriented in one direction during the first hour of capture (Veenhuis *et al.*, 1985a). Veenhuis *et al.* (1985a) suggested that a prerequisite for the fungus to penetrate the nematode against the internal pressure of the intact animal is the firm anchoring by the adhesive. Based on light and transmission electron microscopic observations it was suggested that the cuticle was penetrated—at least largely—by mechanical force (Veenhuis *et al.*, 1985a). The tenacity of the adhesive and its anchoring and sealing function have been further substantiated by video recordings of live material (Nordbring-Hertz *et al.*, 1986). The secretion of a strong adhesive seems to be a prerequisite for the penetration of the live nematode by the fungus (Nordbring-Hertz and Stålhammar-Carlemalm, 1978; Veenhuis *et al.*, 1985a). Recently it has been shown that an interaction between individual cells within the same trap network can also take place (Dowsett *et al.*, 1984; Veenhuis *et al.*, 1985c). Also in this case the interaction was preceded by a secretion of adhesive. The interaction was mediated by the living nematode and could not be achieved by mechanical means (Veenhuis *et al.*, 1985c). Thus, the adhesive in nematode–fungus interaction plays a decisive role for the infection process. Its connection with the lectin–carbohydrate binding is an unsolved problem.

5.2.7. Localization of the Lectin

To understand the function of the lectin in the *A. oligospora*–nematode interaction its location on or in the trap is crucial. In an attempt to solve this problem, antibodies against the trap lectin of *A. oligospora* were raised and partly purified (C. A. K. Borrebaeck, B. Mattiasson, and B. Nordbring-Hertz, unpublished results). The subcellular localization of the lectin was studied immunocytochemically using the protein A/gold method in Lowicryl K4M embedded traps that had captured nematodes. On thin sections of these samples gold particles were largely confined to the cell wall of the fungus and absent on the nematode including its cuticle. Furthermore, labeling was mainly on the trap cell wall and relatively low on walls of normal hyphae (M. Veenhuis, K. Sjollema, and B. Nordbring-Hertz, unpublished results). These results suggest that the trap lectin is indeed developmentally regulated, as it is present almost exclusively in the trap cell wall. Moreover, it is interesting to note that no specific labeling was observed on the adhesive layer present between the trap and the captured nematode. This suggests that the adhesive is distinct from the lectin.

It was mentioned previously that the function of lectins in nature is still obscure. The nematophagous fungus–nematode system provides an example where the probable function of lectin is in prey recognition, since the lectin is indeed present on the surface of the structure responsible for the capture of nematodes and not in ordinary hyphae. Further investigations are required to determine the specific function of the lectin.

6. Future Developments

The nematode–nematophagous fungus system offers an excellent model for studies of interaction mechanisms between two organism groups. Detailed laboratory investigations have generated much information concerning host–parasite interactions that might be of general biological significance. The research discussed in this paper relates both to the fungal and the nematode counterpart and originates from studies by mycologists and biochemists, microbial ecologists, and plant and animal pathologists. The cooperation of these groups of scientists will be extremely important in the future for two reasons. First, an understanding of the background of the activities of these fungi touches many fields of competence, and second, development of biocontrol methods is dependent on basic knowledge of organism and cell biology, population dynamics, and environment manipulation.

It has been stressed many times (Cooke, 1968; Mankau, 1980; Kerry, 1984) that in-depth studies of the abundance, distribution, and population dynamics of nematophagous fungi under natural conditions are badly needed. One difficulty that has often been stressed is the laborious and time-consuming work that seems to be inevitably connected with such studies (Gray, 1983, 1984a–c, 1985; Dackman et al., 1987). Nevertheless, continued work is necessary on detection and quantification of nematophagous fungi in the natural habitat, especially in such habitats where disease seems to be controlled by natural means (Kerry, 1980; Stirling et al., 1979).

In early rhizosphere studies, Katznelson and co-workers (e.g., Peterson and Katznelson, 1965) studied the development of nematode-trapping fungi along with bacteria, actinomycetes, saprotrophic fungi, and the soil fauna in rhizospheres of different plants. Unfortunately, these studies were not continued. Present knowledge of rhizosphere biology stresses the mutual influence of all nutritional groups on crop productivity [cf. reviews by Bowen (1982), Cooke and Rayner (1984), and Whipps and Lynch (1986)]. With this knowledge it should be possible to design experiments to elucidate the factors controlling establishment of the fungi in a highly competitive environment. Such in-depth studies of the interactions in the rhizosphere would also constitute a basis for development of novel methods for biocontrol where interference with nematode behavior and nematode adhesion mechanisms could also be investigated.

108 B. Nordbring-Hertz

The study of plant lectins has led to a better understanding of interactions between cells, and the use of plant lectins as analytical and experimental tools is widespread. Nevertheless, the function of lectins in the intact plant is still largely obscure and so is their function in interactions with other organisms (Lis and Sharon, 1981). In recent years an increasing interest in microbial lectins has developed, and the properties and biological activities of lectins from bacteria, viruses, fungi, protozoa, and algae were reviewed recently (Mirelman, 1986). The possible function of microbial lectins in cell–cell or host recognition is suggested in several systems.

The biological function of the fungal lectins in nematophagous fungi will remain a challenging problem. The fact that the trap lectin of *A. oligospora* is localized to the surface of the traps (C. A. K. Borrebaeck *et al.,* unpublished results) points to a function connected with recognition of prey, as suggested previously from biological and biochemical investigations (Nordbring-Hertz and Mattiasson, 1979; Borrebaeck *et al.,* 1984). Although a signal substance has not been identified, it is an attractive hypothesis to assume that the lectin binding induces a series of signal–response reactions. These might result in secretion of an adhesive capable of holding the live nematode and of hydrolytic enzymes functioning in penetration and/or digestion of the nematode (Veenhuis *et al.,* 1985a).

There exists an increasing body of biochemical and ultrastructural evidence on the importance of surface properties of both fungi (Borrebaeck *et al.,* 1984, 1985; Veenhuis *et al.,* 1985a,c) and nematode species (Zuckerman and Jansson, 1984, and references therein). But the genetics and molecular biology of the fungi in such systems remain uninvestigated. A molecular approach would therefore certainly benefit the development of the applied aspects of this fascinating subject.

ACKNOWLEDGMENTS. I am grateful to Prof. G. L. Barron, Department of Environmental Biology, University of Guelph, Canada, and to Dr. H.-B. Jansson, Microbial Ecology, and Dr. C. A. K. Borrebaeck, Biotechnology, University of Lund, Sweden, for helpful criticism and valuable suggestions on earlier versions of this review.

References

Barron, G. L., 1977, *The Nematode-Destroying Fungi,* Canadian Biological Publications, Guelph, Ontario, Canada.
Barron, G. L., 1978, Nematophagous fungi: Endoparasites of *Rhabditis terricola, Microb. Ecol.* 4:157–163.
Barron, G. L., 1981, Predators and parasites of microscopic animals, in: *Biology of Conidial*

Fungi (G. T. Cole and B. Kendrick, eds.), Vol. 2, pp. 167–200, Academic Press, New York.

Barron, G. L., 1982, Nematode-destroying fungi, in: *Experimental Microbial Ecology* (R. G. Burns, and J. H. Slater, eds.), pp. 533–552, Blackwell, Oxford.

Bone, L. W., and Bottjer, K. P., 1985, Cuticular carbohydrates of three nematode species and chemoreception by *Trichostrongylus colubriformis, J. Parasitol.* **71:**235–238.

Borrebaeck, C. A. K., Mattiasson, B., and Nordbring-Hertz, B., 1984, Isolation and partial characterization of a carbohydrate-binding protein from a nematode-trapping fungus, *J. Bacteriol.* **159:**53–56.

Borrebaeck, C. A. K., Mattiasson, B., and Nordbring-Hertz, B., 1985. A fungal lectin and its apparent receptors on a nematode surface, *FEMS Microbiol. Lett.* **27:**35–39.

Bowen, G. D., 1982, The root–microorganism ecosystem, in: *Biological and Chemical Interactions in the Rhizosphere* (Proceedings of a Symposium of Swedish Natural Science Research Council 1981), pp. 3–42, Sundt Offset, Stockholm.

Carlile, M. J., 1980, Positioning mechanisms—The role of motility, taxis and tropism in the life of microorganisms, in: *Contemporary Microbial Ecology* (D. C. Ellwood, J. W. Hedger, M. J. Latham, J. M. Lynch, and J. H. Slater, eds.), pp. 55–74, Academic Press, London.

Chet, I., and Mitchell, R., 1976, Ecological aspects of microbial chemotactic behavior, *Annu. Rev. Microbiol.* **30:**221–239.

Clarke, A. E., and Knox, R. B., 1978, Cell recognition in flowering plants, *Quart. Rev. Biol.* **53:**3–53.

Cooke, R. C., 1963, Ecological characteristics of nematode-trapping Hyphomycetes, *Ann. Appl. Biol.* **52:**431–437.

Cooke, R., 1968, Relationships between nematode-destroying fungi and soil-borne phytonematodes, *Phytopathology* **58:**909–913.

Cooke, R. C., and Godfrey, B. E. S., 1964, A key to the nematode-destroying fungi, *Trans. Br. Mycol. Soc.* **47:**61–74.

Cooke, R. C., and Rayner, A. D. M., 1984, *Ecology of Saprophytic Fungi,* Longman, London.

Crump, D. H., and Kerry, B. R., 1981, A quantitative method for extracting resting spores of two nematode parasitic fungi, *Nematophthora gynophila* and *Verticillium chlamydosporium,* from soil, *Nematologica* **27:**330–339.

Dackman, C., and Nordbring-Hertz, B., 1985, Fungal parasites of the cereal cyst nematode *Heterodera avenae* in southern Sweden, *J. Nematol.* **17:**50–55.

Dackman, C., Olsson, S., Jansson, H.-B., Lundgren, B., and Nordbring-Hertz, B., 1987, Quantification of predatory and endoparasitic nematophagous fungi in soil, *Microb. Ecol.* **13:**89–93.

Dowe, A., 1966, Untersuchungen zur Frage der Wirtzspezifität nematodenfangender Pilze (Hyphomycetes), *Wiss. Z. Univ. Rostock* **15**(2):261–264.

Dowe, A., 1972, *Räuberische Pilze* (Die Neue Brehm-Bücherei), A. Ziemsen Verlag, Wittenberg Lutherstadt.

Dowsett, J. A., and Reid, J., 1979, Observations on the trapping of nematodes by *Dactylaria scaphoides* using optical, transmission and scanning-electron-microscopic techniques, *Mycologia* **71:**379–391.

Dowsett, J. A., Reid, J., and vanCaeseele, L., 1977, Transmission and scanning electron microscope observations on the trapping of nematodes by *Dactylaria brochopaga, Can. J. Bot.* **55:**2945–2955.

Dowsett, J. A., Reid, J., and Hopkin, A. A., 1984, Microscopic observations on the trapping of nematodes by the predaceous fungus *Dactylaria cionopaga, Can. J. Bot.* **62:**674–679.

Drechsler, C., 1941, Predaceous fungi, *Biol. Rev. Camb. Phil. Soc.* **16:**265–290.

Duddington, C. L., 1962, Predacious fungi and the control of eelworms, in: *Viewpoints of Biology* (J. D. Carthy and C. L. Duddington, eds.), pp. 151–200, Butterworths, London.

Dürschner, U., 1983, Pilzliche Endoparasiten an beweglichen Nematodenstadien, *Mitteilungen aus der Biologischen Bundesanstalt für Land- und Forstwirtschaft,* **217**:1–83, Berlin-Dahlem.

Eayre, C. G., Jaffee, B. A., and Zehr, E. J., 1983, Suppression of *Criconemella xenoplax* by the fungus *Hirsutella rhossiliensis, Phytopathology* **73**:500 (Abstract).

Eren, J., and Pramer, D., 1965, The most probable number of nematode-trapping fungi in soil, *Soil Sci.* **99**:285.

Eren, J., and Pramer, D., 1966, Application of immunofluorescent staining to studies of the ecology of soil microorganisms, *Soil Sci.* **101**:39–45.

Eren, J., and Pramer, D., 1978, Growth and activity of the nematode-trapping fungus *Arthrobotrys conoides* in soil, in: *Microbial Ecology* (M. W. Loutit and J. C. R. Miles, eds.), pp. 121–127, Springer-Verlag, Berlin.

Field, J. I., and Webster, J., 1977, Traps of predacious fungi attract nematodes, *Trans. Br. Mycol. Soc.* **68**:467–469.

Forrest, J. M. S., and Robertson, W. M., 1986, Characterization and localization of saccharides on the head region of four populations of the potato cyst nematode *Globodera rostochiensis* and *G. pallida, J. Nematol.* **18**:23–26.

Friman, E., Olsson, S., and Nordbring-Hertz, B., 1985, Heavy trap formation by *Arthrobotrys oligospora* in liquid culture, *FEMS Microbiol. Ecol.* **31**:17–21.

Goldstein, I. J., Hughes, R. C., Monsigny, M., Osawa, T., and Sharon, N., 1980, What should be called a lectin?, *Nature* **285**:66.

Gray, N. F., 1983, Ecology of nematophagous fungi: Distribution and habitat, *Ann. Appl. Biol.* **102**:501–509.

Gray, N. F., 1984a, The effect of fungal parasitism and predation on the population dynamics of nematodes in the activated sludge process, *Ann. Appl. Biol.* **104**:143–149.

Gray, N. F., 1984b, Ecology of nematophagous fungi: Comparison of the soil sprinkling method with the Baermann funnel technique in the isolation of endoparasites, *Soil Biol. Biochem.* **16**:81–83.

Gray, N. F., 1984c, Ecology of nematophagous fungi: Methods of collection, isolation and maintenance of predatory and endoparasitic fungi, *Mycopathologica* **86**:143–153.

Gray, N. F., 1985, Ecology of nematophagous fungi: Effect of soil moisture, organic matter, pH and nematode density on distribution, *Soil Biol. Biochem.* **17**:449–507.

Gray, N. F., and Bailey, F., 1985, Ecology of nematophagous fungi: Vertical distribution in a deciduous woodland, *Plant Soil* **86**:217–233.

Gray, N. F., and Lewis Smith, R. I., 1984, The distribution of nematophagous fungi in the maritime Antarctic, *Mycopathologia* **85**:81–92.

Hayes, W. A., and Blackburn, F., 1966, Studies on the nutrition of *Arthrobotrys oligospora* Fres. and *A. robusta* Dudd. II. The predacious phase, *Ann. Appl. Biol.* **58**:51–60.

Heintz, C. E., and Pramer, D., 1972, Ultrastructure of nematode-trapping fungi, *J. Bacteriol.* **110**:1163–1170.

Jaffee, B. A., and Zehr, E. I., 1982, Parasitism of the nematode *Criconemella xenoplax* by the fungus *Hirsutella rhossiliensis, Phytopathology* **72**:1378–1381.

Jaffee, B. A., and Zehr, E. J., 1985, Parasitic and saprophytic abilities of the nematode-attacking fungus *Hirsutella rhossiliensis, J. Nematol.* **17**:341–345.

Jansson, H. B., 1982a, Attraction of nematodes to endoparasitic nematophagous fungi, *Trans. Br. Mycol. Soc.* **79**:25–29.

Jansson, H. B., 1982b, Predacity by nematophagous fungi and its relation to the attraction of nematodes, *Microb. Ecol.* **8**:233–240.

Jansson, H. B., and Nordbring-Hertz, B., 1979, Attraction of nematodes to living mycelium of nematophagous fungi, *J. Gen. Microbiol.* **112**:89–93.

Jansson, H. B., and Nordbring-Hertz, B., 1980, Interactions between nematophagous fungi and plant-parasitic nematodes: Attraction, induction of trap formation and capture, *Nematologica* **26**:383–389.

Jansson, H. B., and Nordbring-Hertz, B., 1983, The endoparasitic nematophagous fungus *Meria coniospora* infects nematodes specifically at the chemosensory organs, *J. Gen. Microbiol.* **129**:1121–1126.

Jansson, H. B., and Nordbring-Hertz, B., 1984, Involvement of sialic acid in nematode chemotaxis and infection by an endoparasitic nematophagous fungus, *J. Gen. Microbiol.* **130**:39–43.

Jansson, H. B., von Hofsten, A., and von Mecklenburg, C., 1984a, Life cycle of the endoparasitic nematophagous fungus *Meria coniospora:* A light and electron microscopic study, *Antonie Leeuwenhoek J. Microbiol.* **50**:321–327.

Jansson, H. B., Jeyaprakash, A., Damon, R. A., Jr., and Zuckerman, B. M., 1984b, *Caenorhabditis elegans* and *Panagrellus redivivus:* Enzyme-mediated modification of chemotaxis, *Exp. Parasitol.* **58**:270–277.

Jansson, H. B., Jeyaprakash, A., and Zuckerman, B. M., 1985a, Differential adhesion and infection of nematodes by the endoparasitic fungus *Meria coniospora* (Deuteromycetes), *Appl. Environ. Microbiol.* **49**:552–555.

Jansson, H. B., Jeyaprakash, A., and Zuckerman, B. M., 1985b, Control of root-knot nematodes on tomato by the endoparasitic fungus *Meria coniospora, J. Nematol.* **17**:327–329.

Jansson, H.-B., Jeyaprakash, A., Coles, G. C., Marban-Mendoza, N., and Zuckerman, B. M., 1986, Fluorescent and ferritin labelling of cuticle surface carbohydrates of *Caenorhabditis elegans* and *Panagrellus redivivus, J. Nematol.* **18**:570–574.

Jeyaprakash, A., Jansson, H. B., Marban-Mendoza, N., and Zuckerman, B. M., 1985, *Caenorhabditis elegans:* Lectin-mediated modification of chemotaxis, *Exp. Parasitol.* **59**:90–97.

Kerry, B., 1980, Biocontrol: Fungal parasites of female cyst nematodes, *J. Nematol.* **12**:253–259.

Kerry, B. R., 1984, Nematophagous fungi and the regulation of nematode populations in soil, *Helminthol. Abstracts Ser. B Plant Nematol.* **53**:1–14.

Kerry, B. R., and Crump, D. H., 1977, Observations on fungal parasites of females and eggs of the cereal cyst-nematode, *Heterodera avenae,* and other cyst nematodes, *Nematologica* **23**:193–201.

Lis, H., and Sharon, N. 1981, Lectins in higher plants, in: *The Biochemistry of Plants* (A. Marcus, ed.), Vol. 6, pp. 371–447, Academic Press, New York.

Lis, H., and Sharon, N., 1984, Lectins: Properties and applications to the study of complex carbohydrates in solution and on cell surfaces, in: *Biology of Carbohydrates* (V. Ginsburg and P. H. Robbins, eds.), Vol. 2, pp. 1–85, Wiley, New York.

Lysek, G., and Nordbring-Hertz, B., 1981, An endogenous rhythm of trap formation in the nematophagous fungus *Arthrobotrys oligospora, Planta* **152**:50–53.

Lysek, G., and Nordbring-Hertz, B., 1983, Die Biologie der nematodenfangender Pilze, *Forum Mikrobiol.* **6**:201–208.

Mankau, R., 1980, Biological control of nematode pests by natural enemies, *Annu. Rev. Phytopathol.* **18**:415–440.

Mankau, R., 1981, Microbial control of nematodes, in: *Plant Parasitic Nematodes* (B. M. Zuckerman and R. A. Rohde, eds.), Vol. III, pp. 475–494, Academic Press, New York.

Mattiasson, B., Johansson, P. A., and Nordbring-Hertz, B., 1980, Host–microorganism interaction: Studies on the molecular mechanisms behind the capture of nematodes by nematophagous fungi, *Acta Chem. Scand. B* **34**:539–540.

McClure, M., and Zuckerman, B. M., 1982, Localization of cuticular binding sites of concanavalin A on *Caenorhabditis elegans* and *Meloidogyne incognita, J. Nematol.* **14**:39–44.

Mirelman, D., 1986, *Microbial Lectins and Agglutinins: Properties and Biological Activity,* Wiley, New York.

Nigh, E. A., Thomason, I. J., and van Gundy, S. D., 1980, Identification and distribution of fungal parasites of *Heterodera schachtii* eggs in California, *Phytopathology* **70**:884–891.

Nordbring-Hertz, B., 1972, Scanning electron microscopy of the nematode-trapping organs in *Arthrobotrys oligospora, Physiol. Plant.* **26**:279–284.

Nordbring-Hertz, B., 1973, Peptide-induced morphogenesis in the nematode-trapping fungus *Arthrobotrys oligospora, Physiol. Plant.* **29**:223–233.

Nordbring-Hertz, B., 1977, Nematode-induced morphogenesis in the predacious fungus *Arthrobotrys oligospora, Nematologica* **23**:443–451.

Nordbring-Hertz, B., 1983, Dialysis membrane technique for studying microbial interaction, *Appl. Environ. Microbiol.* **45**:290–293.

Nordbring-Hertz, B., 1984, Mycelial development and lectin–carbohydrate interactions in nematode-trapping fungi, in: *The Ecology and Physiology of the Fungal Mycelium* (D. H. Jennings and A. D. M. Rayner, eds.), pp. 419–432, Cambridge University Press, Cambridge.

Nordbring-Hertz, B., and Chet, I., 1986, Fungal lectins and agglutinins, in: *Microbial Lectins and Agglutinins: Properties and Biological Activity* (D. Mirelman, ed.), pp. 393–408, Wiley, New York.

Nordbring-Hertz, B., and Jansson, H. B., 1984, Fungal development, predacity, and recognition of prey in nematode-destroying fungi, in: *Current Perspectives in Microbial Ecology* (M. J. Klug and C. A. Reddy, eds.), pp. 327–333, Proceedings of the Third International Symposium Microbial Ecology, August 1983.

Nordbring-Hertz, B., and Mattiasson, B., 1979, Action of a nematode-trapping fungus shows lectin-mediated host–microorganism interaction, *Nature* **281**:477–479.

Nordbring-Hertz, B., and Odham, G., 1980, Determination of volatile nematode exudates and their effects on a nematode-trapping fungus, *Microb. Ecol.* **6**:241–251.

Nordbring-Hertz, B., and Stålhammar-Carlemalm, M., 1978, Capture of nematodes by *Arthrobotrys oligospora,* an electron microscope study, *Can. J. Bot.* **56**:1297–1307.

Nordbring-Hertz, B., Friman, E., and Mattiasson, B., 1982, A recognition mechanism in the adhesion of nematodes to nematode-trapping fungi, in: *Lectins, Biology, Biochemistry and Clinical Biochemistry* (T. C. Bøg-Hansen, ed.), Vol. 2, pp. 83–90, de Gruyter, Berlin.

Nordbring-Hertz, B., Veenhuis, M., and Harder, W., 1984, Dialysis membrane technique for ultrastructural studies of microbial interactions, *Appl. Environ. Microbial.* **47**:195–197.

Nordbring-Hertz, B., Zunke, U., Wyss, U., and Veenhuis, M., 1986, Trap formation and capture of nematodes by *Arthrobotrys oligospora,* Film No. C 1622, produced by Institut fur den Wissenshaftlichen Film, Göttingen, Germany.

Persson, Y., Veenhuis, M., and Nordbring-Hertz, B., 1985, Morphogenesis and significance of hyphal coiling by nematode-trapping fungi in mycoparasitic relationships, *FEMS Microbiol. Ecol.* **31**:283–291.

Peterson, E. A., and Katznelson, H., 1965, Studies on the relationships between nematodes and other soil microorganisms. IV. Incidence of nematode-trapping fungi in the vicinity of plant roots, *Can. J. Microbiol.* **11**:491–495.

Premachandran, D., and Pramer, D., 1984, Role of N-acetylgalactosamine-specific protein in trapping of nematodes by *Arthrobotrys oligospora, Appl. Environ. Microbiol.* **47:**1358–1359.

Pramer, D., and Kuyama, S., 1963, Symposium on biochemical bases of morphogenesis in fungi: II Nemin and the nematode-trapping fungi, *Bacteriol. Rev.* **27:**282–292.

Rosenzweig, W. D., and Ackroyd, D., 1983, Binding characteristics of lectins involved in the trapping of nematodes by fungi, *Appl. Environ. Microbiol.* **46:**1093–1096.

Rosenzweig, W. D., and Ackroyd, D., 1984, Influence of soil microorganisms on the trapping of nematodes by nematophagous fungi, *Can. J. Microbiol.* **30:**1437–1439.

Rosenzweig, W. D., and Pramer, D., 1980, Influence of cadmium, zinc and lead on growth, trap formation, and collagenase activity of nematode-trapping fungi, *Appl. Environ. Microbiol.* **40:**694–696.

Rosenzweig, W. D., Premachandran, D., and Pramer, D., 1985, Role of trap lectins in the specificity of nematode capture by fungi, *Can. J. Microbiol.* **31:**693–695.

Saikawa, M., 1982, An electron microscope study of *Meria coniospora,* an endozoic nematophagous fungus, *Can. J. Bot.* **60:**2019–2023.

Saikawa, M., Totsuka, J., and Morikawa, C., 1983, An electron microscope study of initiation of infection by conidia of *Harposporium oxycoracum,* an endozoic nematophagous hyphomycete, *Can. J. Bot.* **61:**893–898.

Schenck, S., Chase, T., Jr., Rosenzweig, W. D., and Pramer, D., 1980, Collagenase production by nematode-trapping fungi, *Appl. Environ. Microbiol.* **40:**567–570.

Spiegel, Y., and Cohn, E., 1982, Lectin binding to *Meloidogyne javanica* eggs, *J. Nematol.* **14:**406–407.

Spiegel, Y., Cohn, E., and Spiegel, S., 1982, Characterization of sialyl and galactosyl residues on the body wall of different plant parasitic nematodes, *J. Nematol.* **14:**33–39.

Spiegel, Y., Robertson, W. M., Himmelhoch, S., and Zuckerman, B. M., 1983, Electron microscope characterization of carbohydrate residues on the body wall of *Xiphinema index, J. Nematol.* **15:**528–534.

Stirling, G. R., McKenry, M. V., and Mankau, R., 1979, Biological control of root knot nematode on peach, *Phytopathology* **69:**806–809.

Sturhan, D., and Schneider, R., 1980, *Hirsutella heteroderae,* ein neuer nematodenparasitärer Pilz, *Phytopathol. Z.* **99:**105–115.

Tzean, S. S., and Estey, R. H., 1978, Nematode-trapping fungi as mycopathogens, *Phytopathology* **68:**1266–1270.

Veenhuis, M., Nordbring-Hertz, B., and Harder, W., 1984, Occurrence, characterization and development of different types of microbodies in the nematophagous fungus *Arthrobotrys oligospora, FEMS Microbiol. Lett.* **24:**31–38.

Veenhuis, M., Nordbring-Hertz, B., and Harder, W., 1985a, An electron microscopical analysis of capture and initial stages of penetration of nematodes by *Arthrobotrys oligospora, Antonie Leeuwenhoek J. Microbiol.* **51:**385–398.

Veenhuis, M., Nordbring-Hertz, B., and Harder, W., 1985b, Development and fate of electron dense microbodies in trap cells of the nematophagous fungus *Arthrobotrys oligospora, Antonie Leeuwenhoek J. Microbiol.* **51:**399–407.

Veenhuis, M., Nordbring-Hertz, B., and Harder, W., 1985c, An ultrastructural study of cell–cell interactions in capture organs of the nematophagous fungus, *Artrobotrys oligospora, FEMS Microbiol. Lett.* **30:**93–98.

Whipps, J. M., and Lynch, J. M., 1986, The influence of the rhizosphere on crop productivity, in *Advances in Microbial Ecology* (K. C. Marshall, ed.), Vol. 9, pp. 187–244, Plenum Press, New York.

Wimble, D. B., and Young, T. W. K., 1983, Capture of nematodes by adhesive knobs in *Dactylella lysipaga, Microbios* **36:**33–39.

Wimble, D. B., and Young, T. W. K., 1984, Ultrastructure of the infection of nematodes by *Dactylella lysipaga, Nova Hedwigia* **40**:9–29.

Zopf, W., 1888, Zur Kenntnis der Infektions-Krankheiten niederer Tiere und Pflanzen, *Nova Acta Leop. Carol.* **52**:314–376.

Zuckerman, B. M., 1983, Hypothesis and possibilities of intervention in nematode chemotaxis, *J. Nematol.* **15**:175–182.

Zuckerman, B. M., and Jansson, H. B., 1984, Nematode chemotaxis and possible mechanism of host/prey recognition, *Annu. Rev. Phytopathol.* **22**:95–113.

Zuckerman, B. M., and Kahane, I., 1983, *Caenorhabditis elegans:* Stage specific differences in cuticle surface carbohydrates, *J. Nematol.* **15**:535–538.

4

Fungal Communities in the Decay of Wood

A. D. M. RAYNER and LYNNE BODDY

1. Wood as a Venue for Community Studies

Nowhere, we believe, can the presence of fungal communities, their structure, dynamics, and diversity, be more explicit and susceptible to direct analysis than in decaying wood. In consequence, wood provides an excellent venue, both for the study of community interactions, and for the development of a conceptual framework within which they can be rationalized. Three outstanding characteristics of wood, as a resource for exploitation by heterotrophs, account for this belief: its bulk, its spatial definition, and its durability.

By contrast with unicellular microorganisms, the mycelial form of many fungi is especially suitable for invasion of solid, bulky, spatially determinate resource units. During such invasion, available colonizable space (in the sense of *domain* or *territory;* that is, space containing nutrients within the immediate sphere of influence of the thallus) progressively becomes depleted as different thalli expand and achieve contact with one another (Rayner and Todd, 1979; see also Bull and Slater, 1982). A general principle is that as the bulk of a spatially determinate resource increases, so the available surface from which colonization proceeds decreases proportionately in relation to volume; hence, there is the potential for individual thalli to occupy increasingly larger volumes. This

A. D. M. RAYNER • School of Biological Sciences, University of Bath, Bath BA2 7AY, United Kingdom. LYNNE BODDY • Department of Microbiology, University College, Cardiff CF2 1TA, United Kingdom.

probably contributes to the situation in decaying wood, where one or a few individual mycelia may sometimes be found occupying 1 m^3 or more, and where differently sized units (e.g., twigs, branches, trunks) contain similar numbers of individuals (Rayner and Webber, 1984).

That we regard the durability of wood as an advantage may cause some surprise, and, indeed, it is a double-edged sword. Without doubt, many potential investigators must have been discouraged from experimental studies by the often protracted periods over which wood decay processes occur. However, in practical terms, relatively slow rates of change can often facilitate detailed analysis without risk of missing vital stages. Possibly even more important from a theoretical point of view is that studies of systems in which rates of change are rapid, and hence attractive to experimentalists, tend necessarily to focus on exploitation of easily assimilable and/or ephemeral resources, and hence on organisms with a limited individual life span and rapid commitment to reproduction (i.e., those with *r*-selected ecological strategies; see Section 4). This carries the risk that during development of concepts, those organisms with a sustained individual existence and slow commitment to reproduction (*K*-selected) may be omitted from consideration. It is among certain of such organisms that the most dramatic interaction phenomena may both be expected and indeed occur. By giving scope, through its durability, for expression of the full range of ecological strategies encompassed with the *r–K*-selection spectrum, wood provides the basis for a more *complete* understanding of community dynamics. In addition, such expression is amplified by the diversity of circumstances, from twigs to trunks, from the living standing tree to cut and processed timber, under which wood may become decayed. In understanding how community patterns and functioning may be affected by these varying circumstances, a necessary first step is to characterize the effects that a range of environmental variables (biotic and abiotic) associated with wood may have on colonization processes. The following is a brief summary of what is known about such effects.

2. Factors Affecting Colonization

As in all cases where factors affecting microbial community characteristics within a particular resource type are being considered, it is useful to distinguish three interacting categories: those factors specifically associated with the resource or substratum (resource quality), microclimate, and other organisms (Fig. 1).

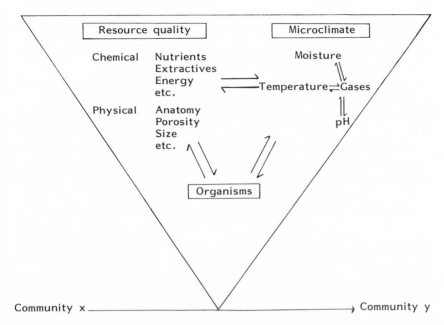

Figure 1. Schematic representation of interactions among the factors affecting colonization and community development in wood.

2.1. Resource Quality

2.1.1. Availability of Organic Nutrients and Their Distribution

Readily accessible, assimilable substrates such as starch, soluble sugars, lipids, peptides, and other primary metabolites occur in relatively small amounts in wood (<10% by dry weight), being found almost exclusively in living or recently dead sapwood parenchyma. They are rapidly depleted during fungal colonization, but enable development of certain fungi and other microorganisms not active in decay of wood structure, and may also be important in aiding *establishment* of decay fungi. Their amount varies according to seasonal and other factors, but may be as high as 7% for starch and 2.5% for lipids, with the amount of starch sometimes being maximal in the middle sapwood, declining to zero at the heartwood boundary (Hillis, 1977).

The dominant available carbon sources are, however, the relatively refractory major structural components of the cell walls, that is, cellulose, lignin, and hemicelluloses. The relative proportions of these can vary considerably both in wood from the same and from different trees (Timell, 1965; Côté, 1977).

Three basic classes of decay fungi have been identified, which differ in the nature of their attack on major cell wall components: white rots, brown rots, soft rots. White rots involve degradation of all components, usually at rates relative to the original amounts present, although lignin is often removed relatively faster, especially in coniferous woods. Typically, the wood acquires a bleached appearance and a fibrous or spongy consistency as decay proceeds. Fungi responsible include many Basidiomycotina, comprising members of all the major taxonomic groups except Hemibasidiomycetes, and certain xylariaceous Ascomycotina (species of *Xylaria, Hypoxylon, Daldinia,* and *Ustulina*).

Brown-rot fungi, which appear to be exclusively Basidiomycotina, utilize hemicelluloses and cellulose, leaving lignin essentially undegraded, although it is slightly modified, as is indicated by demethylation and the accumulation of oxidized polymeric degradation products (Kirk, 1975; Kirk and Adler, 1970). Brown-rotted wood is characterized by its light or dark brown color, and, when decay is advanced, by its crumbly or powdery consistency and cubical cracking pattern, which results from uneven shrinkage during drying. Brown rot typically results in much faster initial loss of weight and tensile strength than white rot, because of the more rapid removal of cellulose from cell walls at considerable distances from hyphae: this may be associated with a partially nonenzymatic process based on an H_2O_2/Fe^{2+} system (Koenigs, 1972 a,b, 1974 a,b).

Soft rot is caused by a wide range of Ascomycotina and Fungi Imperfecti, including species of *Chaetomium, Phialophora,* and *Fusarium*. It is characteristic of wood with a sufficiently high moisture or preservative content to preclude attack by white- and brown-rot fungi, and is frequently superficial. Both cellulose and hemicellulose are degraded, but only in the immediate vicinity of the hyphae, resulting in characteristically shaped cavities in the wood cell walls, and much slower breakdown than with brown rots (e.g., Käärik, 1974; Levy, 1982).

2.1.2. Availability of Mineral Nutrients

Wood is commonly supposed to have a very low nitrogen status, due to the fact that its carbon–nitrogen ratio is commonly as much as 500:1, and may even be as high as 1250:1 (Cowling, 1970). This apparent deficiency may be alleviated in several ways. First, wood-inhabiting fungi may preferentially allocate nitrogen to metabolically active cell constituents such as nucleic acids. Second, they may conserve nitrogen by continual autolysis and consequent reutilization of the nitrogen so released, or alternatively, by lysis of other fungal hyphae present in the wood (Levi *et al.,* 1968; Levi and Cowling, 1969). It might also be relevant that while a high carbon–nitrogen ratio may be unfavorable when either or both the carbon and mineral nutrient sources are soluble, it may be less so when it is insoluble, and concentrated spatially. In the latter situation, a fungus may be able to concentrate nitrogen in certain restricted areas, thus permitting concomitant localized production of hydrolases. In these areas the carbon–nitrogen ratio will be different from, and more favorable than, that of the general substratum, with much of the insoluble carbon-rich material acting as an inert matrix within which growth can take place (Park, 1976; Dowding, 1981).

Typically, more nitrogen is located in sapwood than heartwood; in recent sapwood than in older heartwood; in early wood than in late wood within the same annual ring; in the pith than in adjacent tissues; in inner than outer heartwood; where the proportion of parenchyma cells is highest; and in root wood than in stem wood of gymnosperms (Merrill and Cowling, 1966; Platt *et al.,* 1965). Like nitrogen, other important mineral nutrients, phosphorus and potassium, show a dramatic decrease in concentration by as much as 95% or more during the transition from sapwood to heartwood.

2.1.3. Extractives

Besides the major organic nutrients mentioned previously, wood contains a wide range of extraneous materials or extractives which can be removed using neutral organic solvents or water. They include waxes, fats, organic acids, alkaloids, oils, rosins, resins, phenolics, and many others (Hillis, 1962). Large quantities of extractives are characteristically found in heartwood, but they may also be found in sapwood, either due to impregnation from resin- or gum-filled canals, or following accumulation in discolored tissues associated with wounding and/or decay columns in living trees (Shigo and Hillis, 1973; Hillis, 1977).

Emphasis is usually placed on those extractives with fungitoxic or fungistatic properties, since these are believed to confer resistance to decay (Scheffer and Cowling, 1966). Most are phenolic, including the four major classes, terpenoids, tropolones, flavonoids, and stilbenes. Tannins are prominent in many angiosperms. Evidently these extractives are formed *in situ* from carbohydrate or lipid substrates via the acetate and shikimic acid pathways (Hillis, 1962, 1977). The role of such extractives in decay resistance is indicated not only by their fungitoxicity in agar culture, but by the increased decay rate of extracted heartwood and decreased decay rate of artificially impregnated wood blocks or sawdust (Scheffer and Cowling, 1966). However, assays using impregnated wood or sawdust often show much reduced toxicity compared with tests on nutrient media, and the possibilities of chemical binding to cell walls and other interactions need to be considered (Hart and Shrimpton, 1979). Furthermore, in some instances extractives may actually play a role as specific *stimulants* of certain fungi (Fries, 1973; Rayner and Hedges, 1982).

2.1.4. Wood Anatomy

Access to the nutrients in wood is provided via an interconnected system of voids (lumina) surrounded by solid, relatively impenetrable cell walls (Levy, 1975, 1982). In purely physical terms, the orientation, maintenance of access to, and size of these voids will crucially affect microbial and fungal distribution (Cooke and Rayner, 1984). Longitudinal access is facilitated by the predominantly elongated lumina of xylem elements such as vessels, tracheids, fibers, and axial parenchyma. Radial access will be enhanced by the radially elongated, nutrient-enriched parenchyma of medullary rays, providing that these are dead. However, it will be restricted in nonray tissue by the close-packed xylem elements in the late (autumn) wood of each annual ring. Except where a dead cambial layer facilitates subcortical mycelial spread, tangential access is greatly restricted by the lack of any system, apart in a few cases from trauma-induced resin or gum-filled canals of tangentially elongated lacunae in wood. Any spread in this direction must be effected via numerous woody cell walls, either by direct penetration, formation of bore holes, or through pits. Contiguous large vessels or other elements in spring wood, particularly that which is ring porous, may slightly enhance tangential access in this tissue. On the basis that the effects of these anatomical features operate without interference from other factors, the expected shape of a mycelium arising eccentrically in intact timber is a longitudinally tapering, wedge-shaped column, with slightly corrugated radial margins due to alternation between early and late wood in annual rings (Boddy and Rayner, 1983c; Cooke and Rayner, 1984).

2.2. Microclimate

2.2.1. Moisture Content and Aeration

In addition to providing access routes for entry and growth of microorganisms, the voids in wood similarly act either as reservoirs in which fluids can accumulate or as channels through which they may be interchanged. The principal liquid filling the voids under natural conditions is, of course, water and its content of dissolved solutes: its amount and location are reciprocally related to that of the gaseous phase, and are crucial determinants of fungal activity. In saturated wood, supply of gases required for metabolic function, notably oxygen, will be limited both by their solubility in water and the long diffusion path over which they must travel. Equally, removal of those gases, such as carbon dioxide and ethylene, produced as a result of metabolism will be restricted, so that they are likely to accumulate within the vicinity of any active cells (Boddy, 1986). For fungi, which are predominantly aerobic, this is likely greatly to impede activity, perhaps especially for those dependent on ligninocellulose breakdown, which has been shown to be a highly oxygen-demanding process (Kirk *et al.*, 1978; cited in Kirk and Fenn, 1982). As drying proceeds, water is withdrawn from progressively smaller cavities, which, in consequence, become filled by gases. However, at high moisture contents access of oxygen to this gaseous phase is, for the same reasons as above, likely to be limited and that of carbon dioxide enhanced. Thus, carbon dioxide levels of 10–20% have commonly been recorded in wood (Chase, 1934; Hintikka, 1982) and approaching 100% CO_2 has been recorded in living stems of *Acacia* (Carrodus and Triffett, 1975).

As a generalization, and assuming that they are connected to the exterior and not completely surrounded by smaller voids, voids with radii greater than 5 μm contain no water at matric potentials lower than about 0.03 MPa, osmotic potential being negligible in wood. Micropores, such as pit apertures and pit membrane pores, of radius 0.01–5 μm are emptied at -0.03 to -14 MPa. Inter- and intramolecular spaces in cell walls retain water at less than -14 MPa. At -0.1 MPa voids of less than 1.5 μm are empty (corresponding to a moisture content of about 27–30% of oven dry weight in undecayed wood), and this defines the "fiber saturation point" when all water present can effectively be regarded as "bound" to the cell walls (Griffin, 1977). Below this level water quickly becomes relatively inaccessible to fungal hyphae and their extracellular enzymes, as well as requiring more force to extract. This may account for the fact that the lower general limit for growth of wood-decaying Basidiomycotina appears to be about -4 to -4.5 MPa (Griffin, 1977; Wilson and Griffin, 1979; Boddy, 1983a). In practical terms, undecayed wood is generally considered safe from decay at moisture contents of about 20–24% by dry weight, corresponding with matric potentials of about -7 MPa or below. Even though growth may not be possible below this level, survival cetainly is. In pieces of wood colonized by *Schizopora paradoxa, Fibulopora vaillantii,* and *Lentinus lepideus,* these fungi are still viable after 6 years of storage at less than -100 MPa (Theden, 1961).

2.2.2. *Temperature and pH*

The temperature and gross pH requirements and tolerance of wood inhabiting fungi vary (Wagener and Davidson, 1954; Cartwright and Findlay, 1958), and in some cases there is evidence that this affects their distribution and competitive ability. As an example of possible temperature effects, in a study of the distribution of decay fungi within logging slash in Canada, Loman (1962, 1965) found that the four most common basidiomycetes were remarkably consistent in their spatial location. Thus, *Lenzites saepiaria* was predominant in the central portions of the slash; *Phlebia phlebioides* was isolated mainly from upper portions; and *Stereum sanguinolentum* and *Coniophora puteana* from the lower portions. This distribution may be explained, at least in part, on the basis of their temperature tolerances, optima for growth, and relative decaying abilities. High temperatures lethal to the mycelium of *C. puteana* and *S. sanguiniolentum,* but not to *L. saepiaria* and *P. phlebioides,* occur in the upper and central portions of slash during fine weather. Also, *P. phlebioides* had a much higher growth rate and decay rate at 38°C, which was common in fine weather, than the other three species, and *L. saepiaria* caused more decay than the others at 31°C. At 10°C the greatest decay was caused by *C. puteana.*

In terms of possible pH and related effects, certain elm (*Ulmus* spp.)-inhabiting fungi may be selectively favored by their tolerance of the relatively high pH of elm wood (Rayner and Hedges, 1982), whereas *Laetiporus sulphureus* and *Daedalea quercina* may be favored in heartwood of oak (*Quercus* spp.) by their tolerance of high concentrations of acetic acid (Hintikka, 1969).

2.3. Other Organisms

Besides mycelial fungi, other organisms that may be present, to a greater or lesser extent, in decaying wood include bacteria, yeasts, myxomycetes, and invertebrates, mainly insects, e.g., Isoptera, Coleoptera, Isopoda, Diplopoda, and Diptera, but also Oligochaeta, Acari, and Nematoda.

In some cases these organisms may play a very important or even principal role in decomposition, as where termites are active or in waterlogged wood where bacterial attack predominates. In other cases, interactions with the fungi may considerably affect community dynamics. Such interaction may be direct, as with grazing of fungal mycelium by arthropods and myxomycetes (Swift and Boddy, 1984; Madelin, 1984), antibiosis, and nutrient competition, although there are few cases where the latter have been clearly implicated and quantified. Alternatively, the interactions may be indirect, operating through disturbance, stress alleviation, or stress aggravation, and this will be considered in more detail below (Section 5.2). In terms of direct interactions, the most dramatic and clearly definable are those between mycelial individuals themselves, and this will be the concern of the next major section of this chapter.

3. Mycelial Interactions

3.1. Evidence for Interactions *in Situ:* Observations of Fungal Community Patterns

Concomitant with the nature of fungal mycelia as domain-capturing forms, understanding of fungal community structure in wood (or any other medium) depends on knowledge of their spatial distribution. Failure to appreciate this fundamental concept, combined with a tendency to treat mycelia only as though they were populations of cells, rather than as individuals in their own right, may have been behind many past approaches to studying fungal community structure. Typically these have relied on isolation onto culture media from limited samples of material, often without reference to their spatial origin. In turn, such failure may have resulted from lack of appreciation of the individualistic and essentially territorial nature of mycelia, especially in higher fungi, where somatic recognition and rejection mechanisms operate between different individuals both within and between species (e.g., Rayner and Todd, 1979; Todd and Rayner, 1980; Rayner *et al.,* 1984). As we will discuss further, the occurrence of such territoriality radically affects the conceptional framework within which mycelial interactions must be viewed and contrasted with the behavior of nonmycelial microorganisms.

The most direct means of gaining a first impression of community structure in a piece of decaying wood is simply to saw it into sections and examine the distribution of discolored and/or decayed regions within it, where a variety of patterns may be apparent, depending on circumstances (Fig. 2). Fungi associated with different regions may be identified by correlation with the position of sporophores at the surface, identification of mycelium and sporophores produced by direct incubation under moist conditions, or of cultures obtained by isolation onto culture media: ideally, a combination of all three approaches should be used. In some cases the wood may be generally stained or discolored without clear demarcation of separate regions. This typically occurs where brown- and white-rot fungi are not active and it is common to find a diverse array of microfungi, yeasts, and bacteria present. This implies, at least at the gross level, some element of coexistence or *intermingling* among such microorga-

Figure 2. Community structure in beech logs, 10–20 cm diameter, placed upright with their bases buried in leaf litter: (a, b) a mature decay community (a) before and (b) after incubation. The log has been cut transversely near the top surface and longitudinally for the remaining length to reveal the three-dimensional distribution of columns of decay and discoloration; (c) transverse section of beech log, showing early stages of colonization with general stain plus a column containing *Xylaria hypoxylon;* (d) sections through the base of a beech log, showing the mycelial cords of *Tricholomopsis platyphylla,* which has caused uniform decay. Ab, *Armillaria bulbosa;* Cm, *Chaetosphaeria myriocarpa* (conidial stage) occupying interaction zones and other regions not occupied by active basidiomycete mycelium; Cv, *Coriolus versicolor;* Pv, *Phanerochaete velutina* invading from the base; Sh, *Stereum hirsutum.* [From Coates (1984).]

nisms in the sense of overlapping domains (see Section 1; obviously no two individuals can, by definition, coexist in the sense of occupying exactly the same space). However, it may also be that their domains are so small as not to be distinguishable as separate entities. We will return to this later.

In wood where decay fungi are active, the pattern is very different in that either at one extreme an easily observable mosaic of different regions of decay, separated by distinct boundary regions (zone lines), is present, or at the other extreme the wood may be uniformly decayed. In both cases the decay regions are typically effectively occupied by pure cultures of decay fungi, as can be readily confirmed by isolation onto laboratory media. However, in standing trees we have occasionally found evidence of apparent coexistence of both a basidiomycete and a xylariaceous asco-mycete within uniform decay (Fig. 2d), for example, *Bjerkandera adjusta* and *Hypoxylon nummularium* in beech, even though in other circum-stances these are mutually exclusive. Such apparent coexistence may be enforced by latent invasion mechanisms (Sections 5.1 and 5.2.1) whereby different individuals become irretrievably intermixed.

The origins of zone lines have been discussed by Rayner and Todd (1979, 1982) and it is evident that, whereas some forms may be produced in the presence of a single individual, others are the direct result of con-frontation between individuals of the same or different species: this can readily be confirmed by isolation of mycelia from different sides of a zone line into culture, whence they display an antagonistic interaction when placed adjacent to each other (Fig. 3). Such zone lines are characteristi-cally less decayed and of a different color (usually, but not always, darker) than the surrounding decay. Intriguingly, they appear to provide a habitat for a variety of microfungi, notably dematiaceous species of *Rhinocla-diella, Leptodontium, Catenularia, Phialophora,* and *Endophragmiella.* In wood undergoing very active decay by Basidiomycotina these are often the only microfungi to be found, and the zone lines their only location (Fig. 4).

This pattern is clearly indicative of exclusivity between the domains of different individuals, rather than coexistence, even though the distinc-tion has sometimes been confused (e.g., Swift, 1976). Such exclusivity, which is also implicit in those cases where a single individual occupies a uniformly decayed substratum, can be regarded as the outcome of what has been described as a "deadlock" interaction whereby confronting indi-viduals are unable to enter each other's domains. A situation in which deadlock interactions predominate is, of course, intrinsically stable, but often close examination of decay patterns also reveals evidence of inva-sion, by certain individuals, of domains previously occupied by others. This typically occurs between different species and has been termed *replacement* (Rayner and Todd, 1979). Such evidence can take a variety of forms, including the presence of series of "relic" zone lines left behind by a retreating individual in wood now occupied by another (Fig. 2b), the presence of different fungi within the same decay region, and presence of old sporophores or other structures associated with wood in which the

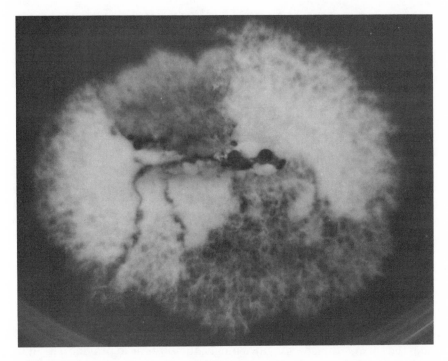

Figure 3. Outgrowth of *Colpoma quercinum* onto malt agar from a piece of oak twig, showing correlation of zone lines in wood with interaction zones between developing colonies. [From Boddy and Rayner (1984).]

fungus responsible is not present. Replacement can also be inferred by examining the change in community patterns with time within a particular resource (Fig. 5) [for more details see Rayner and Todd (1979), Cooke and Rayner (1984), Rayner and Webber, (1984)].

Finally, in wood that has undergone very extensive decomposition, community patterns are often all but destroyed, particularly in association with invasion by animals. White- and brown-rot fungi are often no longer evident, and mucoraceous species and penicillia, as well as a diverse range of other microfungi commonly associated with soil, together with bacteria and yeasts are the only readily isolated forms (Swift and Boddy, 1984).

Clearly, then, there is direct evidence for a wide range of mycelial interactions occurring during community development in wood, and that these may be broadly classified under the headings replacement, deadlock, and intermingling. Such headings do not prejudge the issue of the actual *mechanisms(s)* leading to such *outcomes* (Rayner and Webber,

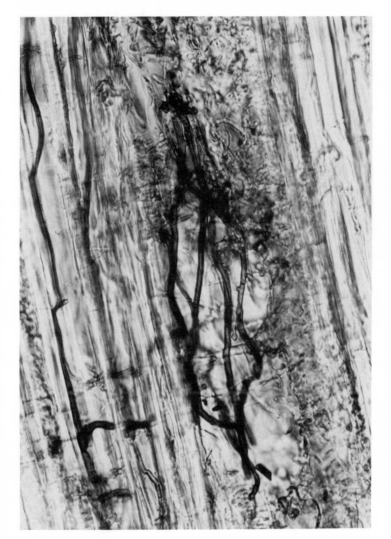

Figure 4. Longitudinal section through a zone line in birchwood colonized by *Coriolus versicolor*, showing a large number of hyaline hyphae of *C. versicolor* and several distinctive dark hyphae. [From Rayner (1976).]

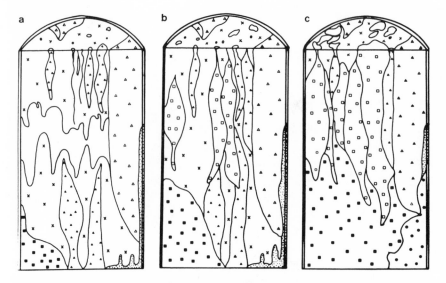

Figure 5. (a–c). Idealized diagrams illustrating a typical pattern of natural decay community development in a cut beech log placed upright with its base buried in the soil and litter of a deciduous woodland site. The community patterns at 6, 12, and 18 months respectively are shown: x, stained or discolored wood containing microfungi and/or the Basidiomycotina *Chondrostereum purpurem* and *Corticium avoluens;* solid triangles, *Xylaria hypoxylon;* open triangles, *Hypoxylon nummularium;* open squares, combative air-borne Basidiomycotina, e.g., *Coriolur versicolor, Bjerkandera adjusta,* and *Stereum hirsutrum;* closed squares, combative early-arriving cord formers, e.g., *Phallus impudicus* and *Tricholomopsis platyphylla;* closed circles, combative late-arriving cord formers, e.g., *Phanerochaete velutina;* stipple, *Armillaria bulbosa.* (After Coates, 1984.)

1984) and, indeed, it is almost impossible to determine this in the opaque natural substratum and under the diverse circumstances that occur. Whereas it may be possible to infer whether the mechanisms are due to *active* confrontation rather than being *passive* due to indirect effects (for example, the deadlock reactions between different mycelial mats shown in Fig. 2c are clearly due to an active mechanism), for more information it is necessary, with all the attendant problems to proceed to studies of laboratory cultures.

3.2. Interactions in Agar Culture

While it is possible to study interactions under seminatural conditions, for example in wood-block or sawdust culture, such experiments fall uncomfortably between the two stools of on one hand knowing that what has been observed has taken place under natural conditions and on the other hand ease of experimentation, observation, and maintenance of environmental conditions (e.g., Rayner and Webber, 1984). For ease of discussion we will therefore restrict ourselves to these two extreme possibilities.

In determining whether or not an interaction seen in culture is relevant to field conditions, an obvious first step, if possible, is simply direct comparison between the two. In fact, especially with white- and brownrot fungi, there is often a remarkably good correlation between laboratory and field observations, and where there is not, it is often possible with experience to determine why: this often is related to understanding the actual mechanism responsible for a particular observed outcome. In terms of such outcomes in culture, it is again possible to recognize three categories of behavior: intermingling, deadlock, and replacement, and these may occur when isolates of the same or different species are adjacent to one another.

3.2.1. Intermingling

3.2.1a. Within Species. Probably the clearest example of an intermingling response between mycelia paired in culture occurs between isolates with the same genotype, at least with respect to factors controlling somatic incompatibility (Rayner *et al.,* 1984). Here, although there may be some initial inhibition as the mycelial fronts come into contact, perhaps related to mechanical effects, nutrient competition, and/or production of inhibitory products, this is normally soon followed by complete coalescence, and, at least in higher fungi where vegetative anastomoses occur, effectively a new mycelial unit is formed. In nature, such coalescence can be expected whenever colonization is effected by genetically identical propagules, such as conidia ultimately derived from the same thallus, or even "sexual" spores (ascospores and basidiospores) if these are produced via self-fertilization or amixis from a haploid homokaryotic parent. For example, in *Stereum sanguinolentum* the basidiocarps are homokaryotic and produce offspring that intermingle in culture. As a result a series of what are effectively clones occur in the natural population; members of the same clone intermingle, but members of different clones are somatically incompatible (Rayner and Turton, 1982). The major circumstance in which coalescence of genetically nonalike mycelia occurs is found between sexually compatible homokaryons of Basidiomycotina, where somatic fusion between different mating types results in production of an independent secondary mycelial or dikaryotic phase. It has been argued that such coalescence is a consequence of override, by the mating system, of the self–nonself somatic rejection reaction that is intrinsic between nonalike genotypes (Rayner et al., 1984).

3.2.1b. Between Species. It is difficult to be certain about the occurrence or otherwise of intermingling between different species in culture, partly because of insufficient work—especially with non-Basidiomycotina—and partly because it is facilitated by culture conditions favoring sparse mycelial development. For example, intermingling may occur where nutrients are either supplied in low overall concentration or in a form or forms inaccessible to one or both participants, so that access to them can only be achieved via complementation. Under conditions favoring dense mycelial growth, coexistence (see Section 3.2.1a) of mycelia, other than as occurs as a result of biotrophic mycoparasitism, seems to be an exceptional event. However, fungi that have constitutively sparse, explorative mycelia, such as certain coenocytic Mucorales, may invade a domain already occupied by another fungus without necessarily deleteriously affecting the latter or being so affected.

How far this type of circumstance may apply in wood is uncertain, but it seems likely that mycelial systems of non-wood-decay species in particular may often be necessarily sparse due to the limited supply of available nutrients, and hence account for the apparent intermingling in discolored but relatively undecayed wood. During active decay, examination of microscope sections or direct incubation reveals that mycelia of white- and brown-rot fungi are often remarkably luxuriant, often appearing just as dense, if not more so, as in culture on 3% malt agar (Fig. 2c). The likelihood of their intermingling thus seems remote, and this correlates with the observation of exclusiveness described above (Section 3.1).

3.2.2. Replacement and Deadlock

As we have implied, replacement and deadlock appear to be the most common outcomes of confrontation in culture. Whereas they may in some cases be due to purely nutritional considerations, in other cases they result from direct physiological challenge, as will be discussed later. Such interactions can be classed as "combative" and are our particular concern here.

3.2.2a. Interspecific Interactions. Interspecific combative interactions have been classified by Rayner and Webber (1984) on the basis of underlying mechanisms as shown in Fig. 6. Two main types are recognized; first, interactions that are mediated at a distance through diffusible or volatile substances (including antibiotics), which result in inhibition or cessation of growth of one of *both* mycelia. Unilateral inhibition may be a prelude to replacement, whereas bilateral inhibition results in dead-

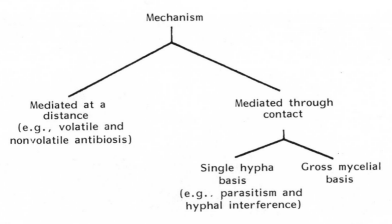

Figure 6. Mechanisms of combat. [From Rayner and Webber (1984).]

lock. Second, very many mycelial interactions only occur following *contact* and may again result in deadlock or replacement. Such contact may either be mediated at the level of recognition responses between individual hyphae, as in hyphal interference and necrotrophic mycoparasitism, or by gross contact between mycelial systems. Antibiosis in its broadest sense, that is, where colony inhibition occurs at a distance, has probably been the most widely reported phenomenon between mycelia in culture and is certainly widespread among wood-inhabiting fungi (e.g., Fig. 7a). Whereas such inhibition may be brought about by a wide range of mechanisms, including production and accumulation of general waste products, changes in pH, etc., the accent has usually been on a single volatile or diffusible substance produced by particular strains. This may have limited the development of understanding of the role of antibiotics in the dynamics of mycelial interactions due to overemphasis of unilateral effects on a particular target organism and on single chemical substances. Thus mutual inhibition (Fig. 7a), which is an extremely common feature, has received little attention. Additionally, the fact that a particular isolate may show antibiotic actions only against certain fungi, but not others, has been interpreted in relation to variation in sensitivity to the antibiotic, rather than differences in elicitation of its production. Mutual inhibition and specificity of action imply a more complex situation, possibly involving a reciprocal exchange of chemical signals between participants (Rayner and Webber, 1984).

One important mechanism resulting in contact inhibition has been termed hyphal interference (Ikediugwu *et al.,* 1970). This characteristically involves mutual or unilateral death of hyphae or hyphal compartments as a result of intimate contact with one another. Death of cyto-

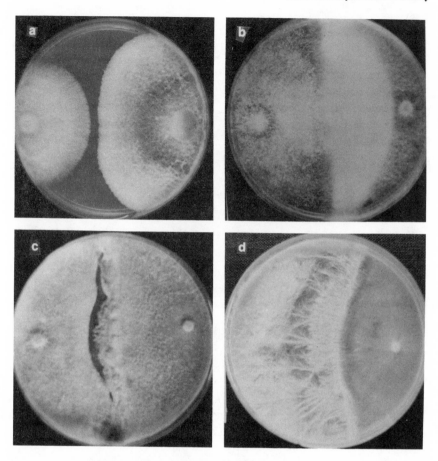

Figure 7. Interspecific interactions in culture. (a) mutual inhibition between *Daedalea quercina* and *Daedaleopsis confragosa* (right), (b) mycoparasitism of *Pseudotrametes gibbosa* (left) on *Bjerkandera adjusta* (right), (c) replacement of *Bjerkandera adjusta* (left) following contact, then lysis, by *Phlebia radiata*, (d) replacement of *Chondrostereum purpureum* (left) by cords of *Phanerochaete velutina* (right). [Parts (c) and (d) from Rayner and Todd (1979).]

plasm in contacted compartments follows changes in membrane permeability, increase of refractility, vacuolation and eventual lysis. Such hyphal interference underlies the biological control of *Heterobasidion annosum* on pine stumps by *Phlebia gigantea* (Rishbeth, 1963; Ikediugwu *et al.,* 1970) and has been studied at the ultrastructural level by Ikediugwu (1976). Affected compartments show a variety of abnormalities, including accumulation of lipid globules, vacuoles of various types, and swelling of mitochondria and nuclei. The swollen mitochondria subsequently become vacuolate, as eventually does the whole cell. At the point of contact with the antagonizing hypha a characteristic extra plasmalemmal zone develops from which various types of vesicles radiate.

It may only be a relatively short step from hyphal interference to the point where one or more hyphae of one fungus, having killed a hypha of another, directly absorbs nutrients from it. Such behavior appears to occur between the white-rot species *Pseudotrametes gibbosa* and *Bjerkandera adusta* (Fig. 7b) associated with a replacement interaction in culture and selective colonization in nature by *P. gibbosa* of wood previously occupied by a *B. adusta* (Rayner, 1978; Rayner and Todd, 1979). Similar behavior occurs between *Lenzites betulina* and *Coriolus* spp., and it may be suggested that both *P. gibbosa* and *L. betulina* can be likened to temporarily socially parasitic ants (Rayner *et al.*, 1987). In the latter, the females cannot found colonies independently, and instead invade those of other species, kill the host queen, and allow their brood to be reared by host workers. The host colony dies out, while the number of social parasites increases until an independently existing colony emerges (Dumpert, 1978).

Observations of basidiomycete interactions in particular suggest that mechanisms involving gross mycelial contact between colonies are important determinants of deadlock and replacement. Very often a dense zone of mycelium is produced at the interaction interface: this may be mutual, in deadlock, or unilateral, in replacement interactions, where a dense wave of mycelium progressively advances across the opposing colony, often preceded by lysis (Fig. 7c). Similarly, aggregated structures, such as cords, are often differentiated prior to replacement (Fig. 7d). Although such replacement may be brought about by lysis and parasitism on a huge scale, the underlying mechanisms of such gross interactions are still largely unknown. In some cases it may be that they simply have a "smothering" effect on the mycelium replaced. Nonetheless, the specificity with which such interactions occur in different combinations, together with the elicitation of dense mycelium, suggest that some form of recognition response is involved (Rayner and Webber, 1984).

3.2.2b. Intraspecific Interactions. As has already been implied, replacement, and especially deadlock, phenomena can also occur between mycelia of the same species, in association with self–nonself recognition reactions. Deadlock is a feature of purely somatic interactions between vegetatively exclusive individuals. Typically it involves formation of a reaction zone (Fig. 8a) containing relatively sparse mycelium, with evidence of hyphal disruption, often associated with production of pigment within the hyphae and/or medium. However, the reaction is very variable, both within and between species, notably in relation to width of the interaction zone, intensity and hue of pigment production, and develop-

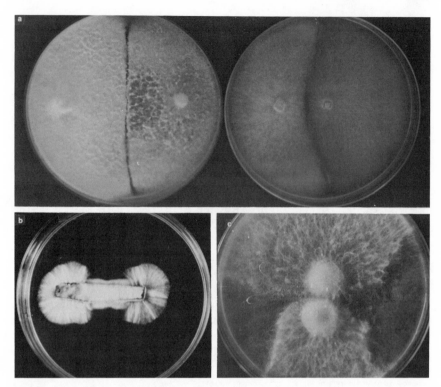

Figure 8. (a) Intraspecific interactions in culture between different wood isolates of *Stereum gausapatum* [from Boddy and Rayner (1982)]. (b) Outgrowth following subculture of a strip of agar plus mycelium from across the interaction zone between sib-related monoascospore isolates of *Hypoxylon serpens* (courtesy of C. G. Dowson). (c) Mating-type incompatible sib-related basidiospores of *Stereum hirsutum,* showing the "bow-tie" interaction.

ment of aerial mycelium on either side of, overarching, or even within the interaction zone. Expression of somatic incompatibility is associated with hyphal fusion, most generally, it seems, being a postfusion event (Rayner *et al.,* 1984). For Ascomycotina (which do not have an independent secondary mycelial phase or mating-type heterokaryon), it occurs directly between homokaryons. In Basidiomycotina expression is typically between heterokaryons, but may also occur between non-mating-compatible homokaryons.

Replacement has sometimes been indicated in association with somatic incompatibility in Ascomycotina. For example, in *Ceratocystis ulmi,* the Dutch elm disease pathogen, a "penetration" reaction occurs, which results in one genotype progressively invading the other across a previously established somatic incompatibility barrier (Brasier, 1984). In pairings between different sibs (ascospore-derived progeny from the same perithecial stroma) of *Hypoxylon serpens,* a ridge of dense, white mycelium characteristically developed in the interaction zone. This ridge widened very gradually and, as it did so, mycelium of the original isolates was destroyed (Fig. 8b) (Dowson, 1982; Rayner *et al.,* 1984). In *Daldinia concentrica* bow-tie interactions similar to those detected in certain Basidiomycotina (see below) have recently been discovered between mycelial isolates from wood, stromata, and single ascospores (Sharland and Rayner, 1986).

In Basidiomycotina replacement of one genotype by another has been detected between homokaryons of *Stereum hirsutum* that are mating-type incompatible but are heteroallelic for a locus conferring what has been termed the "bow-tie" interaction (Fig. 8c) (Coates *et al.,* 1981; Rayner *et al.,* 1984; Coates and Rayner, 1985a). Bow-tie-like interactions have now been detected in a variety of bipolar basidiomycetes and the locus (loci) responsible may well be homologous with the B-locus of tetrapolar species such as *Schizophyllum* (Coates and Rayner, 1985a). There is evidence of a widespread genetic mechanism controlling intraspecific replacement—even more so if the ascomycete interactions prove to be equivalent to those in basidiomycetes. A current idea is that the underlying mechanism involves *access* of donor nuclei into a recipient mycelium and migration via lateral anastomoses (Coates and Rayner, 1985a).

3.3. Classification of Interaction Types

In order not to prejudge the issues, and to introduce the nature of fungal communities in wood at an early stage of discussion, we have so far described mycelial interactions in terms of observable outcomes (i.e., replacement, deadlock, intermingling) of direct proximity between different mycelial thalli. However, in order to produce, as we hope, a conceptual framework of predictive value in understanding community processes, it is necessary to examine how far such observations can be encompassed with existing concepts and classification of interaction types. As we have already hinted, consideration of the territorial characteristics of mycelial systems and of wood, among many other substrata in which fungal communities develop, as a spatially determinate, nonrenewable resource forces us to adopt a terminology at variance with systems of classification preferred by many microbial ecologists [including that described by Bull and Slater (1982)].

The scheme that we prefer (Fig. 9) was developed by Cooke and Rayner (1984) and a full discussion of it, together with the reasons that other terminologies are considered unsatisfactory, is provided by Rayner and Webber (1984). Briefly, the advantages we hope this scheme possesses are

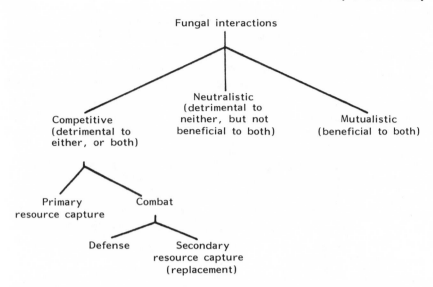

Figure 9. Proposed schema of fungal interactions. [After Cooke and Rayner (1984).]

as follows: (1) It is simple and precise and yet covers a wide range of interaction types *known* to occur in nature, excluding others that may not; (2) it recognizes that interactions are mediated ultimately at the level of *individuals* rather than *populations;* this is in accordance with present views of individual as opposed to group selection (Williams, 1971); (3) accordingly, it recognizes the essential territoriality of mycelia and their potentially indefinite extent; (4) competition is understood in its original Darwinian sense; (5) it can successfully be reconciled with currently developing concepts of ecological strategies and r- and K-selection.

The primary division, according to this scheme, is into competitive, neutralistic, and mutualistic interactions between two participants, depending, respectively, upon whether the outcome is detrimental to either or both, detrimental to neither but not beneficial to both, or beneficial to both.

3.3.1. Competition and Combat

Competition is defined, broadly and simply, as an active demand by two or more individuals of the same or different species for the same resource. It is not necessary to suggest that this resource is in insufficient supply to satisfy the needs of both, and, indeed, in terms of spatiotemporal considerations and ultimate biomass production (as opposed to rate of production) the "needs" of indefinite mycelial systems are hard to define; they are intrinsically greedy! Within competition it is necessary to recognize, as some animal ecologists effectively have, that two distinct aspects exist: primary resource capture and combat.

Primary resource capture describes the process of gaining initial access to, and influence over, an available resource. Since primary resource capture is dependent solely on priority of arrival and sequestration of resources before the occupied domains become contiguous, it does not involve direct challenge between individuals in close proximity, except, perhaps, during initial establishment at the surface of the resource. Success in primary resource capture is therefore determined by such factors as effective dispersal mechanisms, spore germination, mycelial extension rates, possession of suitable enzymes to utilize available substrates, and tolerance of adverse conditions (stress) associated with the resource.

As primary resource capture proceeds, so the domains of different individuals increasingly come into contact. When this happens the three outcomes we have already described (Section 3.2), that is, intermingling, deadlock, and replacement, may follow. The latter two events may, under some circumstances, result from purely nutritional factors; thus, effective uptake of nutrients by one individual may be sufficient to deny access of another individual to its domain. Alternatively, one individual that has exhausted supplies of nutrients that are assimilable to it may then be replaced by another individual capable of utilizing the residue. However, mutual exclusion and replacement also occur as a result of more direct physiological challenge between individuals; for example, access to one individual's domain may be prevented by active defense or achieved via active secondary resource capture mechanisms. This is described as combat. Collectively the mechanisms underlying combat are encompassed by the term antagonism. The combative ability of a fungus is, of course, partly determined by its physiological state, so that changes in physico-chemical or nutritional conditions associated with resource utilization will affect the outcome of interaction between two combatants (Boddy *et al.*, 1985).

Defense and secondary resource capture represent two different aspects of combat. In defense, access to resources gained by primary capture is denied to another individual. Hence, success of a fungus able principally in defense will rest on its ability for effective primary resource capture. In secondary resource capture, access is gained to an already occupied domain, so that fungi able in this aspect of combat need not necessarily be effective in primary resource capture.

3.3.2. Neutralism and Mutualism

Although emphasis in fungal ecology is normally placed on the competitive interactions just described, there are two obvious ways in which fungi may associate noncompetitively in nature, via neutralism and mutualism.

There are two possible kinds of neutralistic associations between fungi. In the first, association results in no discernible benefit or harm to the associated species, interaction being either absent or so minor as to produce no detectable effects. This type of essentially passive coexistence, though it seems to be common between fungi and bacteria, is probably rare between fungi. In the second kind of neutralistic association, one of the associates benefits in some way from the activities of the other without conferring any benefit or harm in return. These two types of neutralism correspond to strict neutralism and commensalism as defined by Bull and Slater (1982). However, since it can be difficult, both theoretically and practically, to decide on whether or not benefit accrues to one of the associates, and indeed whether strict neutralism even exists, we prefer to use a single term to cover the range of possibilities.

Intermingling of mycelia may be evidence for a neutralistic reaction between them, but as we have indicated, the issue is complicated by the requirement for sparse mycelial systems and frequent association with low nutrient availability. It must be accepted that intermingling mycelia may still be depleting the same resource pool and hence be in competition; for associations to be truly neutralistic (as occurs in many animal communities), different portions of the resource pool should be utilized. One way in which this would be achieved is where waste products, exudates, or autolysed remains of one fungus are used by another; contamination of basidiomycete colonies by such fungi as *Penicillium* and *Cladosporium* may involve this type of situation. A characteristic situation where this type of opportunity arises is in the interaction zones between somatically incompatible individuals of decay fungi; accumulation of a wide variety of lytic products would be expected under these regimes, together with the fact that the decay mycelia themselves would tend to be excluded from or be inactivated in the zones. The occurrence of specialized dematiaceous hyphomycetes in such zones (see above; Fig. 4) can be accounted for in this way.

Stimulation of sporulation has been very widely reported as a consequence of mycelial interactions, and represents one way in which an individual may benefit without necessarily bringing about deleterious effects in the other. It may therefore be regarded as one basis for neutralism and for mutualism. However, it is more likely that stimulation of the reproductive mode results from deleterious action on the vegetative mode, and that any advantage accrued must be viewed against the basic life strategy and energetics of the fungus concerned. One example of this type of interaction is provided by what has been termed the *"Trichoderma"* effect, observed among many heterothallic species of *Phy-*

tophthora. The effect is confined to the *A2* compatibility type and, perhaps significantly, is especially marked among species invading roots of woody hosts, such as *P. cinnamomi, P. cambivora,* and *P. palmivora,* where contact with *Trichoderma* species is likely to occur (Brasier, 1975a,b, 1978). The effect involves stimulation of the production of oospores via self fertilization in mycelial cultures exposed to volatile antibiotics from *Trichoderma* strains, variability between *Trichoderma* strains in eliciting the effect being correlated with their production of antibiotics. Since the effect is due to antibiotics, it is evident that it may at least partially function via deleterious effects on the vegetative phase, and indeed *Trichoderma* eventually replaces the *Phytophthora* species. However, the deleterious effect of replacement might be viewed as being counteracted partly by the production of survival structures (the oospores), but also by the stimulation of sexual reproduction in the absence of the *A1* compatibility type. Typically there seems to be a marked imbalance between the compatibility types in *Phytophthora* populations; for example, among a world-wide sample of *P. cinnamomi* isolates, 28 were *A1* and 632 were *A2.* Such an imbalance may be seen either as leading to a requirement for or being a consequence of the *Trichoderma* effect.

Another example of a change of mode brought about by interactions occurs in stimulation of rhizomorph and mycelial cord formation. Thus, rhizomorph development in *Armillaria* is stimulated by the presence of *Aureobasidium pullulans,* apparently via ethanol production (Pentland, 1965, 1967), and in *Sphaerostilbe repens* it is stimulated by a number of fungi, including *Penicillium thomii, Aspergillus niger,* and *A. amstelodami* (Botton and El-Khouri, 1978). Mycelial cords of fungi such as *Phallus impudicus* and *Phanerochaete velutina* appear to be induced to form at a distance by *Penicillium* species, and then may show directed growth toward the latter (Fig. 10) (L. Boddy and A. D. M. Rayner, unpublished observations). As is indicated by Thompson and Rayner (1983), mycelial cord formation in several Basidiomycotina is favored by growth under nonsterile conditions.

Neutralistic interactions may occur when readily assimilable substrates released by extracellular enzyme action of one fungus on a more complex and hence refractory substrate are made available to another, associated species, not itself capable of breaking down the more refractory substrate. Garrett (1963) had this situation in mind when he postulated the occurrence of so-called "secondary sugar fungi" associating with cellulolytic and lignolytic species in decomposing plant residues. Whereas the existence of such fungi has been widely assumed (Hudson, 1968), there is little direct evidence for them in wood, except in special situations, such as interaction zone fungi (see Section 3.1).

Figure 10. Directed growth of cords of *Phallus impudicus* to *Penicillium* colonies. [From Rayner and Webber (1984).]

It has also been pointed out that removal of reducing sugars by an associated sugar fungus may limit catabolite repression, and hence enhance cellulase production by a cellulolytic fungus (Hulme and Stranks, 1970). Also, since cellulase is an enzyme complex, mixtures of cellulases from different fungi growing together may be more efficient than those of an individual (Wood, 1969; Hulme and Shields, 1975). Whereas these examples illustrate synergism in cellulose breakdown, they should not automatically be regarded as mutualistic associations; enhanced cellulose breakdown may not necessarily be in keeping with the particular strategies of the participants. Nevertheless, such examples do illustrate how complementary physiological activities can provide a basis for mutualism.

Complementarity can also be found in culture by growing fungi on media lacking vital growth substances. Thus, *Nematospora gossypii* and *Bjerkandera adjusta* can grow in mixed culture on a synthetic medium lacking biotin, inositol, and thiamin, but not individually. Whereas *B. adjusta* can synthesize biotin and inositol, it cannot synthesize thiamin; and *N. gossypii* can synthesize thiamin, but not biotin and inositol (Kogl and Fries, 1937).

4. Ecological Strategies

One of the ways in which the concept of niche may be understood is to consider the means (ecological strategies) whereby various types of niche may be filled. This has been applied successfully in plant ecology, and it is now evident that it is also of considerable value in fungal ecology (Grime, 1977, 1978, 1979; Pugh, 1980; Andrews and Rouse, 1982; Andrews, 1984a,b; Cooke and Rayner, 1984). The following discussion is based on that of Cooke and Rayner (1984) and Rayner, Boddy, and Dowson (1987b). Underlying the concept is the idea that the lifestyles of all organisms, including fungi, are influenced by three main facets of their environment: incidence of competitors, stress, and disturbance. In turn, these contribute to the development of three primary strategies: combative (this term is preferred to competitive, to avoid ambiguity), stress-tolerant, and ruderal (Fig. 11).

Disturbance describes a process by which uncolonized resource is made suddenly available either by *enrichment* (input of uncolonized material without destruction of resident organisms) or by partial or complete destruction of the resident organisms within a resource.

Stress is caused by the presence in a habitat, on a relatively long-term or permanent basis, of external abiotic environmental constraints that limit the production of biomass by the majority of the organisms in question. For fungi, such factors may be nutritional, with available substrates being intractable or in scarce supply, or they may be due to microenvironmental features, such as high or low temperature and moisture conditions, presence of allelopaths, or unfavorable pH.

In the absence of significant stress conditions, disturbance by providing a newly available resource will initially favor fungi able, by dint of their effective dispersal, germination, and rapid uptake of easily assimilable nutrients, in rapid primary resource capture. Such fungi are *r*-selected and can be described as ruderals, or as having ruderal strategies: their abiding characteristics are that they are ephemeral, noncombative, often capable only of utilizing easily assimilable resources, and have rapid and sometimes total commitment to reproduction. Many mucoraceous fungi, for example, possess these characteristics.

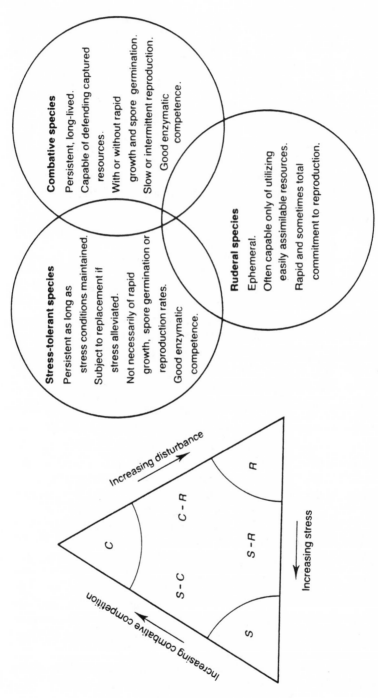

Figure 11 (a) Model of location of primary and secondary strategies in relation to the relative importance of combative competition, stress, and disturbance. Primary strategies: C, combative; S, stress-tolerant; R, ruderal. Secondary strategies: C-R, combative ruderal; S-R, stress-tolerant ruderal; S-C, stress-tolerant combative. [From Cooke and Rayner (1984); modified from Grime (1977).] (b) Summary of expected attributes of saprotrophic fungi in relation to the three primary ecological strategies. [From Cooke and Rayner (1984).]

Stress has two distinct effects: in itself it will tend to cause long-term exploitation (and hence K-selection) of a resource, often (but not necessarily) associated with slow increase in biomass and commitment to reproduction; also, by imposing strong selection pressures, it tends to limit the incidence of competitors. Stress-tolerant fungi accordingly have the general characteristics of physiological adaptation, persistence, frequent lack of rapid extension, germination, and reproduction rates, lack of combative ability, and the enzymic capacity to utilize refractory substrates.

Under conditions of low stress and absence of disturbance, the domains of fungi will increasingly overlap to produce a "closed" community (see Section 5.2.2) and under these circumstances, persistence (K-selection) in a resource will increasingly depend on combative ability, which is either employed in the *defense* of territory obtained by primary resource capture or in invasion and secondary capture. Fungi possessing such characteristics are described as combative and in addition exhibit the following features: rapid or slow germination and growth rates (defenders must be effective in primary capture, invaders need not be so), competence in decomposition, and slow or intermittent commitment to reproduction.

Thus, it may be seen how the interactive properties of fungi can be expected to be inextricably linked with their ecological strategies and hence with their roles in community development. However, before progressing, it is vital to appreciate that, as indicated in Fig. 11, the three primary strategies represent only extreme points in a continuum and perhaps the majority of fungi combine different attributes of the primary strategies in secondary or even tertiary strategies. Furthermore, in some cases, related to the fact that the mycelium operates in a series of overlapping modes associated with establishment, exploitation, exploration, and defense or secondary capture of resources, different operational phases may be characterized by distinctive interactive and other mycelial characteristics. These functional qualities can transcend taxonomic division, as demonstrated by the behavior of the white-rotting species *Phlebia radiata* and *Phlebia rufa* (Boddy and Rayner, 1983a). Here the mycelium develops in culture on malt agar as a rapidly extending broad front of appressed coenocytic hyphae followed by septate mycelium with extensive aerial proliferation and, in secondary mycelia, clamp connections (Fig. 12). The coenocytic front is noncombative and readily replaced by other fungi, whereas contact of other mycelia with the septate mycelium typically results in a dense zone of aerial mycelium, which advances via successive waves of lysis (Fig. 7c). It is as though exploration is effected via a "ruderal" coenocytic phase, whereas consolidation and combat are mediated by the combative septate mycelium.

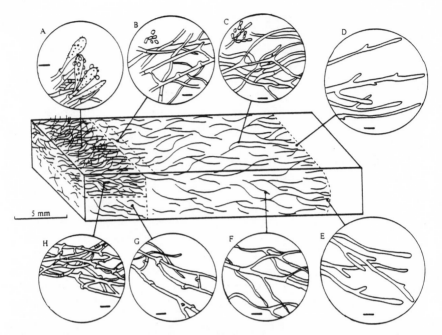

Figure 12. Three-dimensional colony characteristics of a representative dikaryotic culture of *Phlebia radiata* grown on 0.5% malt agar at room temperature. The colony margin is to the right. (A–H) Hyphal characteristics of different regions in (A–D) the surface and (E–H) submerged mycelium; bars represent 10 μm. (A) Mycelium with clamp connections and abundant "cystidioles" with external crystals or liquid globules: (B) heterogeneous mycelium with some septate hyphae, anastomoses, pseudoclamps, and oidia; (C) heterogeneous, mostly coenocytic mycelium with oidia; (D, E) wide, coenocytic marginal hyphae; (F) heterogeneous, mostly coenocytic mycelium; (G) loosely packed mycelium with clamp connections and anastomoses; (H) as (G), but closely packed. [From Boddy and Rayner (1983b).]

Finally, we must point out that for any particular resource type, it is inconceivable that no stress factors should be operating: all substrata available to fungi are characterized by some form of stress factors, making them distinct from others and leading to communities with different biotic components. Thus, wood, in general, is characterized by stress factors including relative lack of assimilable organic and mineral nutrients, lignification, and so on; all its fungal inhabitants must be stress-selected to this extent. The importance of the strategy theory is not that it provides hard and fast rules, but that it enables, for any particular resource type, the classification of fungal communities and their constituents into distinctive components. Hence a predictive understanding of community development processes can be achieved.

5. Patterns of Community Development

Fungal communities in wood are static in neither time nor space. In the past, many ecological studies have focused, often without adequate rigor, on the temporal changes in species composition that occur during what has been described as succession. However, particularly in relation to what is known of the dynamics of mycelial thalli, *spatio*temporal changes are of paramount importance and these can only be understood within the context of overall community development [for more detail of these arguments see Rayner and Todd (1979), Bull and Slater (1982), and Cooke and Rayner (1984)]. This begins when the first participants arrive and establish at the wood surface.

5.1. Setting the Stage: Patterns of Arrival and Establishment

There are two distinct forms in which fungi may arrive at the surface of woody substrata prior to colonization: vegetative mycelium and propagules of some form, usually spores. Establishment processes from these are accordingly distinctive.

Spores and propagules may either be dispersed directly to a suitable surface or may be dormant, either at their point of formation or following a period of dispersal, until a suitable substratum becomes available. A very wide variety of dispersal mechanisms exist (Ingold, 1971), but with the exception of zoospores, many propagules play an essentially passive role in their own dispersal, depending on air or water currents, rain splash, aerosols, or animal vectors to carry them to their destination. Accordingly, contact with a receptive surface depends largely on chance, but propagules carried by animals may not only be taken to specific sites, but may also be inserted within the wood, hence avoiding the often hostile surface environment (Swift and Boddy, 1984).

In addition to their generally chance arrival, spores typically contain limited supplies of endogenous nutrients, so that successful establishment from them at hostile surfaces is dependent on their encountering, again by chance, locally favorable conditions or perhaps by synergistic effects with neighbors. In turn, the possibility of synergism is, in higher fungi, dependent on whether they are genetically similar or different, and hence, often, whether they are sexually or asexually derived (see Sections 3.2 and 3.3). Where genetic differences occur, somatic incompatibility phenomena militate against synergism (Rayner *et al.,* 1984).

By contrast, not only is synergism possible between hyphae in a mycelial inoculum, but the mycelium itself may be capable of bringing in, via translocation, vital resources, including nutrients and water, obviating the requirement to absorb these from the immediate environment during establishment (Jennings and Rayner, 1984). In many instances arrival by mycelium is mediated in soil, litter, and nonnutritive environments by production of aggregated hyphal systems, cords, and rhizomorphs and evidence is accumulating that these may be capable of directed growth responses toward woody substrata, avoiding wastage of biomass (Thompson and Rayner, 1983).

Following arrival, a variety of transitions may occur in the form by which a fungus develops in the wood which affects its ultimate success in establishment (Rayner, Boddy and Dowson, 1987b). One such transition, which appears likely to play an important role, has been termed the "module–mycelium transition" and is associated with a process of "latent invasion" of the tissues (Cooke and Rayner, 1984). The essential feature of this transition is that under conditions of stress that militate against mycelial development, the fungus develops instead as a series of "colonization units" (modules), which may take the form of spores, budding cells, mycelial knots, or fragments. Following alleviation of the stress conditions, mycelial development occurs, and by coalescence between genetically similar mycelia very effective primary resource capture is effected (see below for possible examples).

Other transitions may be associated with changes in the functional form and genetic status of the mycelium. An example of the former might be the explorative–combative transition in *Phlebia* species mentioned previously. In regard to the latter, the transition between primary (homokaryotic) and secondary (heterokaryotic/dikarotic) mycelia that follows successful mating in Basidiomycotina provides a dramatic, but little studied example. This transition is important because, depending on its duration, the homokaryon may be principally responsible for primary resource capture.

The final distribution of mutually antagonistic secondary mycelia may therefore not be so dependent on their own intrinsic properties as on the growth rates and patterns of nucleus exchange of homokaryons. One way of modeling this experimentally is to inoculate petri dishes with arrays of sexually compatible homokaryons and to examine the spatial development of the resulting secondary mycelia. Sample results for *Coriolus versicolor* obtained using non-sib-related monokaryons indicated that differences between the areas occupied by the resultant dikaryons were relatively slight, taking into account the timing of the inoculations (Williams *et al.,* 1981). However, studies using *Stereum hirsutum,* in which arrays of sib- and non-sib-related homokaryons were inoculated, revealed that the domains subsequently occupied by non-sib-composed secondary mycelia were substantially greater than those of sib-composed secondary mycelia (Coates, 1984).

Duration of the homokaryotic phase is obviously, therefore, a crucial determinant of population structure. Observations on colonization of cut beech logs (Coates and Rayner, 1985b–d) indicated that homokaryons of common decay-causing basidiomycetes could be isolated at depth up to 6 months after exposure, but rarely thereafter. Monokaryons of *Coriolus versicolor* directly inoculated on short lengths of doweling into birch logs were dikaryotized 12 months later, but the number of dikaryotization events was low, perhaps one or two per inoculum (Williams *et al.,* 1981). This may be related to the fact that dikaryotization of a single homokaryon rapidly precludes further dikaryotization.

5.2. Subsequent Changes: Community Development Pathways

The limitation of scope in the successional approach to fungal ecology has in the past resulted from attempts to regard change as being due to simple, single underlying causes (Rayner and Todd, 1979; Bull and Slater, 1982). In reality a wide variety of influences may be responsible, and the problem has been how to rationalize these within a suitable conceptual framework. In attempting to do this, Cooke and Rayner (1984) have devised a scheme, a simplified version of which is shown in Fig. 13. The approach is based on the fundamental dichotomy between communities initiated under stress conditions and those initiated in the *relative* absence of stress factors: these result in communities *characterized* by the presence of S-tolerant and ruderal pioneers, respectively. Thereafter, changes in community structure are attributed to four basic processes or factors: stress aggravation, stress alleviation, disturbance, and intensification of combat. The latter two have been defined elsewhere. Stress aggravation involves imposition of new or additional stress, and stress alleviation the opposite.

As will become evident, stress alleviation and aggravation themselves are often closely associated with disturbance, but unlike the latter, which is a sudden event, they are gradually imposed and often follow disturbance. Our intention now is to examine, in a general way, patterns of community development initiated under stress and nonstress conditions, thereafter to determine how the observed changes may be accounted for in relation to the four driving forces. We must stress that for the sake of brevity, and to emphasize what we believe are salient principles, our discussion is deliberately very selective and should in no way be taken as an indication of the current state of knowledge of wood decay processes.

5.2.1. Stress-Initiated Patterns

5.2.1a. In the standing tree. In the absence of man's intervention, fungal colonization leading to decomposition of wood must predominantly be initiated whilst it is still part of the standing tree. Such coloni-

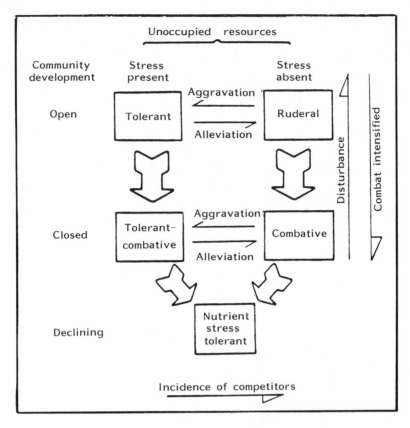

Figure 13. Diagram of possible community development pathways from colonization of a totally unoccupied resource, through an open community stage with still unoccupied resources available for primary capture, to a closed community with all initial primary capture completed. Culmination is a declining community stage characterized by severe nutrient stress. In the absence of competitors, developing tolerant communities may progress directly to declining tolerant communities without an intermediate combative stage. [From Rayner and Webber (1984); simplified from Cooke and Rayner (1984).]

probably arises largely under conditions that impose severe stress, although mechanical damage to sapwood, due to high winds, lightning strikes, frost, heat, drought, and animals, may result in the non-stress-initiated patterns of development described in the subsequent section. The stress conditions may be due to one or more of a variety of factors: (1) relative absence of easily assimilable nutrients and high water content in functional sapwood, (2) inhibitory extractives, (3) inaccessibility. In addition it has been argued that living sapwood is capable of a resistance response to invasion by fungi and microorganisms associated with the laying down of physical and chemical barriers (e.g., Shigo, 1979). The arguments for and against this are discussed in detail by Boddy and Rayner (1983c), and we will not mention them further, beyond saying that we believe that the three features listed above adequately account, as primary factors for many observed patterns of decay in trees.

Where appropriate studies have been made, indications are that stress-initiated decay communities are typically dominated by extensive individuals of relatively few white- or brown-rot species; often a single individual may come to occupy virtually entire trunks and branches (Fig. 14). Also, it is common to find that the decay species exhibit marked selectivity towards particular tree species. These features are all to be expected given the selective influence of stress factors and the resulting lowered incidence of competitors. Latent invasion mechanisms associated with module–mycelium transitions may further facilitate formation of extensive individuals and may lead to enforced coexistence. With regard to stress-initiated colonization, distinctive development patterns may be distinguished based upon whether initial colonization is effected in sapwood or heartwood.

Colonization of true heartwood leads to the development of heartrot, which, for a long period following the pioneering studies of Hartig in the last century, was believed to be the major cause of decay in standing trees (Manion and Zabel, 1979). Perhaps associated with the long periods over which it develops, little is known of the initiation of heartrot. It is generally assumed that colonization occurs via wounds large enough to expose the heartwood (e.g., Peace, 1962; Boyce, 1961). However, it has been suggested that *Echinodontium tinctorium* may establish via what amounts to latent invasion from twigs of a few millimeters diameter (Etheridge and Craig, 1976). *Stereum gausapatum* regularly becomes established in heartwood following initial development of decay in sapwood of branches that may or may not be large enough themselves to contain heartwood (Cartwright and Findlay, 1958; Boddy and Rayner, 1982). In *Phaeolus schweinitzii,* there is evidence that initial establishment is from roots infected with *Armillaria* spp. (Barrett, 1970). Whatever their mechanism of establishment, and whether or not this involves interaction with other microorganisms, heartrot fungi frequently form very extensive individuals, associated with often massive sporophores, and show strong selectivity for particular species of tree. For example, in Britain, heartrot of oak (*Quercus* spp.) is most commonly due to *Laetiporus sulphureus, Fistulina hepatica,* and *Stereum gausapatum,* which cause top-rots, and *Grifola frondosa* and *Inonotus dryadeus,* which cause butt-rots. Each of these species in turn colonizes only a narrow range of other trees. A further feature of heartrot fungi is that many appear to advance very slowly

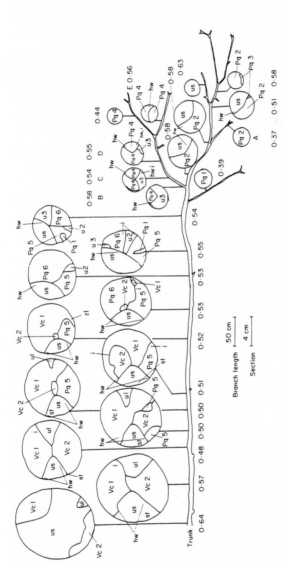

Figure 14. Decay community structure in an attached oak branch. Different basidiomycete individuals were demarcated by interactive zone lines (i) in the wood. The predominant species were *Vuilleminia comedens* (two individuals; Vc 1–2) and *Peniophora quercina* (six individuals; Pq 1–6). Three other unidentified basidiomycetes (u 1–3) occupied relatively smaller volumes. Both *V. comedens* and one of the *P. quercina* individuals occurred in the main stem, possibly originating from small branch stubs or by direct entry. It is likely that Pq 2 originated in the branchlet from which section A was taken and progressed to the main fork, whence it grew in both available directions. In the distal sections B–E heartwood wings (hw) were present in the absence of living tissue (us) and separated decay columns occupied by either the same or two different individuals. In the latter case, an interactive zone line was also present. Presumably the fungi were initially confined to regions between the heartwood wings, with living tissue outside. Subsequently, possibly following some traumatic event, the fungi were able, by some means, to penetrate or bypass the wings, gaining access to the living wood and colonizing it without further wing formation. Values are given for relative density (g cm^{-3}). [From Boddy and Rayner (1983a).]

through woody tissues, so that large individuals are probably many years, if not decades, old.

Colonization of uninjured sapwood characteristically occurs in trees under some form of stress (e.g., drought) and in attached branches which die during outward spread of the canopy. In addition, a relatively few fungi probably colonize by active pathogenesis. The latter include *Armillaria* species and *Heterobasidion annosum*, which colonize roots via ectotrophic mycelial spread, which, combined with their pathogenicity, probably provides them with a selective advantage over, for example, nonpathogenic cord-forming species, hence allowing them to occupy considerable volumes of wood (e.g., Garrett, 1970; Thompson and Boddy, 1983).

Fungi colonizing trunks of stressed trees include many xylariaceous Ascomycotina, notably species of *Hypoxylon, Daldinia,* and *Ustulina.* Many of these show strong tree-species-selectivity; for example, *Hypoxylon cohaerens, H. fragiforme,* and *H. nummularium* show marked preferences for beech *(Fagus sylvatica),* whereas *Daldinia concentrica* and *Hypoxylon rubiginosum* are especially frequent on ash *(Fraxius excelsior).* Although confirmatory studies have yet to be performed, all these fungi appear to develop as very extensive, often single individuals within a trunk, which, by contrast with heartrot fungi, appear very rapidly, often apparently within a single year. This might be due to latent invasion mechanisms (Boddy and Rayner, 1983c; Cooke and Rayner, 1984).

Recently a detailed study has been made of the colonization of attached branches of oak trees in Britain (Boddy and Rayner, 1981, 1982, 1983a–c). The decay communities, which often contained very few, but extensive individuals (e.g., Fig. 14), were characterized by 12 species of Basidiomycotina, each with a distinctive ecological role. *Stereum gaussapatum, Vuilleminia comedens,* and *Phellinus ferreus* often occurred in wood adjacent to living sapwood, and appeared to be directly responsible for initiation of decay, there being no evidence of involvement of any other fungi or microorganisms. Their decay columns were often several meters long and bounded by regions of prematurely formed heartwood ("heartwood wings"), which had probably been induced via aeration of nonfunctional sapwood at the junction with functional sapwood (Fig. 14). Lack of annual rings formed subsequent to the decay columns in abutting functional sapwood suggested that the individuals had developed within a single year. Again latent invasion is implicated. *Peniophora quercina* and *Phlebia rufa* had similar characteristics, but the former was generally confined to distal portions of branches and the latter to branches weakened by light suppression or other factors. *Exidia glandulosa* appeared principally to cause cambial death and loosening, leading to white rot in the wood. *Coriolus versicolor* and *Phlebia radiata* were secondary colonizers only, replacing the pioneers. *Hyphoderma setigerum* and *Schizopora paradoxa* occurred regularly, but their role was less clear: *H. setigerum* tended to be associated with weakened branches, whereas *S. paradoxa* was apparently favored by desiccating conditions, particularly following loss of bark cover; both were closely associated with insect activity. Finally, *Peniophora lycii* and *Stereum hirsutum* were only found occasionally, the former being associated with desiccating conditions. Less detailed studies of branches of ash and beech (Boddy *et al.*, 1985; L. Boddy and A. D. M. Rayner, unpublished results) indicate similar community patterns, although species composition and numbers of individuals are qualitatively and quantitatively different, probably associated with lack of heartwood, thin bark (in beech), and relative rapidity of desiccation.

Stress-initiated development in sapwood also occurs in connection with the establishment of vascular wilt fungi and certain other pathogens, together with fungi associated with certain wood and bark-boring insects. Vascular wilts are a specialized group, notably including certain species of *Ceratocystis (C. ulmi, C. fagacearum)* and *Verticillium (V. albo-atrum, V. dahliae)*, which are able to gain access to conducting xylem, directly or indirectly leading to its blockage and consequent development of wilt symptoms. They appear to be tolerant of the stress conditions, notably oxygen depletion, in functional sapwood and are often spread through the sapstream in the form of detached propagules. Thus, they may represent an extreme example of latent invasion, development being enhanced by early pathogenesis. They may gain access to sapwood either via fine rootlets, as with *Verticillium* species, or via wounds, as with the *Ceratocystis* species. Another pathogen, *Chondrostereum purpureum,* certainly does appear to colonize via wounds. In addition to bringing about silverleaf disease in *Prunus* and other fruit trees, this fungus appears very rapidly in cut stems and stumps of many deciduous trees, where its capacity for longitudinal and radial spread appears far greater than that of other decay fungi (Rayner, 1979). Its early occurrence has been attributed to its invasion only of wood in which parenchyma containing reserve materials is still alive (Guinier, 1933) and pathogenesis, involving unlocking of nutrients sequestered in living parenchyma, together with tolerance of low oxygen levels, might play an important role in its success.

Bark and wood-boring beetles characteristically produce tunnels in bark and wood of trees that are either stressed in some way or recently felled; they play an important role in the establishment of certain fungi that act as pioneers in colonization of the wood. Thus, in conifers, bark-beetles play an important role in establishment of blue-stain fungi, notably *Ceratocystis* spp., which are effectively inoculated into the galleries either by the beetles themselves or by other insects, including Diptera, which occupy the galleries (Dowding, 1984). Success of these fungi is thus due to their being given early access to the timbers, and their distribution is critically affected by moisture content; as the latter is lowered, so the blue-stain fungi develop lenticular colonies of a size related to the form of the brood chambers (Dowding, 1984).

The relatively casual relationship between blue-stain fungi and bark beetles is extended to mutualism in the case of fungi associated with ambrosia beetles (Scolytidae principally of the tribe Xyleborini) and siricid wood wasps. In the former case "ambrosia fungi," of which the major genus appear to be *Fusarium* (Norris, 1979), are actively cultivated in the galleries of the beetles, and disappear when vacated, being replaced by other fungi. In the latter case species of *Stereum* and *Amylostereum* are introduced into wood at oviposition: hence, they are given priority of access to the timber; their success is further dependent on stress tolerance in the mycetangia of the insects (e.g., Cooke and Rayner, 1984).

5.2.1b. In Felled and Processed Timber. Felling and processing of timber remove the problem of inaccessibility that characterizes wood in the standing tree. However, various treatments and the locations in which wood is put to service may themselves often impose considerable selective stresses, and hence colonization occurring under such circumstances may be considered to be stress-initiated.

Change in moisture content following felling and/or processing is probably one of the major sources of stress. Thus, decay of structural timbers in buildings has long been recognized to be strongly affected by moisture availability. The dry-rot fungus *Serpula lacrimans* provides a classic example of a fungus that is selectively favored in relatively dry timber by its capacity to mobilize water from a colonized base through to the advancing margin of its mycelium (Jennings, 1982). Thus, the fungus can establish extensive mycelium under poorly ventilated conditions, provided there is a source of dampness aiding its establishment. However, the niche of the fungus appears to have very narrow limits, and this is manifest in its rarity in natural woodland outside buildings. By contrast, *Coniophora puteana,* which has much wider tolerance and is favored in relatively wet timbers is, together with *S. lacrimans,* a major cause of decay of structural timbers, but is also widespread in nature. This species appears to show an unusual combination of ruderal, stress-tolerant, and combative characteristics, which may account for its success in a wide range of conditions and timbers (A. D. M. Rayner and L. Boddy).

Treatment with preservatives provides another source of stress in processed timber. In this case such treatments are often successful in preventing establishment of white- and brown-rot fungi, but are less effective against soft-rot fungi, perhaps due to the characteristic growth of the latter within secondary cell walls where preservatives cannot accumulate

(Dickinson, 1982). Natural preservatives occur in heartwood, but there is generally little information about development of decay in felled, fallen, or processed heartwood. It is, however, clear that fungi responsible for heartrot in the standing tree often do not remain active after felling, and do not invade felled heartwood (e.g., Highley and Kirk, 1979). Species selectivity is, however, again evident, as in oak, where *Daedalea quercina* and *Hymenochaete rubiginosa* are characteristically associated with decay after felling.

5.2.2. Non-Stress-Initiated Patterns

Non-stress-initiated patterns of community development characteristically occur when sapwood is rendered accessible for colonization, either by major wounds in standing trees or following felling and processing. The latter features effectively cause enrichment disturbance, and very similar community development patterns have been observed in association with them. For reasons already given, these patterns may not be very representative of community development processes in undisturbed woodland. The characteristic pattern is for developing communities at first to be dominated by microorganisms and microfungi that are not active in decay rots, then for increasing dominance of white- and brown-rot fungi. This occurs regardless of whether colonization is effected at wounds in standing trees (e.g., Shigo, 1966; Shortle and Cowling, 1978), in sapwood stakes (e.g., Clubbe, 1980) or in cut logs and stumps (e.g., Rayner, 1977a,b; Coates and Rayner, 1985b–d).

A detailed study of population and community development in cut beech logs has been made by Coates and Rayner (1985b–d), which we believe is particularly apposite in relation to the principles we have discussed, the more so because of the emphasis placed on determination of *spatiotemporal* characteristics. In this investigation, natural colonization was followed in several hundred beech logs, 10–20 cm diameter and 30–40 cm long, cut from freshly felled trees and placed upright, with their bases buried up to 10 cm deep, in the floor of a mixed deciduous woodland. A typical sequence in the development of community patterns such as might occur in a single log over an 18-month period is shown in Fig. 5. Substantially different patterns were associated with colonization from the base as opposed to the aerially exposed cut surfaces. Mycelial penetration from the latter was initially slow, achieving average depths of 3 mm within the first 6 weeks after exposure, but then accelerating, so that by 24 weeks the average depth reached was 140 mm. The fungi most frequently isolated at first were non-Basidiomycotina, these fungi being gradually eliminated as decay communities developed. The Basidiomycotina *Chondrosterium purpureum* and *Corticium evolvens* and homokaryons of *Coriolus versicolor* and *Bjerkandera adusta* were isolated after 3 months, and after 6 months the two former species were widespread, together with secondary mycelia of *Coriolus versicolor, Bjerkandera adusta,* and *Stereum hirsutum.* However, there was little evidence of decay column formation by the latter three species, and numerous individuals of them were sometimes obtained by isolation from very small volumes of discolored wood. For example, 14 individuals of *Bjerkandera adusta* were obtained from a 1.92 cm^3 sample and seven individuals of *Stereum hirsutum* in 0.06 cm^3 (Coates, 1984). By 12 months all three species had begun to form substantial decay columns, and *Chondrostereum purpureum* and *Corticium evolvens* were already in decline, being virtually eliminated by 18 months. Two xylariaceous Ascomycotina regularly formed decay columns. *Xylaria hypoxylon* formed narrow, tapering columns, apparently arising from spores, and these were manifest from 3 months onward. By contrast, *Hypoxylon nummularium* formed very extensive, parallel-sided decay columns, which after 6 months were often virtually continuous between the top and bottom cut surfaces.

Colonization from the bases occurred much more readily, and without an obvious lag phase, than from aerial surfaces, and discoloration was soon evident. Decay-causing Ascomycotina and Basidiomycotina including the rhizomorphic *Armillaria bulbosa,* the cord-forming species *Tricholomopsis platyphylla* and *Phallus impudicus,* and *Xylaria hypoxylon* were quick to arrive, all being present within 3 months. *Armillaria bulbosa* and *Xylaria hypoxylon* were virtually ubiquitous throughout the site. Isolates of the former intermingled in culture, suggesting that a single extensive mycelial type was present. However, *Xylaria hypoxylon* produced numerous individual decay columns, which suggested establishment via ascospores from the soil and litter. The distribution of cord-forming species was sporadic, with isolated groups of logs in the same vicinity all being colonized, indicating the presence of isolated mycelial types. When present, they markedly inhibited colonization by *Armillaria bulbosa,* and whereas the latter was, apart from some direct penetration by rhizomorphs, restricted to the outermost and subcortical regions, they were otherwise capable of extensive penetration and decay of the interior of the logs. They were aggressive combatants, replacing *Xylaria hypoxylon* and *Hypoxylon nummularium* as well as those fungi arising from the aerial cut surface.

5.2.3. Driving Forces for Change

5.2.3a. Stress Aggravation. The predominant cause of stress aggravation during community development in wood is probably that of nutrient stress aggravation whereby the limited supplies of readily assimilable nutrients in decayed wood are rapidly depleted. Associated with this, fungi incapable of utilizing cell wall components become replaced by decay fungi. However, it is debatable whether such replacement is due solely to nutrient stress aggravation or is perhaps simply enhanced by the process. It seems quite probable, from observation of persistence of stain in nondecayed wood, that in the absence of decay species, nondecay fungi could persist for almost indefinite periods, although at a low level of activity, perhaps supported by autolysis and leachates. More active replacement processes are therefore probably also involved.

Another type of stress aggravation may be associated with changes in moisture conditions due to waterlogging or desiccation. Thus, wood in ground contact rotted by *Armillaria* species characteristically becomes water-soaked, and this may enhance persistence of *Armillaria,* which is probably tolerant of such conditions. In standing elms (*Ulmus* spp.) killed by *Ceratocystis ulmi,* desiccation and/or extremely fluctuating moisture conditions develop in the upper portion of trunks; here *Rhodotus palmatus,* which produces large numbers of chlamydospores during growth, is characteristic and appears to be tolerant of these conditions. In lower parts of boles, other fungi, such as *Flammulina velutipes* and *Pleurotus cornucopiae,* are often prevalent (Gibbs and Gulliver, 1977; Rayner and Hedges, 1982). In attached oak branches, detachment of bark as a result of decay leads to desiccating conditions in the wood. This appears to favor *Schizopora paradoxa,* which again is chlamydosporic, allowing it to replace fungi such as *Stereum gausapatum* with which its normal mycelial interaction is deadlock (Boddy and Rayner, 1983a).

As wood becomes increasingly decomposed, it becomes more susceptible to colonization by invertebrates, particularly microarthropods. Entrance of such animals may, as discussed below, result in temporary disturbance, but thereafter they can cause physicochemical stress aggravation, either directly through grazing on fungal mycelium or indirectly via increasing exposure to microenvironmental fluctuations. For example, invasion by tipulid larvae of wood undergoing decay by Basidiomycotina results in disappearance of the latter concomitant with an increase in isolation frequency of microfungi, bacteria, and yeasts (Swift and Boddy, 1984). Stress aggravation is also evident in relation to the activities of fungus-culturing insects, such as the ambrosia beetles and higher termites, which sustain specialized communities of ambrosia fungi on one hand and *Termitomyces* and *Xylaria* spp. on the other.

5.3.2b. Stress Alleviation. Natural nutrient stress alleviation in wood can occur via three distinct routes: improvement in accessibility of nutrient supplies, concentration, and importation. Improvement of accessibility can occur through death of living tissues, for example, following wounding, infection, or physiological stress of standing trees. Whereas wounding also involves enrichment disturbance, the latter processes are more gradually imposed and particularly seem to favor fungi colonizing via latent invasion, hence the development of extensive individuals described above. Indeed, latent invasion can be regarded as involving a stress-tolerant strategy followed by immediate capitalization on stress alleviation.

Alternatively, improvement in accessibility can result from lignocellulolytic extracellular enzyme action and consequent provision of assimilable breakdown products. The possibility that this can lead to neutralistic associations, as perhaps with interaction zone microfungi, has already been mentioned (Section 3.3.2). Additionally, where decay fungi are lost following disturbance or stress aggravation, assimilable nutrients released via their previous activities may become available. Occurrence of mucoraceous fungi at late stages of decomposition may be partially explained in this way.

Animals, by causing wounds in standing trunks or by collecting and fragmenting material, can also increase accessibility of nutrients. However, as indicated earlier in this section, their subsequent activity may also often result in stress aggravation.

Concentration occurs particularly with respect to mineral nutrients, notably nitrogen and phosphorus, which are not significantly depleted by heterotrophy. Thus, the nitrogen concentration (by weight) of decomposing wood may progressively be expected to increase (Swift, 1977; Swift and Boddy, 1984).

Importation can occur via external mycelium, perhaps particularly mycelial cords and rhizomorphs, capable of active translocation. Similarly, animal invasion can result in importation. There is also evidence that woody substrata in contact with soil may be enriched with nitrogen and that such effects are particularly important near their surfaces, leading—in preserved wood—to the development of soft rot (King et al., 1981). In general, fluxes associated with evapotranspiration may be expected to lead to importation of nutrients from solution.

As with stress aggravation, physicochemical stress alleviation in wood is often associated with changes in moisture regime from extremes of high or low moisture content. In the standing tree, alleviation of high moisture can, as with nutrient stress alleviation, result from wounding, infection, or physiological stress, and with similar consequences in relation to community development patterns and latent invasion. Alleviation of low moisture content may often accompany trunk or branch fall, whence previously permanently or intermittently desiccated wood comes into ground contact. At this stage, stress-tolerant residents often become susceptible to replacement by combative fungi invading from the forest floor, perhaps particularly mycelial-cord and rhizomorph-formers, which arrive as vegetative mycelium. In the same way, alleviation of low moisture content is a familiar cause of decay in structural timbers.

Removal or inactivation of allelopathic chemicals may often play a significant part in stress alleviation. Thus, inactivation of phenolic inhibitors by tolerant microorganisms has been implicated in the "succession" that leads to dominance of Basidiomycotina in decay originating from wounds in standing trees (Shigo, 1979; Shortle and Cowling, 1978). Similarly, leaching of toxic extractives from heartwood may eventually render it susceptible to decay.

Since fungus-culturing insects cause stress aggravation, their exit will in turn lead to stress alleviation and replacement of their tolerant communities. This regularly occurs.

5.2.3c. Disturbance. That felling and wounding of standing trees results in enrichment disturbance should already be clear, together with the way in which this leads to initial establishment of a species-rich ruderal community followed by a combative one. In the case of previously established heartrot fungi, felling can also act as a destructive disturbance, as indicated by their disappearance and/or replacement by other fungi.

Another cause of destructive disturbance of resident fungal communities in wood is fire. This both wholly or partially removes the resident microbiota and introduces new environmental stresses associated, for example, with changes of pH (usually toward alkalinity), moisture, and temperature regimes. Accordingly, burned timber is often characterized by communities of apparently specialized fungi that occur rarely, if at all, in other habitats.

Animals can cause destructive disturbance of resident fungal communities in wood, chiefly via comminution of the material. A general consequence of comminution, with the exception of the fungus gardens of ants and termites, is the progressive depletion of particle size, obviating the mycelial advantage in resource capture and hence a deflection toward bacterially dominated processes (Cooke and Rayner, 1984; Swift and Boddy, 1984).

5.2.3d. Intensification of Combat. In the absence of sufficient stress to prevent establishment of a potentially rich community, combat inevitably intensifies under undisturbed conditions as domains progressively come into contact. In this circumstance combative interactions attain a predominant role in the maintenance of equilibrium, via deadlock, or the generation of change, via replacement. In turn, species with combative strategies will be selectively favored. This situation characteristically occurs where community development follows enrichment disturbance or following stress alleviation, and adequately accounts for many of the observed patterns of community change described in Sections 5.2.1 and 5.2.2.

6. Community Functioning

It remains to consider the role of wood decay fungi in relation to ecosystem processes, that is, how their activities contribute to overall rates of decomposition and nutrient flux. This involves the vital bridging of the gap that has traditionally developed between "black-box" decomposition ecology and identification of community structure and dynamics. This schism is particularly unfortunate, since it is obvious that understanding of the one is dependent on knowledge of the other. All we can do at this stage is to point to various problems that require resolution and indicate ways in which, we believe, knowledge of community structure and interaction is likely to produce insight into processes.

Where studies of wood decomposition have been made in a generalized way, without reference to the fungal communities present, it has been evident that the rate and pattern of decay can be quite variable, but on average very slow. For example, estimates of half-lives for wood in the forest floor are frequently as high as 15–20 years or more (e.g., Boddy and Swift, 1984). This leads to substantial accumulations of woody material, which can be regarded as an important reservoir of nutrients in many woodland ecosystems. Observations of active fungal communities, as in the beech logs described in Section 5.2.2, suggest that decomposition rates can be much faster, with virtually complete decomposition of units 1000 cm^3 or more in volume within under 5 years. The cause of such variations must surely lie in the actual nature of decay communities, and in the fact that conditions are rarely optimal for their development. This raises the question as to what the optimal conditions are and how the nature of a fungal community can affect its overall activity. Crucial issues here concern moisture content and fluctuation, patterns of arrival of fungi and other organisms at the wood surface, and the extent to which interactions can inhibit or accelerate decomposition processes.

Moisture content is of particular importance because of the relatively narrow range of moisture conditions over which optimal decomposition occurs: relatively small departures from this range toward either high or low levels swiftly impede activity of decay fungi (Boddy, 1983c, 1986). Thus, logs with their bases buried in soil, fallen branches covered by leaf litter, or partially living branches and trunks with bark cover are likely to have a moisture regime within the appropriate limits, probably for protracted periods (Boddy, 1983b). By contrast, standing dead timber, dead timber not in contact with the ground, or submerged timber will not.

With regard to pattern of arrival, contact with the ground may again be important with respect to mycelial-cord and rhizomorph-forming fungi. Where conditions are favorable, such fungi can cause very rapid decay indeed (Thompson, 1982). By the same token, where conditions are not suitable for such fungi, as in mull soils, where extensive comminution by earthworm activity in particular disrupts their activity or prevents buildup of an appreciable leaf litter layer, decay of wood in ground contact may be significantly delayed.

Finally, there is the question as to how interfungal interactions may themselves affect community functioning. Here, although there has been some suggestion of synergistic interaction, for example, between different cellulase complexes (Hulme and Shields, 1975), the fact that most interactions appear to be competitive suggests that overall they may inhibit decomposition processes. There are two levels at which this may apply. In the one case, an individual relatively inactive in decay may, by primary capture and successful subsequent defense, prevent establishment or access by a more active decomposer. For example, several *Hypoxylon* species appear capable of restricting colonization by more active decomposers such as *Coriolus versicolor* and *Stereum hirsutum* (Rayner and Todd, 1979). Further, some microfungi, such as species of *Trichoderma*, *Cryptosporiopsis,* and *Scytalidium* and the Ascomycotina species *Ascocoryne sarcoides,* have been proposed as biological control agents, preventing decay via antibiosis or preemption of nutrients (Rayner and Todd, 1979).

Alternatively, the proximity of large numbers of antagonistic individuals might also be expected to lead to decay inhibition. Some indicatin of this has been found in cut beech logs (see Section 5.2.2), where inoculation of basidiospore suspensions of *Bjerkandera adusta, Coriolus versicolor, Stereum hirsutum,* and *Hypholoma fasciculare* prevented decay column formation and inhibited production of sporophores by these fungi (Coates and Rayner, 1985a–c; Rayner *et al.,* 1984). The most rapid decay often seems to occur when extensive individuals of decay fungi become established. This raises the intriguing conundrum that conditions favoring ascospore and basidiospore establishment may inhibit subsequent resource exploitation—an unusual example of autonomous population regulation.

References

Andrews, J. H., 1984a, Life history strategies of plant parasites, *Adv. Plant, Pathol.* **2:**105–130.

Andrews, J. H., 1984b, Relevance of *r*- and *K*-theory to the ecology of plant pathogens, in:

Current Perspectives in Microbial Ecology (M. J. Klug and C. A. Reddy, eds.), pp. 1–7, American Society for Microbiology, Washington, D.C.

Andrews, J. H., and Rouse, D. I., 1982, Plant pathogens and the theory of *r*- and *K*-selection, *Am. Nat.* **120**:283–296.

Barrett, D. K., 1970, *Armillaria mellea* as a possible factor predisposing roots to infection by *Polyporus schweinitzii*: *Trans. of Brit. Mycol. Soc.* **55**:459–462.

Boddy, L., 1983a, The effect of temperature and water potential on the growth rate of wood-rotting basidiomycetes, *Trans. Br. Mycol. Soc.* **80**:141–149.

Boddy, L., 1983b, Microclimate and moisture dynamics of wood decomposing in terrestrial ecosystems, *Soil Biol. Biochem.* **15**:149–157.

Boddy, L., 1983c, Carbon dioxide release from decomposing wood: Effect of water content and temperature, *Soil Biol. Biochem.* **15**:501–510.

Boddy, L., 1986, Water and decomposition processes in terrestrial ecosystems, in: *Water, Fungi and Plants* (P. G. Ayres and L. Boddy, eds.), pp. 375–398, Cambridge University Press, Cambridge.

Boddy, L., and Rayner, A. D. M., 1981, Fungal communities and formation of heartwood wings in attached oak branches undergoing decay, *Ann. Bot.* **47**:271–274.

Boddy, L., and Rayner, A. D. M., 1982, Population structure, inter-mycelial interactions and infection biology of *Stereum gausapatum*, *Trans. Br. Mycol. Soc.* **78**:337–351.

Boddy, L., and Rayner, A. D. M., 1983a, Ecological roles of basidiomycetes forming decay communities in attached oak branches, *New Phytol.* **93**:77–88.

Boddy, L., and Rayner, A. D. M., 1983b, Mycelial interactions, morphogenesis and ecology of *Phlebia radiata* and *P. rufa* from oak, *Trans. Br. Mycol. Soc.* **80**:437–448.

Boddy, L., and Rayner, A. D. M., 1983c, Origins of decay in living deciduous trees: The role of moisture content and a reappraisal of the expanded concept of tree decay, *New Phytol.* **94**:623–641.

Boddy, L., and Swift, M. J., 1984, Wood decomposition in an abandoned beech and oak coppiced woodland in south-east England. III. Decay rate and turnover time of twigs and branches. *Holarc. Ecol.* **7**:229–238.

Boddy, L., Gibbon, O. M., and Grundy, M. A., 1985, Ecology of *Daldinia concentrica:* Effect of abiotic variables on mycelial extension and interspecific interactions, *Trans. Br. Mycol. Soc.* **85**:201–211.

Botton, B., and El-Khouri, M., 1978, Synnema and rhizomorph production in *Sphaerostilbe repens* under the influence of other fungi, *Trans. Br. Mycol. Soc.* **70**:131–136.

Boyce, J. S., 1961, *Forest Pathology*, 3rd ed. McGraw-Hill, New York.

Brasier, C. M., 1975a, Stimulation of sex organ formation in *Phytophthora* by antagonistic species of *Trichoderma*. I. The effect *in vitro, New Phytol.* **74**:183–194.

Brasier, C. M., 1975b, Stimulation of sex organ formation in *Phytophthora* by antagonistic species of *Trichoderma*. II. Ecological implications, *New Phytol.* **74**:195–198.

Brasier, C. M., 1978, Stimulation of oospore formation in *Phytophthora* by antagonistic species of *Trichoderma* and its ecological implications, *Ann. Appl. Biol.* **89**:135–138.

Brasier, C. M., 1984, Inter-mycelial recognition systems in *Ceratocystis ulmi:* Their physiological properties and ecological importance, in *The Ecology and Physiology of the Fungal Mycelium* (D. H. Jennings and A. D. M. Rayner, eds.), pp. 451–497, Cambridge University Press, Cambridge.

Bull, A. T., and Slater, J. H., 1982, Microbial interactions and community structure, in: *Microbial Interactions and Communities* (A. T. Bull and J. H. Slater, eds.). pp. 13–44, Academic Press, London.

Carrodus, B. B., and Triffett, A. C. K., 1975, Analysis of composition of respiratory gases in woody stems by mass spectrometry, *New Phytol.* **74**:43–246.

Cartwright, K. St. G., and Findlay, W. P. K., 1958, *Decay of Timber and its Prevention*, 2nd ed. HMSO, London.

Chase, W. W., 1934, The Composition, Quantity and Physiological Significance of Gases in Tree Stems, University of Minnesota Agricultural Experiment Station, *Technical Bulletin*, No. 99, pp. 1–51.

Clubbe, C. P., 1980, Colonisation of wood by micro-organisms, Ph.D. Thesis, University of London.

Coates, D., 1984, The biological consequences of somatic incompatibility in wood decaying basidiomycetes and other fungi, Ph.D. Thesis, University of Bath.

Coates, D., and Rayner, A. D. M., 1985a, Genetic control and variation in expression of the 'bow-tie' reaction between homokaryons of *Stereum hirsutum, Trans. Br. Mycol. Soc.* **84:**191–205.

Coates, D., and Rayner, A. D. M., 1985b, Fungal population and community development in beech logs. I., Establishment via the aerial cut surface, *New Phytol.* **101:**153–171.

Coates, D., and Rayner, A. D. M., 1985c, Fungal populations and community development in beech logs. II., Establishment via the buried cut surface, *New Phytol.* **101:**173–181.

Coates, D., and Rayner, A. D. M., 1985d, Fungal pupulation and community development in beech logs. III., Spatial dynamics, interactions and strategies, *New Phytol.* **101:**183–198.

Coates, D., Rayner, A. D. M., and Todd, N. K., 1981, Mating behaviour, mycelial antagonism and the establishment of individuals in *Stereum hirsutum, Trans, Br. Mycol. Soc.* **76:**41–51.

Cooke, R. C., and Rayner, A. D. M., 1984, *The Ecology of Saprotrophic Fungi*, Longman, London.

Côté, W. A. Jr., 1977, Wood ultrastructure in relation to chemical composition, in: *The Structure, Biosynthesis and Degradation of Wood* (F. A. Loewus and V. C. Runeckels, eds.), pp. 1–44, Plenum Press, New York.

Cowling, E. B., 1970, Nitrogen in forest trees and its role in wood deterioration, *Acta Univ. Upsal. Diss. Sci.* **164.**

Dickinson, D. J., 1982, The decay of commercial timbers, in: *Decomposer Basidiomycetes* (J. C. Frankland, J. N. Hedger, and M. J. Swift, eds.), pp. 179–190, Cambridge University Press, Cambridge.

Dowding, P., 1981, Nutrient uptake and allocation during substrate exploitation by fungi, in: *The Fungal Community* (D. T. Wicklow and G. C. Carroll, eds.), pp. 621–635, Dekker, New York.

Dowding, P., 1984, The evolution of insect–fungus relationships in the primary invasion of forest timber, in: *Invertebrate–Microbial Interactions* (J. M. Anderson, A. D. M. Rayner, and D. W. H. Walton, eds.), pp. 133–154, Cambridge University Press, Cambridge.

Dowson, C. G., 1982, Mycelial ecology of the Xylariaceae, Project report, University of Bath.

Dumpert, K., 1978, *The Social Biology of Ants*, Pitman, Boston.

Etheridge, D. E., and Craig, H. M., 1976, factors influencing infection and initiation of decay by the Indian paint fungus *(Echinodontium tinctorium)* in western hemlock, *Can. J. For. Res.* **6:**299–318.

Fries, N., 1973, Effects of volatile organic compounds on the growth and development of fungi, *Trans. Br. Mycol. Soc.* **60:**1–21.

Garrett, S. D., 1963, *Soil and Soil Fertility*, Pergamon Press, Oxford.

Garrett, S. D., 1970, *Pathogenic Root-Infecting Fungi*, Cambridge University Press, Cambridge.

Gibbs, J. N., and Gulliver, C. C., 1977, Fungal decay of dead elms, *Eur. J. For. Pathol.* **7:**193–200.

Griffin, D. M., 1977, Water potential and wood decay fungi, *Annu. Rev. Phytopathol.* **15:**319–329.

Grime, J. P., 1977, Evidence for the existence of three primary strategies in plants and its relevance to ecological and evolutionary theory, *Am. Nat.* **111:**1169–1194.

Grime, J. P., 1978, Competition and the struggle for existence, *Symp. Br. Ecol. Soc.* **20:**123–139.

Grime, J. P., 1979, *Plant Strategies and Vegetation Processes,* Wiley, London.

Guinier, P., 1933, Sur la biologie de deux champignons lignicoles, *C. R. Soc. Biol. Nancy* **112:**1363.

Hart, J. H., and Shrimpton, D. M., 1979, Role of stilbenes in resistance of wood to decay, *Phytopathology* **69:**1138–1143.

Highley, T. L., and Kirk, T. K., 1979, Mechanisms of wood decay and the unique features of heartrots, *Phytopathology* **69:**1151–1157.

Hillis, W. E. (ed.), 1962, *Wood Extractives,* Academic Press, New York.

Hillis, W. E., 1977, Secondary changes in wood, in: *The Structure, Biosynthesis, and Degradation of Wood* (F. A. Loewus and V. C. Runeckles, eds.), pp. 247–309, Plenum Press, New York.

Hintikka, V., 1969, Acetic acid tolerance in wood and litter-decomposing Hymenomycetes, *Karstenia* **10:**177–183.

Hintikka, V., 1982, The colonization of litter and wood by basidiomycetes in Finnish forests, in: *Decomposer Basidiomycetes: Their Biology and Ecology* (J. C. Frankland, J. N. Hedger, and M. J. Swift, eds.), pp. 227–239, Cambridge University Press, Cambridge.

Hudson, H. J., 1968, The ecology of fungi in plant remains above the soil, *New Phytol.* **67:**837–874.

Hulme, M. A., and Shields, J. K., 1975, Antagonistic and synergistic effects for biological control of decay, in: *Biological Transformation of Wood by Micro-organisms* (W. Liese ed.), pp. 52–63, Springer-Verlag, New York.

Hulme, M. A., and Stranks, D. W., 1970, Induction and the regulation of production of cellulase by fungi, *Nature* **226:**469–470.

Ikediugwu, F. E. O., 1976, Ultrastructure of hyphal interference between *Coprinus heptemerus* and *Ascobolus crennlatus, Trans. Br. Mycol. Soc.* **66:**281–290.

Ikediugwu, F. E. O., Dennis, C., and Webster, J., 1970, Hyphal interference by *Pueniophora gigantea* against *Heterobasidion annosum, Trans. Br. Mycol. Soc.* **54:**307–309.

Ingold, C. T., 1971, *Fungal Spores; Their Liberation and Dispersal,* Clarendon Press, Oxford.

Jennings, D. H., 1982, The movement of *Serpula lacrimans* from substrate to substrate over nutritionally inert surfaces, in: *Decomposer Basidiomycetes: Their Biology and Ecology* (J. C. Frankland, J. N. Hedger, and M. J. Swift, eds.), pp. 91–108, Cambridge University Press, Cambridge.

Jennings, D. H., and Rayner, A. D. M., 1984, *The Ecology and Physiology of the Fungal Mycelium,* Cambridge Unviversity Press, Cambridge.

Käärik, A. A., 1974, Decomposition of wood, in: *Biology of Plant Litter Decomposition,* (C. H. Dickinson and G. J. F. Pugh, eds.), pp. 129–174, Academic Press, London.

King, B., Smith, G. M., Baecker, A. A. W., and Bruce, A., 1981, Wood nitrogen control of toxicity of copper chrome arsenic preservatives, *Mat. Org.* **16:**105–118.

Kirk, T. K., 1975, Effects of a brown-rot fungus, *Lenzites trabea,* on lignin in spruce wood, *Holzforschung* **29:**99–107.

Kirk, T. K., and Adler, E., 1979, Methoxyl-deficient structural elements in lignin of sweetgum decayed by a brown-rot fungus, *Acta Chem. Scand. B* **24:**3379–3390.

Kirk, T. K., and Fenn, P., 1982, Formation and action of the ligninolytic system in basidiomycetes, in: *Decomposer Basidiomycetes: Their Biology and Ecology* (J. C. Frankland,

J. N. Hedger, and M. J. Swift, eds.), pp. 67–90, Cambridge University Press, Cambridge.

Kirk, T. K., Schulz, E., Connors, W. J., Lorenz. L. F., and Zeikus, J. G., 1978, Influence of culture parameters on lignin metabolism by *Phanerochaete chrysosporium*, *Arch; Microbiol.* **117**:277–285.

Koenigs, J. W., 1972a, Effects of hydrogen peroxide on cellulose and its susceptibility to cellulase, *Mat. Org.* **7**:133–147.

Koenigs, J. W., 1972b, Production of extra-cellular hydrogen peroxide and peroxidase by wood-rotting fungi, *Phytopathology* **62**:100–110.

Koenigs, J. W., 1974a, Hydrogen peroxide and iron: A proposed system for decomposition of wood by brown-rot basidiomycetes, *Wood Fibre* **6**:66–79.

Koenigs, J. W., 1974b, Production of hydrogen peroxide by wood-decaying fungi in wood and its correlation with weight loss, depolymerisation and pH changes, *Arch. Microbiol.* **99**:129–145.

Kogl, F., and Fries, N., 1937, Uber den einfluss von biotin, aneurin und meso-inosit auf das wachstum verschieden pilzarlen, *Z. Physiol. Chem.* **249**:23–110.

Levi, M. P., and Cowling, E. B., 1969, Role of nitrogen in wood deterioration. VII., Physiological adaptation of wood-destroying and other fungi to substrates deficient in nitrogen, *Phytopathology* **59**:460–468.

Levi, M. P., Merrill, W., and Cowling, E. B., 1968, Role of nitrogen in wood deterioration. VI., Mycelial fractions and model nitrogen compounds as substrates for growth of *Polyporus versicolor* and other wood-destroying and wood-inhabiting fungi, *Phytopathology* **58**:626–634.

Levy, J. F., 1975, Colonisation of wood by fungi, in: *Biological Transformation of Wood by Micro-organisms* (W. Liese, ed.), pp. 16–23, Springer-Verlag, Berlin.

Levy, J. F., 1982, The place of basidiomycetes in the decay of wood in contact with the ground, in: *Decomposer Basidiomycetes: Their Biology and Ecology* (J. C. Frankland, J. N. Hedger, and M. J. Swift, eds.), pp. 161–178, Cambridge University Press, Cambridge.

Loman, A. A., 1962, The influence of temperature on the location and development of decay fungi in lodgepole pine logging slash, *Can. J. Bot.* **40**:1545–1559.

Loman, A. A., 1965, The lethal effect of periodic high temperatures on certain lodgepole slash decaying basidiomycetes, *Can. J. Bot.* **43**:334–338.

Madelin, M. F., 1984, Myxomycetes, micro-organisms and animals: A model of diversity in animal–microbial interactions, in: *Invertebrate–Microbial Interactions* (J. M. Anderson, A. D. M. Rayner, and D. W. H. Walton, eds.), pp. 1–33, Cambridge University Press, Cambridge.

Manion, P. D., and Zabel, R. A., 1979, Stem decay perspectives—An introduction to the mechanisms of tree defense and decay patterns, *Phytopathology* **69**:1136–1138.

Merrill, W., and Cowling, E. B., 1966, Role of nitrogen in wood deterioration: Amounts and distribution of nitrogen in tree stems, *Can. J. Bot.* **44**:1555–1580.

Norris, D. M., 1979, The mutualistic fungi of Xyleborini beetles, in: *Insect–Fungus Symbiosis* (L. R. Batra, ed.), pp. 53–63, Allanheld, Osmun, Montclair, New Jersey.

Park, D., 1976, Carbon and nitrogen levels as factors influencing fungal decomposers, in: *The Role of Terrestrial and Aquatic Organisms in Decomposition Processes* (J. M. Anderson and A. Macfadyen, eds.), pp. 41–59, Blackwell, Oxford.

Peace, T. R., 1962, *Pathology of Trees and Shrubs,* Clarendon Press, Oxford.

Pentland, G. D., 1965, Stimulation of rhizomorph development of *Armillaria mellea* by *Aureobasidium pullulans* in artificial culture, *Can. J. Microbiol.* **11**:345–350.

Pentland, G. D., 1967, Ethanol produced by *Aureobasidium pullulans* and its effect on growth of *Armillaria mellea, Can. J. Microbiol.* **13**:1631–1639.

Platt, W. D., Cowling, E. B., and Hodges, C. S., 1965, Comparative resistance of coniferous root wood and stem wood to decay by isolates of *Fomes annosus, Phytopathology* **55:**1347–1353.

Pugh, G. J. F., 1980, Strategies in fungal ecology, *Trans. Br. Mycol. Soc.* **75:**1–14.

Rayner, A. D. M., 1976, Dematiaceous hyphomycetes and narrow dark zones in decaying wood. *Transactions of the British Mycological Society* **67:**546–549.

Rayner, A. D. M. 1977a, Fungal colonization of hardwood stumps from natural sources, I., Non-basidiomycetes, *Trans. Br. Mycol. Soc.* **69:**291–302.

Rayner, A. D. M., 1977b, Fungal colonization of hardwood stumps from natural sources, II., Basidiomycetes, *Trans. Br. Mycol. Soc.* **69:**303–312.

Rayner, A. D. M. 1978, Interactions between fungi colonizing hardwood stumps and their possible role of determining patterns in colonization and succession, *Ann. Appl. Biol.* **89:**131–134.

Rayner, A. D. M., 1979, Internal spread of fungi inoculated into hardwood stumps, *New Phytol.* **82:**505–517.

Rayner, A. D. M., and Hedges, M. J., 1982, Observations on the specificity and ecological role of basidiomycetes colonizing dead elm wood, *Trans. Br. Mycol. Soc.* **78:**370–373.

Rayner, A. D. M., and Todd, N. K., 1979, Population and community structure and dynamics of fungi in decaying wood, *Adv. Bot. Res.* **7:**333–420.

Rayner, A. D. M., and Todd, N. K., 1982, Population structure in wood-decomposing basidiomycetes, in: *Decomposer Basidiomycetes: Their Biology and Ecology* (J. C. Frankland, J. N. Hedger, and M. J. Swift, eds.), pp. 109–128, Cambridge University Press, Cambridge.

Rayner, A. D. M., and Turton, M. N., 1982, Mycelial interactions and population structure in the genus *Stereum: S. rugosum, S. sanguinolentum* and *S. rameale, Trans. Br. Mycol. Soc.* **78:**438–493.

Rayner, A. D. M., and Webber, J. F., 1984, Interspecific mycelial interactions—An overview, in: *The Ecology and Physiology of the Fungal Mycelium* (D. H. Jennings and A. D. M. Rayner, eds.), pp. 383–417. Cambridge University Press, Cambridge.

Rayner, A. D. M., Boddy, L., and Dowson, C. G., 1987a, Temporary parasitism of *Coriolus* spp. by *Lenzites betulina:* A strategy for domain capture in wood decay fungi, *FEMS Microbiol. Ecol.* **45:**53–58.

Rayner, A. D. M., Boddy, L., and Dowson, C. G., 1987b, Genetic interactions and developmental versatility during establishment of decomposer basidiomycetes in wood and tree litter. In: *Ecology of Microbial Communities* (M. Fletcher, T. Gray, and J. G. Jones, eds.), pp. 83–123, Cambridge University Press, Cambridge.

Rayner, A. D. M., Coates, D., Ainsworth, A. M., Adams, T. J. H., Williams, E. N. D., and Todd, N. K., 1984, The biological consequences of the individualistic mycelium, in *The Ecology and Physiology of the Fungal Mycelium* (D. H. Jennings and A. D. M. Rayner, eds.), pp. 509–540, Cambridge University Press, Cambridge.

Rishbeth, J., 1963, Stump protection against *Fomes annosus,* III., Inoculation with *Peniophora gigantea, Ann. Appl. Biol.* **52:**63–77.

Scheffer, T. C., and Cowling, E. B., 1966, Natural resistance of wood to microbial deterioration, *Annu. Rev. Phytopathol.* **4:**147–170.

Sharland, P. R., and Rayner, A. D. M., 1986, Mycelial interactions in *Daldinia concentrica, Trans. Br. Mycol. Soc.* **86:**643–649.

Shigo, A. L., 1966, Decay and Discoloration following Logging Wounds on Northern Hardwoods, U. S. Department of Agriculture, Forestry Service Research Paper NE-43.

Shigo, A. L., 1979, Tree Decay: An Expanded Concept, U. S. Department of Agriculture, Forestry Service Agricultural Information Bulletin 419.

Shigo, A. L., and Hillis, W. E., 1973, Heartwood, discoloured wood and microorganisms in living trees, *Annu. Rev. Phytopathol.* **11**:197–222.

Shortle, W. C., and Cowling, E. B., 1978, Interaction of live sapwood and fungi found in discolored and decayed wood, *Phytopathology* **68**:617–623.

Swift, M. J., 1976, Species diversity and the structure of microbial communities in terrestrial habitats, in: *The Role of Terrestrial and Aquatic Organisms in Decomposition Processes* (J. M. Anderson and A. Macfadyen, eds.), pp. 185–222, Blackwell, Oxford.

Swift, M. J., 1977, The role of fungi and animals in the immobilization and release of nutrient elements from decomposing branch-wood, in: *Soil Organisms as Components of Ecosystems* (U. Lohm and T. Persson, eds.), pp. 193–202, Swedish National Science Research Council, Stockholm.

Swift, M. J., and Boddy, L., 1984, Animal–microbial interactions during wood decomposition, in: *Invertebrate–Microbial Interactions* (J. M. Anderson, A. D. M. Rayner, and D. W. H. Walton, eds.), pp. 89–131, Cambridge University Press, Cambridge.

Theden, G., 1961, Untersuchungen uber die Fahigkeit holzzerstorender Pilze zur Trockenstrarre, *Angew. Bot.* **35**:131–145.

Thompson. W., 1982, Biology and ecology of mycelial cord-forming basidiomycetes in deciduous woodlands, Ph.D. Thesis, University of Bath.

Thompson, W., and Boddy, L., 1983, Decomposition of suppressed oak trees in even-aged plantations, II., Colonization of tree roots by cord and rhizomorph producing basidiomycetes, *New Phytol.* **93**:277–291.

Thompson, W., and Rayner, A. D. M., 1983, Extent, development and function of mycelial cord systems in soil, *Trans Br. Mycol. Soc.* **81**:333–345.

Timell, T. E., 1965, Wood and bark polysaccharides, in: *Cellular Structure of Woody Plants* (W. A. Côté, Jr., ed.), pp. 127–156, Syracuse University Press, Syracuse, New York.

Todd, N. K., and Rayner, A. D. M., 1980, Fungal individualism, *Sci. Prog. (Oxford)* **66**:331–354.

Wagener, W. W., and Davidson, R. W., 1954, Heart rots in living trees, *Bot. Rev.* **20**:61–134.

Williams, G. C. (ed.), 1971, *Group Selection,* Aldine, Atherton, Chicago.

Williams, E. N. D., Todd, N. K., and Rayner, A. D. M., 1981, Spatial development of populations of *Coriolus versicolor, New Phytol.* **89**:307–319.

Wilson, J. M., and Griffin, D. M., 1979, The effect of water potential on the growth of some soil basidiomycetes, *Soil Biol. Biochem.* **11**:211–212.

Wood, T. M., 1969, Relation between cellulolytic and pseudocellulolytic microorganisms, *Biochim. Biophys. Acta* **192**:531–534.

5

Phagotrophic Phytoflagellates

ROBERT W. SANDERS and KAREN G. PORTER

1. Introduction

Phytoflagellates are known to be important contributors to aquatic primary production; however, their role as consumers has been largely overlooked by ecologists. This is despite the many incidental observations and laboratory studies of algal phagotrophy reported in the literature (Table I). Close phylogenetic relationships exist between the groups classically known as algae and protozoa (Margulis and Schwartz, 1982; Corliss, 1983). Mixotrophic phytoflagellates, which photosynthesize, ingest particulate matter, and absorb dissolved organic matter, illustrate the functional overlap of these groups. The apochlorotic microflagellates, in particular, have close taxonomic affinities with pigmented flagellates (Table II). This led us to propose that the pigmented forms in groups with unpigmented phagotrophs, such as the dinoflagellates, cryptophytes, coccolithophores, chrysophytes, euglenoids, and flagellated chlorophytes, had the potential for mixotrophy (Porter *et al.,* 1985; Sanders *et al.,* 1985; Porter, 1987).

 Current interest by ecologists in the phenomenon of algal phagotrophy has been stimulated by recent conceptual and methodological developments in the area of microbial ecology. The microbial food loop is now recognized as an important component of planktonic ecosystems (Pomeroy, 1974; Azam *et al.,* 1983). Procaryotic picoplankton represent a major part of this loop and grazing is considered to be the most important factor controlling their abundances. Heterotrophic microflagellates are

ROBERT W. SANDERS and KAREN G. PORTER • Department of Zoology, University of Georgia, Athens, Georgia 30602.

Table I. Reports of Particle Ingestion by Phytoflagellates

Particle type	Reference
Division Chrysophyta	
Class Chrysophyceae		
Catenochrysis hispida — Polystyrene beads	Bird and Kalff (1987)
Chromulina sp. — "Phagotrophic"	Hutner and Provasoli (1951)
Chromulina sp. — Diatoms	J. R. Pratt, personal communication
Chromulina elegans — Polystyrene beads	Bird and Kalff (1987)
Chrysamoeba spp. — Bacteria	Estep *et al.* (1986)
Chrysamoeba radians — Bacteria	Hibberd (1971)
Chrysococcus cystophorus — Polystyrene beads	Bird and Kalff (1987)
Chrysosphaerella longispina[a] — Polystyrene beads	Bird and Kalff (1987)
Chrysostephanosphaera globulifera — Polystyrene beads	Present work (Table III)
Cyrtophora pedicellata — "Animal nutrition"	Pascher (1911)
Dinobryon sp. — Beads, starch, bacteria	K. G. Porter, R. W. Sanders, and S. J. Bennett, unpublished results
Dinobryon sp. — "Small organic particles"	Pascher (1943)
Dinobryon bavaricum — Polystyrene beads	Bird and Kalff (1986)
Polystyrene beads	Present work (Table III, Fig. 3)
Dinobryon cylindricum — Polystyrene beads, bacteria	Bird and Kalff (1986)
Polystyrene beads	Present work (Table III, Fig. 3)
Dinobryon eurystoma — Polystyrene beads	Bird and Kalff (1987)
Dinobryon sertularia — Unidentified particles	Wujek (1969)
Polystyrene beads	Bird and Kalff (1986)
Dinobryon sociale v. *americanum* — Polystyrene beads	Bird and Kalff (1986)
Ochromonas sp. — *Anacystis* (cyanobacteria)	Daley *et al.* (1973)
Ochromonas sp.[b] — Bacteria, cannibalism	Fenchel (1982a,b)
Ochromonas spp. — Bacteria	Estep *et al.* (1986)
Ochromonas sp. — *Spumella,* small algae	J. R. Pratt, personal communication
Polystyrene beads	Present work (Table III)
Ochromonas danica — Bacteria	Aaronson and Baker (1959)
Bacteria	Schuster *et al.* (1968)
India ink, bacteria	Aaronson (1973a,b)
Anacystis	Daley *et al.* (1973)
Bacteria, yeast	Aaronson (1974)
Microcystis, bacteria	Cole and Wynne, 1974
Beads, flagellate	Porter (1987)
Bacteria	Present work (Fig. 2)

Table I. (*Continued*)

	Particle type	Reference
Ochromonas *malhamensis*[c]	Bacteria, *Chlorella* *vulgaris,* *Ankistrodesmus* *angustus,* Fungal conidia, cannibalism, oil, casein, paint particles	Pringsheim (1952) Aaronson and Baker (1959)
	bacteria, *Escherichia coli,* cannibalism *Aerobacter* *aerogenes,* latex beads	Stoltze *et al.* (1969) Dubowsky (1974)
Ochromonas *tuberculatus*	Algae, graphite particles	Hibberd (1970)
Ochromonas miniscula	Polystyrene beads	Bird and Kalff (1987)
Ochromonas minuta	Polystyrene beads	Porter (1988)
Ochromonas monicis	Bacteria	Doddema and van der Veer (1983)
Palatinella cyrtophora	*Cryptomonas,* diatoms "Animal nutrition"	Lauterborn (1906) Pascher (1911)
Pedinella sp.	Bacteria	Lee (1980)
Pedinella hexacostata	Bacteria, graphite, India ink	Swale (1969)
	Bacteria	Conrad (1926)
	Bacteria	Vӯsotskiï (1888) (cited in Swale, 1969)
Phaeaster pascheri	Graphite particles	Belcher and Swale (1971)
Poterioochromonas *malhamensis*	Polystyrene beads, starch, bacteria, *Microcystis*	Porter (1987) present work (Table IV, Fig.2)
Poterioochromonas *stipitata*	Not described	Pitelka (1963)
	Ferritin molecules	Tsekos (1973)
Pseudopedinella elastica	"Food particles"	Skuja (1948) (cited in Swale, 1969)
Pseudopedinella erkensis	"Food particles"	Skuja (1948) (cited in Swale, 1969)
Pseudopedinella *rhizopodiaca*	"Mixotrophic"	Schiller (1952) (cited in Swale, 1969)
Synochromonas pallida	*Chromatium,* "various smaller organisms"	Korshikov (1929)
Uroglena americana	Bacteria Polystyrene beads, bacteria	Kimura and Ishida (1985) Bird and Kalff (1986)
Uroglena conradii	Polystyrene beads	Bird and Kalff (1986)
Uroglenopsis sp.	Bacteria	Wujek (1976)
Unidentified species[d]	Polystyrene beads	Bird and Kalff (1987)
Class Prymnesiophyceae		
Chrysochromulina alifera	Graphite particles, bacteria	Parke *et al.* (1956)
Chrysochromulina *brevifilum*	Graphite particles, bacteria, algal cells ($<5\ \mu$m)	Parke *et al.* (1955)

(*continued*)

Table I. (*Continued*)

Particle type		Reference
Chrysochromulina chiton	Graphite particles, bacteria, scales	Parke *et al.* (1958)
Chrysochromulina ephippium	Graphite particles, bacteria, algal cells (< 2.5 μm)	Parke *et al.* (1956)
Chrysochromulina ericina	*Oicomonas, Stichococcus* spp., *Chlorella*, small *Nitzschia*, bacteria	Parke *et al.* (1956)
Chrysochromulina kappa	Graphite particles, bacteria, algal cells (<5 μm)	Parke *et al.* (1955)
Chrysochromulina megacylindra	Diatoms, detritus	Manton (1972)
Chrysochromulina minor	Graphite particles, bacteria, algal cells (<5 μm)	Parke *et al.* (1955)
Chrysochromulina spinifera	Algal cells	Pienaar and Norris (1979)
Chrysochromulina strobilus	Graphite particles, bacteria, algal cells (<4 μm)	Parke *et al.* (1959)
Coccolithus pelagicus	Graphite particles	Parke and Adams (1960)
	Class Xanthophyceae	
Chlorochromonas polymorpha	"Animal nutrition"	Gavaudan (1931)
	Division Pyrrhophyta	
Class Dinophyceae		
Amphidinium steini	Food vacuoles	Kofoid and Swezy (1921)
Ceratium lunula	*Peridinium* sp.	Norris (1969)
Ceratium hirundinella[e]	Bacteria, dinoflagellates, diatoms, *Chlamydomonas* bacteria, blue-green algae, diatoms	MacKinnon and Hawes (1961) Dodge and Crawford (1970)
Gymnodinium agile	"Food vacuoles"	Kofoid and Swezy (1921)
Gymnodinium cnecoides	"Solid food"	Harris (1940)
Gymnodinium discoidale	"Solid food"	Harris (1940)
Gymnodinium flavum	"Food vacuoles"	Kofoid and Swezy (1921)
Gymnodinium fulgens	"Food vacuoles"	Kofoid and Swezy (1921)
Gymnodinium herbaceum	"Food vacuoles"	Kofoid and Swezy (1921)
Gymnodinium ravenescens	"Food vacoules"	Kofoid and Swezy (1921)

Table I. (*Continued*)

	Particle type	Reference
Gyrodinium sp.	Smaller dinoflagellate	D. Jacobson, personal communication
Gyrodinium pavillardi	*Strombidium* (ciliate)	Biecheler (1952)
Gyrodinium melo	"Food vacuoles"	Kofoid and Swezy (1921)
Gyrodinium vorax	Dinoflagellates	Biecheler (1952)
Massartia hyperxantha	"Partly holozoic"	Harris (1940)
Massartia hyperxanthoides	"Partly holozoic"	Harris (1940)
	Division Cryptophyta	
Cryptomonas sp.	Small flagellates	Pratt and Cairns (1985)
Cryptomonas borealis[f]	Cannibalism	Wawrik (1970)
Cryptomonas erosa	Polystyrene beads	Porter (1987)
	Bacteria	Porter *et al* (1985)
Cryptomonas ovata	Polystyrene beads	Porter (1987)

[a]We have also observed beads to be ingested by cells in colonies of *C. longispina*. However, the cells that ingested beads were morphologically different and did not autofluoresce. We believe these to be heterotrophic microflagellates attached to the autotrophic colony.
[b]Strain had a reduced chloroplast.
[c]*Ochromonas malhamensis* has been reassigned to the genus *Poterioochromonas* (see text).
[d]Similar to *Chrysostephanosphaera*.
[e]Hofeneder (1930) saw ingestion by *Ceratium hirundinella* with reduced chloroplasts.
[f]This could be an observation of a vegetative division process (Gantt, 1980).

major consumers of the picoplankton (Fenchel, 1982c; Sieburth and Davis, 1982; Güde, 1986). Recent field studies have shown that certain pigmented flagellates can also have a large grazing impact on the picoplankton. Phagotrophic phytoflagellates can exert up to 79% of the total grazing pressure on bacterioplankton (Fig. 1) (Bird and Kalff, 1986; Porter, 1987). Phytoflagellates in turn are the major food sources for higher trophic levels, such as *Daphnia* and other crustacean zooplankton (Porter, 1973).

The majority of phytoflagellates are phototrophic, whereas some routinely use organic compounds obtained through osmotrophy or phagotrophy. When carbon is acquired by photosynthesis plus particle ingestion, we use the term mixotrophic. There has already been considerable work on the heterotrophic uptake of dissolved organic compounds by algae [for reviews see Pringsheim (1963), Aaronson (1980), and Antia (1980)], and it is not emphasized here. We report the name of an organism as used by each author. This should be kept in mind, since *Ochro-*

Table II. Examples of Morphologically Similar Flagellates that Possess or Lack Plastids

Class	Plastidic	Aplastidic
Cryptophyceae	*Cryptomonas*	*Chilomonas*
Dinophyceae	*Gymnodinium*	*Gymnodinium*
	Gyrodinium	*Gyrodinium*
Euglenophycea	*Euglena*	*Astasia*
	Phacus	*Hyalophacus*
Chrysophyceae	*Dinobryon*	*Stokesiella*
	Ochromonas	*Spumella* (= *Monas*)
	Mallomonas	*Paraphysomonas*
	Chromulina	*Oikomonas*
Prymnesiophycea	*Chrysochromulina*	*Chrysochromulina*
Chlorophyceae	*Chlamydomonas*	*Polytoma*
	Chlorogonium	*Hyalogonium*

monas malhamensis is currently known as *Poterioochromonas malhamensis* (Pringsheim) Peterfi (see Leedale and Hibberd, 1985). In our tables, we have followed the classification system used by Bold and Wynne (1985). The classes have nearly exact equivalents as orders in recent classifications of protists (Leedale and Hibberd, 1985).

2. Reports of Phagotrophy in Pigmented Algae

The species of algae that are reported to be phagotrophic and the particles that were ingested are listed in Table I. All were reported to contain apparently active chloroplasts at the same time that phagocytized particles were observed and can thus be considered mixotrophic. Chrysophytes and dinoflagellates, which have numerous nonphotosynthetic species (Leedale and Hibberd, 1985), also have the greatest number of photosynthetic members known to utilize particulate food. Many members of the coccolithophorid genus *Chrysochromulina* are also phagotrophic. Some *Cryptomonas* species have recently been observed to take up particulate matter into the gullet. Extreme environmental conditions may be necessary to induce feeding by autotrophic cryptomonads. Interestingly, particle ingestion by euglenoids with chlorophyll has not been reported, even though there are colorless forms and some chlorophytic species that can be induced to become permanently apochloritic (Hutner and Provasoli, 1951).

The presence of particles inside the algae does not necessarily imply phagotrophy. For example, *Euglena proteus,* a colored species, has been

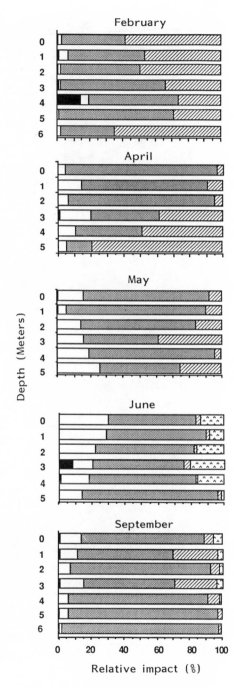

Figure 1. Percentage of total grazing by planktonic groups on bacteria-sized particles in Lake Oglethorpe, Georgia, in 1986. Grazers include cladoceran crustaceans (solid), ciliates (open), heterotrophic microflagellates (stippled), mixotrophic flagellates (hatched), and rotifers (chevrons).

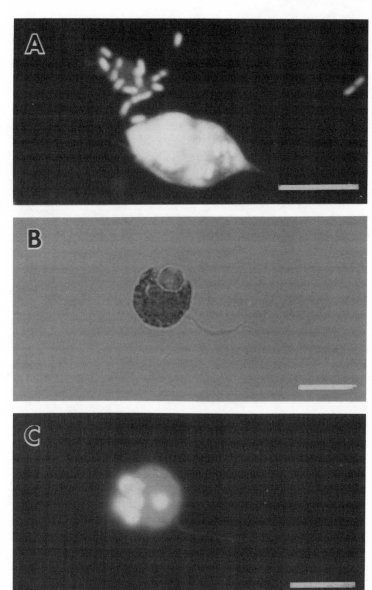

Figure 2. (A) *Ochromonas danica* with ingested bacteria that were prestained with acridine orange. The cell was stained with primulin and photographed using epifluorescent light excitation. (B) *Poterioochromonas malhamensis* with ingested starch particle. (C) *P. malhamensis* with ingested 1.17 μm microspheres. Scale bars are 5 μm.

observed to contain *Escherichia coli* when the bacteria were introduced after *Euglena proteus* had been maintained for 24 hr in a low-nutrient medium in the dark (H. King, personal communication). However, when we offered the same *Euglena* clone a variety of other food particles (starch, fluorescent beads, *Microcystis aeruginosa,* and *Aerobacter aerogenes*) under similar conditions, no ingestion was observed (K. G. Porter, R. W. Sanders, and S. J. Bennett, unpublished data). In this case, the strain of bacteria may be an infective agent that was able to invade the *Euglena* cells (H. King, personal communication).

Phytoflagellates are known to have associated endo- and episymbiotic bacteria (e.g., Geitler, 1948; Kochert and Olson, 1970; Leedale, 1969; Wujek, 1978; Klaveness, 1982; Doddema and van der Veer, 1983). These reports illustrate a possible danger of attributing internalized particles to feeding when they are in fact due to a symbiotic relationship. However, the uptake of several types of food particles by many of the species listed in Table I (see also Fig. 2) would suggest that these algae are actually phagotrophic. Furthermore, some species have been microscopically observed while actively feeding (Pringsheim, 1952; Biecheler, 1952; Doddema and van der Veer, 1983; M. Boraas, personal communication, J. Pratt, personal communication).

3. Environmental Distribution

Phagotrophic phytoflagellates are found in a variety of marine and freshwater habitats. They may be found free-swimming and attached to or temporarily associated with a variety of substrata. These include suspended particles, aquatic plants, and sediments. It is likely that some of the planktonic species also can graze on surfaces and enter flocculent, decomposing material where particulate food would be concentrated. Although we are not aware of any reports of ingestion by pigmented flagellates in the soil, moist terrestrial environments may also harbor these protists.

The phagotrophic chrysomonads are common in the freshwater plankton and occur in all bodies of water, from small pools and streams to large lakes. There are also planktonic marine species, and they may be more abundant in this environment than was previously suspected (Estep *et al.,* 1986). Most are free-swimming single cells or colonies. The pedinellid chrysomonads, which have a peduncle used to attach temporarily to a substratum, are a small group usually found in brackish water. Dinoflagellates are widely distributed in marine, brackish, and fresh waters. Mixotrophic dinoflagellate species have been reported in each of these planktonic environments. Most species of the Prymnesiophyceae, which includes the coccolithophorids, are marine and planktonic. The crypto-monads are common in the plankton and also on natural substrates (Pratt and Cairns, 1985). Phagotrophy by pigmented cryptomonads has been reported only for freshwater species.

4. Feeding Mechanisms

There are a variety of means by which flagellates are known to capture particulate food. Several orders have species containing trichocysts, which Aaronson (1973a) suggests may function to paralyze or entrap prey. *Noctiluca* can use mucus to entrap particles (Uhlig and Sahling, 1985). Euglenoid flagellates and a chrysophyte, *Prymnesium parvum,* also produce muciferous bodies near their surface, which are extruded, whereas *Ochromonas danica* excretes membranous vesicles, both of which may serve to entrap particles (Aaronson, 1973a, 1974). The holozoic dinoflagellate *Gymnodinium fungiforme* ingests the cytoplasm of its prey through an extensible peduncle (Spero and Morée, 1981). Some colorless euglenoids also use special rods or tubes in feeding (Leedale, 1967). The process of pseudopodial feeding has been observed in several dinoflagellates, including a photosynthetic species (Biecheler, 1936, 1952; Gaines and Taylor, 1984; Lessard and Swift, 1985). Not all of these mechanisms have been observed for phytoflagellates, but they are potential means of incorporating particulate food.

Engulfment of prey items into food vacuoles (phagocytosis) has been observed for several algal species. The process is best described for several *Ochromonas* species. Aaronson (1973a) observed that *O. danica* ingested bacteria, yeast, cyanobacteria, and other *Ochromonas* via phagocytosis. Bacteria were apparently ingested at any point of the cell surface forming primary vacuoles, which coalesced into a larger one near the posterior of the cell. Cole and Wynne (1974) reported similar observations when *O. danica* fed on *Microcystis,* except that the food was generally ingested only at a point just posterior to flagella. *Ochromonas malhamensis* captures food with the aid of the flagella, which causes a particle to spin rapidly at the anterior of the cell until it suddenly stops movement and is immediately engulfed by short pseudopodia (Pringsheim, 1952). *Ochromonas monicis,* a marine species (Doddema and van der Veer, 1983), and an as yet unidentified freshwater *Ochromonas* sp. (K. Estep, personal communication) also use flagella to aid in the capture of food.

Several phytoflagellates excrete exoenzymes (Aaronson, 1973a). These may be utilized for external digestion, especially of particles concentrated by such means as flocculation (Porter, 1987). Uptake of dissolved organics from the prey items by pinocytosis or some form of osmotrophy would then serve the same function as phagocytosis. At least one species, *Poterioochromonas stipulata,* can apparently take up large molecules by phagocytosis (Tsekos, 1973).

5. Grazing Experiments

While most of the literature about phytoflagellate feeding can be considered as incidental, some laboratory and field experiments have focused on the feeding process. These address three basic questions: (1) what are the feeding rates, (2) does phagotrophy enhance growth, and (3) what induces particle feeding? Factors that elicit feeding in a mixotroph may also affect ingestion and growth rates.

5.1. Feeding Rates

Determination of accurate feeding rates in protists has been largely limited by the techniques available. The indirect method of determining ingestion by the disappearance of prey items over time does not always take into account the growth of prey populations. Other methods, such as metabolic inhibition (Sanders and Porter, 1986) and uptake of radioactively labeled bacteria (Hollibaugh *et al.,* 1980), are frequently unable to identify individual species as grazers. Recently, the use of tracer particles (e.g., fluorescent microspheres), which can be observed directly within individual cells after addition to natural populations, has shown that phytoflagellates can be the dominant consumers of bacteria-sized particles in lakewater (Bird and Kalff, 1986; R. W. Sanders *et al.,* in preparation: "Seasonal patterns of bacterivery by freshwater flagellates, ciliates, rotifers, and crustaceans"). This is illustrated in Fig. 1 for tracer experiments using 0.6-μm Polysciences "Fluoresbright" microspheres in Lake Oglethorpe, Georgia. The autotrophic grazers at those times were *Dinobryon cylindricum, Dinobryon bavaricum, Ochromonas* sp., and *Chrysostephanosphaera globulifera.*

Clearance rates (the volume of water that would need to be swept clear of the known density of microspheres) were determined from average individual ingestion rates (Table III). If bacteria are assumed to be ingested without discrimination and in the same manner as the spheres, then ingestion rates for bacteria can be determined from the clearance rates. From 5×10^3 to 5×10^4 bacteria were removed hourly per milliliter of lakewater by these algal species. Bacterial density was about 3×10^6 cells/ml during these experiments, so that phytoflagellates alone have the potential to remove a large proportion of the bacterial standing stock daily. *Dinobryon bavaricum* had a higher average clearance rate and

Table III. Clearance Rates of Phytoflagellates Determined Using 0.6-μm
Microspheres and Calculated Removal of Bacteria from Experiments in Water
from Lake Oglethorpe, Georgia[a]

Species	Clearance rate (nl cleared flagellate^{-1} hr^{-1})	Bacteria ingested (flagellate^{-1} hr^{-1})
Dinobryon bavaricum	8.4	24
Dinobryon cylindricum	3.5	10
Ochromonas sp.	2.5	7
Chrysostephanosphaera[b]	26.0	77

[a]R. Sanders (unpublished results).
[b]Per colony.

ingested more microspheres per cell than did *D. cylindricum* (Table III,
Fig. 3). However, *D. cylindricum* was much more abundant and removed
more bacteria as a population than did any other single species of bacter-
ivore. Bird and Kalff (1986) observed similar clearance rates for *Dinobryon* (5.8 nl flagellate^{-1} hr^{-1}) in Lake Memphremagog (Quebec–Ver-
mont). They reported ingestion rates for *Uroglena* spp. that were
approximately 18% that of *Dinobryon*.

We have also determined grazing rates for several phytoflagellates
from cultures (Table IV) (Porter, 1987). Rates were determined by offer-
ing known concentrations of microspheres and other particles to axenic
cultures. Ingested particles (e.g., Fig. 2) were counted and clearance rates
determined as for the field experiments. All culture experiments were run
at 20°C. These rates are lower than those we measured in field experiments, but are still as high as for some heterotrophic microflagellates
(Sherr *et al.,* 1983; Cynar and Sieburth, 1986; Porter, 1987). Other fluorescent particles that can be distinguished from the autofluoresence of
the algal cells (e.g., *Microcystis,* acridine orange-stained bacteria) are also
ingested by these species and can be used in an analogous manner for
feeding rate determination.

Cole and Wynne (1974) found *Ochromonas danica* ingested the
cyanobacterium *Microcystis aeruginosa* at a rate of >2 cells
Ochromonas^{-1} hr^{-1} when both predator and prey were present at approximately 10^6 cells ml^{-1} (calculated from their Fig. 8). Grazing was calculated from the disappearance of *Microcystis* when ingestion was highest
during the first 9 min after combining the two cultures. This is equivalent
to a clearance rate of approximately 2.3 nl flagellate^{-1} hr^{-1}. The decrease
in ingestion rate with time may have been due to the decreasing numbers
of *Microcystis* left in suspension after a period of grazing, since the rate
of endocytosis was apparently a factor of encounter rate and timing of
digestive processes (Cole and Wynne, 1974).

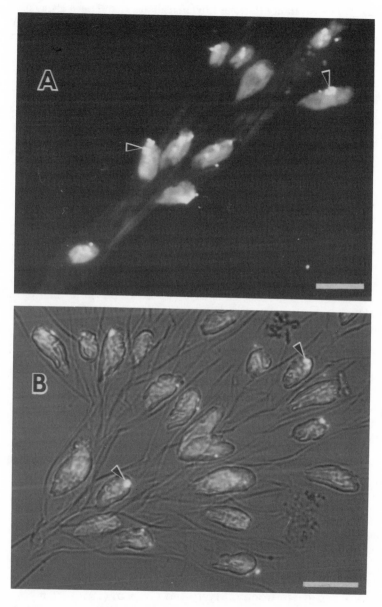

Figure 3. Light micrographs of (A) *Dinobryon bavaricum* and (B) *Dinobryon cylindricum* with ingested 0.57 μm fluorescent microspheres. Arrows indicate some of the ingested beads. Note that *D. bavaricum* ingested more beads per cell during the same experiment. Scale bars are 10 μm.

Table IV. Clearance Rates of Phytoflagellates Determined Using 0.6-μm
Fluorescent Microspheres in Axenic Cultures

Species	Particle density (number ml^{-1})	Clearance rate (nl cleared flagellate^{-1} hr^{-1})	Particles ingested (flagellate^{-1} hr^{-1})
Ochromonas danica	7.7×10^6	0.13	1.0
Ochromonas minuta	5.0×10^6	0.16	0.8
Poterioochromonas malhamensis	3.8×10^6	0.40	1.5

Fenchel (1982b) calculated a maximum clearance rate of 10 nl flagellate^{-1} hr^{-1} based on cell yield in cultures for an *Ochromonas* sp. feeding on bacteria. This species, which was originally isolated from a *Paramecium* culture, had a very reduced chloroplast and was incapable of autotrophic growth (Fenchel, 1982a). The maximum calculated uptake was 190 bacteria cell^{-1} hr^{-1}.

Grazing is frequently considered to be responsible for the relative constancy of bacterial numbers in both marine and freshwater planktonic systems (e.g., Sieburth and Davis, 1982; Wright and Coffin, 1984; Sanders and Porter, 1986). The heterotrophic flagellates have been proposed as the major grazers of bacteria. Yet, grazing rates for some phytoflagellates are as high as those measured for many nonphotosynthetic flagellates. Up to 20% of the flagellates ingesting bacteria-sized particles in experiments in Kaneohe Bay, Hawaii, were also photosynthetic (M. Pace, personal communication). Clearly, the role of phytoflagellates in planktonic food webs needs reevaluation.

5.2. Enhancement of Growth Due to Mixotrophy

Growth experiments tend to indicate that several species of algae have better growth when organic compounds are present in either dissolved or particulate form. The possible importance of phagotrophy is also supported by the inability to maintain axenic cultures of some species (Parke *et al.,* 1955; Kimura and Ishida, 1985; R. R. L. Guillard, personal communication). In the spectrum of nutrition between photoautotrophy and heterotrophy, Myers and Graham (1956) even concluded that *Ochromonas malhamensis* lies closer to the heterotrophic end of the scale. They found photosynthesis to make only a marginal contribution to the nutrition of this species. Pringsheim (1952) found that media that supported the growth of *Ochromonas malhamensis* in the light also did so in the dark, and concluded that photosynthesis was important for cell maintenance, but not growth in this species. Aaronson and Baker (1959), however, reported good phototrophic growth of both *O. malhamensis* and *O. danica*. They suggested that photosynthesis in these species may require organic H-donors from the environment.

It has been proposed that phagocytosis may be a means of incorporating nitrogen or phosphorus when dissolved forms are depleted (Doddema and van der Veer, 1983). Dissolved amino acids can also be the sole nitrogen source for several chrysomonads, dinoflagellates, and euglenoids (Aaronson, 1973a). Although these species have not all been observed to be phagotrophic, it does suggest a function for mixotrophy in general. Many phytoflagellates are auxotrophic—vitamins that they cannot produce are required for growth (Table V). Bacteria could supply essential vitamins (Pringsheim, 1952; Aaronson and Baker, 1959). Both *O. malhamensis* and *O. danica* can obtain their obligate requirement for biotin by ingestion of the bacterium *Thiobacillis* (Aaronson and Baker, 1959), indicating at least a potential for growth enhancement by phagotrophy. Doddema and van der Veer (1983) disagree and maintain that "if sufficient vitamin-producing bacteria are present to prey upon, the amount of vitamin dissolved in the seawater will be sufficient to sustain the growth of the algae." The question of whether phagocytosis enhances growth or increases ecological fitness in mixotrophic phytoflagellates requires more investigation.

Table V. Vitamins Required by Phagotrophic Phytoflagellates when Grown Axenically in Culture

	Species	Vitamins	Reference
Chrysophyceae	*Ochromonas danica*	Biotin, thiamine	Provasoli (1958)
	Ochromonas minuta	None	Aaronson (1980)
	Ochromonas malhamensis	B_{12}, biotin, thiamine	Hutner *et al.* (1953)
	Poterioochromonas stipitata	B_{12}, biotin, thiamine	Hutner *et al.* (1953)
Prymnesiophyceae	*Chrysochromulina brevifilum*	B_{12}, thiamine	Aaronson (1980)
	Chrysochromulina kappa	B_{12}, thiamine	Aaronson (1980)
	Chrysochromulina strobilus	B_{12}	Aaronson (1980)
	Coccolithus sp.	B_{12}	Aaronson (1980)
Dinophyceae[a]	*Gymnodinium* spp.	B_{12}	Provasoli (1958)
	Gyrodinium spp.	B_{12}	
Cryptophyceae	*Cryptomonas ovata*	B_{12}	Provasoli (1958)

[a]Most dinoflagellates appear to have a requirement for B_{12} (Provasoli, 1958; Loeblich, 1967).

5.3. Elicitation of Feeding Behavior

Some experiments would indicate that many of the ochromonads are obligate mixotrophs (i.e., they require some sort of dissolved or particulate organic substrate). *Dinobryon* and *Chrysochromulina* also resist axenic culturing and may require organic matter. If the impact of phytoflagellate grazing is to be evaluated, the factors determining ingestion need further investigation. Available light, nutrient and vitamin requirements, and need of a carbon source are possible factors regulating ingestion.

We have observed in an organic-free medium that phagocytosis by *Poterioochromonas malhamensis* is greater in low light versus high light (Porter, 1987). Ingestion rate by *P. malhamensis* increased steadily as light and, consequently, primary production decreased. Conversely, Aaronson (1974) reported a relative constancy of the percentage of *Ochromonas danica* eating in short-term experiments in both light and dark, suggesting that light is not an important regulator of feeding. However, the uptake rates of individual flagellates may change even when the percentage with ingested particles remains constant.

Grazing rates of *Dinobryon* spp. in Lac Gilbert, Quebec, correlated well with temperature and not with light (Bird and Kalff, 1987). A relative constancy of grazing from day to night by *Dinobryon* was observed in two other Canadian lakes (Bird and Kalff, 1987). We observed that clearance rates for *Dinobryon cylindricum* tended to decrease slightly with depth in the shallow Lake Oglethorpe, Georgia, during a February bloom (R. Sanders, S. Bennett, A. DeBiase, and K. Porter, unpublished results). Temperature and light decreased with depth and both had high positive correlations with ingestion rate for this species. As with any correlational relationships, this does not necessarily mean that either of these factors caused the change in feeding rates. However, temperature is likely to affect the overall metabolism of protists and thus their ingestion. Aaronson (1974) found that a greater number of *Ochromonas danica* were phagotrophic at higher temperatures (Fig. 4).

The needs to satisfy nitrogen, phosphorus, and vitamin requirements are possible reasons for phagotrophy (see Section 5.2). Yet, little attention has been given to the possibility that phagotrophy can be an important carbon source for phytoflagellates. Ingestion of bacteria may provide from 30 to 70% of the carbon assimilation for *Dinobryon sertularia* at some depths in Lac Cromwell, Quebec (Bird and Kalff, 1987). Particle ingestion by *Poterioochromonas malhamensis* was inhibited by increasing dissolved glucose concentrations in laboratory experiments (Porter, 1987). F. R. Pick (personal communication) has observed ephemeral

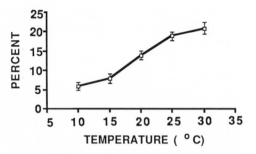

Figure 4. Percentage of *Ochromonas danica* cells that showed evidence of phagotrophy as temperature was increased. Error bars show 95% confidence intervals. [Graphed from data in Aaronson (1974).]

peaks of the chrysophytes *Dinobryon* and *Mallomonas* in zones of high detritus in Canadian shield lakes. These experiments and observations imply that one function of mixotrophy may be to obtain carbon. Phagotrophy could be an important source of carbon when dissolved organics are low and light levels are suboptimal for photosynthesis. Both organics and light vary with depth in the water column. If these factors are determinants of ingestion, then the impact of phytoflagellate grazing is also expected to change with depth. Bird and Kalff (1987) suggest that deep chlorophyll layers composed of algae with phagotrophic abilities probably subsist by ingesting bacteria.

6. Digestive Processes

Digestion in phytoflagellates has received a fair amount of attention, although most of the work has focused on the genus *Ochromonas* [see review by Aaronson (1973a)]. There is evidence for both intracellular digestion, with formation of food vacuoles, and of extracellular digestion.

6.1. Intracellular Digestion

Phagotrophy in several *Ochromonas* spp. has some similarity. After a primary vacuole is formed around the food particle, it migrates to the posterior of the cell and fuses with a secondary vacuole. There is controversy about the amount of digestion that takes place in these vacuoles (endosomes). Cole and Wynne (1974) suggest that at least the mucilaginous sheath of a cyanobacterium is removed in the primary vacuole of *Ochromonas danica,* and further digestion takes place after fusion with the secondary vacuole. Ingestion in this species apparently stops when the secondary endosome is full and the contents are then digested simultaneously. Lysosome enzymes for digestion in *Ochromonas danica* may be synthesized to some degree by the endosomes, although acid phosphatase activity has been observed in the Golgi cisternae (Cole and Wynne, 1974).

Acid phosphatase activity occurs in both the large "leucosin" vacuole and the smaller food vacuoles (primary endosomes) of *Ochromonas malhamensis* (Stoltze *et al.*, 1969). Schuster *et al.* (1968) also observed acid phosphatase activity in food vacuoles of *O. danica*, but did not report activity in the larger posterior vesicle. Daley *et al.* (1973) suggested that cells ingested by an *Ochromonas* species were first sequestered in the large (leucosin) vacuole, with final stages of digestion and elimination occurring in the smaller vacuoles. Aaronson (1974) believed that the large leucosin vacuole was compressed by coalescing food vacuoles, but that they did not actually fuse. Elimination apparently occurs at undifferentiated sites on the membrane (Dubowsky, 1974).

There is also some controversy as to whether digestive enzymes are released upon ingestion of any particle by a protist. Ricketts (1971) maintained that only useful food elicited acid phosphatase release in the ciliate *Tetrahymena*, whereas Müeller *et al.* (1965) found activity when latex beads were ingested. Enzyme activity was also observed in *Ochromonas malhamensis* when either polystyrene particles or bacteria were eaten (Dubowsky, 1974).

6.2. Extracellular Digestion

Several phytoflagellates, including *Ochromonas* and *Amphidinium*, produce extracellular hydrolases (Aaronson, 1973a). *Ochromonas danica* can secrete a variety of hydrolases into the environment (Meyer and Aaronson, 1973). Many protists reclaim enzymes before material is egested. But the explusion of food vacuoles that apparently still contain hydrolytic enzymes has been described for an *Ochromonas* sp. (Dubowsky, 1974). This could explain the occurrence of extracellular enzymes in some algal cultures, although the extent of this process is unknown.

Extracellular hydrolases may have a role in algal nutrition. Breakdown of particulate food could be enhanced by attachment to the algal cell with concomitant release of hydrolases. However, unless there is some mechanism for preventing dilution of the enzymes in the aquatic environment, the release without attachment of particles would seem useless (Aaronson, 1973a). Alkaline phosphatases do appear to be firmly bound near the cell surface of some phosphorus-limited algae (Kuhl, 1974). These may allow the cells to use phosphorylated compounds present in the water. Analogously, other enzymes could serve to hydrolyze large molecules into smaller ones that are more easily taken up by the algae (Aaronson, 1973a).

7. Selective Feeding

There is some controversy about the ability of protists to feed selectively. Many will ingest nonnutritive particles as readily as nutritive ones, whereas others may select specific food if given a choice (Müeller *et al.*, 1965; Rapport *et al.*, 1972; Fenchel, 1980; Stoecker *et al.*, 1981). As shown in Table I, many phagotrophic phytoflagellates are not always qualitatively selective about particles they ingest. *Ochromonas malhamensis* ingested several paint pigment particles, cyanobacteria, fungi, bacteria, and latex beads. It did not ingest sand grains, calcium sulfate, calcium carbonate, or powdered glass (Pringsheim, 1952). Pringsheim attributed this to the different weights of the particles. Particles that the flagellum could not press against the cell surface were not incorporated.

Dubowsky (1974) observed polystyrene particles and bacteria within the same food vacuole when both were offered simultaneously to *Ochromonas malhamensis*. We have made the same observation for *O. danica, O. minuta,* and *Poterioochromonas malhamensis*. Several unidentified phytoflagellate species in lakewater also ingest both beads and bacteria (S. Bennett, personal communication). However, there may be quantitative differences in ingestion. Bird and Kalff (1986) found that *Dinobryon* species ingested 1.3 bacteria for each bead. Phytoflagellates in Kaneohe Bay, Hawaii, may also select bacteria over latex beads (M. Pace, personal communication). A freshwater *Ochromonas* sp. consumed *Chlorella pyridenosa,* but not bacteria or plastic beads <2 μm in diameter (Boraas, personal communication).

If selective feeding is a common phenomenon, the estimates of grazing rates using tracer cells are not true rates on natural particles. However, any difference is likely to be an underestimate of grazing and calculated rates would be conservative. There may also be species-specific differences in feeding selectivity in phytoflagellates, but this requires further investigation.

8. Ecological Significance

Phytoflagellates are usually the most abundant eucaryotes in the plankton of freshwater and marine systems. They can be important grazers of bacteria (Bird and Kalff, 1986; R. W. Sanders *et al.,* in preparation), cyanobacteria (Daley *et al.,* 1973; Cole and Wynne, 1974), and other algae (K. Estep and M. Boraas, personal communications). Grazing by phytoflagellates will have some effects on the biotic community similar to that of completely heterotrophic organisms. These include changes in the population dynamics of the prey, and possible competition for prey with other grazers. If phytoflagellates cause clumping of prey cells, as reported for *Ochromonas* (Aaronson, 1973b), then the grazing of other predators that are size-selective feeders would be affected. This could include both crustaceans and ciliates.

Cole and Wynne (1974) suggest a possible role for *Ochromonas* in the biological control of problem cyanobacterial blooms. This is feasible if a strain or species could be isolated that feeds preferentially on these procaryotes. Another ecological effect of phagotrophic algae may be to harbor pathogenic bacteria. This can protect the pathogens from chlorination in water treatment plants (H. King, personal communication), as has already been demonstrated for ciliates (Fields *et al.*, 1984).

The ecological impact of phagotrophic phytoflagellates is probably more complicated. Phytoplankton exudates can be a dominant substrate for bacterial growth (Coveney, 1982; Larsson and Hagström, 1979), and grazing by microplankton also releases carbon that is available for bacterioplankton (Taylor *et al.*, 1985). Yet species of bacterivorous phytoflagellates bring some of this organic carbon back into the phytoplankton component. Mixotrophic algae are primary producers as well as grazers and their effect on nutrient recycling is unclear.

Mixotrophy may serve as a source of essential nutrients and vitamins to some algal cells. It may also supply a large proportion of the carbon requirements of the cells. Even if phagotrophy does not always provide for all components necessary for growth (Pringsheim, 1952), it could still supply enough nutrition for cell maintenance. This is especially true where light is limiting. It is possible that phagotrophy and uptake of dissolved organics are the processes responsible for maintaining the large numbers of chrysophytes frequently found at metalimnetic peaks in lakes.

9. Evolutionary Significance

Phagotrophy has evolutionary implications for phytoflagellates, as well as for eucaryotes in general. Invasion of procaryotes by other procaryotes may have led to the first eucaryotes, whereas ingestion of photosynthetic procaryotes without digestion may have led to the first plastidic algae (Margulis, 1981). Phagotrophy should thus be considered a primitive trait in algal cells, and the ultrastructure and method of ingestion could shed light on the exact progression from colorless to colored forms of flagellates. Groups such as dinoflagellates and chrysophytes, which have the greatest incidence of phagotrophic species, would be predicted to be primitive. The dinoflagellates and chrysophytes have a fossil record going back to the early Paleozoic (Margulis and Schwartz, 1982).

10. Concluding Remarks

Many phytoflagellates are morphologically similar to nonphotosynthetic flagellates (Table II). Although the extent of phagotrophy by phytoflagellates is not well documented, the field experiments reported here suggest that it is pervasive. It is becoming clear that both colored and colorless groups can play an important role as consumers and as links between the picoplankton and larger grazers.

The phenomenon of mixotrophy in flagellates, as well as the retention of chloroplasts or algal endosymbionts by ciliates and amoeba (Taylor, 1982; Laval-Peuto and Febvre, 1986; McManus and Fuhrman, 1986), raises questions about the utility of trophic-level definitions. This concept may be of little use in modeling microbial food webs, where many of the "players" could be considered to be functioning on at least two trophic levels. Plankton surveys in which the categories "phytoplankton" and "microzooplankton" are used to create budgets to estimate material flow in the planktonic food web are likely to be inaccurate if they do not account for photosynthesizing ciliates (McManus and Fuhrman, 1986) and phagotrophic phytoplankton.

ACKNOWLEDGMENTS. We have had access to several manuscripts and appreciate the generosity of our colleagues in sharing their unpublished work with us. We thank S. Bennett, D. Bird, M. Boraas, K. Estep, R. Guillard, D. Jacobson, J. Kalff, H. King, M. Pace, F. R. Pick, and J. R. Pratt for their communications. D. Klaveness and S. Hutner offered insights, and numerous others supplied reprints and encouragement. R. McDonough and S. Bennett provided technical and photographic assistance, respectively. This work was supported by NSF grant BSR-8407928 to K. G. P. and is contribution No. 29 of the Lake Oglethorpe Limnological Association.

References

Aaronson, S., 1973a, Digestion in phytoflagellates, in: *Lysosomes in Biology and Pathology,* Vol. 3 (J. T. Dingle, ed), pp. 18–37, North-Holland, Amsterdam.

Aaronson, S., 1973b, Particle aggregation and phagotrophy by *Ochromonas, Arch. Mikrobiol.* 92:39–44.

Aaronson, S., 1974, The biology and ultrastructure of phagotrophy in *Ochromonas danica* (Chrysophyceae: Chrysomonadida), *J. Gen. Microbiol.* 83:21–29.

Aaronson, S., 1980, Descriptive biochemistry and physiology of the Chrysophyceae (with some comparisons to Prymesiophyceae), in: *Biochemistry and Physiology of Protozoa,* Vol. 3, 2nd ed. (M. Levandowsky and S. H. Hutner, eds.), pp. 117–169, Academic Press, New York.

Aaronson, S., and Baker, H., 1959, A comparative biochemical study of two species of *Ochromonas, J. Protozool.* **6**:282–284.

Antia, N. J., 1980, Nutritional physiology and biochemistry of marine Cryptomonads and Chrysomonads, in: *Biochemistry and Physiology of Protozoa,* Vol. 3, 2nd ed. (M. Levandowsky and S. H. Hutner, eds.), pp. 67–115, Academic Press, New York.

Azam, F., Fenchel, T., Field, J. G., Gray, J. S., Meyer-Reil, L. A., and Thingstad, F., 1983, The ecological role of water-column microbes in the sea, *Mar. Ecol. Prog. Ser.* **10**:257–263.

Belcher, J. H., and Swale, E. M. F., 1971, The microanatomy of *Phaeaster pasheri* Scherffel (Chrysophyceae), *Br. Phycol. J.* **6**:157–169.

Biecheler, B., 1936, Des conditions et du mécanisme de la prédation chez un dinoflagellé à enveloppe tabulée, *Peridinium gargantua* n. sp., *C. R. Seances Soc. Biol. Fil.* **121**:1054–1057.

Biecheler, B., 1952, Recherches sur les peridiniens, *Bull. Biol. Fr. Belg.* **36**:1–149.

Bird, D. F., and Kalff, J., 1986, Bacterial grazing by planktonic lake algae, *Science* **231**:493–495.

Bird, D. F., and Kalff, J., 1987, Algal phagotrophy: Regulating factors and importance relative to photosynthesis in *Dinobryon* (Chrysophyceae), *Limnol. Oceanogr.* **32**:277–284.

Bold, H. C., and Wynne, M. J., 1985, *Introduction to the Algae,* 2nd ed., Prentice-Hall, Englewood Cliffs, New Jersey.

Cole, G. T., and Wynne, M. J., 1974, Endocytosis of *Microcystis aeruginosa* by *Ochromonas danica, J. Phycol.* **10**:397–410.

Conrad, W., 1926, Recherches sur les flagellates de nos eaux saumâtres. 2: Chrysomonadines, *Arch. Protistenkd.* **56**:167–231.

Corliss, J. O., 1983, Consequences of creating new kingdoms of organisms, *BioScience* **33**:314–318.

Coveney, M. F., 1982, Bacterial uptake of photosynthetic carbon from freshwater phytoplankton, *Oikos* **38**:8–20.

Cynar, F. J., and Sieburth, J. McN., 1986, Unambiguous detection and improved quantification of phagotrophy in apochlorotic nanoflagellates using fluorescent microspheres and concomitant phase contrast and epifluorescent microscopy, *Mar. Ecol. Prog. Ser.* **32**:61–70.

Daley, R. J., Morris, G. P., and Brown, S. R., 1973, Phagotrophic ingestion of a blue-green alga by *Ochromonas, J. Protozool.* **20**:58–61.

Doddema, H., and van der Veer, J., 1983, *Ochromonas monicis* sp. nov., a particle feeder with bacterial endosymbionts, *Cryptogamie Algologie* **4**:89–97.

Dodge, J. D., and Crawford, R. M., 1970, The morphology and fine structure of *Ceratium hirundinella* (Dinophyceae), *J. Phycol.* **6**:137–149.

Dubowsky, N., 1974, Selectivity of ingestion and digestion in the chrysomonad flagellate *Ochromonas malhamensis, J. Protozool.* **21**:295–298.

Estep, K. W., Davis, P. G., Keller, M. D., and Sieburth, J. McN., 1986, How important are oceanic algal nanoflagellates in bacterivory? *Limnol. Oceanogr.* **31**:646–650.

Fenchel, T., 1980, Suspension feeding in ciliated protozoa: Functional response and particle size selection, *Microb. Ecol.* **6**:1–11.

Fenchel, T., 1982a, Ecology of heterotrophic microflagellates. I. Some important forms and their functional morphology, *Mar. Ecol. Prog. Ser.* **8**:211–223.

Fenchel, T., 1982b, Ecology of heterotrophic microflagellates. II. Bioenergetics and growth, *Mar. Ecol. Prog. Ser.* **8**:225–231.

Fenchel, T., 1982c, Ecology of heterotrophic microflagellates. IV. Quantitative occurrence and importance as bacterial consumers, *Mar. Ecol. Prog. Ser.* **9**:35–42.

Fields, B. S., Shotts, E. B., Jr., Feeley, J. C., Gorman, G. W., and Martin, W. T., 1984, Proliferation of *Legionella pneumophila* as an intracellular parasite of the ciliated protozoam *Tetrahymena pyriformis*, *Appl. Environ. Microbiol.* **47**:467–471.

Gaines, G., and Taylor, F. J. R., 1984, Extracellular digestion in marine dinoflagellates, *J. Plank. Res.* **6**:1057–1061.

Gantt, E., 1980, Photosynthetic cryptophytes, in: *Phytoflagellates* (E. R. Cox, ed.), pp. 381–405, Elsevier/North-Holland, New York.

Gavaudan, P., 1931, Quelques remarques sur *Chlorochromonas polymorpha,* spec. nov., *Botaniste* **23**:277–300.

Geitler, L., 1948, Symbiosen zweischen Chrysomonaden und knospenden bakterienartigen Organismen sowie Beobachtungen über Organisationseigentümlichkeiten der Chrysomonaden, *Öst. Bot. Z.* **95**:300–324.

Güde, H., 1986, Loss processes influencing growth of planktonic bacterial populations in Lake Constance, *J. Plank. Res.* **8**:795–810.

Harris, T. M., 1940, A contribution to the knowledge of the British freshwater Dinoflagellata, *Proc. Linn. Soc.* **152**:4–33.

Hibberd, D. J., 1970, Observations on the cytology and ultrastructure of *Ochromonas tuberculatus* sp. nov. (Chrysophyceae), with special reference to the discobolocysts, *Br. Phycol J.* **5**:119–143.

Hibberd, D. J., 1971, Observations on the cytology and ultrastructure of *Chrysamoeba radians* Klebs (Chrysophyceae), *Br. Phycol J.* **6**:207–223.

Hofeneder, H., 1930, Über die animalische Ernährung von *Ceratium hirundinella* O.F. Muller und über die Rolle des kernes bei dieser Zellfunktion, *Arch. Protistenkd.* **71**:1–32.

Hollibaugh, J. T., Fuhrman, J. A., and Azam, F., 1980, Radioactively labelling of natural assemblages of bacterioplankton for use in trophic studies, *Limnol. Oceanogr.* **25**:172–181.

Hutner, S. H., and Provasoli, L., 1951, The phytoflagellates, in: *Biochemistry and Physiology of Protozoa*, Vol. 1 (A. Lwoff, ed.), pp. 27–128, Academic Press, New York.

Hutner, S. H., Provasoli, L., and Filfus, J., 1953. Nutrition of some phagotrophic freshwater chrysomonads, *Ann. N. Y. Acad Sci.* **56**:852–862.

Kimura, B., and Ishida, Y., 1985, Photophagotrophy in *Uroglena americana,* Chrysophyceae, *Jpn. J. Limnol.* **46**:315–318.

Klaveness, D., 1982, The *Cryptomonas–Caulobacter* consortium: Facultative ectocommensalism with possible taxonomic consequences?, *Nord. J. Bot.* **2**:183–188.

Kochert, G., and Olson, L. W., 1970, Endosymbiotic bacteria in *Volvox carteri, Trans. Am. Microscop. Soc.* **89**:475–478.

Kofoid, C. A., and Swezy, O., 1921, The free-living unarmored dinoflagellata, *Mem. Univ. Calif.* **5**:1–562.

Korshikov, A. A., 1928, Studies on the chrysomonads. I. *Arch. Protistenkd.* **67**:253–290.

Kuhl, A., 1974, Phosphorus, in: *Algal Physiology and Biochemistry* (W. D. P. Stewart, ed.), pp. 636–654, University of California Press, Berkeley.

Larsson, U., and Hagström, A., 1979, Phytoplankton exudate release as an energy source for the growth of pelagic bacteria, *Mar. Biol.* **52**:199–206.

Lauterborn, V. R., 1906, Eine neue Chrysomonadinen-Gattung (*Palatinella cyrptophora* nov. gen. nov. spec.), *Zool. Anz.* **30**:423–428.

Laval-Peuto, M., and Febvre, M., 1986, On plastid symbiosis in *Tontonia appendiculariformis* (Cilophora, Oligotrichina), *BioSystems* **19**:137–158.

Lee, R. E., 1980, *Phycology,* Cambridge University Press, Cambridge.

Leedale, G. F., 1967, *Euglenoid Flagellates,* Prentice-Hall, Englewood Cliffs, New Jersey.

Leedale, G. F., 1969, Observations on endonuclear bacteria in euglenid flagellates, *Ost. Bot. Z.* **116**:279–294.

Leedale, G. F., and Hibberd, D. J., 1985, Class 1. Phytomastigophorea Calkins, 1909, in: *Illustrated Guide to the Protozoa* (J. J. Lee, S. H. Hutner, and E. C. Bovee, eds.), pp. 18–105, Society of Protozoologists, Lawrence, Kansas.

Lessard, E. J., and Swift, E., 1985, Species-specific grazing rates of heterotrophic dinoflagellates in oceanic waters, measured with a dual-label radioisotope technique, *Mar. Biol.* **87**:289–296.

Loeblich, A. R., III, 1967, Aspects of the physiology and biochemistry of Pyrrhophyta, *Phykos* **5**:216–235.

MacKinnon, D. L., and Hawes, R. S. J., 1961, *Introduction to the Study of Protozoa*, Oxford University Press, Oxford.

Manton, I., 1972, Observations on the biology and micro-anatomy of *Chrysochromulina megacylindra* Leadbeater, *Br. Phycol. J.* **7**:235–248.

Margulis, L., 1981, *Symbiosis in Cell Evolution*, Freeman, San Francisco.

Margulis, L., and Schwartz, K. V., 1982, *Five Kingdoms*, Freeman, San Francisco.

McManus, G. B., and Fuhrman, J. A., 1986, Photosynthetic pigments in the ciliate *Laboea strobila* from Long Island Sound, USA, *J. Plank. Res.* **8**:317–327.

Meyer, D. H., and S. Aaronson, 1973, Evidence for secretion by *Ochromonas danica* of an acid hydrolase into its environment, *J. Phycol.* **9**(Suppl.):20.

Müeller, M., Röhlich, P., and Törö, I., 1965, Studies on the feeding and digestion of protozoa. VII. Ingestion of polystyrene latex particles and its early effect on acid phosphatase in *Paramecium multinucleatum* and *Tetrahymena pyriformis*, *J. Protozool.* **12**:27–34.

Myers, J., and Graham, J., 1956, The role of photosynthesis in the physiology of *Ochromonas*, *J. Cell. Comp. Physiol.* **47**:397–414.

Norris, D. R., 1969, Possible phagotrophic feeding in *Ceratium lunula* Schimper, *Limnol. Oceanogr.* **14**:448–449.

Parke, M., and Adams, I., 1960, The motile (*Crystallolithus hyalinus* Gaardner and Markali) and the non-motile phases in the life history of *Coccolithus pelagicus* (Wallich) Schiller, *J. Mar. Biol. Assoc. U. K.* **39**:263–274.

Parke, M., Manton, I., and Clarke, B., 1955, Studies on marine flagellates. II. Three new species of *Chrysochromulina*, *J. Mar. Biol. Assoc.* **34**:579–609.

Parke, M., Manton, I., and Clarke, B., 1956, Studies on marine flagellates. III. Three further species of *Chrysochromulina*, *J. Mar. Biol. Assoc.* **35**:387–414.

Parke, M., Manton, I., and Clarke, B., 1958, Studies on marine flagellates. IV. Morphology and microanatomy of a new species of *Chrysochromulina*, *J. Mar. Biol. Assoc.* **37**:209–228.

Parke, M., Manton, I., and Clarke, B., 1959, Studies on marine flagellates. V. Morphology and microanatomy of *Chrysochromulina strobilus* sp. nov., *J. Mar. Biol. Assoc.* **38**:169–188.

Pascher, A., 1911, *Cyrtophora*, eine neue tentakeltragende Chrysomonade aus Franzensbad und ihre Verwandten, *Ber. Deutsch. Bot. Ges.* **29**:112–125.

Pascher, A., 1943, Zur Kenntnis verschiedener Ausbildungen der planktontischen Dinobryen, *Int. Rev. Gesamten. Hydrobiol.* **43**:110–123.

Pienaar, R. N., and Norris, R. E. 1979, The ultrastructure of the flagellate *Chrysochromulina spinifera* (Fournier) comb. nov. (Prymnesiophyceae) with special reference to scale production, *Phycologia* **18**:99–108.

Pitelka, D. R., 1963, *Electron-Microscopic Structure of Protozoa*, Pergamon Press, Oxford.

Pomeroy, L. R., 1974, The ocean's food web, a changing paradigm, *Bioscience* **24**:499–504.

Porter, K. G., 1973, Selective grazing and differential digestion of algae by zooplankton, *Nature,* **244**:179–180.

Porter, K. G., 1987, Phagotrophic phytoflagellates in microbial food webs, *Hydrobiologia,* in press. (Special volume, *The Role of Microorganisms in Aquatic Food Webs* [T. Berman, ed.])

Porter, K. G., Sherr, E. B., Sherr, B. F., Pace, M., and Sanders, R. W., 1985, Protozoa in planktonic food webs, *J. Protozool.* **32**:409–415.

Pratt, J. R. and Cairns, J., Jr., 1985, Functional groups in the protozoa: Roles in differing ecosystems, *J. Protozool.* **32**:415–423.

Pringsheim, E. G., 1952, On the nutrition of *Ochromonas, Q. J. Microscop. Sci.* **93**:71–96.

Pringsheim, E. G., 1963, *Farblose Algen,* G. Fischer, Jena.

Provasoli, L., 1958, Nutrition and ecology of protozoa and algae, *Annu. Rev. Microbiol.* **12**:279–308.

Rapport, D. J., Berger, J., and Reid, D. W. B., 1972, Determination of food preference of *Stentor coeruleus, Biol. Bull.* **142**:103–109.

Ricketts, T. R., 1971, Endocytosis in *Tetrahymena pyriformis, Exp. Cell. Res.* **66**:49–58.

Sanders, R. W., and Porter, K. G., 1986, Use of metabolic inhibitors to estimate protozooplankton grazing and bacterial production in a monomictic lake with an anaerobic hypolimnion, *Appl. Environ. Microbiol.,* **52**:101–107.

Sanders, R. W., Porter, K. G., and McDonough, R. J., 1985, Bacterivory by ciliates, microflagellates and mixotrophic algae: Factors influencing particle ingestion, *Eos* **66**:1314.

Schiller, J., 1952, Neue Mikrophyton aus dem Neusiedler See und benachbarter gebiete, *Öst. Bot. Z.* **99**:100–117.

Schuster, F. L., Hershenov, B., and Aaronson, S., 1968, Ultrastructural observations on aging of stationary cultures and feeding in *Ochromonas, J. Protozool.* **15**:335–346.

Sherr, B. F., Sherr, E. B., and Berman, T., 1983, Grazing, growth, and ammonium excretion rates of a heterotrophic microflagellate fed with four species of bacteria, *Appl. Environ. Microbiol.* **45**:1196–1201.

Sieburth, J. McN., and Davis, P. G., 1982, The role of heterotrophic nanoplankton in the grazing and nuturing of planktonic bacteria in the Sargasso and Caribbean Sea, *Ann. Inst. Oceanogr.* **58** (Suppl.):285–296.

Skuja, H., 1948, Taxonomie des Phytoplanktons einiger Seen in Uppland, Schweden, *Symb. Bot. Upsal.* **9**:1–399.

Spero, H. J., and Morée, M. D., 1981, Phagotrophic feeding and its importance to the life cycle of the holozoic dinoflagellate *Gymnodinium fungiforme, J. Phycol.* **17**:43–51.

Stoecker, D., Guillard, R. R. L., and Kavee, R. M., 1981, Selective predation by *Favella ehrenbergii* (Tintinnia) on and among dinoflagellates, *Biol. Bull.* **160**:136–145.

Stoltze, H. J., Lui, N. S. T., Anderson, O. R., and Roels, O. A., 1969, The influence of the mode of nutrition on the digestive system of *Ochromonas malhamensis, J. Cell. Biol.* **43**:90–104.

Swale, E. M. F., 1969, A study of the nannoplankton flagellate *Pedinella hexacostata* Vȳsotskiï by light and electron microscopy, *Br. Phycol. J.* **4**:65–86.

Taylor, F. J. R., 1982, Symbioses in marine microplankton, *Ann. Inst. Oceanogr. Paris* **58**(S):61–90.

Taylor, G. T., Iturriaga, R., and Sullivan, C. W., 1985, Interactions of bactivorous grazers and heterotrophic bacteria with dissolved organic matter, *Mar. Ecol. Prog. Ser.* **23**:129–141.

Tsekos, I., 1973, Licht- und electronenmikroskopische Untersuchunger über die Stoffaufnahme durch *Poterioochromonas stipulata, Protoplasma* **77**:397–409.

Uhlig, G., and Sahling, G., 1985, Blooming and red tide phenomenon in *Noctiluca scintillans, Bull. Mar. Sci.* **37**:780.

Vȳsotskiĭ, A. V., 1888 (1887), Mastigophora i Rhizopoda, naigenȳya v 'Veisovom' i, R(ĕ)pnom 'ozerakh', *Tr. Obshch. Ispȳt. Prir. Imp. Khar'kov. Univ.* **21**:119–140.

Wawrik, F., 1970, Mixotrophie bei *Cryptomonas borealis* Skuja, *Arch. Protistenkd.* **112**:312–313.

Wright, R. T., and Coffin, R. B., 1984, Measuring microzooplankton grazing by its impact on bacterial production, *Microb. Ecol.* **10**:137–150.

Wujek, D. E., 1969, Ultrastructure of flagellated chrysophytes. I. *Dinobryon, Cytologia* **34**:71–79.

Wujek, D. W., 1976, Ultrastructure of flagellated chrysophytes. II. *Uroglena* and *Uroglenopsis, Cytologia* **41**:665–670.

Wujek, D. E., 1978, Ultrastructure of flagellated chrysophytes. III. *Mallomonas caudata, Trans. Kans. Acad. Sci.* **81**:327–335.

6

The Microbial Ecology of the Dead Sea

AHARON OREN

> . . . a barren land, bare waste. Vulcanic lake, the dead sea: no
> fish, weedless, sunk deep in the earth. No wind would lift those
> waves. Brimstone they called it raining down: Sodom, Gomor-
> rah, Edom. All dead names. A dead sea in a dead land, grey
> and old.
>
> —James Joyce, Ulysses (1922)

1. Introduction

The last decade has shown a great revival in the study of halophilic
microorganisms. In part this interest has been caused by the discovery of
properties interesting from a theoretical point of view, such as mechanisms of osmotic adjustment, the functioning of enzymes in the presence
of high salt concentrations, and the possession of retinal pigments, such
as bacteriorhodopsin and halorhodopsin in a number of *Halobacterium*
strains, representing simple mechanisms of converting light energy into
biologically available energy (Stoeckenius and Bogomolni, 1982). Moreover, accumulation of valuable products, such as glycerol and (in certain
strains) β-carotene, in the halotolerant unicellular green alga *Dunaliella*
has industrial potential (Ben-Amotz and Avron, 1983).

However, our understanding of the ecology of the hypersaline environments in which these interesting micoorganisms thrive is very lim-

AHARON OREN • Division of Microbial and Molecular Ecology, Institute of Life Sciences, Hebrew University of Jerusalem, Jerusalem 91904, Israel.

ited. To exemplify this, the properties of bacteriorhodopsin-containing purple membrane occurring in certain *Halobacterium* strains have been intensively studied since the discovery of the pigment in 1971 (Oesterhelt and Stoeckenius, 1971), but it took until 1981 to demonstrate that this pigment also occurs in halobacterial communities in nature (Oren and Shilo, 1981). One of the reasons for our limited knowledge of the ecology of hypersaline water bodies is the geographic isolation of most of these biotopes; moreover, growth of extremely halophilic microorganisms is relatively slow, and requires special techniques.

The properties of hypersaline lakes as biotopes for microorganisms are determined to a great extent by the ionic composition of the brines. Thus, alkaline soda lakes in Egypt and Kenya lack measurable concentrations of divalent cations, and the microorganisms thriving in them do not require addition of Ca^{2+} or Mg^{2+} to their growth media (Tindall *et al.*, 1980); the ionic composition of the Great Salt Lake, Utah, resembles that of seawater, and Great Salt Lake microorganisms are well adapted to the chemical properties of their environment (Post, 1977). Dead Sea brines are characterized by extremely high Mg^{2+} and Ca^{2+} concentrations (about 1.8 and 0.4 M, respectively); thus, the organisms living in the Dead Sea should be able to tolerate these concentrations.

The biology of the Dead Sea, one of the most saline biotopes existing, has been reviewed before (Volcani, 1944; Nissenbaum, 1975; Larsen, 1980). These reviews were mainly based on qualitative studies on the nature of the microorganisms isolated from the lake, quantitative data being scarce, not enabling an understanding of the factors determining the community sizes, growth rates, and death rates of the different microorganisms dominating in the lake, and the factors limiting their development.

The recent renewal of interest in the Dead Sea was triggered by the planning of a water carrier connecting the lake with the Mediterranean, which should enable the exploitation of the energy potential of the difference in elevation between the two water bodies of more than 400 m (Steinhorn and Gat, 1983; Weiner, 1985). Thorough studies were initiated, not only of the physics and chemistry of the lake, but also of its biology. This chapter will review these and earlier studies of the biology of the Dead Sea.

2. Physicochemical Properties of the Dead Sea

An in-depth discussion of the geological, physical, and chemical properties of the Dead Sea is outside the scope of this chapter; interested readers can be referred to review papers by Neev and Emery (1966, 1967) and Steinhorn and Gat (1983). An extensive bibliography on the subject was published recently (Arad et al., 1984).

The Dead Sea is the lowest geographic feature at the surface of the earth, 404.5 m below mean sea level* (November 1983). Its origin can be traced in terms of plate tectonics; the Dead Sea is part of the Syrio-African rift valley formed between two plates that have been sliding past each other. At present the lake is about 50 km long and 17 km wide, and its maximum depth is about 320 m.

2.1. Hydrographic Properties: The Physical Structure of the Water Column

The physical and chemical properties of the Dead Sea water column are a result of its geographic situation: the Dead Sea is a terminal lake; thus, its water level is determined by the balance between water inflow (from the Jordan river, springs near the shore of the lake and below the water level, and rainfloods from the catchment area) and evaporation. Evaporation is high because of the high air temperature (averages are in the range 16–34°C, depending on the season). From the beginning of this century the water balance has been negative, especially since the 1950s, when the National Water Carrier was completed, and the outlet of the Sea of Galilee to the Jordan river was closed. The drop in water level (on an average about 0.5 m/year during the past 15 years) changed the shape of the lake: the shallow basin south of the Lisan peninsula became detached from the deep northern basin, and has since ceased to exist; the water present today in the southern part of the lake has been pumped from the northern basin to fill the evaporation ponds of the Dead Sea Works potash plant.

The drop in water level caused a drastic change in the physical structure of the water column. A survey in the years 1959–1960 (Neev and Emery, 1966, 1967) showed the lake to be meromictic, with a stable pycnocline at a depth of 40–80 m. The lower water mass (below a depth of 80–100 m) consisted of "fossil" water that had not been in contact with the surface for approximately 260–280 years, as shown by studies of the distribution of ^{226}Ra in the water column (Stiller and Chung, 1984). The drop in water level and the accompanying increase in salinity of the surface water weakened the pycnocline (Steinhorn et al., 1979) and caused the end of the "permanent" stratification in February 1979 (Carmi et al., 1984; Stiller and Kaufman, 1984). This complete mixing of the water col-

---

*All Dead Sea elevations mentioned in this review refer to the leveling grid utilized by the Dead Sea Works, Ltd. They are 2.46 m lower than the precise level determination of the Survey of Israel; thus, −404.50 m in this chapter equals 402.04 m below mean sea level.

umn caused the disappearance of the sulfide that had been present in the lower water layers (Nissenbaum and Kaplan, 1976), and O_2 penetrated to the bottom (Levy, 1980). Unusually abundant winter floods in the winter of 1979–1980 (a total freshwater inflow of 1.5 × 10^9 m^3) reinduced stratification, which lasted for 3 years (Stiller et al., 1984a). An overturn of the water column in December 1982 was followed by another overturn in December 1984 (D. A. Anati, Solmat Systems, Ltd., unpublished data).

The temperature of the upper water layers varies seasonally from lows of around 18°C to highs of up to 36°C (Neev and Emery, 1967; Oren and Shilo, 1982).

2.2. Salt Concentrations and Ionic Composition

The Dead Sea can be described as a hypersaline chloride lake. The variable hydrographic properties of the lake discussed above caused long-term changes in the salt concentrations, as exemplified in Table I.

The specific gravity of the water (with 340 g/liter dissolved salts) is 1.234 g/cm^3 at 25°C (Stiller et al., 1984a).

During recent years the lake has been saturated with respect to NaCl, and massive precipitation of NaCl has been reported during certain periods (Steinhorn, 1983); another 20 million tons of NaCl are precipitated annually in the Dead Sea Works evaporation pans at the southern end of the lake (Weiner, 1985).

From a biological point of view the concentrations of nutrients are important. Standard nutrient analyses are often unreliable, as the high salt concentrations interfere with the commonly used analytical methods. Therefore, data by different authors, using different analytical techniques, may be difficult to compare, especially when these data were collected

Table I. Average Salt Composition (in g/liter) of the Upper and Lower Water Masses in the Period 1959–1960, and Average Salt Composition of Northern Basin Brines, March 1977[a]

	Total dissolved salts	Na^+	Mg^{2+}	Ca^{2+}	K^+	Cl^-	Br^-	Na^+/Mg^{2+}
1959–1960, upper water mass	299.89	38.51	36.15	16.38	6.50	196.94	4.60	1.07
1959–1960, lower water mass	332.06	39.70	42.43	17.18	7.59	219.25	5.27	0.94
March 1977	339.6	40.1	44.0	17.2	7.65	224.9	5.3	0.91

[a]Data taken from Beyth (1980).

during different periods in a lake whose properties are constantly changing.

Reported phosphate concentrations in the waters of the Dead Sea are very low; Kaplan and Friedmann (1970) found less than 1 μg/liter dissolved phosphate (less than 0.01 μM). More recent estimates of M. Stiller (Weizmann Institute of Science, Rehovot, unpublished data) were around 1 μM. Inorganic nitrogen is present in high concentrations, not to be expected to limit microbial development: ammonium concentrations were between 1.9 and 7.5 mg/liter, nitrate between 16 and 20 μg/liter, and dissolved organic nitrogen between 0.33 and 2.9 mg/liter (Neev and Emery, 1967). Sulfate is present in Dead Sea water at an average concentration of 0.45 g/liter (March 1977) (Beyth, 1980), its concentration being limited by the high Ca^{2+} concentration in the lake. At times sulfate precipitates as gypsum ($CaSO_4 \cdot 2H_2O$) (Neev and Emery, 1967). Iron is found mainly in its trivalent form as particulate iron; its concentration is quite variable (2–22 μM, generally in the range 5–10 μM; March 1977–May 1980) (Nishry and Stiller, 1984). Before the overturn of the lake in February 1979 dissolved Fe^{2+} was present in the lower water mass at a concentration of about 5 μM.

The high calcium concentration of the water limits not only the solubility of sulfate, but also that of inorganic carbon. Sass and Ben-Yaakov (1977) found a concentration of 2.5–2.6 meq/liter total dissolved inorganic carbon. Dissolved organic carbon concentrations of 4.2–8.3 μg/liter have been reported (Neev and Emery, 1967).

The solubility of O_2 in hypersaline brines is limited; a study of dissolved O_2 concentrations in the Dead Sea in 1980 (Levy, 1980) showed values generally between 0.7 and 0.9 ml/liter, reaching highs of up to 2 ml/liter after floods by dilution, and again high values in August 1980 (1.8 ml/liter), probably as a result of biological activity (see Section 6.1).

Reported concentrations of trace metals such as Cu^{2+} are inconsistent: Nissenbaum (1977) measured a concentration range of 300–500 μg/liter, while Stiller et al. (1984b) found (not in the same period) concentrations as low as 2 μg/liter. High concentrations of other trace metals were reported by Nissenbaum, e.g., 308–330 mg/liter Sr, 3.1–8 mg/liter Mn, 500 μg/liter Zn, and 120–300 μg/liter Pb; Stiller et al. (1984b) also reported low concentrations of Co and Hg (1.3 and 1.2 μg/liter, respectively).

The pH values of Dead Sea water are relatively low (Ben-Yaakov and Sass, 1977; Sass and Ben-Yaakov, 1977). They range from 5.95 (deep waters, before the overturn) to 6.56 (at the surface). The pH values increase with increasing dilution: for example, Dead Sea brine of pH 6.46, diluted 50% with distilled water, yielded a solution of pH 7.42 (Sass and Ben-Yaakov, 1977).

3. The Search for Life in the Dead Sea

The Dead Sea has long been considered as an environment hostile to life, and the opinion that no living organism can thrive in it was widespread until only a few decades ago (Nissenbaum, 1979). Greek and Roman writers of the antique world mentioned the fact that no living fish is found in the lake; therefore, it is remarkable that the lifelessness of the Dead Sea is never stressed in the ancient Jewish literature: the Old Testament refers to the lake as "Yam Ha Melakh," the Salt Sea (e.g., *Genesis* **14:3**), the Sea of the Arabah (e.g., *Deuteronomy* **3:17**, *Joshua* **3:16**) or the Eastern Sea (*Joel* **2:20**), while the Talmud uses designations such as the Salt Sea or the Sea of Sodom. The term Dead Sea *(Mare Mortuum)* was used for the first time by Greek and Roman writers of the second century A.D.

Interest in the chemical composition of the Dead Sea developed in the 18th and 19th centuries. The search for microscopic forms of life started as early as the beginning of the 19th century, when Louis Joseph Gay-Lussac in 1819 examined Dead Sea water samples microscopically; he did not detect microscopic organisms. A similar observation was reported by M. Barrois, a zoologist of Lille University, who, in the mid 1880s, traversed the Dead Sea in search for lower animals. The French microbiologist M. L. Lortet, having heard of Barrois' findings, thought of looking for useful applications for the "sterile" brines by using them as an aseptic liquid. For that purpose he investigated mud samples collected by Barrois and inoculated them into media, prepared with fresh water, designed for the growth of bacteria. To his surprise, he obtained vigorous growth of pathogenic anaerobic bacteria of the genus *Clostridium,* which are the etiological agents of tetanus and gas gangrene (Lortet, 1892). These bacteria are unable to multiply in highly saline environments, but they belong to the few bacterial types producing endospores that resist adverse conditions such as extremes of temperature, desiccation, and salinity. These endospores cannot be expected to have originated in the Dead Sea, but may have been brought there by the river Jordan, by flood waters, or from the air; they may have survived immersion in the Dead Sea brines until inoculation into suitable growth media enabled them to resume their development.

At the beginning of the 20th century it was discovered that hypersaline environments, even those saturated with salt, are generally inhabited by a variety of microorganisms tolerating, and even requiring, extremely high salt concentrations. These organisms include red halophilic bacteria [often imparting a deep red color to salterns or to lakes like the Great Salt Lake in Utah (Post, 1977)], cyanobacteria ("blue-green algae"), unicellular green algae, such as *Dunaliella,* protozoa, and even macroscopically visible animals, such as the brine shrimp, *Artemia salina*. In view of these findings, a renewed study of the biology of the Dead Sea was initiated in the 1930s by Benjamin Elazari-Volcani (previously named Wilkansky), and in 1936 the first of a series of reports on the isolation and characterization of microorganisms from the Dead Sea was published (Wilkansky, 1936; Elazari-Volcani, 1940, 1943a,b, 1944; Volcani, 1944). From a number of water samples from the Dead Sea, most of them at a site about 3–4 km from the mouth of the river Jordan, and at depths of up to 7 m, Volcani isolated by means of enrichment cultures a variety of microorganisms: bacteria, cyanobacteria, unicellular green algae, and even protozoa. He also obtained samples of anaerobic sediments from great depth, from which he was able to grow anaerobic bacteria (Elazari-Volcani, 1943a).

Since Volcani's pioneering work, studies on the microbiology of the lake have been few and far between: quantitative measurements were performed in 1963–1964 by Kaplan and Friedmann (1970), a few measurements were made in 1973 by Kritzman (1973), and from 1980 onward systematic sampling and monitoring of the sizes of the microbial communities have been performed by our group (Oren, 1981, 1983b, 1985; Oren and Shilo, 1982).

4. The Microorganisms of the Dead Sea

Studies on the microbiology of the Dead Sea have demonstrated the presence of a variety of microorganisms adapted to the high salt concentrations in the lake: unicellular algae, cyanobacteria, aerobic and anaerobic chemoorganotrophic bacteria, and protozoa. Thus far, no higher organisms have been reported from the Dead Sea. Even the brine shrimp, *Artemia salina,* which is found in many hypersaline environments at salt concentrations of up to saturation, such as the Great Salt Lake, Utah (Post, 1977), has never been found in the Dead Sea. Possible explanations for its absence are the high concentrations of ions of the alkaline earth metals and/or the low pH values, which are much below those of the water bodies in which *Artemia* commonly thrives.

4.1. Phototrophic Microorganisms

All reports on the microbiota of the Dead Sea published since the days of Volcani mention the presence of the unicellular green alga *Dunaliella* (e.g., Elazari-Volcani, 1940; Kaplan and Friedmann, 1970; Kritzman, 1973; Oren, 1981). Its cells are pear-shaped, and possess two polar flagella and a pink eye-spot, and lack a rigid cell wall. Members of the genus *Dunaliella* are found in hypersaline environments all over the world; some of them are red as a result of a high carotenoid content. Red types have never been observed in the Dead Sea itself, but red cells resembling *Dunaliella bardawil* (Ben-Amotz *et al.*, 1982) appeared in a few of our simulation experiments in ponds at the northern end of the lake (Oren and Shilo, 1985) (see Section 6.1). Some confusion exists concerning the species designation of the green strain(s?) occurring in the Dead Sea. Volcani classified his strain as *Dunaliella viridis,* and reported as its dimensions 5–7.3 by 11–15 μm (Elazari-Volcani, 1940; Volcani, 1944). Kaplan and Friedmann (1970) described the dominant type in the lake as "similar to that described by Volcani as *D. viridis,*" but reported as its size 10 by 20 μm. The cells observed by us in the Dead Sea during the *Dunaliella* bloom in 1980 (see Section 6.1) were only 8–10 μm in length. Whether these differences in cell size are significant remains unclear; at least theoretically a possibility exists that during the time, a succession occurred of different types with different properties (see also Section 8). Later taxonomic revisions of the genus *Dunaliella* no longer recognized the name *D. viridis* (Lerche, 1937; Butcher, 1959), and replaced it by *D. parva*. Ben-Amotz and Ginzburg (1969) identified their Dead Sea isolate as such, and in a recent report we did the same (Oren and Shilo, 1982).

Occasionally other algal types have been reported from the Dead Sea. Volcani (1944) described the isolation of a small green flagellate, motile by means of a single flagellum arising from one side of the cell. The strain grew in a wide range of salt concentrations, from 1% to 27%; the isolate has unfortunately been lost. Furthermore, Volcani grew *Chlorella* sp. from sediment samples by enrichment in media with a low salt content (0.5%); as this isolate may not be able to grow at elevated salt concentrations, it may not be indigenous to the Dead Sea.

Often remnants of freshwater organisms originating from the river Jordan or reaching the lake with rain floods from the catchment area can be found in the Dead Sea. Volcani (1944) already reported the presence of diatom frustules in mud samples from the bottom of the lake, and a similar observation was reported by Nissenbaum (1979). Kritzman (1973) photographed and described a number of cyanobacteria, green algae, diatoms, and Xanthophyceae from Dead Sea water. Insofar as his photographs show living organisms and not artefacts, it remains to be determined whether these are able to withstand the high salt concentrations prevailing in the lake; as long as they have not been characterized further, they cannot be considered indigenous to the Dead Sea. The high numbers of "marine" types of *Coccolithus fragilis* and of the dinoflagellate *Exuviella* reported by Bernard (1957) in Dead Sea surface water in 1956, up to 800 and 5500 cells/ml, respectively, (numbers exceeding by far their population density in the Mediterranean), probably should be considered as artefacts.

Several types of cyanobacteria ("blue-green algae") have been observed by Volcani in enrichment cultures from water and sediment samples (Elazari-Volcani, 1940; Volcani, 1944): an obligately halophilic *Aphanocapsa,* growing optimally in media of 9–18% salt, *Phormidium* sp. and *Nostoc* sp., growing from 0 to 12% salt, and *Microcystis* sp. (?).

4.2. Bacteria of the Dead Sea

4.2.1. Aerobic Bacteria

4.2.1a. The Archaebacteria: Genera Halobacterium and Halococcus. From enrichment cultures designed for the isolation of denitrifying bacteria, using medium with 24% salt, Elazari-Volcani (1940) isolated a pink bacterium, described as a nonmotile short rod, measuring 0.5 by 1.3–3 μm. The strain, named *Halobacterium ("Flavobacterium") marismortui,* did not grow at salt concentrations below 15%. Though described as a rod-shaped bacterium, a micrograph of the strain (Elazari-Volcani, 1940) is strongly reminiscent of the irregular pleomorphic cells observed by direct microscopic examination of Dead Sea water samples by Kaplan and Friedmann (1970) (though these were reported to display irregular fast movements), and of the pleomorphic strain described as *Halobacterium volcanii* (Mullakhanbhai and Larsen, 1975). Volcani's strain of *H. marismortui* has been lost, but a new isolate by Ginzburg (Ginzburg *et al.,* 1970; Kirk and Ginzburg, 1972) resembles the original strain in many respects, such as its ability to ferment different sugars and the reduction of nitrate to N_2 (Werber and Mevarech, 1978). No valid species description of *Halobacterium marismortui* has been published.

Mullakhanbhai and Larsen (1975) isolated, from bottom sediment at 1 m depth close to the northern shore of the Dead Sea, a pleomorphic pink *Halobacterium,* differing from *"H. marismortui"* by its inability to evolve gas from nitrate. The strain was designated *Halobacterium volcanii;* it grows optimally at NaCl concentrations of 1.7–2.5 M and 0.2–0.5 M $MgCl_2$. Other similar (but distinctly different) strains have been isolated by us (Oren, 1981).

All the above strains do not possess the regular rod shape of most other representatives of the genus *Halobacterium,* but rather look irregular and flat, and have been described as "cup-shaped" or "disk-shaped" (Mullakhanbhai and Larsen, 1975). Recently, extremely halophilic bacteria have been isolated from salt flats in the Sinai peninsula and in California, displaying "square" or "box-shaped" morphology (Walsby, 1980; Javor *et al.,* 1982). The box-shaped strains have tentatively been called

"Haloarcula" (Javor *et al.,* 1982). *"Halobacterium marismortui"* (the Ginzburg strain) may be a close relative of the "box-shaped" halobacteria and of *H. vallismortis,* isolated from salt pools in Death Valley, California (Gonzalez *et al.,* 1978), as appears from a comparison of the nucleotide sequences in their 5S ribosomal RNA (Nicholson and Fox, 1983).

Kaplan and Friedmann (1970) observed long, rod-shaped halobacteria in the Dead Sea by means of direct microscopic observation. This type was probably isolated by Kritzman (1973), but his isolates have been lost. A rod-shaped type has again been isolated by Oren, using a medium containing starch (Oren and Shilo, 1981; Oren, 1983c). The organism measures 0.5 by 2.5–5 μm, and is motile by means of a tuft of polar flagella. No gas is evolved from nitrate. This strain, designated *Halobacterium sodomense,* grows optimally at 1.7–2.5 M NaCl in addition to 0.8 M $MgCl_2$. Starch is hydrolyzed to glucose during growth (Oren, 1983c,d), but could not be replaced by glucose or other sugars. Starch, however, could be replaced by clay minerals, such as bentonite; thus, starch, in addition to its properties as a substrate, may serve as a scavenger of toxic metabolites or of toxic components of the medium. When incubated at low O_2 tensions and in the light, cultures of *H. sodomense* turn purple due to the formation of purple membrane (Oren and Shilo, 1981; Oren, 1983c). The potential for making purple membrane has never been demonstrated in *"H. marismortui"* or in *H. volcanii,* and seems to be restricted to a few *Halobacterium* strains only. The importance of bacteriorhodopsin for the bacterial community in the Dead Sea will be discussed in Section 6.2.1.

Further aerobic obligately halophilic isolates from the Dead Sea include *"Halobacterium trapanicum"* (now *species incertae sedis*) isolated by Volcani (1944), a nonmotile rod measuring 0.45–0.55 by 1.5–15 μm, able to reduce nitrate to gas, and distinguishable from *"H. marismortui"* by its inability to produce acid from carbohydrates.

None of the halobacteria reported from the Dead Sea possess gas vacuoles; strains like *H. halobium* and *H. salinarium* often contain refractile gas vacuoles, easily seen microscopically. It must be kept in mind, however, that in part of the studies microscopic examination of Dead Sea water samples was preceded by high-speed centrifugation, a treatment collapsing gas vacuoles.

From enrichment cultures for nitrate-reducers, Volcani (1944) isolated spherical bacteria described as *Micrococcus (Halococcus?) morrhuae,* which did not grow in salt concentrations below 6%, reduced nitrate to nitrite, and did not produce acid from carbohydrates.

4.2.1b. Halophilic Aerobic Eubacteria. A number of aerobic, chemoorganotrophic halotolerant bacteria were isolated from the Dead Sea by Volcani (1944). They are *"Flavobacterium halmephilum,"* nonmotile short rods, 0.5–0.6 by 1.5–2 μm, forming yellowish colonies, and *"Pseudomonas halestorgus,"* pleomorphic rods, 0.5 by 1.3–3.4 μm, motile by means of a single polar flagellum, and forming characteristic colonies, which are colored brown in the center and surrounded by blue, brown-gray, blue-gray, and finally yellowish zones. The two last species names do not appear on the presently recognized list of approved bacterial names. However, at least some of the original isolates have been preserved, and are awaiting a taxonomic reevaluation. Additional strains have been isolated by later workers, such as strain Bal isolated by Rafaeli-Eshkol (1968), and other strains have been isolated by us, growing well in a medium containing 2.1 M NaCl + 0.25 M MgCl$_2$, and differing with respect to morphology, motility, and pigmentation (A. Oren, unpublished data).

Several studies on the microbiology of the Dead Sea included enrichment cultures for different physiological groups of bacteria, using media with different carbon and energy sources. In many of these cultures bacterial growth was observed, but in most cases the bacteria were not isolated and characterized further. Positive enrichment cultures were reported of cellulose-decomposing bacteria, protein-decomposers, denitrifiers, sulfur- and thiosulfate-oxidizers, hydrocarbon-oxidizers, and organisms degrading starch to glucose (Volcani, 1944; Kaplan and Friedmann, 1970; Kritzman, 1973). It is possible that the number of different bacteria appearing in these cultures was actually smaller than the number of different incubation conditions used, as a single organism versatile with respect to its nutritional demands may have developed in more than one of the cultures.

4.2.2. Anaerobic Bacteria

An anaerobic, chemoheterotrophic, very long, thin, and flexible bacterium, measuring about 0.35 by 10–20 μm and longer, and evolving a large amount of gas was isolated from an enrichment culture designed for the isolation of sulfate-reducing bacteria. The enrichment medium consisted of 80% Dead Sea water and 20% distilled water, enriched with pyruvate, sulfate, yeast extract, ascorbate, and thioglycollate, and was inoculated with mud collected from 60 m depth. The strain has been isolated in pure culture, and grows well with glucose and yeast extract as carbon and energy sources. The organism has been named *Halobacteroides halobius* (Oren *et al.*, 1984b). The strain is obligately anaerobic, and does not reduce sulfate. Its growth is very fast, with generation times as short as 1 hr under optimal conditions (37–42°C, and in the presence of 1.4–2.8 M NaCl). Below 1 M NaCl no growth occurred and the bacteria lysed. They tolerate fairly high magnesium concentrations: excellent growth is possible at magnesium concentrations as high as 1 M, and fair growth is observed at 1.5 M (in the presence of 1.5 M NaCl). The bacterium is able to ferment a variety of carbohydrates; fermentation products from glucose are ethanol, acetate, H_2, and CO_2. *Halobacteroides halobius*-like cells are present in Dead Sea bottom sediments and in anaerobic sediments at the shore of the lake in high numbers: values of 10^3–10^5 viable cells per g of sediment have been reported (Oren *et al.*, 1984b).

From an enrichment culture set up for the isolation of sulfate-reducing bacteria as above, but using lactate instead of pyruvate, we grew an anaerobic, endospore-forming bacterium, which also does not reduce sulfate. It was therefore classified in the genus *Clostridium*, and described as *Clostridium lortetii* (Oren, 1983e)—named for M. L. Lortet, who had isolated clostridia from the lake as early as 1892 (Lortet, 1892). It grows in a medium containing 1–2 M NaCl and other salts, enriched with glutamate, casamino acids, and yeast extract. Among the fermentation products are acetate, propionate, and butyrate. Glucose is used, but only in older cultures, after other, more suitable substrates are depleted. In aging cells gas vacuoles appear concomitantly with the formation of the endospore, and these remain attached to the spore after degeneration of the vegetative cell. Gas vacuole formation occurs very rarely in the genus *Clostridium;* a group of Russian scientists (Krasil'nikov *et al.*, 1971) isolated a number of gas-vacuole-containing, nonhalophilic *Clostridium* strains from soil, some of which morphologically resemble our isolate. Also, the sulfate-reducing spore-former *Desulfotomaculum acetoxidans* produces gas vacuoles (Widdel and Pfennig, 1981).

Analysis of 16S ribosomal RNA oligonucleotide sequences showed no relationship of *Clostridium lortetii* with the other members of the genus *Clostridium;* moreover, *C. lortetii* has a Gram-negative-type cell envelope, whereas the true clostridia are typically Gram-positive. Thus, this strain may better be reclassified in another, yet to be created genus. Analysis of 16S ribosomal RNA also showed that *Halobacteroides halobius,* though obviously belonging to the eubacterial kingdom, does not belong to any of the recognized eubacterial families, and therefore a new family was created, the Haloanaerobiaceae, to classify this organism (Oren *et al.*, 1984a); the anaerobic Dead Sea bacterium *Clostridium lortetii* also belongs to this family (Oren, 1986b).

Halobacteroides halobius was found to contain high intracellular Na^+ and K^+ concentrations, rather than organic osmotic solutes; moreover, its proteins contain an excess of acidic amino acids. Thus, its mode of osmotic adjustment resembles that of the halophilic archaebacteria, rather than that of the aerobic halotolerant eubacteria (Oren, 1986a).

Bacteria fermenting glucose, bacteria fermenting lactose, and denitrifying bacteria were grown by Elazari-Volcani (1940, 1943a; Volcani, 1944) in anaerobic enrichment cultures containing 25% salt and inoculated with mud from the bottom sediments of the Dead Sea, but these bacteria have not been characterized further.

The existence of bacterial sulfate reduction in the bottom sediments of the lake is well established (see Section 6.3), but the nature of the organism(s) responsible for the process is still unknown. Enrichment cultures for sulfate-reducing bacteria by Kaplan and Friedmann (1970) yielded *Vibrio*-like organisms, small, rod-shaped bacteria, or spore-forming rods, resembling *Desulfotomaculum ("Clostridium") nigrificans,* and not growing at NaCl concentrations above 5%. Likewise, Nissenbaum (1975) reported the existence of sulfate-reducers similar to *Desulfotomaculum nigrificans* in Dead Sea sediments. Kritzman (1973) observed small, rod-shaped, motile bacteria and spheroids in his enrichments for sulfate-reducers. None of the above types exist in culture.

4.3. Protozoa

Elazari-Volcani (1943b, 1944) isolated two halophilic protozoa from the Dead Sea. An ameba was isolated from mud samples from a depth of 330 m, which developed in enrichment cultures with 9–18% salt, in association with flagellates and cyanobacteria. The organism required at least 6% salt, grew optimally between 15 and 18%, and even tolerated 33% salt. Under certain conditions the organism developed flagella, and it was identified as a dimastigameba. The cells are 7–54 μm in width and 17–80 μm in length; they contain many vacuoles, are fairly motile, also in the nonflagellate stage, and are able to form cysts.

From a pebble of flint collected at a depth of 1.5 m, Elazari-Volcani grew a ciliate, developing in an enrichment culture in which the organism appeared together with *Dunaliella,* a cyanobacterium, and an ameba, after an incubation period of more than 3 years. The ciliate was ovally shaped, measuring 8–13 by 16–24 μm, showed four longitudinal ridges and two posterior spines, and was reported to grow well at salt concentrations between 15 and 21%. No cultures of the two protozoa have been preserved.

Later studies have not revealed the presence of protozoa in the Dead Sea, but amebae abound in a hypersaline sulfur spring on the western shore of the lake near Ein Gedi (A. Oren, unpublished results).

5. Adaptations of Dead Sea Microorganisms to Their Environment

Microorganisms living in hypersaline environments are exposed to concentrated salt solutions exerting extremely high osmotic pressures. In order to withstand this physical force, the organisms have evolved mechanisms enabling them to maintain their intracellular space at osmotic values of the same order of magnitude as the surrounding medium. The details of the mechanisms vary in the different groups of organisms:

halophilic archaebacteria (genera *Halobacterium* and *Halococcus*) accumulate high concentrations of KCl inside the cells, and their enzymes are active in the presence of high salt concentrations; halophilic aerobic eubacteria regulate their intracellular osmotic pressure by the accumulation of betaine (Imhoff and Rodriguez-Valera, 1984), the halophilic green flagellate *Dunaliella* accumulates glycerol (Ben-Amotz and Avron, 1983), and at least five different organic osmotic solutes have been identified in halophilic and halotolerant cyanobacteria (Mackay *et al.,* 1984). An in-depth discussion of the different strategies of osmotic adjustment is outside the scope of this chapter; it suffices to refer to the recent reviews on the subject (e.g., Kushner, 1978). Dead Sea microorganisms do not differ greatly from their counterparts from other hypersaline biotopes with respect to the general mechanisms used to osmotically equilibrate the cell contents with the external medium; earlier studies on the adaptation of Dead Sea microorganisms to high salinities have been reviewed by Nissenbaum (1975).

However, to be able to thrive in the Dead Sea, microorganisms must withstand and tolerate, in addition to the high salinity of their environment and the extremely high osmotic pressure it exerts on the cells, the specific composition of its waters. In contrast to better known environments in which halobacteria develop massive blooms, such as salterns (Javor, 1984) and the Great Salt Lake, Utah (Post, 1977), and in which sodium chloride is the dominant salt, the Dead Sea contains extremely high concentrations of ions of the alkaline earth metals magnesium and calcium, which together account for more than 50% of the cation sum of the water (Beyth, 1980), and such an ionic composition does not allow growth of most strains of halophilic organisms isolated from biotopes in which monovalent cations dominate. A discussion of the distinguishing properties of Dead Sea microorganisms, relating to their adaptations to the specific conditions in the Dead Sea, is necessarily limited to the *Halobacterium* group, since very few data are available on the relationships of the aerobic and the anaerobic, moderately halophilic bacteria and of the algae from the Dead Sea toward divalent cations. The different halobacterial types thriving in the lake not only display a high tolerance toward the high divalent cation concentrations, but also require these ions in much higher concentrations than do other bacteria, including halobacteria from other biotopes.

5.1. Tolerance toward High Concentrations of Divalent Cations

Representatives of the genus *Halobacterium* can be found in hypersaline environments of greatly differing ionic composition, each strain being adapted relatively well to the biotope in which it is found. Thus, alkaliphilic halobacteria have been isolated from African soda lakes, such as Lake Magadi in Kenya and Wadi Natrun in Egypt, and these grow with only trace amounts of divalent cations such as are present in their natural habitat, as high pH values and abundance of divalent cations are mutually exclusive (Tindall *et al.,* 1980, 1984; Soliman and Trüper, 1982); moreover, magnesium concentrations as low as 10 mM exert an inhibitory effect (Tindall *et al.,* 1980). The commonly studied strains *H. halobium* and *H. salinarium,* isolated from such environments as salterns or dried salted fish, tolerate and also require somewhat higher divalent cation concentrations. Dead Sea strains, as exemplified by *H. volcanii* (Mullakhanbhai and Larsen, 1975), are especially well adapted to the salt composition of the Dead Sea waters. Their NaCl requirement is relatively low: instead of at 4–5 M, the concentration generally reported to be optimal for the halobacteria, *H. volcanii* grows optimally at 1.7 M NaCl (at 30°C, 2.5 M at 40°C), concentrations that are of the same order of magnitude as those actually existing in the Dead Sea (Table I). On the other hand, the strain is extremely tolerant toward high magnesium concentrations, and at a Mg^{2+} concentration as high as 1.4–1.5 M growth is still possible at half of the maximal rate (Mullakhanbhai and Larsen, 1975). Other isolates resembling *H. volcanii* (Oren, 1981) equally show a high magnesium tolerance. The rod-shaped *H. sodomense,* isolated from the Dead Sea, displays the highest reported requirement for and tolerance toward divalent cations: it grows optimally at 0.6–1.2 M $MgCl_2$ in the presence of 2.1 M NaCl; lowering the magnesium concentration causes a reduction in growth rate and a change in cell morphology from rods to irregular spheres, even at elevated monovalent cation concentrations. Calcium ions can at least partially replace magnesium. At lowered NaCl concentrations (0.4–0.9 M), fair growth was obtained even at $MgCl_2$ concentrations above 2 M (Oren, 1983c).

The low requirement for NaCl of the Dead Sea halobacteria precludes their classification as extremely halophilic organisms when using the commonly accepted definitions, according to which an extremely halophilic organism grows best in 2.5–5.2 M salt (Kushner, 1978). Thus, Mullakhanbhai and Larsen (1975) correctly designated their *H. volcanii* isolate as a moderate halophile. Likewise, *H. sodomense* grows at NaCl concentrations as low as 0.5 M, and maximal growth rates are found at 1.7–2.5 M NaCl (Oren, 1983c).

A method to compare different microorganisms with respect to their requirement for and tolerance toward monovalent and divalent cations has been proposed by Edgerton and Brimblecombe (1981), who plotted the salt concentrations enabling growth in the "environmental space" of the different possible combinations of monovalent and divalent cations. In this type of plot, an example of which is found in Fig. 1, the X axis represents the mole fraction of monovalent cations, the Y axis the total ionic strength of the medium. Compared in the figure are the combina-

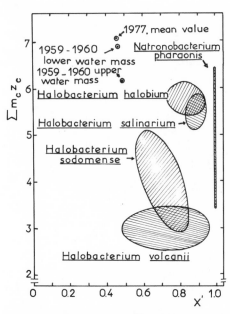

Figure 1. Cation composition of media enabling optimal growth of five *Halobacterium* strains: the pleomorphic Dead Sea strain *H. volcanii*, the rod-shaped Dead Sea strain *H. sodomense*, *H. halobium*, *H. salinarium*, and the alkaliphilic strain *Halobacterium (Natronobacterium) pharaonis* SP-1. The ionic composition of growth media enabling optimal growth was plotted into the "environmental space" according to Edgerton and Brimblecombe (1981). The X axis represents the mole fraction of monovalent cations $(m\mathrm{Na}^+ + m\mathrm{K}^+)/(m\mathrm{Na}^+ + m\mathrm{K}^+ + m\mathrm{Mg}^{2+} + m\mathrm{Ca}^{2+})$; the Y axis represents the total charge concentration $(m\mathrm{Na}^+ + m\mathrm{K}^+ + 2m\mathrm{Mg}^{2+} + 2m\mathrm{Ca}^{2+})$ $(m$ = concentration in molal). Data were derived from Oren (1983c), Tindall *et al.* (1980), and Edgerton and Brimblecombe (1981). For comparison the composition of Dead Sea brines (upper water mass, 1959–1960; lower water mass, 1959–1960; and the average composition in 1977; all derived from Table I) are plotted in the figure. [Modified from Cohen *et al.*, 1983 and Oren, 1983c. By permission.]

tions of monovalent and divalent cation concentrations optimal for growth of two Dead Sea halobacteria (the pleomorphic *H. volcanii* and the rod-shaped *H. sodomense*) with halobacterial types from other biotopes: *H. halobium, H. salinarium,* and the alkaliphilic strain *Halobacterium (Natronobacterium) pharaonis* isolated from soda lakes (Tindall *et al.,* 1980, 1984; Soliman and Trüper, 1982). In the same figure data on the composition of the Dead Sea brines have been plotted. It is shown that the Dead Sea strains, though well adapted to the ratio of monovalent and divalent cations in the Dead Sea water, grow optimally at a total ionic strength much below that found in the Dead Sea. This observation has important implications for the dynamics of the halobacterial communities in the lake, as will be discussed in Section 6.2. Different isolates of the *H. volcanii*-type strains differ in their tolerance toward extremely high magnesium concentrations (Oren, 1981). The importance of this finding will be discussed in Section 8.

5.2. Requirement for High Concentrations of Divalent Cations

Dead Sea halobacteria display their high degree of adaptation toward life at high divalent cation concentrations not only by their tolerance to extremely high Mg^{2+} concentrations, but also by their specific requirement for divalent cations in relatively high concentrations; thus, *Halobacterium volcanii* requires 0.2 M Mg^{2+} for optimal growth (Mullakhanbhai and Larsen, 1975; Cohen *et al.*, 1983). *Halobacterium sodomense* even requires magnesium concentrations as high as 0.6–1.2 M for optimal growth (Oren, 1983c). At suboptimal concentrations (0.05–0.075 M Mg^{2+}), *H. volcanii* cells lose their native morphology and become spherical, but are still able to multiply. Pleomorphic Dead Sea halobacteria such as *H. volcanii* are unique in the respect that they cannot even survive temporary withdrawal of divalent cations from their medium. Even in the presence of saturating concentrations of sodium chloride in media lacking divalent cations, cells lose their native morphology and become spherical within seconds, and within a few minutes they lose their viability (Cohen *et al.*, 1983).

In the above discussion on the effects of divalent cations on halophilic bacteria from the Dead Sea we referred to their concentrations in the external medium only; nothing is known about their internal Mg^{2+} and Ca^{2+} concentrations. We are also unaware of the existence of published measurements of internal divalent cation concentrations of halobacterial strains isolated from other habitats. A study on the moderately halophilic *Vibrio costicola* (not a Dead Sea isolate) reported a higher Mg^{2+} concentration inside the cell than in the medium (20–40 and 0.41 mM, respectively) (Shindler *et al.*, 1977). This concentration gradient of magnesium ions was thought to be due not to active accumulation, but rather to its binding to macromolecules such as nucleic acids and ribosomes within the cell, and to cell envelopes. It is unknown whether in media containing extremely high concentrations of divalent cations, such as the Dead Sea brines, these cations do enter the cells freely or are efficiently excluded. In any case, biochemical studies on a number of enzymes isolated from "*H. marismortui*" (see, e.g., Werber and Mevarech, 1978; Pundak and Eisenberg, 1981) did not reveal an exceptionally high magnesium requirement of the enzymes.

6. Life in the Dead Sea: Quantitative Aspects

Not all organisms described in Section 4 contribute significantly to the biomass of the Dead Sea; some of the organisms isolated from the lake proved unable to tolerate high salinities, and thus cannot be expected to reach high densities in the lake. Most of the strains have been isolated from enrichment cultures, and no quantitative conclusions may be drawn from the development of organisms in such cultures, as the presence of even one viable cell able to grow in the enrichment medium under the conditions used can give rise to massive growth in the laboratory.

Comparison of quantitative data on the microbiota of the Dead Sea is complicated by the fact that quantitative relations may differ from site to site in the lake. Thus, near the mouth of the Jordan River where Volcani collected most of his samples, moderately halophilic and even non-halophilic organisms may be more important quantitatively than in the center of the lake. Furthermore, the evolution of the sea itself, with its rapid changes in chemical and physical properties, makes it difficult to compare quantitative data collected in different periods.

In the few quantitative studies on the microbiology of the water column three types of microorganisms together represented more than 99% of the total biomass; these are the green flagellate alga *Dunaliella* and two types of halobacteria: the pleomorphic *H. volcanii*-like cells (80–90% of the total halobacterial community in surface water) and the rod-shaped halobacteria (Kaplan and Friedmann, 1970; Oren, 1981, 1983b; Oren and Shilo, 1982, 1985). In the next two sections we will present data on the sizes of the communities of these organisms and on the factors determining their growth and death rates in the Dead Sea.

6.1. Population Dynamics of *Dunaliella* in the Dead Sea

Though most studies on the biology of the Dead Sea mention the occurrence of the unicellular green alga *Dunaliella,* quantitative data on its distribution in the lake before 1980 are scarce. The first enumerations of algae were made by Kaplan and Friedmann (1970), who reported in 1964 (sampling date and site not specified) a density of 4×10^4 *Dunaliella* cells/ml in surface water; at a depth of 50 m the numbers were reduced by two orders of magnitude, and at a depth of 100 m no algae were found. S. Garlick (unpublished results) counted 140 *Dunaliella* cells/ml of surface water in the summer of 1978. The above numbers are quite high, but are surpassed by data from other hypersaline environments such as the Great Salt Lake, where *Dunaliella* densities as high as 2×10^5 cells/ml are not uncommon (Post, 1977).

At the beginning of 1980 we started a systematic survey of the spatial and temporal distribution of *Dunaliella* in the Dead Sea (Oren, 1981; Oren and Shilo, 1982) and part of the results are shown in Fig. 2. Before May 1980 no *Dunaliella* cells were observed, but in the summer of 1980 a bloom of algae developed, reaching a peak density of 8800 cells/ml measured in surface water on July 10, 1980 (Oren and Shilo, 1982). Counts in surface samples taken from different stations in the northern basin during the same period did not differ significantly. The *Dunaliella* population was initially restricted to the upper 5–10 m of the water column, but later in the summer algae were found as deep as 25 m. Generally the population densities were similar at all depths down to the lower

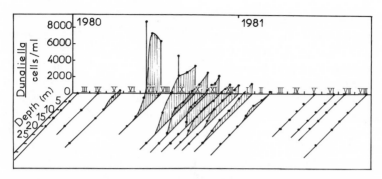

Figure 2. Distribution of *Dunaliella* in the Dead Sea, March 1980–August 1981. Data are derived in part from Oren and Shilo (1982). [Modified from Oren and Shilo (1982), with permission.]

boundary of the algal distribution, and no diel vertical migration of the cells could be demonstrated. During the period May–August 1980, population densities at the water surface were slightly lower than those at a depth of 5 m, an effect that may have been due to supraoptimal light intensities at the surface. The vertical distribution of the algae appeared to be closely linked to the physical structure of the lake: in the summer of 1980 the lake was stratified as a result of abundant rain floods in the winter of 1979–1980 (see Section 2.1). From May to September 1980 two pycnoclines were present, separating three distinct water layers, and *Dunaliella* was restricted to the upper layer of 5–10 m depth. With the decrease of freshwater inflow and the increased evaporation rates in summer, the salinity of the upper layer finally equalled that of the intermediate one, resulting in mixing of the upper two layers. From then on, *Dunaliella* was distributed evenly above the remaining single pycnocline, and the maximal depth at which the algae were found closely followed the variations in depth—from 15 to 25 m below the surface—of this pycnocline, due to internal seiches.

From October 1980 onward the *Dunaliella* population in the Dead Sea declined. When analyzing the causes of this decline we must remember that no protozoa were found in any of the water samples, and thus the quantitative role of ciliate and ameboid protozoa (Elazari-Volcani, 1943b, 1944) in controlling microbial communities must have been negligible.

Possible causes of the observed decline in the algal population are starvation for phosphate (see later in this section) and/or starvation by prolonged darkness. The upper water layer of the Dead Sea in the summer of 1980 was extremely turbid, partially because of the high community densities of halobacteria (up to 2×10^7 cell/ml), and at a depth of 2 m the penetrating light intensity amounted to only about one-tenth of that at the surface. The low light penetration was also demonstrated in an experiment in which bottles filled with a culture of *Dunaliella* and enriched with $^{14}CO_2$ were lowered to different depths in the water column; no net photosynthesis could be demonstrated at depths below 5–10 m (Oren and Shilo, 1982). As a result of the deepening of the upper mixed water layer to depths of 20 m and below, a great part of the *Dunaliella* population was found in darkness. As *Dunaliella* can survive continuous darkness for a limited time only (laboratory experiments suggest periods of no longer than 2 weeks), those cells that do not reach the photic surface layer as a result of mixing with the upper water layer may die. This hypothesis was supported by changes observed in the depth distribution of the algae after the new winter floods of December 1980–January 1981, which caused an additional layer of less saline water to be formed on top of the water column. A new pycnocline was formed at a depth of 5–10 m, preventing the *Dunaliella* cells below from reaching the euphotic surface layer. A short time after these floods most of the cells in depths below 10 m lacked intracellular starch and had lost motility, and at the end of February 1981 *Dunaliella* cells had completely disappeared from depths below 15 m. The dead cells probably had sunk to the bottom of the lake; large amounts of chlorophyll *a*, probably originating from *Dunaliella* blooms in the past, could be found in the sediments (Nissenbaum *et al.*, 1977). Chlorophyll *a* remained intact in the sediments for long periods as the high magnesium concentrations of Dead Sea water protect the molecule from loss of its magnesium. As no chlorophyll *b* could be demonstrated in the sediment, it was suggested that this molecule is less stable than chlorophyll *a* under the conditions present.

Summer blooms of *Dunaliella* are not an annually recurring phenomenon, and no such blooms were observed in 1981–1985. On the contrary, the decline in the algal population size that started in the autumn of 1980 continued, and from a density of around 1000 cells/ml in the surface layers in December 1980 only about 100 remained in May 1981, and less than 10 in November 1981 (Fig. 2).

Only few measurements of primary production have been performed in Dead Sea waters (Oren, 1981, 1983a). Such measurements may be complicated because of the chemical composition of Dead Sea water: the water is often saturated with respect to calcium carbonate, which may be present as crystals in the form of calcite or aragonite (Neev and Emery, 1967; Sass and Ben-Yaakov, 1977). Thus, the $^{14}CO_2$ added in the standard primary production assays may precipitate, making part of the labeled CO_2 unavailable for the algae. A few measurements have been performed by us *in situ* by suspending light and dark glass bottles with Dead Sea water enriched with $H^{14}CO_3^-$ in the sea at the depth of sampling; other incubations were performed with constant illumination (5 W/m^2 at 25°C) in the laboratory. Calculations of the results were based on a concentration of 2.5 meq/liter total dissolved inorganic carbon (Sass and Ben-Yaakov, 1977). Values of net primary production varied from 6.9 mg C/m^3 hr at the height of the algal bloom (August 1980, at a depth of 5 m, and determined by incubation in the laboratory), to values obtained *in situ* of 1.95 mg C/m^3 day in July 1981, and less than 0.02 mg C/m^3 day in December 1981. Calculations over the whole water column per unit of surface yielded values varying from 15 mg C/m^2 day in July 1981 to less than 1 mg C/m^2 day in December 1981. Such values are low compared, for example, to those measured in the Great Salt Lake: up to 6 g C/m^2 day or 0.7 g C/m^3 day (Stephens and Gillespie, 1976). Unfortunately no *in situ* primary production measurements were performed at the height of the algal bloom in the summer of 1980. Part of the light-dependent CO_2 assimilation in the Dead Sea may be due to bacterial activity rather than to algal photosynthesis (Oren, 1983a), as will be discussed in Section 6.2.1.

The presence of algal blooms in the Dead Sea only in certain years raises the question as to what are the factors limiting the development of *Dunaliella* in the Dead Sea. Enrichment culture experiments demonstrated that, in addition to the availability of light, as described above, two conditions must be fulfilled before mass development of *Dunaliella* can take place in the Dead Sea: phosphate must be present, and the salinity must be sufficiently low. When dilutions of Dead Sea water were incubated in the light, *Dunaliella* developed only when phosphate was added, and the rate of division increased with decreasing salinity. Without phosphate no growth occurred, and the final cell yield was proportional to the phosphate concentration added (Oren and Shilo, 1985). Extrapolating these data [Fig. 4B in Oren and Shilo (1985)], one can infer that the bloom of 8800 cells/ml in the upper 10 m of the water column must signify an input of at least 1000 tons of phosphate with rain floods or by the Jordan river in the winter of 1979–1980 (the total amount of allochthonous material annually brought to the Dead Sea by the Jordan river and by rain floods has been estimated to vary between 2×10^5 and 3×10^5 tons (Nishry and Stiller, 1984).

In water of a specific gravity ρ_{20} of 1.22 or higher, growth is extremely slow; thus, at these salinities no mass development of algae can be expected in the Dead Sea, even in the presence of phosphate. This explains the absence of algal blooms in 1981–1985: in the period April–June 1980 (the year of the summer bloom of *Dunaliella*) ρ_{20} values were 1.2–1.215; in the corresponding period in 1981 values were much higher (1.215–1.221), while in subsequent summers the specific gravity of the water exceeded 1.23. Thus, salinities in the Dead Sea are generally higher than the optimal values for growth of *Dunaliella*, a situation resembling that existing in the Great Salt Lake (Brock, 1975).

Results of simulation experiments performed in 5.6-m³ fiberglass tanks at the northern shore of the lake (Oren and Shilo, 1985) confirmed the conclusion that salinity and availability of phosphate are the important factors determining the possibility of development of a *Dunaliella* bloom: *Dunaliella* was seen to develop (often in numbers up to 10^5 cells/ml) in ponds containing Dead Sea water diluted with Mediterranean water (75–65% Dead Sea water, 25–35% Mediterranean water, not in more concentrated brines), and then only when phosphate was also added.

6.2. Community Dynamics of Halobacteria in the Dead Sea

As is the case with the alga *Dunaliella*, only a few quantitative data were collected on the density of the bacterial communities in the Dead Sea before 1980. The first bacterial enumerations were reported by Kaplan and Friedmann (1970): densities of cells in surface water increased from 2.3×10^6 cells/ml in March 1964 to 8.8×10^6 in July 1964, remaining approximately constant until their last reported measurement in November 1964 (8.9×10^6 cells/ml). During this period the surface water had a reddish color (Neev and Emery, 1966). Most of the cells belonged to the pleomorphic, cup-shaped type, resembling *Halobacterium volcanii* (Mullakhanbhai and Larsen, 1975), a minor component consisted of long, motile, red rods resembling *H. sodomense* (Oren, 1983c). Bacterial numbers decreased with depth: at 50 m depth, $(2–3) \times 10^6$ cells/ml; at 100 m, $(6–8) \times 10^5$; and at 250 m, 2×10^4; with increasing depth the relative contribution of the rod-shaped cells increased. Kritzman (1973) and Kritzman *et al.* (1973) reported $(1–6) \times 10^4$ bacteria/ml by direct microscopic enumeration (sampling site and date not specified), and $(1.7–4.5) \times 10^4$ by colony counting on plates. A single determination by S. Garlick (unpublished data) in the summer of 1978 yielded 4.4×10^5 cells/ml in water from 40 to 100 m depth.

A systematic survey of bacterial community densities from 1980 onward (Oren, 1983b, 1985) showed the development of a bloom of halobacteria in the summer of 1980 (Fig. 3), paralleling the *Dunaliella* bloom in the same period (Fig. 2). A maximal community density of 1.9×10^7 cells/ml was recorded in July 1980. The bloom consisted of 80–90% of pleomorphic halobacteria, with the remainder rod-shaped red halobacteria. As a result of the color of these bacteria, caused by red-orange carotenoid pigments and by the purple pigment bacteriorhodopsin (see Section 6.2.1), the bacteria once more imparted a reddish color to the water. All

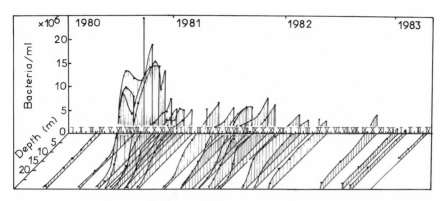

Figure 3. Distribution of bacteria in the Dead Sea as determined by direct microscopic enumeration, April 1980–April 1983. Data are derived in part from Oren (1983b). [Modified from Oren (1983b), with permission.]

quantitative data available on bacterial communities relate to these two *Halobacterium* types, and no data are available on the spatial and temporal distribution of other extremely or moderately halophilic bacteria. Like the *Dunaliella* bloom, the bacterial bloom was also restricted to the water layer above the (upper) pycnocline, and closely followed depth variations of this pycnocline. Horizontal variation of bacterial numbers was not significant. Below the pycnocline, bacterial community densities were invariably low: (4–7) $\times$ 10^5 bacterialike particles/ml was enumerated in samples collected from all depths below the pycnocline, down to 200 m. Bacterial densities of up to 1.9 $\times$ 10^7 cells/ml are very high if compared, for example, with those in the oceans and most freshwater bodies, but are not uncommon in hypersaline water bodies, such as the Great Salt Lake north arm, where bacterial densities have been reported to vary generally from 4 $\times$ 10^7 to 10^8 cells/ml, except near the sediment and after algal blooms, when counts ranged from 10^8 to as high as 2.4 $\times$ 10^8 cells/ml (Post, 1977).

Enrichment cultures with Dead Sea water samples, set up to determine the factor(s) limiting bacterial development in the Dead Sea (Oren, 1983b), showed that mass development of bacteria was possible when the water samples were supplemented with phosphate and with a suitable source of organic carbon. The bacterial types developing in the enrichment cultures were generally the same as those that dominated the bacterial community in the lake. A single carbon source was sufficient to give bacterial yields of more than 2 $\times$ 10^9 cells/ml; suitable carbon sources were, for example, glucose, glycerol, casamino acids, and glutamate. The carbon source that enabled the summer bloom of bacteria in 1980 may have been glycerol, which is present in large concentrations inside the

Dunaliella cells, serving as an osmoregulator and compatible solute. Under normal conditions *Dunaliella* is able to retain most of the glycerol produced within the cells, but even under optimal conditions up to 5% of the glycerol was found outside the cells (Wegmann *et al.*, 1980). The finding that single carbon sources could give rise to development of halobacteria in enrichment cultures was unexpected, as most halobacteria have complex nutritional requirements. Recently, however, halobacterial types resembling the Dead Sea strain *H. volcanii* have been described as having simple nutritional requirements, being able to grow on single carbon sources (Rodriguez-Valera *et al.*, 1980; Javor, 1984). As mass development of bacteria in the Dead Sea depends on the presence of algal blooms supplying organic carbon sources, the bacterial community did not increase in the summer of 1981, when no new algal bloom developed (Fig. 3).

Dilution of the Dead Sea water samples with distilled water had a stimulating effect on the growth of enrichment cultures in which organic carbon sources and phosphate were added. This can be explained by the finding (Mullakhanbhai and Larsen, 1975; Oren, 1983c) that salinities (and more specifically the divalent cation concentrations) in the Dead Sea are supraoptimal for the growth of Dead Sea halobacteria (Fig. 1). The effect of salinity, the presence of phosphate, and the presence of different carbon sources (glycerol, glucose, or the presence of a *Dunaliella* bloom) were also examined under field conditions in outdoor ponds (Oren, 1985; Oren and Shilo, 1985). Under suitable conditions (presence of phosphate and either of the carbon sources, in a relatively diluted brine) bacterial densities reached values of up to 2×10^8 cells/ml. Also, slow bacterial development was observed at relatively low temperatures (14–17°C).

In the period October–December 1980 the bacterial community was found to decrease almost threefold (Fig. 3). When trying to explain the cause of this decrease it should be kept in mind that the quantitative role of protozoa, if these were present at all, must have been negligible (see Section 6.1). However, we cannot rule out a potential role of delicate microflagellates that may escape detection with the techniques used. The possibility that halophilic bacteriophages (Torsvik and Dundas, 1974) were responsible for the decline has not been investigated. It is not probable that part of the bacteria reached the deep waters by mixing, as stable thermoclines and pycnoclines were present during this period (Stiller *et al.*, 1984a), and no sign of massive mixing of the water layers was found. A steady precipitation of cells through the pycnocline to the deep layers is theoretically possible, but in that case it is difficult to explain the stability of the community size after this short decline period (Fig. 3). Another possibility, at least theoretically, is that bacteria served as seeds around which crystals of salts such as aragonite or gypsum (Neev and

Emery, 1966, 1967) were formed, which afterward precipitated to the bottom of the lake, or that the bacteria adhere to crystals formed by other means and are sedimented along with them. Whatever the mechanism may be, part of the bacteria developing in the water column finally do reach the bottom sediments, and large amounts of characteristic compounds, such as phytanic acid (up to 520 μg/kg sediment), dihydrogenphytol (up to 800 μg/kg sediment), and phytane (up to 17 μg/kg sediment), have been found in the sediment (Kaplan and Baedecker, 1970; Nissenbaum et al., 1977). These substances are thought to be derived from the phosphatidyl glycerophosphate lipid with isoprenoid side chains characteristic of the halobacteria, and also found in the Dead Sea strains *H. volcanii* (Mullakhanbhai and Francis, 1972) and *"H. marismortui"* (Evans et al., 1980). The carotenoid pigment of the halobacteria, α-bacterioruberin, however, was not found in the sediments, probably because it undergoes rapid transformation (Nissenbaum et al., 1977).

From the beginning of 1981 onward the bacterial community size remained at a constant level of around 5×10^6 cells/ml surface water during a period of almost 2 years (Fig. 3). Measurements of phosphate turnover rates during this period (Oren, 1983b) suggest that this stability in community size represented a stable equilibrium in which cell multiplication was extremely slow. Phosphate, though a limiting factor for bacterial development, and though present in very low concentrations, underwent very slow turnover (turnover times as long as 240–1200 days), in spite of the high bacterial densities present. During the period discussed a significant part of the bacterial community proved viable (Oren, 1983b). These observations suggest that the Dead Sea halobacteria may survive starvation conditions for long periods. The presence of purple membrane in at least part of the halobacterial community in the Dead Sea may have enabled the use of light as a source of maintenance energy, thus prolonging viability under nutrient starvation (Brock and Petersen, 1976). Therefore, a bacterial bloom, once formed, can remain in the Dead Sea water for extremely long periods, long after the *Dunaliella* cells that had enabled its development have disappeared (compare Figs. 2 and 3).

The new overturn of the water column in December 1982 was accompanied by a sudden decline in the bacterial community size in the surface waters (Oren, 1985); as no new *Dunaliella* blooms developed in the subsequent 3 years, bacterial community sizes have also remained small.

6.2.1. Bacteriorhodopsin in the Dead Sea

A special feature of the halobacterial community of the Dead Sea in 1980–1982 was the presence of large amounts of purple-membrane-containing bacteriorhodopsin (Oren and Shilo, 1981). Certain strains of halobacteria are able to produce a purple pigment, bacteriorhodopsin, localized in special parts of the cell membrane [for a review see, e.g., Stoeckenius and Bogomolni (1982)]. Light excitation of bacteriorhodopsin causes the extrusion of protons from the cell to the outer medium, and the proton gradient thus formed can be used for the generation of ATP (Danon and Stoeckenius, 1975). The possession of bacteriorhodopsin can be of great ecological advantage, as bacteriorhodopsin-containing halobacteria may survive starvation in the light for prolonged periods (Brock and Petersen, 1976). However, purple membrane is synthesized only by a small number of halobacterial strains, and only under special conditions: in the presence of light and low O_2 concentrations. The first biotope in which the presence of bacteriorhodopsin was reported was the Dead Sea, where concentrations as high as 0.6 nmole/liter (0.4 nmole/mg protein) were found at the beginning of 1981, surprisingly in the presence of quite high O_2 tensions (around 1.2 ml/liter) (Oren and Shilo, 1981). Until now the ability to synthesize bacteriorhodopsin among the Dead Sea halobacteria has been demonstrated only in the rod-shaped *H. sodomense* (Oren, 1983c).

Light energy absorbed by bacteriorhodopsin can be used for many purposes, one of them being the fixation of CO_2: Danon and Caplan (1977) showed light-dependent CO_2 assimilation in bacteriorhodopsin-containing *H. halobium,* and as a possible mechanism they suggested the reaction

$$CO_2 + propionyl\text{-}CoA + 2[H] \rightarrow \alpha\text{-ketobutyrate} + CoA$$

The reducing power needed for this reaction was thought to be supplied by reversed electron transport driven by the proton gradient generated by light excitation of bacteriorhodopsin. The α-ketobutyrate formed can then be used for the biosynthesis of amino acids such as isoleucine (Buchanan, 1969). In our studies on the primary production in the Dead Sea from November 1981 to June 1982, a period in which almost no algal cells were present in the lake, some light-dependent CO_2 assimilation was measured, which, by means of the use of different inhibitors and specific wavelengths of exciting light, was shown to be bacteriorhodopsin-mediated rather than chlorophyll-dependent (Oren, 1983a).

6.3. Sulfate Reduction in the Dead Sea

Dead Sea bacteria are believed to play important roles in a number of geochemical processes, though the nature of the bacteria involved, their numbers, and the rates of the processes are generally unknown. The occurrence of active sulfate reduction in the bottom sediments was first indicated by the finding by Neev and Emery (1967) that a large amount of gypsum was collected in their sediment traps, whereas only little gypsum accumulated on the bottom of the lake. The biological nature of the sulfate reduction was suggested by data on the isotopic composition of the sulfate and sulfide present in the sediments and in the lower water mass before the 1979 overturn of the lake: sulfide in the lower water mass was relatively depleted of the heavy stable isotope ^{34}S ($\delta^{34}S = -19.6$ to $-21.7‰$) whereas the heavy isotope was abundant in the sulfate ($\delta^{34}S = +14.1$ to $+15.6‰$). In the interstitial waters of the reduced sediments these values were $-17‰$ and $+16‰$, respectively (Nissenbaum and Kaplan, 1976). Isotopic fractionations like these are characteristic for bacterial dissimilatory sulfate reduction. Lerman (1967) calculated the rate of sulfate reduction in Dead Sea sediments from the depth distribution of gypsum, and his estimations yielded low rates of around 7.8×10^{-7} mole S/cm^3 year.

In the period before the 1979 overturn of the lake, part of the sulfide from the lower water mass reached the aerobic water layers and was oxidized there. Whether this process was due to chemical oxidation, or whether chemolithotrophic bacteria were involved, is unknown.

Activity of chemoorganotrophic bacteria in the lake's sediments may be low, as evidenced by large amounts of accumulated organic carbon (0.23–0.4% by weight). Even substrates that are easily degraded by many microorganisms, such as amino acids, were found in large concentrations (in oxidized sediments, 60–120 mg amino acids/kg dry sediment; in reduced sediments, 750–790 mg/kg, which is 8–12% of the total organic carbon, or 16–24% when humic and fulvic acids are not taken into account) (Nissenbaum et al., 1977). In reducing sediments the preservation of organic matter is better than under oxidizing conditions.

6.4. Other Processes in the Dead Sea Mediated by Microorganisms

Another process in which microorganisms may be involved is the deposition of manganese-rich crusts at certain sites along the Dead Sea shore (Garber et al., 1981; Nishry, 1984). Samples collected from soils and springs near the Dead Sea have yielded Mn^{2+}-oxidizing bacteria (Gram-positive spore-formers and Gram-negative rods). These are not extreme halophiles and are unable to grow in Dead Sea water. They can use Mn^{2+} as a source of energy, but they do not grow autotrophically (Ehrlich and Zapkin, 1983). However, in view of the low salt tolerance of these bacteria, Garber et al. (1981) suggested an abiogenic origin of the manganese-rich excrustations.

7. The Fate of Nonhalophilic Bacteria Entering the Dead Sea

One of the categories of bacteria isolated from the Dead Sea was termed "haloresistant" by Volcani (Elazari-Volcani, 1940; Volcani, 1944). These were organisms that were unable to grow in media containing high salt concentrations. Upon accidentally entering the Dead Sea water, e.g., with flood waters, the river Jordan, or from the air, they were able to survive exposure to the concentrated brines until transferred to a suitable growth medium with a low salt concentration. Many of these isolates belong to the genus *Bacillus,* whose members produce endospores that are particularly resistant to adverse conditions such as excessive heat or salinity. Volcani obtained most of his water samples from the northern part of the lake, close to the location of the entrance of fresh water from the Jordan river, and nonhalophilic bacteria were especially easy to isolate from zones of relatively fresh water floating on top of the concentrated brines before mixing with them. The pathogenic *Clostridium* types isolated by Lortet (1892) from mud from the Dead Sea must similarly belong to this category.

The fate of nonhalophilic bacteria, such as *Escherichia coli* entering the Dead Sea with sewage discharged into the lake, was studied by Kritzman (1973) and by Oren and Vlodavsky (1985). Kritzman counted viable *E. coli* cells at different distances from a sewage discharge site in the southern part of the lake; initial numbers of 2×10^4 viable cells/ml were reduced by 90% at a distance of 70 m, while no viable cells were recovered at 90 m distance. Similarly, high numbers of coliforms were counted as far as 20 m from a sewage discharge site at the western shore of the northern part of the lake (Oren and Vlodavsky, 1985). A laboratory study showed that numbers of viable *E. coli* and *Vibrio harveyi* (a marine, luminescent bacterium) cells decrease by 90% in less than 1 hr to several hours upon suspension in Dead Sea water. The rate of dieoff depends on the salinity of the water (a dilution enables a prolonged survival), on temperature (at higher temperatures dieoff is faster), and on light intensity (the higher the irradiation, the faster the decrease in viability). The toxic principle for *E. coli* in Dead Sea water is the combination of high Ca^{2+} and Mg^{2+} concentrations: high NaCl concentrations only, or NaCl together with either $CaCl_2$ or $MgCl_2$ at the concentrations present in the Dead Sea, are much less toxic (Oren and Vlodavsky, 1985).

8. The Biology of the Dead Sea: Past, Present, and Future

The physical and chemical properties of the Dead Sea determining its properties as a biotope are far from constant. In addition to the seasonal fluctuations in salinity and temperature, whose effects we have discussed in previous sections, the composition of the Dead Sea water also undergoes long-term changes. Lake Lisan, the Pleistocene precursor of the Dead Sea, was probably much less saline than the present-day lake, and it was accordingly inhabited by a much greater variety of forms of life. Fossils found in geological strata originating from Lake Lisan (about 60,000 to 18,000 years B.P.) include a variety of freshwater and euryhaline diatoms, freshwater gastropods (mainly in the northern basin of Lake Lisan), nonmarine ostracods, and euryhaline fish, *Aphanius* sp. (Begin *et al.*, 1974). The fossil record suggests an increase in salinity with time. All these organisms disappeared from the lake when its salinity became too high.

Also in more recent periods long-term changes in salinity have taken place. Data on the water level of the lake since the beginning of the 19th century (Klein, 1961) show a positive water balance until the end of the 19th century, followed by excess evaporation, which caused a drop in water level from about −391 m in 1890 to −397 m in 1960 and −402 m in 1978. Part of the effect must have been due to climatic changes, and similar dramatic changes in water level may also have occurred in earlier periods (Klein, 1982). Since the completion of the National Water Carrier in the 1950s, the amount of water flowing in the River Jordan has been greatly reduced, and the water level of the Dead Sea has dropped by 0.5 m/year on average during recent years, causing an increase in salt concentrations (Table I). Even the salinity of the bottom layers is no longer constant: after having been isolated for about 300 years as "fossil water," the lower water layers have been involved in several overturn events. After the second overturn (December 1982) the lake was more saline by 1.0 g/kg than after the first (February 1979) (Stiller *et al.*, 1984a).

In addition to the general increase in salinity, the relative concentration of Mg^{2+} increased in comparison with Na^+ (Table I), signifying supersaturation of at least part of the lake and, at least during certain periods between 1960 and 1977, with respect to sodium chloride, which precipitated as halite (Steinhorn, 1983). It has been shown (Oren, 1983b; Oren and Shilo, 1985; see also Fig. 1) that halobacteria live in the Dead Sea at the upper limit of tolerance of magnesium concentrations. Thus, an increase in salinity, and additionally a specific increase in the concentration of magnesium ions in the upper water layers (22% in less than 20 years) has made the Dead Sea less suited as a biotope for these bacteria. Salinities such as found in the first half of this century were not extremely high for the development of *Dunaliella;* during recent years salinities have been so high that *Dunaliella* development was possible only after a drastic dilution of the upper water layers by flood waters.

The microbial communities may be capable of adapting themselves to a certain extent to the long-term changes in the physical and chemical properties of the sea. Different *Halobacterium* isolates from the Dead Sea show significant differences in magnesium tolerance. Nothing is known about the extent of adaptability of halophilic microorganisms to supraoptimal salt concentrations, but it is tempting to speculate that the increase in salinity, and more specifically in magnesium concentrations, caused a succession of bacterial types with increasing magnesium tolerance. This hypothesis is difficult to prove because of the small number of strains isolated in the past that have survived. However, the process of drying out of the lake is expected to continue during the years to come, and the process of selection and evolution may continue together with the increase in overall salinity of the lake and the changes in its salt composition.

The complete mixing of the water column during the overturn of the lake in February 1979 had important biological implications: first, the less saline upper water layer, in which microorganisms less well adapted to the most extreme salinities could find a refuge, ceased to exist, thus making conditions in the Dead Sea still more extreme as a biotope. Furthermore, with the mixing of the water layers, the whole water column became aerobic (Nishry and Stiller, 1984), and the anoxic, sulfide-containing lower layer (Neev and Emery, 1967) ceased to exist. Thus, activity of anaerobic microorganisms became restricted to the reduced layers of the sediments. During the following 3 years of meromixis, lasting until December 1982 (Stiller et al., 1984a), no new accumulation of sulfide occurred in the lower layers.

The drop in the water level of the Dead Sea and the resulting increase in salinity of the water are expected to continue in the years to come. To change this situation, a water carrier connecting the Dead Sea with the Mediterranean has been planned (Steinhorn and Gat, 1983; Weiner, 1985),* and its possible biological implications have been examined. Our understanding of the biology of the Dead Sea in its present state allows us to make some predictions regarding the biology of the lake when large amounts of water from the Mediterranean will enter it, causing at first a rise in water level, followed by a stabilization. The physical properties of the lake will change, and these changes will also have a profound influence on the biology of the lake. The following factors are expected to affect the biology of the Dead Sea as a result of the exploitation of the water carrier.

1. Salinity effects. The inflow of large quantities of Mediterranean water will probably cause the formation of a layer of less saline water floating on top of the heavier Dead Sea brine, at least locally near the place of discharge of Mediterranean waters and during the first years after

---

*Implementation of the planned Mediterranean–Dead Sea hydroelectric project has been suspended for financial reasons.

the completion of the project. The depth of this upper layer and its salinity depend on many factors, such as the rate of entrance of Mediterranean water and the amount of mixing by wind. As discussed earlier, microorganisms live in the Dead Sea at salinities (or, more specifically, at divalent cation concentrations) much above their optimum. Thus, when the availability of nutrients is not limiting, bacterial and algal growth will be much faster in the more dilute waters than in almost saturated brines. The stimulating effect of dilution on the growth of microorganisms in Dead Sea water has been demonstrated for the alga *Dunaliella* (Oren and Shilo, 1982, 1985) as well as for halobacteria (Oren, 1983b, 1985; Oren and Shilo, 1985). In addition to a lowering of the salt concentrations, changes in salt composition are also expected to occur in the Dead Sea as a result of mixing of Mediterranean waters with Dead Sea brines, and these changes may also influence the biology of the lake. One of the changes due to occur is the formation of crystals of gypsum ($CaSO_4 \cdot 2H_2O$) as a result of the excess of calcium in Dead Sea water and the excess of sulfate in Mediterranean water (Katz *et al.*, 1981). Whether the gypsum crystals will precipitate quickly to the bottom of the lake or will remain suspended for prolonged periods in the upper water layer is not yet clear.

2. The water carrier may add nutrients that previously limited development of microorganisms. As we have seen, lack of phosphate often limits the development of algae and bacteria in the lake; thus any amount of phosphate originating from the Mediterranean waters or added to them on their way to the Dead Sea may cause an increase in microbial biomass in the lake. Phosphate concentrations in Mediterranean water are generally low (around 1 μeq/liter), and thus cannot be expected to contribute significantly to the phosphorus concentration in Dead Sea water. However, any addition of phosphate may easily give rise to algal and bacterial blooms in the lake, especially if the lowered salinity of the upper water layer enables high growth rates.

Addition of organic nutrients to the Dead Sea through the water carrier may enhance the activity of chemoorganotrophic bacteria. The effect of eutrophication has been studied by Kritzman (1973) at a site in the southern part of the lake where domestic sewage was released into the sea. Nonhalophilic bacteria were killed soon after their entrance into the Dead Sea, and no viable coliforms could be detected at a distance of 90 m from the site of discharge (see Section 7); no halobacteria were observed at distances closer than 50 m to the point of entrance of the wastewater, but dense halobacterial communities of around 10^6 cells/ml were observed at a distance of 120 m. The heterotrophic activity of the bacterial community, as monitored by the rate of $^{14}CO_2$ evolution from [1-^{14}C]glucose, showed two maxima, one at 0–50 m, due to nonhalophilic

microorganisms, and the other at 130–250 m, due to halophilic chemoorganotrophic bacteria.

3. The temperature of part of the Dead Sea water may increase: when
a thin layer of relatively fresh water overlies more concentrated salt solutions, these may warm up as a result of the "solar pond effect" (Tabor,
1966; Assaf, 1976). As halobacteria isolated from the Dead Sea grow optimally around 40°C (Mullakhanbhai and Larsen, 1975; Oren, 1983c), an
increase in water temperature may give rise to increases in microbial
growth rates.

During the first years of the operation of the water carrier the Dead
Sea upper water layers may become relatively diluted according to the
preliminary planning, and forms of life not found in the lake at present
may get a chance to develop. The salinity of this upper layer is not
expected to reach values lower than half the present ones; thus, typically
Mediterranean organisms, requiring salt concentrations of around 3.5%,
will not be able to survive. In the long run, after the achievement of a
stable water level, the salinity of the upper water layer will rise again, and
only microorganisms adapted to the extreme salinities and high divalent
cation concentrations will be able to survive. Thus, the Mediterranean–
Dead Sea water carrier cannot be expected to enable the growth of fish in
the Dead Sea, such as prophesized by the prophet Ezekiel (*Ezekiel* **47**:8–
10):

> This water flows toward the eastern region and goes down into the Arabah;
> and when it enters the stagnant waters of the Sea, the water will become fresh.
> And wherever the river goes, every living creature which swarms shall live,
> and there will be very many fish; for this water goes there, that the waters of
> the sea may become fresh; so everything will live where the river goes. Fish-
> ermen shall stand beside the sea; all the way from En-ge'di to En-eg'laim it will
> be a place for the spreading of nets; its fish will be of very many kinds, like the
> fish of the Great Sea.

ACKNOWLEDGMENTS. I thank M. Shilo and M. Kessel for critically reading
the manuscript.

Research on the microbiology of the Dead Sea in my laboratory has
been supported by grants from the Israeli Ministry of Energy and Infrastructure and from the Mediterranean–Dead Sea Company, Ltd.

References

Arad, V., Beyth, M., and Bartov, Y., 1984, The Dead Sea and Its Surroundings. Bibliography
of Geological Research, Geological Survey of Israel, Special Publication No. 3.
Assaf, G., 1976, The Dead Sea: A scheme for a solar lake, *Solar Energy* **18**:293–299.
Begin, Z. B., Ehrlich, A., and Nathan, Y., 1974, Lake Lisan. The Pleistocene Precursor of
the Dead Sea, Bulletin No. 63, State of Israel, Ministry of Commerce and Industry,
Geological Survey.

Ben-Amotz, A., and Avron, M., 1983, Accumulation of metabolites by halotolerant algae and its industrial potential, *Annu. Rev. Microbiol.* **37**:95–119.

Ben-Amotz, A. and Ginzburg, B. Z., 1969, Light-induced proton uptake in whole cells of *Dunaliella parva, Biochim. Biophys. Acta* **183**:144–154.

Ben-Amotz, A., Katz, A., and Avron, M., 1982, Accumulation of β-carotene in halotolerant algae: Purification and characterization of β-carotene-rich globules from *Dunaliella bardawil* (Chlorophyceae), *J. Phycol.* **18**:529–537.

Ben-Yaakov, S., and Sass, E., 1977, Independent estimate of the pH of Dead Sea brines, *Limnol. Oceanogr.* **22**:374–376.

Bernard, F., 1957, Présence de flagellés marins *Coccolithus* et *Exuviella* dans le plancton de la Mer Morte, *C. R. Acad. Sci. (Paris)* **245**:1754–1756.

Beyth, M., 1980, Recent evolution and present stage of Dead Sea brines, in *Hypersaline Brines and Evaporitic Environments* (A. Nissenbaum, ed.), pp. 155–165, Elsevier, Amsterdam.

Brock, T. D., 1975, Salinity and the ecology of *Dunaliella* from Great Salt Lake, *J. Gen. Microbiol.* **89**:285–292.

Brock, T. D., and Petersen, S., 1976, Some effects of light on the viability of rhodopsin-containing halobacteria, *Arch. Microbiol.* **109**:199–200.

Buchanan, B. B., 1969, Role of ferredoxin in the synthesis of α-ketobutyrate from propionyl coenzyme A and carbon dioxide by enzymes from photosynthetic and nonphotosynthetic bacteria, *J. Biol. Chem.* **244**:4218–4223.

Butcher, R. W., 1959, *An Introductory Account of the Smaller Algae of British Coastal Waters. Part 1: Introduction and Chlorophyceae, HMSO,* London.

Carmi, I., Gat, J. R., and Stiller, M., 1984, Tritium in the Dead Sea, *Earth Planet. Sci. Lett.* **71**:377–389.

Cohen, S., Oren, A., and Shilo, M., 1983, The divalent cation requirement of Dead Sea halobacteria, *Arch. Microbiol.* **136**:184–190.

Danon, A., and Caplan, S. R., 1977, CO_2 fixation by *Halobacterium halobium, FEBS Lett.* **74**:255–258.

Danon, A., and Stoeckenius, W., 1975, Photophosphorylation in *Halobacterium halobium, Proc. Natl. Acad. Sci. USA* **71**:1234–1238.

Edgerton, M. E., and Brimblecombe, P., 1981, Thermodynamics of halobacterial environments, *Can. J. Microbiol.* **27**:899–909.

Ehrlich, H. L., and Zapkin, M. A., 1981, Mn^{2+} oxidizing bacteria from the Dead Sea region in Israel, Abstract N-60, Annual Meeting of the American Society for Microbiology.

Elazari-Volcani, B., 1940, Studies on the microflora of the Dead Sea, Ph. D. Thesis, Hebrew University of Jerusalem (in Hebrew).

Elazari-Volcani, B., 1943a, Bacteria in the bottom sediments of the Dead Sea, *Nature* **152**:274–275.

Elazari-Volcani, B., 1943b, A dimastigamoeba in the bed of the Dead Sea, *Nature* **152**:301–302.

Elazari-Volcani, B., 1944, A ciliate from the Dead Sea, *Nature* **154**:335.

Evans, R. W., Kushwaha, S. C., and Kates, M., 1980, The lipids of *Halobacterium marismortui,* an extremely halophilic bacterium in the Dead Sea, *Biochim. Biophys. Acta* **619**:533–544.

Garber, R. A., Nishry, A., Nissenbaum, A., and Friedman, G. M., 1981, Modern deposition of manganese along the Dead Sea shore, *Sed. Geol.* **30**:267–274.

Ginzburg, M., Sachs, L., and Ginzburg, B. Z., 1970, Ion metabolism in a *Halobacterium.* I. Influence of age of culture on intracellular concentrations, *J. Gen. Physiol.* **55**:187–207.

Gonzalez, C., Gutierrez, C., and Ramirez, C., 1978, *Halobacterium vallismortis* sp. nov. An amylolytic and carbohydrate-metabolizing, extremely halophilic bacterium, *Can. J. Microbiol.* **24**:710–715.

Imhoff, J. F., and Rodriguez-Valera, F., 1984, Betaine is the main compatible solute of halophilic eubacteria, *J. Bacteriol.* **160**:478–479.

Javor, B. J., 1984, Growth potential of halophilic bacteria isolated from solar salt environments: Carbon sources and salt requirements, *Appl. Environ. Microbiol.* **48**:352–360.

Javor, B. J., Requadt, C., and Stoeckenius, W., 1982, Box-shaped halophilic bacteria, *J. Bacteriol.* **151**:1532–1542.

Kaplan, I. R., and Baedecker, M. J., 1970, Biological productivity in the Dead Sea. Part II. Evidence for phosphatidyl glycerophosphate lipid in sediment, *Isr. J. Chem.* **8**:529–533.

Kaplan, I. R., and Friedmann, A., 1970, Biological productivity in the Dead Sea. Part I. Microorganisms in the water column, *Isr. J. Chem.* **8**:513–528.

Katz, A., Starinsky, A., Taitel-Goldman, N., and Beyth, M., 1981, Solubilities of gypsum and halite in the Dead Sea and in its mixtures with sea water, *Limnol. Oceangr.* **26**:709–716.

Kirk, R. G., and Ginzburg, M., 1972, Ultrastructure of two species of *Halobacterium, J. Ultrastructure Res.* **41**:80–94.

Klein, C., 1961, On the Fluctuations of the Level of the Dead Sea Since the Beginning of the 19th Century, Hydrological Paper No. 7, Ministry of Agriculture, Hydrological Service of Israel.

Klein, C., 1982, Morphological evidence of lake level changes, western shore of the Dead Sea, *Isr. J. Earth Sci.* **31**:67–94.

Krasil'nikov, N. A., Duda, V. I., and Pivovarov, G. E., 1971, Characteristics of the cell structure of soil anaerobic bacteria forming vesicular caps on their spores, *Microbiology* **40**:592–597.

Kritzman, G., 1973, Observations on the microorganisms in the Dead Sea, M. Sc. Thesis, Hebrew University of Jerusalem (in Hebrew).

Kritzman, G., Keller, P., and Henis, Y., 1973, Ecological studies on the heterotrophic extreme halophilic bacteria of the Dead Sea, in: *Abstracts of the 1st International Congress Bacteriology,* Vol. II, p. 242, Jerusalem.

Kushner, D. J., 1978, Life in high salt and solute concentrations: Halophilic bacteria, in: *Microbial Life in Extreme Environments* (D. J. Kushner, ed.), pp. 317–368, Academic Press, London.

Larsen, H., 1980, Ecology of hypersaline environments, in: *Hypersaline Brines and Evaporitic Environments* (A. Nissenbaum, ed.), pp. 23–39, Elsevier, Amsterdam.

Lerche, W., 1937, Untersuchungen über Entwicklung und Fortpflanzung in der Gattung *Dunaliella,* Arch. Protistenkd. **88**:236–268.

Lerman, A., 1967, Model of chemical evolution of a chloride lake—The Dead Sea, *Geochim. Cosmochim. Acta* **31**:2309–2330.

Levy, Y., 1980, Seasonal and Long Range Changes in Oxygen and Hydrogen Sulfide Concentration in the Dead Sea, Report MG/9/80, Ministry of Energy and Infrastructure, Geological Survey of Israel.

Lortet, M. L., 1892, Researches on the pathogenic microbes of the mud of the Dead Sea, *Palestine Exploration Fund* **1892**:48–50.

Mackay, M. A., Norton, R. S., and Borowitzka, L. J., 1984, Organic osmoregulatory solutes in cyanobacteria, *J. Gen. Microbiol.* **130**:2177–2191.

Mullakhanbhai, M. F., and Francis, G. W., 1972, Bacterial lipids. 1. Lipid constituents of a moderately halophilic bacterium, *Acta Chem. Scand.* **26**:1399–1410.

Mullakhanbhai, M. F., and Larsen, H., 1975, *Halobacterium volcanii* spec. nov., a Dead Sea halobacterium with a moderate salt requirement, *Arch. Microbiol.* **104**:207–214.

Neev, D., and Emery, K. O., 1966, The Dead Sea, *Science J.* **2**:50–55.

Neev, D., and Emery, K. O., 1967, The Dead Sea. Depositional Processes and Environments of Evaporites, Bulletin No. 41, State of Israel, Ministry of Development, Geological Survey.

Nicholson, D. E., and Fox, G. E., 1983, Molecular evidence for a close phylogenetic relationship among box-shaped halophilic bacteria, *Halobacterium vallismortis,* and *Halobacterium marismortui, Can. J. Microbiol.* **29**:52–59.

Nishry, A., 1984, The geochemistry of manganese in the Dead Sea, *Earth Planet. Sci. Lett.* **71**:415–426.

Nishry, A., and Stiller, M., 1984, Iron in the Dead Sea, *Earth Planet Sci. Lett.* **71**:405–414.

Nissenbaum, A., 1975, The microbiology and biogeochemistry of the Dead Sea, *Microb. Ecol.* **2**:139–161.

Nissenbaum, A., 1977, Minor and trace elements in Dead Sea water, *Chem. Geol.* **19**:99–111.

Nissenbaum, A., 1979, Life in a Dead Sea—Fables, allegories and scientific search, *Bioscience* **29**:153–157.

Nissenbaum, A., and Kaplan, I. R., 1976, Sulfur and carbon isotopic evidence for biogeochemical processes in the Dead Sea ecosystem, in: *Environmental Biochemistry* (J. O. Nriagu, ed.), Vol. 1, pp. 309–325, Ann Arbor Scientific, Ann Arbor, Michigan.

Nissenbaum, A., Baedecker, M. J., and Kaplan, I. R., 1977, Organic geochemistry of Dead Sea sediments, *Geochim. Cosmochim. Acta* **36**:709–727.

Oesterhelt, D., and Stoeckenius, W., 1971, Rhodopsin-like protein from the purple membrane of *Halobacterium halobium, Nature* **233**:149–152.

Oren, A., 1981, Approaches to the microbial ecology of the Dead Sea, *Kieler Meeresforsch. Sonderh.* **5**:416–424.

Oren, A., 1983a, Bacteriorhodopsin-mediated CO_2 photoassimilation in the Dead Sea, *Limnol. Oceanogr.* **28**:33–41.

Oren, A., 1983b, Population dynamics of halobacteria in the Dead Sea water column, *Limnol. Oceanogr.* **28**:1094–1103.

Oren, A., 1983c, *Halobacterium sodomense* sp. nov., a Dead Sea halobacterium with an extremely high magnesium requirement, *Int. J. Syst. Bacteriol.* **33**:381–386.

Oren, A., 1983d, A thermophilic amyloglucosidase from *Halobacterium sodomense,* a halophilic bacterium from the Dead Sea, *Curr. Microbiol.* **8**:225–230.

Oren, A., 1983e, *Clostridium lortetii* sp. nov., a halophilic obligatory anaerobic bacterium producing endospores with attached gas vacuoles, *Arch. Microbiol.* **136**:42–48.

Oren, A., 1985, The rise and decline of a bloom of halobacteria in the Dead Sea, *Limnol. Oceanogr.* **30**:911–915.

Oren, A., 1986a, Intracellular salt concentrations of the anaerobic halophilic eubacteria *Haloanaerobium praevalens* and *Halobacteroides halobius, Can. J. Microbiol.,* **32:** 4–9.

Oren, A., 1986b, The ecology and taxonomy of anaerobic halophilic eubacteria, *FEMS Microbiol. Rev.* **39:** 23–29.

Oren, A., and Shilo, M., 1981, Bacteriorhodopsin in a bloom of halobacteria in the Dead Sea, *Arch. Microbiol.* **130**:185–187.

Oren, A., and Shilo, M., 1982, Population dynamics of *Dunaliella parva* in the Dead Sea, *Limnol. Oceang.* **27**:201–211.

Oren, A., and Shilo, M., 1985, Factors determining the development of algal and bacterial blooms in the Dead Sea: A study of simulation experiments in outdoor ponds, *FEMS Microbiol. Ecol.* **31**:229–237.

Oren, A., and Vlodavsky, L., 1985, Survival of *Escherichia coli* and *Vibrio harveyi* in Dead Sea water, *FEMS Microbiol. Ecol.* **31**:365–371.

Oren, A., Paster, B. J., and Woese, C. R., 1984a, Haloanaerobiaceae: A new family of moderately halophilic, obligatory anaerobic bacteria, *Syst. Appl. Microbiol.* **5**:71–80.

Oren, A., Weisburg, W. G., Kessel, M., and Woese, C. R., 1984b, *Halobacteroides halobius* gen. nov., sp. nov., a moderately halophilic anaerobic bacterium from the bottom sediments of the Dead Sea, *Syst. Appl. Microbiol.* **5**:58–70.

Post, F. J., 1977, The microbial ecology of the Great Salt Lake, *Microb. Ecol.* **3**:143–165.

228   A. Oren

Pundak, S., and Eisenberg, H., 1981, Structure and activity of malate dehydrogenase from the extreme halophilic bacteria of the Dead Sea. 1. Conformation and interaction with water and salt between 5 M and 1 M NaCl concentration, *Eur. J. Biochem.* **118:**463–470.

Rafaeli-Eshkol, D., 1968, Studies on halotolerance in a moderately halophilic bacterium. Effect of growth conditions on salt resistance of the respiratory system, *Biochem. J.* **109:**679–685.

Rodriguez-Valera, F., Ruiz-Berraquero, F, and Ramos-Cormenzana, A., 1980, Isolation of extremely halophilic bacteria able to grow in defined inorganic media with single carbon sources, *J. Gen. Microbiol.* **119:**535–538.

Sass, E., and Ben-Yaakov, S., 1977, The carbonate system in hypersaline solutions: Dead Sea brines, *Mar. Chem.* **5:**183–199.

Shindler, D. B., Wydro, R. M., and Kushner, D. J., 1977, Cell-bound cations of the moderately halophilic bacterium *Vibrio costicola*, *J. Bacteriol.* **130:**698–703.

Soliman, G. S. H., and Trüper, H. G., 1982, *Halobacterium pharaonis* sp. nov., a new, extremely haloalkaliphilic archaebacterium with low magnesium requirement, *Zentralbl. Bakteriol. Hyg. I Abt. Orig. C* **3:**318–329.

Steinhorn, I., 1983, *In situ* salt precipitation at the Dead Sea, *Limnol. Oceanogr.* **28:**580–583.

Steinhorn, I., and Gat, J. R., 1983, The Dead Sea, *Sci. Am.* **249**(4):102–109.

Steinhorn, I., Assaf, G., Gat, J. R., Nishry, A., Nissenbaum, A., Stiller, M., Beyth, M., Neev, D., Garber, R., Friedman, G. M., and Weiss, W., 1979, The Dead Sea: Deepening of the mixolimnion signifies the overture to overturn of the water column, *Science* **206:**55–57.

Stephens, D. W., and Gillespie, D. M., 1976, Phytoplankton production in the Great Salt Lake, Utah, and a laboratory study of algal response to enrichment, *Limnol. Oceanogr.* **21:**74–87.

Stiller, M., and Chung, Y. C., 1984, Radium in the Dead Sea: A possible tracer for the duration of meromixis, *Limnol. Oceanogr.* **29:**574–586.

Stiller, M., and Kaufman, A., 1984, ^{210}Pb and ^{210}Po during the destruction of stratification in the Dead Sea, *Earth Planet. Sci. Lett.* **71:**390–404.

Stiller, M., Gat, J. R., Bauman, N., and Shasha, S., 1984a, A short meromictic episode in the Dead Sea: 1979–1982 *Verh. Int. Verein. Limnol.* **22:**132–135.

Stiller, M., Mantel, M., and Rapaport, M. S., 1984b, The determination of trace elements (Co, Cu, and Hg) in the Dead Sea by neutron activation followed by X-ray spectrometry and magnetic deflection of beta ray interference, *J. Radioanalyt. Nucl. Chem.* **83:**345–352.

Stoeckenius, W., and Bogomolni, R. A., 1982, Bacteriorhodopsin and related pigments of halobacteria, *Annu. Rev. Biochem.* **52:**587–616.

Tabor, H. Z., 1966, Solar ponds, *Sci. J.* **1966**(June):66–71.

Tindall, B. J., Mills, A. A., and Grant, W. D., 1980, An alkalophilic red halophilic bacterium with a low magnesium requirement from a Kenyan soda lake, *J. Gen. Microbiol.* **116:**257–260.

Tindall, B. J., Ross, H. N. M., and Grant, W. D., 1984, *Natronobacterium* gen. nov. and *Natronococcus* gen. nov., two new genera of haloalkaliphilic archaebacteria, *Syst. Appl. Microbiol.* **5:**41–57.

Torsvik, T., and Dundas, I. D., 1974, Bacteriophage of *Halobacterium salinarium*, *Nature* **248:**680–681.

Volcani, B. E., 1944, The microorganisms of the Dead Sea, in: *Papers Collected to Commemorate the 70th Anniversary of Dr. Chaim Weizmann*, pp. 71–85, Collective Volume, Daniel Sieff Research Institute, Rehovoth.

Walsby, A. E., 1980, A square bacterium, *Nature* **283**:69–71.

Wegmann, K., Ben-Amotz, A., and Avron, M., 1980, Effect of temperature on glycerol retention in the halotolerant algae *Dunaliella* and *Asteromonas, Plant Physiol.* **66**:1196–1197.

Weiner, D., 1985, The Dead Sea. Past, present, future, *Interdisc. Sci. Rev.* **10**:151–158.

Werber, M. M., and Mevarech, M., 1978, Induction of dissimilatory reduction pathway of nitrate in *Halobacterium* of the Dead Sea. A possible role for the 2 Fe-ferredoxin isolated from this organism, *Arch. Biochem. Biophys.* **186**:60–65.

Widdel, F., and Pfennig, N., 1981, Sporulation and further nutritional characteristics of *Desulfotomaculum acetoxidans, Arch. Microbiol.* **129**:401–402.

Wilkansky, B., 1936, Life in the Dead Sea, *Nature* **138**:467.

7

Biogeochemistry and Ecophysiology of Atmospheric CO and H₂

RALF CONRAD

1. Introduction

Hydrogen and carbon monoxide are just trace constituents in our environment, but their cycles are nevertheless of great importance for life on earth. Two different cycles may be distinguished, the cycling between the biosphere and the atmosphere (atmospheric cycle) and the cycling and turnover within individual ecosystems of the biosphere (biospheric cycle). For H₂, it is the biospheric cycle in anoxic environments that is of special interest, since H₂ is an important intermediate in the decomposition of organic matter and functions as a regulator for the whole mineralization process. The role of H₂ in these environments has been described and discussed in a number of reviews on methane production and sulfate reduction in anoxic ecosystems (Zehnder, 1978; Nedwell, 1984; Zeikus, 1983) and thus will not be the subject of this review. The biospheric cyles of CO and H₂ in anoxic environments are of relatively little importance for the atmospheric budgets of CO and H₂ (see Section 6). However, they are of great importance for the budget of atmospheric CH₄ (Seiler, 1984), which is an indirect source for atmospheric CO and H₂ (see Section 7). Biospheric cycles of CO and H₂ are also operative in oxic environments. There, they may function as more or less closed cycles or they may result in significant exchange with the atmosphere and

RALF CONRAD • Max-Planck-Institut für Chemie, D-6500 Mainz, Federal Republic of Germany. Present address: Universität Konstanz, Fakultät für Biologie, D-7750 Konstanz 1, Federal Republic of Germany.

Table I. Most Important Processes Involved in the Biospheric Emission
and/or Deposition of Atmospheric CO and H_2

Part of biosphere	CO	H_2
Biospheric production		
Vegetation	Photochemical oxidation of cell organic carbon	Unknown
Ocean and freshwater	Photochemical oxidation of dissolved organic carbon	N_2 fixation by cyanobacteria
Soil	Chemical oxidation of humus	N_2 fixation in root nodules
Anoxic environments	Methanogenesis, CO emitted together with CH_4	Fermentation of organic matter, H_2 emitted together with CH_4
Biospheric consumption		
Vegetation	Oxidation in mitochondria, assimilation via CO_2	Unknown
Soil and water	(1) Oxidation by unknown microorganisms with high efficiency, (2) Cooxidation by NH_4^+-oxidizers	(1) Oxidation by abiontic hydrogenases, (2) Oxidation by unknown microorganisms with high efficiency
Anoxic environments	Methanogenesis, sulfidogenesis, acetogenesis	Methanogenesis, sulfidogenesis, acetogenesis

influence the atmospheric cycles of CO and H_2 (see Section 2). These atmospheric cyles, especially that of CO, have a key function in air chemistry and thus influence the earth's climate (see Section 3).

This review will concentrate on the biogeochemistry of those processes that are of significance for atmospheric H_2 and CO and will further concentrate on the ecophysiology of microbial activities and of other processes involved in the cycling of atmospheric H_2 and CO (see Sections 4–8). Our knowledge on the biogeochemistry of CO and on microorganisms producing or consuming CO was summarized by Nozhevnikova and Yurganov (1978). Since that time, large numbers of field and laboratory studies have been done and resulted in new perspectives that have helped, and will continue to help, in the design of the appropriate experimental studies necessary to obtain a valid qualitative and quantitative description of the atmospheric cycles of trace gases. An overview of the processes that seem to be most important for the exchange of CO and H_2 between biosphere and atmosphere is given in Table I. These processes include chemical reactions (CO production), reactions catalyzed by abiontic enzymes (H_2 consumption), and metabolic processes in which the gases are either a major metabolite (e.g., H_2 in anoxic environments), a side product (e.g., H_2 production during N_2 fixation), or a cosubstrate (e.g., CO consumption).

New perspectives include the awareness that processes involved in consumption or production of atmospheric trace gases in nature are not necessarily the same as those that are known from studies of microbial gas metabolism at much higher gas concentrations. Thus, it becomes more and more evident that the chemolithoautotrophic bacteria that are able to grow on CO and H_2 as sole energy source cannot be responsible for destruction of atmospheric CO and H_2, but must occupy alternative ecological niches (see Sections 9 and 10).

Another perspective concerns the concept that fluxes between biosphere and atmosphere are the net result of simultaneously operating production and consumption processes that are individually controlled by environmental parameters. Turnover may take place within one species of microorganisms, within particular oxic or anoxic compartments, and between these compartments. The turnover, together with the physical transport processes, results in the establishment of concentration gradients that by themselves greatly influence the magnitude of consumption rates, which usually follow first-order kinetics.

A simplified scheme of the relations among different parts of the biosphere is shown in Fig. 1. The scheme was drawn from a pragmatic point

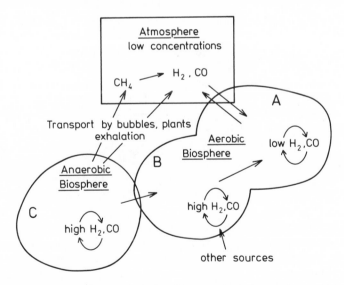

Figure 1. Hypothetical scheme of biospheric–atmospheric interactions with respect to atmospheric CO and H_2.

of view, since it has turned out so far that the magnitude of the trace gas concentration in the biospheric compartments as well as the (an)aerobiosis are features that determine the ecophysiological character of the biological processes involved. The biospheric compartments that are most important for the atmospheric cycle of CO and H_2 are those that are oxic and exhibit CO and H_2 concentrations close to the ambient atmospheric concentrations (compartment A). Anoxic environments (compartment C) contribute to a lesser extent, usually by direct emission via exhalation, bubbles, or plant-mediated transport, so that the gases produced escape oxidation in surrounding oxic environments. Aerobic biospheric compartments or interfaces (compartment B) with high concentrations of H_2 and CO, i.e., concentrations much higher than ambient atmospheric, only exist in special biotopes, e.g., geothermal environments. However, these compartments most probably constitute important ecological niches for chemolithotrophic bacteria oxidizing CO or H_2.

2. Global Budget of Tropospheric CO and H_2

The budgets of H_2 and CO are summarized in Table II. The sources include anthropogenic, biospheric, and chemical processes. Anthropogenic emissions essentially are due to combustion of fossil fuels and to burning of biomass in forest clearing and shifting cultivation. Biospheric emissions are due to processes in plants, oceans, and soils that will be described below. Chemical formation of H_2 and CO occurs by oxidation of tropospheric CH_4 and nonmethane hydrocarbons initiated by reaction with OH. On the other hand, reaction with OH is also a sink for H_2 and CO. Since a concentration gradient of CO exists at the tropopause, the stratosphere is a sink for tropospheric CO (Seiler and Warneck, 1972). This is not the case for H_2, since there is no significant gradient between troposphere and stratosphere (Schmidt, 1978).

Soils are also a sink for tropospheric H_2 and CO. Biological reactions in soil constitute >90% and >15% of the total sink strength for H_2 and CO, respectively. Direct emission of H_2 and CO from the biosphere into the atmosphere makes up only 6–11% of the total sources. It must be pointed out, however, that a further 30–40% of the tropospheric H_2 and CO are photochemically formed from tropospheric CH_4 and other hydrocarbons, predominantly isoprene and terpene. Since isoprene and terpene are mainly emitted by plants (Zimmerman et al., 1978) and ~60% of the CH_4 is of biogenic origin (Seiler, 1984), the biosphere is a significant indirect source of tropospheric H_2 and CO.

Table II. Budgets of Atmospheric CO and $H_2{}^a$

	Amount (Tg/year^{-1})	
	CO	H_2
Sources		
Anthropogenic		
Industry, traffic, heating	640 ± 200	20 ± 10
Biomass burning, shifting cultivation	1000 ± 600	20 ± 10
Total anthropogenic	1640 ± 800	40 ± 20
Biospheric		
Vegetation	75 ± 25	< 0.1
Ocean	100 ± 80	4 ± 2
Soil	17 ± 15	3 ± 2
Freshwater, anoxic environments	< 0.2	< 0.2
Total biospheric	192 ± 120	7 ± 4
Chemical		
Oxidation of methane	600 ± 300	15 ± 5
Oxidation of NMHC^b	900 ± 500	25 ± 10
Total chemical	1500 ± 800	40 ± 15
Total sources	3332 ± 1720	87 ± 39
Sinks		
Chemical oxidation	2000 ± 600	8 ± 3
Diffusion into stratosphere	110	
Soil	390 ± 140	90 ± 20
Total sinks	2500 ± 740	98 ± 23
Residence time (years)	$0.1 - 0.4$	$1.5 - 4.0$

^{a}Compiled from Seiler and Conrad (1987).
^{b}NMHC, Nonmethane hydrocarbons.

The residence time τ of H_2 and CO in the troposphere is given by the equation

$$\tau = M/Q = M/S$$

where M is the total mass of H_2 or CO in troposphere, Q is the total source strength, and S is the total sink strength. The equation assumes that the cycles of H_2 and CO are in steady state. Changes of the total source or sink strengths automatically result in changes of the mixing ratios. Since sink processes generally are first-order reactions with respect to the atmospheric mixing ratio, the sink strength will adapt to changing atmospheric mixing ratio until a new steady state is reached. The time constant of this adaptation increases with the residence time.

The residence time of CO is about ten times shorter than that of H_2 (Table II) and this explains the significantly higher variability of tropospheric CO mixing ratios compared to tropospheric H_2 mixing ratios. Hydrogen is well mixed within the troposphere and does not show significant gradients with latitude, altitude, or season (Schmidt, 1974, 1978; Seiler, 1978). Exceptions are local emissions by strong anthropogenic activities, e.g., in urban areas (Seiler and Zankl, 1975; Scranton et al., 1980) or local depositions by strong biospheric activities, e.g., in the humid tropics (Seiler and Conrad, 1987).

In contrast to H_2, CO mixing ratios are significantly higher in the Northern than in the Southern Hemisphere (Seiler, 1974, 1978; Heidt et al., 1980; E. Robinson et al., 1984; Seiler and Fishman, 1981) and are strongly influenced by local anthropogenic or natural emissions (Seiler and Zankl, 1976; Marenco and Delaunay, 1980; Crutzen et al., 1985) and by photochemistry (Fishman and Seiler, 1983). These provide a reason for the strong seasonal periodicity of the tropospheric CO mixing ratio (Seiler et al., 1984c).

3. Importance of CO and H_2 for Tropospheric Chemistry

Hydrogen and CO are atmospheric trace gases with average tropospheric mixing ratios of 550 and 110 ppbv, respectively (Seiler, 1978). The term mixing ratio (volume per volume) given in parts per million (ppmv = 10^{-6} v/v) or parts per billion (ppbv = 10^{-9} v/v) usually is preferred by air chemists, since in contrast to concentration terms (i.e., mass per volume), it is independent of temperature and pressure. The troposphere, i.e., the lower 8–12 km of atmosphere, contains 75% of the total atmospheric mass of 5.2×10^{21}g. The troposphere is separated from the stratosphere by the tropopause, which is characterized by a temperature inversion. This inversion is the reason that gas exchange between troposphere and stratosphere on the average takes 1–2 years. Troposphere and stratosphere are well-separated entities; the following discussion will deal exclusively with the tropospheric cycle of H_2 and CO.

The exchange of H_2 and CO between biosphere and atmosphere influences the atmospheric mixing ratios of H_2 and CO. The magnitude of their mixing ratios has a significant impact on the chemical reactions in the atmosphere. Tropospheric chemistry is very complex and can only be treated with elaborate models (Logan *et al.*, 1981; Crutzen, 1983). The OH radical plays a central role in tropospheric chemistry by initiating chemical reactions with a multitude of atmospheric compounds. The mixing ratio of OH is the result of its production and destruction rates. Production of OH occurs in the second reaction of the following chain of reactions:

$$O_3 + h\nu \rightarrow O(^1D) + O_2 \quad (\lambda \leq 310 \text{ nm})$$
$$O(^1D) + H_2O \rightarrow 2OH$$
$$O(^1D) + M \rightarrow M + O$$

where $O(^1D)$ is the energetically excited O atom and M is the inert reaction partner, usually N_2 or O_2. The production rate of OH is a function of the O_3 and H_2O mixing ratios as well as of temperature and radiation. The precursor O_3 reaches the troposphere by diffusion from the stratosphere or originates as a product of other chemical reactions, e.g., chemical oxidation of CO or CH_4 (see later in this Section).

The destruction of OH occurs by its reaction with other compounds X according to the following equation:

$$\frac{dm_{OH}}{dt} = \frac{dm_X}{dt} = -k(T,p)\, m_{OH}\, m_X$$

where m_X is the mixing ratio of X, m_{OH} is the mixing ratio of OH, and $k(T,p)$ is the reaction constant k as function of temperature T and pressure p. Many compounds react with OH. Important for tropospheric chemistry are those compounds that influence the tropospheric mixing ratio of OH. This influence depends on the magnitude of m_X and of k. Values of $k(T = 298$ K; $p = 1$ bar) and m_X, as well as the average lifetime τ_{OH} of the OH radical, calculated by

$$\tau_{OH} = 1/k\, m_X$$

are summarized in Table III for the most important tropospheric compounds.

The shortest lifetime of OH is caused by the reaction with CO, followed by CH_4, formaldehyde, H_2 and O_3. Hence, the OH mixing ratio in the troposphere is predominantly controlled by the mixing ratio of CO and to a lesser extent of CH_4. The OH mixing ratio can be calculated by using the mixing ratios of CO and CH_4 and their reaction constants with

Table III. Reaction Constants and Average Lifetimes of OH and Other Compounds in the Troposphere[a]

Compound	k (cm^3 molecule^{-1} sec^{-1})	Mixing ratio m_X (ppbv)	Lifetime τ_{OH} (sec)	τ_X (hr)
CO	2.8×10^{-13}	110	1.3	2,000
CH$_4$	7.7×10^{-15}	1700	3.1	72,000
HCOH	1.3×10^{-11}	0.5	6.2	43
H$_2$	7.5×10^{-15}	550	9.8	74,000
O$_3$	7.9×10^{-14}	40	12.8	7,000

[a]Values of k are from Hampson (1980).

OH, resulting in an average value of 5×10^5 molecules/cm^3 of OH (Singh, 1977; Crutzen, 1982).

Also shown in Table III is the lifetime τ_X of the individual compounds X calculated for an OH mixing ratio of 5×10^5 molecules/cm^3 by

$$\tau_X = 1/k \, m_{OH}$$

Because of their reaction constants, formaldehyde has the shortest lifetime (~ 2 days), followed by CO and ozone (2–10 months), and then CH$_4$ and H$_2$ (8–9 years).

The reactions of H$_2$ or CO with OH result in the formation of HO$_2$ (Crutzen, 1979):

$$H_2 + OH \rightarrow H_2O + H$$
$$CO + OH \rightarrow CO_2 + H$$
$$H + O_2 + M \rightarrow M + HO_2$$

The fate of HO$_2$ mainly depends on the mixing ratio of NO, a product of combustion processes, and N-transformation reactions in soil (Crutzen, 1979). The critical NO mixing ratio is ~ 5 pptv (5×10^{-12} v/v). At NO mixing ratios > 5 pptv, HO$_2$ decomposition results in the net formation of O$_3$ in addition to the regeneration of OH:

$$HO_2 + NO \rightarrow OH + NO_2$$
$$NO_2 + h\nu \rightarrow NO + O$$
$$\underline{O + O_2 + M \rightarrow M + O_3}$$
$$\text{net:} \quad HO_2 + O_2 \rightarrow OH + O_3$$

At NO mixing ratios <5 pptv, HO$_2$ decomposition results in the net consumption of O$_3$:

$$HO_2 + O_3 \rightarrow OH + O_2$$

or does not involve ozone at all:

$$HO_2 + HO_2 \rightarrow H_2O_2 + O_2$$
$$H_2O_2 + h\nu \rightarrow 2OH$$

Hence, H$_2$ and CO not only influence the mixing ratio of OH, but also that of O$_3$ in the troposphere. Because of their reaction constants with OH, only CO, but not H$_2$, is of significance for tropospheric O$_3$ (compare τ_X in Table III). The influence of CO on tropospheric O$_3$ was confirmed by measuring the distribution of CO and O$_3$ in the troposphere, using aircraft for sampling (Fishman and Seiler, 1983). The importance of CO (and, to a virtually insignificant extent, of H$_2$) for tropospheric chemistry rests in its impact on O$_3$ and OH, which themselves influence the chemical decomposition of other tropospheric constituents. Thus, the recently recognized increase of the atmospheric abundance of CH$_4$ by ~1.5%/year (Rasmussen and Khalil, 1981; Khalil and Rasmussen, 1983) was explained by a possible increase of atmospheric abundance of CO (Khalil and Rasmussen, 1984), although an increase of CO was debated and still awaits confirmation by long-term measurements from baseline stations (Seiler, 1985).

4. Emission of CO from the Biosphere into the Atmosphere

Carbon monoxide is emitted from (1) vegetation, (2) ocean, (3) freshwater, and (4) soils. In all these parts of the biosphere, the predominant CO-producing processes are chemical rather than metabolic. In other words, CO is produced from biogenic material, but by chemical reactions.

4.1. Vegetation and Phototrophic Microorganisms

The vegetation cover of the earth, with a source strength of 50–100 Tg/year, is an important source in the global CO budget. This figure is based on field studies of CO emission from trees carried out by Bauer *et al.* (1979), Seiler *et al.* (1978), and Seiler and Giehl (1977). During these measurements, the authors generally observed emission but never deposition of atmospheric CO. The CO uptake by plants can only be observed by utilizing ^{14}CO or high CO mixing ratios. When the uptake is measured by using ^{14}CO, a simultaneous production of ^{12}CO is not observable even if much stronger than uptake. At increasing CO mixing ratios, the CO uptake rate increases and eventually may reach values exceeding those of the simultaneous CO production rate. At ambient CO mixing ratios, however, this does not happen and plants would show net emission of CO.

By using ^{14}CO, several authors actually demonstrated that plants are able to oxidize CO to CO_2 and to assimilate CO into cell material (Bidwell and Fraser, 1972; Bidwell and Bebee, 1974; Bzdega et al., 1981; Peiser et al., 1982). In the light, CO is mainly incorporated into serine (Bidwell and Bebee, 1974), whereas in darkness CO is first oxidized to CO_2, which then is partially incorporated into malate, citrate, and aspartate (Peiser et al., 1982). In the light, too, CO fixation may occur via CO_2, especially in C_4 plants (Bidwell and Fraser, 1972). Green algae and cyanobacteria are also able to oxidize CO (Chapelle, 1962). The rate of CO utilization is strongly dependent on the CO concentration and the temperature, but only to a small extent on light intensity. Light only seems to be necessary for assimilation, but not for oxidation. The CO_2 concentration also seems to affect CO assimilation, but not CO oxidation (Peiser et al., 1982). The oxidation reaction only operates in presence of O_2, but a concentration of 1% is sufficient for full activity (Chapelle, 1962; Bzdega et al., 1981). The CO-oxidizing activity is localized in the mitochondria and exhibits an extremely low K_m for CO (about 7 nM) (Peiser et al., 1982). On the other hand, a K_m of 0.3 mM CO has been observed for assimilation of CO in Scenedesmus (Chapelle, 1962). In higher plants (nine species studied), too, CO assimilation apparently has a K_m of >200 nM, since the rate of incorporation of the radioactive CO up to 100 ppmv was proportional to the CO mixing ratio in the gas phase (Bidwell and Bebee, 1974). Future research should find out whether the different affinities for CO are due to different CO-converting enzymes. It is also unclear whether the phototrophic organism gains any advantage in possessing CO-utilizing capacity. One might speculate that CO is a central metabolite in phototrophs and that oxidation or assimilation of CO serves in detoxification or in the gain of energy, reduction equivalents, or cell carbon.

In fact, CO is a metabolic product during synthesis of phycobilins from heme precursors (Troxler, 1972; Troxler and Dokos, 1973). Carbon monoxide originates from the oxidation of a methin bridge in the porphyrin ring. A similar CO production process is most probably involved in the degradation of other prophyrins, e.g., chlorophyll, cytochrome, and cobalamine. However, experimental evidence of CO production from these compounds only exists for animal cells (Tenhunen *et al.,* 1969; Yoshida et al., 1982) and for heterotrophic bacteria (Engel *et al.,* 1972, 1973).

Carbon monoxide production during photorespiration of plants has been proposed by Fischer and Lüttge (1978, 1979; Lüttge and Fischer, 1980). These authors observed that (1) CO production increases relative to net CO_2 assimilation as soon as the ratio of O_2/CO_2 concentration increases, (2) the relative CO production is higher in C_3 than in C_4 plants, and (3) the CO production increases linearly with light intensity, whereas CO_2 assimilation becomes light-saturated. The authors assumed that CO is formed via formate by oxidation of the glycolate formed during photorespiration. It is as yet unknown, however, whether a CO formation can occur by oxidation of glycolate.

Bauer *et al.* (1979, 1980) made similar observations using trees, *Chlorella,* and phototrophic bacteria. However, these authors presented a different interpretation, i.e., formation of CO by photooxidation rather than by photorespiration. A reaction mechanism was postulated in which reactive oxygen (O_2^*) is formed by photosensitization of chlorophyll:

$$Chl + h\nu \rightarrow Chl^*$$
$$Chl^* + O_2 \rightarrow Chl + O_2^*$$
$$O_2^* + cell\ C \rightarrow CO$$

This reaction would be consistent with the following observations (Wilks, 1959; Loewus and Delwiche, 1963; Krinsky, 1978; Wolff and Bidlack, 1976; Bauer *et al.,* 1979, 1980): (1) CO production was almost independent of temperature because of the low activation energy of photooxidative processes, was independent of the metabolic integrity of cells as long as a photosensitizer was present, and was stimulated by blue light. (2) CO emission from tree leaves increased linearly with light, but was more or less independent of the season (Fig. 2). Similar rates were observed even when leaves discolored in fall. (3) Reactive oxygen is photodynamically produced by irradiation of chlorophyll, if the sensitized state of chlorophyll is not quenched by carotenoids. The CO production was higher in cells with an altered arrangement of the pigments and thus with a presumably decreased quenching efficiency. (4) Reactive oxygen may result in CO formation by peroxidation of lipids. In some experiments with cells having a low chlorophyll content, CO production was much too high to be explained by oxidation of the methin bridge of the porphyrin ring of chlorophyll.

A contribution of metabolic CO formation by photorespiration and cleavage of porphyrin rings to the production and emission of CO from

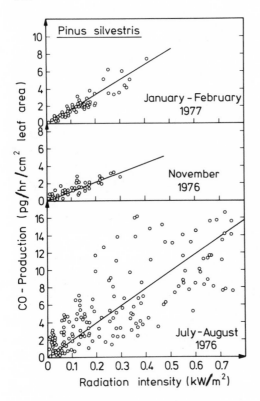

Figure 2. Influence of season and light intensity on the emission of CO from pine trees into the atmosphere. The measurements were conducted under field conditions by enclosing a tree branch in a glass box that was permanently flushed with ambient air. [After Bauer *et al.* (1979).]

phototrophic organisms into the environment cannot be excluded. However, it is questionable to what extent they contribute and whether they may exceed CO production by purely photochemical mechanisms.

4.2. Ocean

It is a general finding that ocean water is supersaturated with respect to the atmospheric CO mixing ratio. This observation has repeatedly been confirmed by measurements during ship expeditions (Swinnerton *et al.*, 1970; Swinnerton and Lamontagne, 1974; Seiler and Schmidt, 1974; Seiler, 1978; Conrad *et al.*, 1982). Since the CO concentration of surface water changes by a diel rhythm and increases with light intensity, and, furthermore, correlates with abundance of phytoplankton, particulate organic carbon, primary productivity, and chlorophyll *a*, it has been suggested that CO is produced by photometabolism of phytoplankton (Swinnerton *et al.,* 1977). Other experiments, however, indicate that CO is produced by photooxidative processes (Bullister *et al.,* 1982; Wilson *et al.,* 1970; Conrad and Seiler, 1980c; Conrad *et al.,* 1982). These experiments in particular demonstrated the following: (1) CO production increases linearly with the light intensity and is dependent on the presence of oxygen. (2) CO production rates do not decrease when algae are removed by filtration through 3-μm filters. In contrast, CO production even increases and is further stimulated after filtration through 0.2-μm filters, which remove bacteria. (3) CO production is not abolished by boiling or by treatment with metabolic inhibitors (Fig. 3). (4) CO production is observed in every water sample, even in those from >2000 m depth, where photometabolism does not occur, as soon as the water is exposed to light. In darkness, however, the CO concentration decreased, demonstrating that CO-consuming processes also exist in ocean water (Fig. 3).

The dependence of CO production on the presence of light and oxygen and its independence on the presence of plankton or bacteria indicates that CO production is a photochemical rather than a photometabolic process. Plankton and bacteria apparently contribute to CO consumption rather than to CO production. The substrates for photooxidative CO formation are most probably dissolved organic substances. A likely candidate is "Gelbstoff," which is a humuslike material common

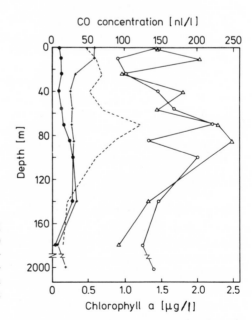

Figure 3. Vertical profiles of dissolved CO in the deep Atlantic Ocean. (+) Immediately analyzed; (●) 24 hr incubation in darkness; (○) 24 hr incubation in daylight; (△) 24 hr incubation in daylight after addition of NaN$_3$; (- -) Chlorophyll a.

in ocean water and is most probably a degradation product of algal excretions (Ehrhardt, 1984). Humuslike materials can be photosensitized, stabilize free radicals, and can be photooxidized (Choudry, 1984). It has been shown that superoxide radicals and singlet oxygen radicals are formed in humus-rich water (Baxter and Carey, 1983; Haag et al., 1984). It appears that CO is formed by either the photosensitized "Gelbstoff" molecules or by reactions initiated by the photoproduction of oxygen or other radicals (Zafiriou et al., 1984).

Carbon monoxide concentrations in ocean water were found to be correlated to chlorophyll a (Swinnerton et al., 1977; Conrad et al., 1982). This correlation does not prove that CO is produced by phytoplankton, but indicates that CO production is related to the presence of plankton. The plankton-related CO production is most likely caused by the greater abundance of algal excretion products, "Gelbstoff," or other dissolved organic substrates in these waters. It is not excluded, however, that some CO may also be produced directly from plankton cells by mechanisms similar to those described for plants and other phototrophic organisms. Even in this case, however, a large part of the produced CO may originate from photooxidative decomposition of cell constituents rather than from photometabolic processes.

Sometimes, CO production has also been found in the dark zone of ocean water, i.e., below 100 m depth. High CO concentrations were often observed at the upper boundary layer of water with higher salinity, where organic matter may accumulate because of the increase in water density (the same for H_2 concentration; see Section 5.1) (Seiler and Schmidt, 1974). However, deep ocean water generally is slightly supersaturated with CO (Conrad et al., 1982). This supersaturation has been explained by CO production by heterotrophic bacteria. Junge et al. (1971, 1972) isolated strains of Alginomonas, Brevibacterium, and Agrobacterium from deep ocean water and showed that these bacteria were able to produce traces of CO. Other experiments did not show an increase of dissolved CO when water samples were incubated for 24 hr in darkness (Conrad et al., 1982). Dark CO production in deep ocean water, thus, can only be a very slow process.

It should be emphasized that CO is not only produced in ocean water, but is also consumed. In contrast to CO production, CO consumption apparently is a microbial process (see Section 8.1). The simultaneous production and consumption result in cycling of CO within the water column. A vertical transport of CO is only of minor importance (Conrad et al., 1982), so that almost all of the CO formed in deeper water layers is recycled by microorganisms and does not reach the atmosphere. However, the light-dependent CO production in surface water is so high that oceans function as a net source for atmospheric CO.

The flux of CO from the water into the atmosphere is generally calculated by the laminar film model (Broecker and Peng, 1974; Liss and Slater, 1974). This model assumes that the gas exchange between water and air is limited by molecular diffusion through a hypothetical laminar film. The flux depends on the gradient of the CO partial pressure between water and gas phase and on the thickness of the laminar film. The film thickness decreases with wind speed and, in fact, CO supersaturation in surface water decreases with increasing wind speed (Conrad et al., 1982). It is completely unclear, however, whether the film thickness is also a function of chemical or biological parameters. The film thickness is just a hypothetical entity that is experimentally defined by calculating the flux of the noble gas radon or helium from isotopic analysis (Broecker and Peng, 1974; Peng et al., 1979). On the other hand, the ocean surface is covered by a physically and chemically defined lipid film (Lion and Leckie, 1981; Wangersky, 1976) and by a biologically defined neuston layer (Norkrans, 1980). It cannot be excluded that in this film CO is produced or consumed in completely different quantities than in water samples below. Recent observations comparing fluxes of different trace gases (H_2, CO, CH_4, N_2O) that are measured directly by a box method or are calculated from the supersaturation of the water using the laminar film model suggest that fluxes indeed are influenced by chemical or biological reactions in the surface film (Conrad and Seiler, 1988; Schütz et al., 1988)

4.3. Freshwater

The processes responsible for CO production in freshwater environments should principally be similar to those of the ocean. So far, there has been only one study on CO distribution and reactions involving CO in a freshwater ecosystem, i.e., a eutrophic lake (Conrad et al., 1983b). This work confirmed the importance of photooxidative CO production, but could not exclude the possibility of CO production by phytoplankton. Light-independent aerobic CO production processes, however, such as those described for animals (Wittenberg, 1960; Sjöstrand, 1970), fungi (Simpson et al., 1963; Westlake et al., 1959), yeasts (Radler et al., 1974), or hemolytic bacteria (Engel et al., 1972, 1973), apparently do not play an important role, since incubation of lake water samples in darkness generally resulted in CO consumption. High CO concentrations were observed in the anoxic hypolimnion (Conrad et al., 1983b). This is most probably due to CO production by methanogenic, acetogenic, or sulfidogenic bacteria and is in accordance with the observation of high CO concentrations in gas bubbles in submerged anoxic soils (see Section 6).

4.4. Soil

It is well known that soils act as a sink for atmospheric CO (e.g., Seiler, 1978). It was not generally recognized, however, that CO is also produced in soils. Thus, CO formation is overlooked if experiments are conducted by using radioactive ^{14}CO or using CO concentrations that are much higher than ambient. Such experiments may result in wrong estimates of CO fluxes between soil and atmosphere (Bartholomew and Alexander, 1981; Seiler and Conrad, 1982). Seiler (1978) has pointed out that the flux of CO between soil and atmosphere is the net result of simultaneously operating CO production and consumption reactions. A recently developed technique allows the discrimination of CO production and CO consumption rates under *in situ* conditions (Conrad and Seiler, 1985a,b). These studies showed that soils in different climatic regions act differently. In the relatively humid soils of temperate regions, CO consumption usually exceeds CO production by a factor of 5–10, so that these soils permanently act as a net sink of atmospheric CO (Seiler, 1978; Conrad and Seiler, 1980b). In most cases, CO is consumed within the uppermost soil layers, so that in 1 cm depth, CO mixing ratios are below 10–30 ppbv (Liebl and Seiler, 1976; Seiler *et al.,* 1977). In contrast to humid soils, arid soils show a diel rhythm in acting as a net source or net sink for

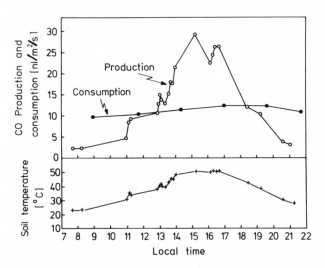

Figure 4. Rates of production and consumption of CO in arid soil (Andalusia, Spain). The measurements were done under field conditions by a modified static box technique. The figures are gross rates; net emission from the soil into the atmosphere is taking place when the production rates exceed consumption rates, otherwise net deposition of atmospheric CO is taking place. [After Conrad and Seiler (1985b).]

atmospheric CO (Fig. 4). This behavior is due to the diel rhythm of CO production, which is a function of the surface soil temperature. The CO consumption, on the other hand, appears to be independent of soil surface temperature and does not change significantly with daytime (Conrad and Seiler, 1985a,b). The reason for this behavior may be the localization of the individual processes within the soil column, i.e., CO production processes being localized in top soil with strong diel temperature changes and CO consumption processes being localized in deep soil where temperature stays relatively constant. It is obvious that the localization of CO production and consumption processes may significantly affect the CO flux between soil and atmosphere. In comparison to savannah soils, desert soils were even more extreme, in that they did not show CO consumption at all unless they were irrigated (Conrad and Seiler, 1982a). Thus, arid desert soils permanently act as a net source of atmospheric CO.

The total source strength of soils was estimated to be <30 Tg/year (Conrad and Seiler, 1985a). It is of interest to determine the processes responsible for this CO production. Laboratory experiments showed that it is a thermally sensitized, abiological process with CO production rates P strictly correlated to temperature according to the Arrhenius equation (Conrad and Seiler, 1980b, 1982a, 1985c):

$$P = A \exp(-E_a /RT)$$

where E_a is the activation energy, R is the gas constant, T is the temperature (kelvins), and A is the Arrhenius constant. Carbon monoxide was also produced from humic acids and a variety of phenolic compounds. In every case, CO production was strongly affected by content of organic matter, moisture, and pH. Both the activation energy and the Arrhenius constant (a measure of activation entropy) decreased with moisture and pH (Table IV). This decrease may be due to the increasing ionization and solubilization of the humic acid polymers.

The exact mechanism of abiotic CO production in soils is unknown. Since humic materials contain and stabilize free radicals (Choudhry, 1984), CO is likely to be produced by a mechanism involving such radicals. Since CO production is stimulated by an oxygen atmosphere and inhibited by a nitrogen atmosphere, CO production seems to involve reactions with O_2. Gohre and Miller (1983) demonstrated a light-dependent production of singlet O_2 in soil. Since addition of quenchers to soil did not show an inhibitory effect (Conrad and Seiler, 1985c), free oxygen radicals appear to be of minor importance for CO production. Carbon monoxide is most probably formed from phenolic moieties in the soil humic materials. It is formed during the autooxidation of alkaline pyro-

Table IV. Activation Energy E_a and Arrhenius Constant A of CO Production in Soil (Eolian Sand)[a]

Addition to soil	E_a (kJ mole^{-1})	A (nl hr^{-1}g^{-1})	CO production rate at 20°C (nl hr^{-1} g^{-1} dry weight)
Dry soil	129	2×10^{21}	0.03
HCl (pH 2.4)	96	2×10^{17}	0.90
H_2O (pH 5.0)	70	3×10^{12}	1.06
NaOH (pH 9.3)	51	5×10^{10}	37.5

[a]Compiled from Conrad and Seiler (1985c).

gallol solution (Calvert *et al.*, 1864) and CO formation was confirmed in alkaline solutions of other phenolic compounds (Miyahara and Takahashi, 1971). It is also formed from mono-, di-, and trihydroxybenzenes, and from gallic acid, tyroxin, *p*-hydroxycoumaric acid, and coniferyl alcohol as soon as water is added. The extent of CO formation apparently is dependent on the relative positions of the hydroxyl groups and other substituents on the benzene ring (Conrad and Seiler, 1985c); however, the molecular mechanism by which CO is released is unknown.

5. Emission of H_2 from the Biosphere into the Atmosphere

In contrast to the case for CO, vegetation does not appear to be a significant source for atmospheric H_2. This conclusion is based on *in situ* measurements on tree leaves analyzing H_2 simultaneously with CO (Seiler *et al.*, 1978). However, light-dependent H_2 production has recently been observed when pieces of rice plants or cotton wood leaves were incubated under water (Schütz *et al.*, 1988). It is unclear by which reactions H_2 was produced and whether it was due to the submerged conditions or not. Since the experiments were not carried out under aseptic conditions, it is even unclear whether H_2 was produced by the plants themselves or whether microorganisms were involved in addition. However, it is remarkable that H_2 production proceeded under highly oxic conditions, since the plant material photosynthetically produced O_2.

It is unclear whether the vegetation may be of significance to the atmospheric H_2 budget. However, H_2 production processes in other oxic environments, i.e., ocean, fresh water, and soil, play a small but significant role in the atmospheric H_2 budget. The H_2 production in these environments seems to be predominantly due to the activity of nitrogenase in N_2-fixing organisms.

5.1. Ocean

The oceans, with a source strength of 2–6 Tg/year are only a relatively small source for atmospheric H_2 (Seiler and Schmidt, 1974). Possibly this figure is even overestimated, since some parts of the oceans, namely the Arctic Sea, turned out to be undersaturated with respect to atmospheric H_2 (Herr et al., 1981; Herr, 1984) (see Section 8.3). The oceans in the tropical, temperate, and antarctic zones, on the other hand, seem to be generally supersaturated with respect to atmospheric H_2 and thus act as a source (Seiler and Schmidt, 1974; Herr and Barger, 1978; Williams and Bainbridge, 1973; Setser et al., 1982; Scranton et al., 1982; Conrad and Seiler, 1986b).

Hydrogen supersaturation principally occurs in the light-penetrated euphotic mixing layer of the ocean, whereas the dark, deep ocean water is often undersaturated with H_2. Similar to the case for CO, there is an exponential decrease of dissolved H_2 from the sea surface to depths of about 100 m (Seiler and Schmidt, 1974; Herr and Barger, 1978; Lilley et al., 1982a; Scranton et al., 1982). In contrast to CO, however, H_2 concentrations in surface water do not exhibit a diel rhythm in general. Although diurnal changes with maxima have been observed at some stations, they have not been observed at others (Bullister et al., 1982; Herr et al., 1984; Setser et al., 1982; Seiler and Conrad, unpublished results). The origin of the H_2 dissolved in ocean water is unclear. This is mainly due to the fact that observations made in a particular water body are often not reproducible in another water body. The same is the case for incubation experiments to test for H_2 production or consumption activities.

There are three major hypotheses to explain H_2 supersaturation in ocean surface water: (1) H_2 production along with the photooxidation of long-chain aldehydes (Herr et al., 1984), (2) H_2 production by anaerobic bacteria present in detritus particles and protozoa (Lilley et al., 1982a), and (3) H_2 production by N_2-fixing cyanobacteria (Scranton, 1983, 1984).

The discussion of H_2 production by anaerobic bacteria in the generally oxic ocean water has a similar basis to that of the origin of CH_4 in the same water bodies. However, it has been pointed out that the number of anaerobic bacteria in fecal pellets is not sufficient to account for the required CH_4 production (Rudd and Taylor, 1980) and that particles must be as large as 500 μm to develop anoxic conditions in their centers (Jörgensen, 1977). The same is the case for H_2. The third explanation, i.e., H_2 production by N_2-fixation, seems to be the most promising, although the experimental evidence is not yet sufficient (Scranton, 1984); thus, the H_2-producing cyanobacterium *Oscillatoria thiebautii* had to be filtered by plankton nets out of large quantities of seawater to obtain detectable activities. The H_2 production rates were not proportional to colony counts and, furthermore, were affected by the volume of the incubation assay.

Hydrogen supersaturation has sometimes also been observed in the dark, deep sea, usually on top of a pycnocline (Seiler and Schmidt, 1974; Scranton *et al.*, 1982), where organic matter tends to be enriched due to the density gradient. Junge *et al.* (1972) were able to isolate some bacterial strains from these water bodies that produced traces of H_2 and CO in laboratory cultures. In fact, supersaturation of CO and CH_4 is usually observed simultaneously with H_2 (Seiler and Schmidt, 1974; Burke *et al.*, 1983; Scranton and Farrington, 1977) and again this has been explained by microbial activity, although the ecophysiological basis for production of these reduced gas species is unknown.

5.2. Freshwater

It has generally been believed that H_2 in freshwater, e.g., lakes, is produced by fermentative bacteria in the anoxic sediment, from which it diffuses into the supernatant water column (e.g., Kuznetsov, 1959; Atlas and Bartha, 1981). Thus, the fate of H_2 would be similar to that of CH_4, which exactly exhibits such a distribution pattern (Rudd and Taylor, 1980; Hanson, 1980). However, *in situ* measurements of dissolved H_2 in lake water refuted this hypothesis. Fermentation-derived H_2 apparently is completely cycled within the sediment and anoxic hypolimnion and does not reach the oxic parts of the water (Conrad *et al.*, 1983a).

Hydrogen production apparently is operating in the oxic, light-penetrated part of the water column, where maximum H_2 concentrations have been observed (Conrad *et al.*, 1983a; Schink and Zeikus, 1984; Dahm *et al.*, 1983; Schütz *et al.*, 1988). These H_2 maxima coincide with maximum numbers of phytoplankton, especially *Oscillatoria rubescens* (Conrad *et al.*, 1983a), and partially coincide with maximum activities in acetylene reduction (Fig. 5). Thus, it is likely that the H_2 is produced by the N_2-fixation process. Paerl (1982, 1983) showed an association of hydrogenase activity (measured by tritium exchange) with the nitrogenase activity in natural aquatic communities. Since this hydrogenase activity is believed to be due to the uptake hydrogenase being associated with the nitrogenase, this observation is indirect evidence for H_2 production by the nitrogen-fixing communities (compare Section 10). Conrad *et al.* (1983a) observed the highest H_2 concentrations in the early morning and the lowest H_2 concentrations in the late afternoon, and concluded that H_2 is produced predominantly during the night. Such behavior excludes photoproduction of H_2 (Hallenbeck and Benemann, 1979) that is dependent on light and at least transient anoxic conditions. Hydrogen production by the nitrogenase system of cyanobacteria, on the other hand, may also be

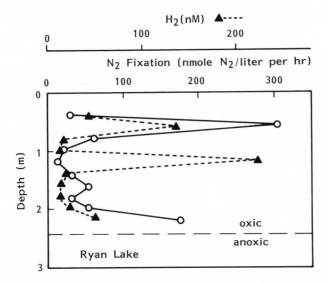

Figure 5. Vertical profiles of dissolved H_2 and of rates of N_2 fixation (acetylene reduction) in Ryan Lake (Washington). [After Dahm *et al.* (1983).]

operative in darkness, although it is usually assayed under illuminated incubation conditions (Bothe *et al.*, 1980; Lambert and Smith, 1981). Dark N_2 fixation is common in unicellular cyanobacteria and most probably also in other nonheterocysteous species, since in this way the bacteria protect their O_2-sensitive nitrogenase from the photosynthetically produced O_2 by temporal separation (Gallon, 1981).

Hydrogen production has also been observed in association with macrophytes *(Ruppia)* in Big Soda Lake (Nevada) (Oremland, 1983). The *Ruppia* leaves were densely colonized with *Anabaena*. Since H_2 production was only observed in darkness and was stimulated by acetylene, the author concluded that H_2 was not produced from the cyanobacteria, but from anaerobic bacteria decomposing dead cells. Hydrogen production has also been observed in a small anoxic salt pond, where H_2 concentrations showed a diel rhythm with maxima during the day (Scranton *et al.*, 1984). In this ecosystem, H_2 production was believed to be due to the activity of phototrophic bacteria.

5.3. Soil

Usually, H_2 is not produced, but is consumed in soils, and, in fact, soils are the predominant sink in the atmospheric budget of H_2 (Seiler, 1978; Conrad and Seiler, 1980a, 1985a). In some soils, however, i.e., those covered with N_2-fixing legumes, H_2 is produced and eventually even emitted into the atmosphere (Conrad and Seiler, 1979a, 1980a). Net emission occurs only during 2–3 months of the vegetative period; during the remainder of the time H_2 production is too low and more than balanced by H_2 consumption reactions in the soil. It is due to this balance that H_2 production can only be observed if the detection of H_2 mixing ratios lower than ambient (i.e., 0.55 ppmv) is analytically possible. As soon as H_2 accumulates, H_2 consumption rates increase in proportion to the H_2 concentration and balance H_2 production at a particular equilibrium value that usually is in the range of <1 ppmv. Hence, it is not unexpected that other researchers using analyzers with H_2 detection limits $>$ 2 ppmv failed to detect H_2 production in soil–legume systems (La Favre and Focht, 1983; Popelier et al., 1985). These authors observed an increase in H_2-oxidation potential in soil surrounding the root nodules and suggested that most of the released H_2 should be consumed in soil before reaching the atmosphere (see Section 9). Conrad and Seiler (1979a, 1980a) calculated from their field data that most of the H_2 that is produced by the N_2-fixation process in root nodules is in fact released into the soil atmosphere. They also showed that the release rates are so high during part of the growing season that they are not balanced by consumption, but result in a net emission of H_2 into the atmosphere (Fig. 6).

Evans and co-workers have demonstrated that H_2 is produced in the root nodules of legumes during N_2 fixation and that about 30% of the electron flow to nitrogenase is diverted to reduce protons (Schubert and Evans, 1976; Evans et al., 1977). The production of H_2 by nitrogenase seems to be an unavoidable event (Simpson and Burris, 1984). Most of the N_2-fixing microorganisms are able to recycle some of the produced H_2 by means of an uptake hydrogenase (Robson and Postgate, 1980). The oxidation of H_2 by uptake hydrogenase may have further advantages by avoiding the inhibition of nitrogenase by H_2 and by removing oxygen from the vicinity of the oxygen-sensitive nitrogenase (Dixon, 1972). Due to the presence of uptake hydrogenases, H_2 release into the soil cannot be expected for most N_2-fixing systems, such as free-living, N_2-fixing microorganisms (e.g., *Azotobacter, Derxia, Xanthobacter*) as well as N_2-fixers associated with grass roots (e.g., *Azospirillum*) or N_2-fixing actinomycetes forming root nodules in trees, such as *Alnus* or *Myrica* (Malik and Schlegel, 1980; Pedrosa et al., 1980, 1982; Lespinat and Berlier, 1981; Walker and Yates, 1978; Benson et al., 1980).

In contrast to these N_2-fixing microorganisms or symbiotic systems, legumes very often seem to have *Rhizobium* strains devoid of uptake hydrogenase and thus are releasing H_2 into the soil environment. An overview of over 1400 commercial isolates of *Rhizobium japonicum* showed that 69% of the strains were devoid of uptake hydrogenase

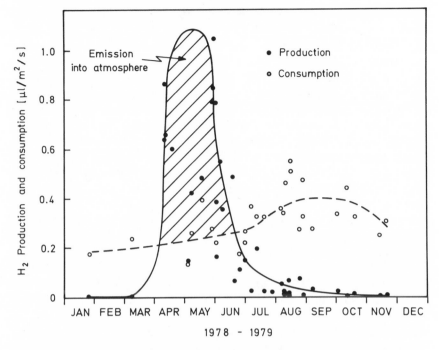

Figure 6. Rates of production and consumption of H$_2$ in soil vegetated with legumes (clover). The measurements were done under field conditions by the static box technique. The data points were gross rates; the soil is emitting H$_2$ into the atmosphere when production rates exceed consumption rates. Otherwise the soil is acting as a sink for atmospheric H$_2$. [After Conrad and Seiler (1980a).]

(Uratsu *et al.*, 1982). A similar survey conducted in China arrived at a similar result (Li *et al.*, 1980). Hence, it is possible that legume fields are the only soil areas able to act as sources for atmospheric H$_2$. It would be interesting to know whether tropical and subtropical trees (e.g., *Acacia*) belonging to the legume family contain *Rhizobium* symbionts with uptake hydrogenase. Release of H$_2$ from N$_2$-fixing symbionts into the environment may also be possible if the hydrogenase activity changes with season by a different pattern than the nitrogenase activity, e.g., as observed for actinomycete symbiosis in *Alnus glutinosa* (Roelofsen and Akkermans, 1979). However, even under optimal conditions for H$_2$ release, the total N$_2$-fixation processes, as summarized by Burns and Hardy (1975), can only account for less than 5 Tg of H$_2$ produced per year, which is a marginal source in the atmospheric H$_2$ budget (Conrad and Seiler, 1979a, 1980a).

6. Anoxic Environments As Sources for Atmospheric CO and H$_2$

In anoxic environments, such as aquatic sediments, H$_2$ is produced by fermentative bacteria, but is kept at a low steady-state concentration by H$_2$-consuming methanogenic bacteria or other species, usually at H$_2$ partial pressures of 1–10 Pa (Conrad *et al.*, 1985a; Lovley *et al.*, 1982; J. A. Robinson and Tiedje, 1982). The maintenance of a low H$_2$ partial pressure is important for the functioning of the anaerobic food chain decomposing organic matter because of thermodynamic reasons (Zehnder, 1978). Apparently CO is also produced in anoxic ecosystems (W. O. Robinson, 1930; Conrad *et al.*, 1983b), most probably by the action of methanogenic, acetogenic, or sulfidogenic bacteria (Conrad and Thauer, 1983; Eikmanns *et al.*, 1985; Lupton *et al.*, 1984; Diekert *et al.*, 1984). The physiological basis of CO production seems to be the interconversion of CO$_2$ and the carboxyl function of acetate or acetyl-CoA or the conversion of pyruvate to acetate during catabolic and anabolic metabolism (Hu *et al.*, 1982; Jansen *et al.*, 1984; Diekert and Ritter, 1983; Kerby *et al.*, 1983; Pezacka and Wood, 1984; Krzycki *et al.*, 1982; Eikmanns *et al.*, 1985). During this interconversion, CO is formed as an intermediate. The intermediate character of CO explains why CO is also consumed by methanogenic, acetogenic, or sulfidogenic bacteria and why the steady-state concentration of CO in anoxic sediments is as low as that of H$_2$.

There is still only a limited number of *in situ* determinations of H$_2$ and CO in anoxic environments, because of analytical limitations (Conrad *et al.*, 1983a,b; 1985a). These data show, however, that the steady-state concentrations of H$_2$ and CO in anoxic sediments are high enough to make these environments supersaturated with respect to the atmosphere. Therefore, H$_2$ and CO could be released into the atmosphere. However, most of the anoxic environments are surrounded by a more or less deep layer of oxygenated water, where H$_2$ and CO are rapidly consumed by aerobic microorganisms and thus do not reach the atmosphere. Hence, H$_2$ and CO only reach the atmosphere if the oxic layer is sufficiently shallow, or if the gas is released in the form of bubbles or by plant-mediated transport. In this respect, littoral zones and paddy fields should be considered as possible sources of these gases.

The role of bubbles and plant-mediated transport has almost exclusively been studied with respect to the emission of CH$_4$. Ebullition of CH$_4$ is important when marine (Martens, 1976; Martens and Val Klump, 1980) or freshwater sediments (Molongoski and Klug, 1980; Strayer and Tiedje, 1978) are heavily eutrophied, so that rates of methanogenesis are high. Ebullition of CH$_4$ is also important in rice paddies. However, plant-mediated transport is even more important and constitutes >90% of the total flux of CH$_4$ from paddy fields into the atmosphere (Seiler *et al.*,

1984b; Holzapfel-Pschorn *et al.*, 1986). Other aquatic plants also mediate the transport of gases from the sediment into the atmosphere (Sebacher *et al.*, 1985). This transport is the side effect of the necessity for plants to transport oxygen into the roots. As soon as CH_4 is transported from the anoxic sediment by bubbles or through plants, one should expect that other gases present in the sediment are transported in the same way. Hydrogen and CO are minor constituents of the CH_4-containing gas bubbles trapped from submerged soils (W. O. Robinson, 1930; Yamane and Sato, 1967; Harrison and Aiyer, 1913; Bell, 1969). The analysis of gas bubbles in submerged soil and of dissolved gas concentrations in lake sediments shows that H_2 and CO concentrations are more than four orders of magnitude lower than CH_4 concentrations (Table V). Therefore, the flux of H_2 and CO into the atmosphere must be much lower than that of CH_4. By comparison with the source strength of wetlands for CH_4 (Seiler, 1984, 1985), less than 0.02 Tg/year of H_2 or CO should be emitted from these environments.

The digestive tract of animals also constitutes an anoxic environment from which fermentatively formed H_2 or CO may escape into the atmosphere without being oxidized by microorganisms. Ruminant digestion makes up 30% of the total CH_4 budget (Seiler, 1984). The source strength of ruminants for atmospheric H_2 or CO is unknown. However, it is evident that most of the produced H_2 is transformed into CH_4, so that H_2 emission should be of minor importance. The same should be the case with the digestive system of termites, which also produce and emit CH_4 into the atmosphere (Breznak, 1982; Zimmerman *et al.*, 1982; Rasmussen and Khalil, 1983; Seiler *et al.*, 1984a). Based on laboratory experiments with termite colonies, Zimmerman *et al.* (1982) estimated a total source strength of 200 Tg H_2/year and 150 Tg CH_4/year. Field measurements on termite mounds, on the other hand, resulted in source strengths of only <7 Tg CH_4/year (Seiler *et al.*, 1984a) and showed that H_2 or CO is not released at all from the termite mounds (W. Seiler and R. Conrad unpublished results). Whereas CH_4 reached mixing ratios of >100 ppmv

Table V. H_2, CO, and CH_4 Mixing Ratios in Gas Bubbles from Submerged Paddy Soils[a]

Location and date	H_2 (ppmv)	CO (ppmv)	CH_4 (%v)	H_2/CH_4	CO/CH_4
Spain, August 82	13	20	42	0.3×10^{-4}	0.5×10^{-4}
Italy, June 84	41	31	56	0.7×10^{-4}	0.5×10^{-4}
Italy, July 85	10	7	12	0.8×10^{-4}	0.6×10^{-4}

[a]Average values of 2–10 determinations. Data are in part from H. Schütz *et al.* (1988).

inside the termite nests and was oxidized to only a small extent before it was released into the atmosphere, H_2 and CO mixing ratios were lower than ambient. If H_2 and CO were produced by termites, the gases were obviously oxidized inside the nest before they could reach the atmosphere.

7. Indirect Biospheric Sources of Atmospheric CO and H_2

Hydrogen and CO are formed as well as decomposed in the troposphere by photochemical and chemical reactions. Important precursors for the (photo)chemical formation of H_2 and CO are methane and non-methane hydrocarbons (NMHC), both of which are mainly emitted from the biosphere (Khalil and Rasmussen, 1983; Seiler, 1984, 1985; Duce et al., 1983). These indirect biospheric sources are even more important for the atmospheric budget of H_2 and CO than the direct biospheric sources.

The chemical decomposition of CH_4 in the troposphere can occur by different pathways that are initiated by reaction with the OH radical, but differ with respect to the further destruction pathway that is mainly dependent on the concentration of NO (see Section 3) (Crutzen, 1979). The intermediate product of CH_4 decomposition is always formaldehyde, which by itself is the direct precursor of H_2 and CO. The formation of H_2 and CO thus is principally dependent on the formation of formaldehyde. The mixing ratios of formaldehyde in the troposphere are very low (0.5–5 ppbv) and the cycle of atmospheric formaldehyde is very poorly understood (Neitzert and Seiler, 1981; Lowe and Schmidt, 1983). Hydrogen and CO are formed from formaldehyde by photolytic cleavage at wavelengths between 270 and 360 nm:

$$H_2CO + h\nu \rightarrow H_2 + CO$$
$$H_2CO + h\nu \rightarrow H + HCO$$
$$HCO + O_2 \rightarrow HO_2 + CO$$
$$H + O_2 \rightarrow HO_2$$

Above 330 nm the quantum yield of the second reaction is so low that H_2 and CO are formed by the first reaction exclusively. Below 330 nm, however, the second reaction is relatively more important, so that H_2 is formed in smaller amounts than CO (Moortgat and Warneck, 1979). Therefore, H_2 and CO formation from CH_4 and NMHC depends not only on the mixing ratios of CH_4 and OH radical, but also on the fate of the intermediate formaldehyde.

In addition to CH_4, large quantities of other hydrocarbons are emitted into the atmosphere. This emission is due not only to anthropogenic sources such as industrial activities and automobiles, etc., but is also due to natural processes in soil, water, and vegetation [review by Duce et al. (1983)]. The largest amounts of NMHC apparently are emitted from vegetation in the form of isoprene and terpenes. Emission of terpenes seems to be a physical process in which the terpene droplets present in leaves evaporate relative to their amount, their partial pressure, the leaf temperature, and the relative humidity. The emission of isoprene, on the other hand, seems to be mainly dependent on the metabolism of the plant. Isoprenes are not a storage product and accumulate in a light-dependent manner. Isoprene emission is restricted to particular plant genera (e.g., *Quercus*). The total emission of isoprene and terpene from vegetation was estimated to 830 Tg C/year (Zimmerman *et al.*, 1978). Additional NMHC is emitted from soils, aqueous environments, and biomass burning.

The residence times of NMHC in the troposphere are very short (<1 day). The destruction mechanisms are still largely unknown. It is assumed, however, that formaldehyde is formed as an intermediate, so that decomposition of NMHC or at least of isoprene and terpenes would finally result in the formation of H$_2$ and CO (Duce *et al.*, 1983).

8. Decomposition of Atmospheric CO and H$_2$

The observation of H$_2$ and CO uptake by soil and water has been interpreted as the result of microbial activity (Schlegel, 1974; Seiler, 1978). The actual nature of these activities, the microorganisms involved, and their ecophysiological characterization is a matter of controversy, speculation, and active research. The microbial and enzymatic processes involved in decomposition of CO and H$_2$ will be reviewed in the following paragraphs. The review will not concentrate on the general capacity of organisms to utilize H$_2$ and CO, but on the capacity for utilization at the extremely low concentrations typical for atmospheric trace gases.

8.1. Capacity of Aerobic Microorganisms to Utilize Atmospheric CO and H$_2$

Carbon monoxide and H$_2$ are present only in traces in the atmosphere. In addition, their solubilities in water are very low. At 20°C and atmospheric pressure only a fraction of about 0.02 dissolves in water (Crozier and Yamamoto, 1974; Douglas, 1967; Winkler, 1901; Schmidt, 1979), so that atmospheric CO and H$_2$ is in equilibrium with an aqueous concentration of <0.5 nM. Hence, microorganisms must be able to utilize CO and H$_2$ at a concentration range of <0.5 nM.

This concentration is extremely low even in comparison to the concentrations of other energy substrates in oligotrophic environments. For example, oligotrophic lakes and water of the open ocean contain 5–100 nM glucose (Vallentyne and Whittaker, 1956; Degens et al., 1964). Despite the low concentration of H_2 and CO, the oxidation with oxygen as electron acceptor yields useful energy:

$$2H_2 + O_2 \rightarrow 2H_2O; \qquad \Delta G^0 = -474.35 \text{ kJ}$$
$$2CO + O_2 \rightarrow 2CO_2; \qquad \Delta G^0 = -514.21 \text{ kJ}$$

The lower concentration level of H_2 and CO at which the Gibbs free energy change of the reaction becomes zero is in the range of 10^{-45} M at an oxygen mixing ratio of 20%. Therefore, the concentration at which H_2 or CO could no longer serve as energy substrate is many orders of magnitude lower than the concentrations in nature.

Although microorganisms can gain energy from oxidation of CO or H_2 even at <0.5 nM concentrations, it is questionable whether the energy yield is sufficient for maintenance and growth or whether the expense of energy in the form of necessary enzyme synthesis, etc., is larger. Conrad (1984) estimated the magnitude of V_{max} and K_m values that a microorganism must have in order to achieve at least its maintenance energy by oxidizing atmospheric CO or H_2 as sole energy substrates. Interestingly, the calculated K_m values were in the same range of 5–80 nM as those measured for decomposition of CO and H_2 in soil and water samples. However, the K_m values of chemolithoautotrophic bacteria for H_2 or CO usually are higher than 0.5 μM (see Section 8.3). Hence, it is unlikely that the latter organisms are able to cover their maintenance energy by oxidizing atmospheric CO or H_2.

There is also direct experimental evidence that indicates that the known laboratory strains of chemolithoautotrophic H_2-oxidizing bacteria (so-called hydrogen bacteria) are unable to utilize atmospheric H_2, although they are able to utilize H_2 at higher concentrations (Conrad and Seiler, 1979b; Conrad et al., 1983c). These experiments were done by incubating bacterial suspensions in soil samples in such a way that the tested strains should have adjusted their metabolic pathways and induced all the necessary enzymes to utilize atmospheric H_2. Although it is very unlikely that these bacteria would behave differently in nature, it must be emphasized that our knowledge of the ecophysiological control of kinetic properties for utilization of H_2 or CO is still marginal. In particular, the effect of one or more additional substrates on the utilization of H_2 and CO should be tested in a rigorous way. It has been shown with various nongaseous substrates that the affinity for a substrate may increase when used in a mixotrophic mode of metabolism (Law and Button, 1977; Gottschal and Kuenen, 1980; Egli et al., 1983).

The inability of chemolithotrophic hydrogen bacteria to utilize atmospheric H_2 apparently is not only due to their K_m for H_2, which generally is higher than 1 μM, but also to the existence of a relatively high threshold concentration (Conrad *et al.*, 1983c). Below this threshold H_2 is only slowly or no longer utilized by these bacteria (Fig. 7). Since the threshold for H_2 appears to be generally higher than the atmospheric H_2 concentration, the traces of H_2 in ambient air cannot be utilized.

In contrast to hydrogen bacteria, the CO-oxidizing chemolithotrophic bacteria (so-called carboxydotrophic bacteria) apparently do not exhibit a threshold for CO and take up atmospheric CO as soon as the density of the bacterial suspension is sufficient (Conrad and Seiler, 1980b). However, the K_m of the carboxydotrophic bacteria for CO is much higher (see Section 8.2) than that of soil or water, so that they cannot be responsible for the decomposition of atmospheric CO (Conrad *et al.*, 1981; Conrad and Seiler, 1982b).

If atmospheric H_2 and CO is not utilized by the chemolithotrophic hydrogen bacteria or carboxydotrophic bacteria, there must exist other microorganisms or enzymatic activities responsible for the degradation that explain the uptake kinetics observed in soil and water. These microorganisms may utilize H_2 or CO as a cosubstrate in a mixotrophic mode

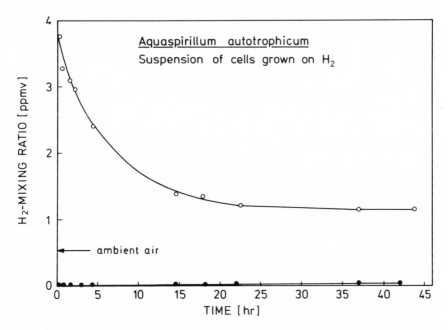

Figure 7. Metabolism of H_2 by a suspension of hydrogen bacteria demonstrating a threshold for H_2 consumption. (O) Consumption of H_2; (●) production of H_2. [After Conrad *et al.* (1983c).]

of metabolism or even may cooxidize the gaseous substrates in a nonutilitarian way. In any case, the kinetic properties of the microorganisms
must fit the kinetic properties of natural soil and water samples. Based
on this prediction, Conrad (1984) estimated the number of bacteria in
soil that would be necessary to account for the decomposition of atmospheric CO and H_2 observed under field conditions. Based on the assumption that the microbial population would exhibit the same K_m as observed
in soil and assuming for the responsible microbes a reasonable specific
enzyme activity for H_2 and CO uptake, microbial numbers of about 10^6–
10^7 cells/g of soil were calculated. This number represents a fairly high
proportion (0.1–1%) of the soil microbial biota, bearing in mind that all
these microbes must possess an enzymatic machinery with high efficiency
for H_2 and CO uptake.

8.2. Decomposition of Atmospheric CO in Soil and Water

The consumption of low concentrations of CO has been characterized in soil (Inman et al., 1971; Heichel, 1973; Ingersoll et al., 1974;
Seiler, 1974; Liebl and Seiler, 1976; Conrad and Seiler, 1980b; Spratt and
Hubbard, 1981; Bartholomew and Alexander, 1979; Nohrstedt, 1984;
Duggin and Cataldo, 1985) as well as in ocean and lake water (Conrad
and Seiler, 1982b). The main characteristics of CO decomposition in soil
are summarized in Table VI. The consumption activity is destroyed by
boiling, autoclaving, or treatment with x-rays. Poisoning with $HgCl_2$,
NaCN, and NaN_3 results in strong inhibition of CO consumption in soil
and water. The CO consumption in soil depends on the moisture content
and shows maximum activities at about 30°C. All these experimental
results are a clear indication for the microbial nature of the CO decom-

Table VI. Properties of Soil Activities Oxidizing
Atmospheric CO and H_2.

Property	CO	H_2
Stimulation by O_2	+	+
Optimum temperature (°C)	25–35	25–35
Range of pH	4–8	4–8
V_{max} (nmole $hr^{-1}g^{-1}$ dry weight)	10–400	10–400
K_m (nM)	5–50	8–50
Inactivation by		
Heat	+	+
Toluene, $CHCl_3$	+	−
Antibiotics	+	−
NaCN, NaN_3	+	−
Drying	+	−

position process. The CO decomposition in soil is an aerobic process, but slow rates persist even in the absence of oxygen.

Many aerobic microorganisms are able to utilize CO, including chemolithoautotrophic carboxydobacteria (Hegeman, 1980; Meyer and Schlegel, 1983; Meyer, 1985) and other species of bacteria, fungi, and algae (Bartholomew and Alexander, 1979; Inman and Ingersoll, 1971; Chapelle, 1962). It is unlikely, however, that these species really can account for the degradation of atmospheric CO in soil and water, because of their low uptake efficiency (see below and Sections 4.1, 8.1, and 10). Other microorganisms, therefore, must be responsible. The contribution of both bacteria and fungi to the decomposition of atmospheric CO in soil was indicated by inhibition experiments using streptomycin and cycloheximide (Conrad and Seiler, 1980b). These antibiotics specifically inhibit protein synthesis in procaryotes and eucaryotes, respectively, and thus make it possible to distinguish between bacterial and fungal activity (Anderson and Domsch, 1973). Although the antibiotics should not inhibit the activity of existing enzymes, the full extent of inhibition was reached after 3–4 days of incubation. This observation indicates that the turnover of CO-oxidizing enzymes or the growth and total turnover of the CO-utilizing microbial population is in the range of a few days. This is a rather rapid turnover compared to that of the total soil microbial biomass, which is of the order of years (Jenkinson and Ladd, 1981), indicating that the CO-oxidizing microbial population is much less dormant than other soil microorganisms. In water samples, similar experiments have not yet been done. However, filtration experiments with ocean and lake water have shown that the CO-utilizing microorganisms are in a size fraction of 0.2–3.0 μm (Conrad et al., 1982; Conrad and Seiler, 1982b), which excludes the eucaryotic phytoplankton population (Williams, 1981).

The CO consumption in soil is predominantly due to oxidation to CO_2 (Bartholomew and Alexander, 1979; Spratt and Hubbard, 1981; Duggin and Cataldo, 1985). Bartholomew and Alexander (1982) showed that assimilation of CO into soil biomass was negligible and that CO_2 assimilation was not stimulated by the presence of CO. These authors therefore concluded that CO is cooxidized in a nonutilitarian way. In contrast to these observations, Spratt and Hubbard (1981) observed a stimulation of CO_2 assimilation by CO and concluded that CO was utilized by chemolithotrophic carboxydobacteria in a utilitarian way. The discrepancy between these experiments is most probably due to the application of different CO mixing ratios (2 and 200 ppmv, respectively). The CO concentration most probably selects for different populations of microorganisms, one that is active at high CO concentrations and one that is active at low concentrations (Duggin and Cataldo, 1985).

This selection is certainly due to the different K_m values of the microorganisms for CO. The bacteria that so far have been isolated by their ability to grow on CO generally exhibit K_m values higher than 0.5 μM (Conrad et al., 1981; Bell et al., 1985; Meyer, 1985). However, the K_m values for CO consumption in soil and water are in the range of 5–50 nM (Conrad and Seiler, 1982b; Conrad et al., 1981; Bartholomew and Alexander, 1981; Jones and Morita, 1983; Spratt and Hubbard, 1981; Duggin and Cataldo, 1985). It is therefore tempting to conclude that only the high CO concentrations are utilized in a utilitarian way by the carboxydotrophic bacteria, whereas the low, atmospheric CO concentrations are utilized in a nonutilitarian way (e.g., cooxidation) by other microorganisms.

Cooxidation of CO is known from methanotrophic bacteria (Ferenci et al., 1975) and NH_4^+-oxidizing nitrifying bacteria (Jones and Morita, 1983). Carbon monoxide obviously is a substrate of methane monooxygenase or ammonium monooxygenase. Pseudomonas methanica has a K_m value for CO of 2.7 μM (Ferenci et al., 1975). Although this K_m is about five times lower than the K_m for the true substrate CH_4 (15 μM), it is still about 100 times higher than the K_m for CO of soil and water. Therefore, cooxidation of atmospheric CO by methanotrophic bacteria is very unlikely. Nitrifying bacteria, on the other hand, have a much lower K_m for CO, in a range below 200 nM. The lowest K_m was 97 nM CO and was observed in Nitrosomonas europaea (Jones and Morita, 1983). These K_m values are not so much higher than those observed in soil and water, and therefore nitrifiers may play a significant role in the decomposition of atmospheric CO. Carbon monoxide is only cooxidized by these bacteria and is apparently not used in a utilitarian way, since it does not support growth or stimulate carbon assimilation (Jones and Morita, 1983).

Morita's group presented another interesting experiment (Jones et al., 1984), in which they used the nitrification inhibitor nitrapyrin to differentiate between CO oxidation by nitrifiers and nonnitrifiers. Some of their data measured along a transect in Yaquina Bay river estuary are shown in Fig. 8. The results indicate a varying degree of contribution by nonnitrifying bacteria to the oxidation of CO in water, and that CH_4 oxidation rates were two orders of magnitude lower. Examining various soil and water samples, Jones et al. (1984) concluded that as much as 28–97% of the CO oxidation activity may be due to nitrifying bacteria. These figures clearly show that nitrifiers are important for oxidation of CO in environmental samples. However, the results also show that a large part of the atmospheric CO is oxidized by nonnitrifying microorganisms. Experiments with Yaquina Bay water indicate that the K_m of the CO-oxidizing nonnitrifiers is lower (8 nM CO) than that of the nitrifiers (42 nM CO). Therefore, the relative importance of the unknown nonnitrifiers for oxidation of CO increases with decreasing CO concentration. In soil, where

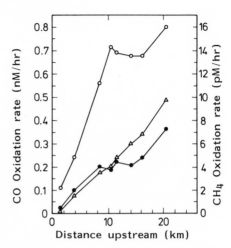

Figure 8. The CO oxidation in water from Yaquina Bay river estuary by nitrifying (○) and nonnitrifying (●) bacteria and (△) rates of CH₄ oxidation. The activity of nitrifiers and nonnitrifiers was differentiated by use of nitrapyrin as inhibitor of NH₄⁺ oxidation. [After Jones *et al.* (1983).]

atmospheric CO mixing ratios results in concentrations of dissolved CO being <0.5 nM, the nonnitrifying population most probably is much more important for CO oxidation than the nitrifying population.

Because of the low K_m of the unknown CO-oxidizers, direct enrichment and isolation by growth on CO as sole substrate is a tedious task; concentrations below 10 nM CO must be applied in order to give the microorganisms a selective advantage over other species with high K_m, such as carboxydotrophic bacteria. Moreover, enrichment would be impossible if the nonnitrifying CO-oxidizers use CO in a nonutilitarian way by cooxidation. However, enrichment cultures supplied with CO at ambient air concentrations resulted in an increase of the CO-oxidizing activity, whereas control cultures without CO did not (Conrad and Seiler, 1982b). This experiment indicates that growth of a microbial population was induced by the presence of atmospheric CO concentrations. Therefore, it is very likely that the unknown CO-oxidizing microorganisms profit to some extent from CO. When the enriched CO-oxidizing microorganisms were inoculated into fresh mineral medium, the CO-oxidizing activity again increased upon gassing with CO-containing ambient air. Further transfers into fresh medium, however, resulted in decrease of activity (R. Conrad, unpublished results). It appears that the unknown microorganisms need further growth factors or substrates, which were not contained in the mineral medium, but were necessary to utilize atmospheric CO. Mixotrophic utilization of atmospheric CO by oligocarbophilic microorganisms might be a possible explanation, but further research is required.

8.3. Decomposition of Atmospheric H_2 in Soil and Water

The consumption of H_2 at low concentrations has been characterized in soil (Schmidt, 1974; Liebl and Seiler, 1976; Seiler, 1978; Conrad and Seiler, 1981; Conrad et al., 1983d; Trevors, 1985; Popelier et al., 1985; La Favre and Focht, 1983) and in lake water (Conrad et al., 1983a). The H_2 consumption in ocean water has not been characterized. In addition to characterization of H_2 decomposition, there is literature on oxidation of tritium in soil and water (Ehhalt, 1973; McFarlane et al., 1978; Fallon 1982a,b; Schink and Zeikus, 1984; Schink et al., 1983). However, the reaction of tritium (T_2) to HTO may be due to an exchange reaction with H_2O rather than a net oxidation, and thus is not necessarily equivalent to the decomposition process of H_2. However, the tritium exchange reaction is a good assay for hydrogenase activity irrespective of whether the hydrogenase is involved in H_2 production or H_2 consumption under in situ conditions (Schink et al., 1983).

The main characteristics of the oxidation of atmospheric H_2 in soil are summarized in Table VI. As with CO, H_2 decomposition activity is destroyed by autoclaving or boiling. In soils, H_2 decomposition reactions, as well as tritium exchange activities (Fallon, 1982a), are dependent on temperature. The existence of an optimum temperature range at 25–35°C indicates that these activities are based on biological rather than chemical processes (Radmer and Kok, 1979). The H_2 decomposition activity is positively correlated with the content of biomass carbon, is dependent on soil moisture content, and slowly decreases when soil is stored in dry conditions. Hydrogen decomposition is an aerobic process, but low rates persist even in the absence of oxygen.

In contrast to CO, H_2 consumption in soil is not completely inhibited by treatment with metabolic inhibitors, antibiotics, or solvents (Conrad and Seiler, 1981; Popelier et al., 1985). These treatments, especially fumigation with toluene or $CHCl_3$, are typically done to assay for soil enzymes (Burns, 1978). Therefore, and because of other unusual characteristics (see below), Conrad and Seiler (1981) concluded that the H_2 consumption activity is due to abiontic soil hydrogenases rather than to soil microorganisms. In this respect it is interesting that the rates of H_2 consumption are similar to the rates of tritium exchange (Fallon, 1982b). This similarity indicates that both reactions are catalyzed by the same type of hydrogenase. Therefore, the H_2 decomposition reaction in soil cannot be specifically associated with H_2-consuming microorganisms, as it is unlikely that H_2-producing microorganisms that also catalyze tritium exchange are lacking in soil samples.

According to the definition by Skujins (1978), abiontic soil enzymes include not only free extracellular enzymes and enzymes bound to inert soil components, but also active enzymes within dead or nonproliferating cells. Hence, the essential difference between soil microorganisms and soil enzymes is that the latter are not associated with proliferating cells. However, they are not necessarily free proteins, but may also be associated with clay minerals, humus colloids, and other soil structures. Popelier et al. (1985) were unable to recover H$_2$ consumption activity in filtrates of soil homogenates unless the soil had been first fumigated with CHCl$_3$. Their conclusion that soil enzymes do not contribute significantly to H$_2$ consumption by soil may be unwarranted, however, as abiontic hydrogenases may have been bound to unfiltrable particles.

Binding of enzymes to clay minerals or humus colloids actually may initiate enzyme modifications that result in a complete alteration of catalytic properties (Burns, 1982). In fact, the catalytic properties of soil for decomposition of atmospheric H$_2$ are quite peculiar. The most interesting phenomenon is the inhibition of uptake of atmosphere H$_2$ by preincubation at high H$_2$ mixing ratios (Conrad and Seiler, 1981). This preincubation results in a decreased V_{max} or in an increased K_m for H$_2$ oxidation (Conrad et al., 1983d). The inhibition can only be observed if tested at low H$_2$ mixing ratios at which V_{max} is not yet reached, i.e., <100 ppmv H$_2$. Incubation of inhibited soil at ambient H$_2$ mixing ratios results in a slow reversion of the inhibition that is most probably due to the release of overreduction of the electron transport chain of the H$_2$-oxidizing enzyme systems. However, knowledge of electron transport in soil systems is meager, since the H$_2$-oxidizing activity has not been isolated from its soil matrix and all results have been obtained from experiments with soil samples containing other enzyme activities (Conrad et al., 1983d). Oxygen is the predominant electron acceptor of the H$_2$-oxidizing activity, but it can be replaced by artificial electron acceptors with redox potentials >80 mV. The fact that artificial electron acceptors can be used in assays of H$_2$ consumption indicates that the responsible activity is freely accessible for redox carriers. This observation again is indicative for abiontic soil hydrogenases, since the membrane of living microbial cells would be impermeable to many artificial redox carriers.

Another interesting feature of H$_2$ consumption in soil is its low K_m for H$_2$. As for CO, the K_m for H$_2$ is in a range of 8–50 nM (Conrad et al., 1983d), which is more than one order of magnitude lower than the K_m values reported for hydrogenases of various microorganisms (Adams et al., 1981). The existence of a low K_m has been confirmed by La Favre and Focht (1983), who found K_m values of 25 ± 18 nM. It is remarkable that the affinity of soil for atmospheric H$_2$ is in the same order of magnitude as that for atmospheric CO, although other characteristics differ (Table VI).

As shown by theoretical considerations (Section 8.1) and discussed in detail by Conrad (1984), the decomposition velocity of atmospheric H_2 and CO in soil can only be explained with activities having a high affinity for H_2 and CO. However, even if microorganisms possess the required kinetic properties, they must exist in rather large numbers in soil, making up 0.1–1% of the total microbial biomass. In the case of H_2, this problem may not exist if H_2 consumption is due to abiontic soil hydrogenases rather than H_2-oxidizing microorganisms. However, it is an important question as to the source from which these abiontic hydrogenases originate. The good correlation of activity with soil biomass carbon and ATP content (Popelier *et al.*, 1985) indicates that the hydrogenases may originate from soil microorganisms.

Very little is known on H_2 decomposition in water samples. Most work has been done by measuring activities at high H_2 concentrations or by applying the tritium exchange technique (Paerl, 1982; 1983; Schink and Zeikus, 1984). Only one study was done by testing activities at the *in situ* concentration range of lake water (Conrad *et al.*, 1983a). This work showed similar low K_m values for H_2 oxidation as in soil and demonstrated that the activity was associated with particles having a size of 0.2–3 μm. Chemolithotrophic hydrogen bacteria, on the other hand, were associated with particles >3 μm and therefore, were obviously not responsible for the observed H_2 consumption. Further characterization of the H_2-oxidizing activity in water has yet to be done, especially to find out whether the activity is due to as yet unknown microorganisms or to abiontic particle-bound hydrogenases as in soil.

Assuming that microorganisms are responsible for the observed rates of H_2 consumption, one can estimate their numbers in water by using the kinetic parameters of H_2 consumption observed in lake water (Conrad *et al.*, 1983a) and making reasonable assumptions on the kinetic properties of the microorganisms. Assuming that the microorganisms would have a rather high specific H_2 oxidation activity of ≤ 100 mmole $hr^{-1} g^{-1}$ cell dry weight at a biomass content of 1 pg dry weight per cell and a rather low K_m value of ≤ 100 nM H_2, one finds that the water must contain 5×10^2 cells/ml to account for the H_2 consumption rate observed *in situ*. These H_2-consuming microorganisms would make up a fraction of >1% of the total bacterial population and would be 10–250 times greater than the population of hydrogen bacteria counted by plating and most probable number techniques under H_2-chemolithoautotrophic growth conditions (Conrad *et al.*, 1983a). Hence, the population size of the hypothetical H_2-consuming microorganisms must be rather large despite the fact that their assumed kinetic properties were much more efficient for H_2 utilization than those published for hydrogen bacteria. Therefore, the alternative hypothesis, that H_2 consumption in lake water is due to abiontic hydrogenases rather than to microorganisms, appears tempting.

Hydrogen consumption processes are also operative in ocean water, since large parts of the oceans are obviously undersaturated with respect to atmospheric H_2 (Herr *et al.*, 1981; Herr, 1984). Because of the relatively low H_2 concentrations of <0.5 nM H_2 in ocean water, the conditions for H_2-consuming microorganisms are similarly restrictive as in the case of the soil environment. Conrad (1984) calculated that highly efficient microorganisms ($K_m < 100$ nM; $V_{max} > 100$ mmole $hr^{-1} g^{-1}$ cell dry weight) must be present in numbers of >1 cell/ml to explain the undersaturation of H_2 observed in the Norwegian Sea (Herr *et al.*, 1981).

9. Ecological Niches of Hydrogen Bacteria

Chemolithoautotrophic hydrogen bacteria can be isolated from almost every soil and water sample (Aragno and Schlegel, 1981). It appears that they are ubiquitous in nature. However, their chemolithotrophic capacities are not sufficient for utilization of atmospheric H_2 (see Sections 8.1 and 8.3). It is unknown how they function in their natural habitats. In principle, it is possible that they grow heterotrophically, since they are facultative autotrophs (Bowien and Schlegel, 1981). If their chemolithotrophic capacities are of any biogeochemical significance, then this must be in those habitats where H_2 emerges in large amounts and at relatively high concentrations.

Habitats where this is the case are found at geothermal sites (Gunter and Musgrave, 1966; Lilley *et al.*, 1982b). Aerobic hot or acid geothermal spring water in Yellowstone Park contained up to 800 μl H_2/liter and in addition contained up to 280 cells of hydrogen bacteria/ml (Schink *et al.*, 1983). Although it has not yet been shown that the H_2 present in geothermal water is oxidized by the resident hydrogen bacteria, it is very likely, since H_2 concentrations are sufficiently high compared to the threshold and affinity of these bacteria. Recently, Conrad *et al.* (1985b) showed that H_2 is consumed under aerobic conditions in hot geothermal spring water and that the K_m of the consumption process is in a range typical for hydrogen bacteria. Hence, H_2 oxidation in geothermal spring water seems to be a suitable ecological niche for thermophilic hydrogen bacteria. In fact, several strains of thermophilic hydrogen bacteria have been isolated from geothermal environments (Kawasumi *et al.*, 1984; Goto *et al.*, 1978; Kristjansson *et al.*, 1985; Schink *et al.*, 1983; Kryukov *et al.*, 1983; Saveleva *et al.*, 1982; Bonjour and Aragno, 1984; McGee *et al.*, 1967).

Syntrophy with H_2-producing cyanobacteria may be another possible ecological niche for hydrogen bacteria (Conrad *et al.*, 1983a; Schink and Zeikus, 1984). Although not yet conclusively proven for *in situ* conditions, N_2-fixing cyanobacteria are a potential source for H_2 in the photic parts of a lake (Conrad *et al.,* 1983a; Oremland, 1983; Paerl, 1982, 1983). Hydrogen bacteria exist in the same water body. Since these bacteria were shown to be associated with particles >3 μm (Conrad *et al.,* 1983a), it might be possible that syntrophic clusters of H_2-producing cyanobacteria and H_2-oxidizing hydrogen bacteria exist. A syntrophic cluster would allow the direct transfer of H_2 between the juxtapositioned partners, avoiding dilution of H_2 and the problems arising from the high threshold and low affinity for H_2 of the chemolithotrophic hydrogen bacteria (Conrad *et al.,* 1983c). The cyanobacteria, on the other hand, would benefit from the removal of H_2 and the consumption of O_2, which both are inhibitory for the nitrogenase activity (Lupton and Marshall, 1981). It is conceivable that part of the hydrogenase activity that is associated with aquatic populations of cyanobacteria (Paerl, 1982, 1983) is due to attached hydrogen bacteria rather than to uptake hydrogenase activity of the cyanobacteria. Recently, juxtapositioning of H_2-producing and H_2-consuming partners was demonstrated for anoxic methanogenic environments (Conrad *et al.,* 1985a). Demonstration of a juxtapositioned interspecies H_2 transfer between aerobic syntrophic partners awaits future experiments.

A syntrophic relation between H_2-producing nitrogen-fixers and H_2-oxidizing chemolithotrophs may also exist in the soil environment. Most N_2-fixing soil microorganisms have a very active uptake hydrogenase and therefore do not release H_2 into the soil environment (see Section 5.3). An exception is the release of H_2 from root nodules of legumes infected by *Rhizobium* strains deficient in uptake hydrogenase (Hup$^-$ strains). An ecological niche for hydrogen bacteria may be in a habitat juxtapositioned to the root nodules where the released H_2 is at a sufficiently high concentration. La Favre and Focht (1983) calculated H_2 concentrations and H_2 oxidation rates in the vicinity of root nodules of pigeon peas *(Cajanus cajan)*. They showed that the population of hydrogen bacteria, as well as the calculated H_2 oxidation rates, decreased with distance from the nodules. However, the observed H_2 oxidation rates had a K_m for H_2 in the low range typical for H_2-oxidizing activity of soil (e.g., abiotic soil hydrogenases) and not in the high range typical for chemolithotrophic hydrogen bacteria. On the other hand, H_2-oxidation rates were higher when the legumes contained Hup$^-$ strains than Hup$^+$ strains of rhizobia (La Favre and Focht, 1983; Popelier *et al.,* 1985). The capacity for H_2 oxidation is apparently stimulated in an environment where H_2 is released from root nodules. Conrad and Seiler (1980a), on the other hand, observed that the uptake rate of atmospheric H_2 was reduced in fields covered with H_2-producing legumes and concluded that this reduction was due to the inhibition of soil hydrogenases by the increased H_2 concentrations (Conrad and Seiler, 1981; Conrad *et al.,* 1983d). The discrepancy between these observations is most probably due to the different assay conditions applied. Whereas the latter authors measured H$_2$ at concentrations lower than or equal to atmospheric, the former authors applied high, saturating H$_2$ concentrations. Therefore, it is possible that the observed effects are the result of the activity of both of the abiontic soil hydrogenases and of the hydrogen bacteria. The former are most active at low atmospheric H$_2$ concentrations, but inhibited at high H$_2$ concentrations, whereas the latter are most active at high H$_2$ concentrations in the vicinity of root nodules, but inefficient for oxidation of atmospheric H$_2$.

A hypothetical scheme of the relationship between H$_2$-releasing root nodules, hydrogen bacteria, abiontic soil hydrogenases, soil air, and ambient atmosphere is shown in Fig. 9. It should be noted that during most of the season there is a net flux of H$_2$ from the atmosphere into the soil, although the root nodules evolve some H$_2$. During some months, however, H$_2$ evolution is so strong that H$_2$ mixing ratios in the soil generally reach values higher than in the ambient atmosphere, so that the soil acts as a net source for atmospheric H$_2$ (see Section 5.3) (Conrad and Seiler, 1979a, 1980a).

Finally, hydrogen bacteria may find an ecological niche at anoxic–oxic interfaces. Hydrogen is produced under anaerobic conditions by fermentative bacteria. In strictly anoxic envionments, such as lake sediments, the produced H$_2$ is almost completely recycled within the anoxic environment. The H$_2$ partial pressure is kept at a very low value so that very little H$_2$ can escape into adjacent oxic zones (see Section 6). The situation is different, however, in environments such as periodically flooded soils, soils with stagnant water table, and heaps of composting solid waste or plant material, where high H$_2$ concentrations may arise at least periodically (W. O. Robinson, 1930; Seiler *et al.,* 1977; Farquhar and Rovers, 1973; Glauser *et al.,* 1988) and where a certain degree of aeration is guaranteed despite the abundance of anoxic sites. Composting plant material as a suitable habitat for hydrogen bacteria has been proposed by Schlegel (1974). Submerged soils release relatively large amounts of H$_2$ at high concentrations immediately after flooding. The release of H$_2$ usually persists until the onset of methanogenesis and is most probably due to the rapidly initiated fermentation of carbohydrates, whereas the release of CH$_4$ is based on the accumulation of the fermentation products H$_2$ and acetate (Yamane and Sato, 1963, 1964, 1967; Bell, 1969). Hence, periodically flooded soils sometimes release H$_2$ into the adjacent oxic environments, whereas permanently anoxic environments only release CH$_4$ or H$_2$S.

A similar situation may arise at the sediment surface of holomictic

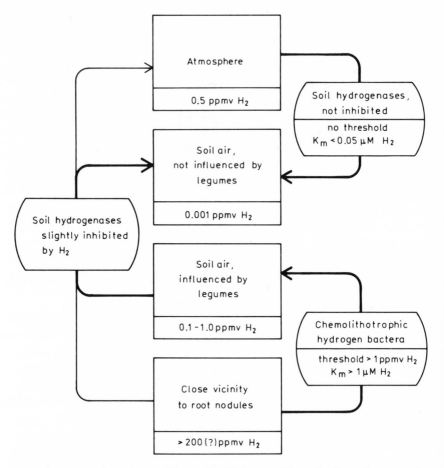

Figure 9. Hypothetical scheme of the relationship between H_2 production and consumption processes in soil and the mixing ratio of H_2 in the soil and atmosphere.

lakes during winter. M. Aragno (personal communication) observed in the surface sediment of Lake Loklat, Switzerland, a hundredfold increase of numbers of hydrogen bacteria after fall turnover, when the water of the hypolimnion changed from anoxic to oxic conditions. Schink and Zeikus (1984), on the other hand, observed a hundredfold decrease in the holomictic Lake Mendota, Wisconsin. Both research groups, however, reported much higher numbers of hydrogen bacteria in the surface sediment than in the water column, even when the hypolimnion and the sediment were completely anoxic. Schink and Zeikus (1984) speculated that the bacteria may accumulate by sedimentation from the oxic water layers and may survive the adverse stratification period by fermentative metabolism as described for *Alcaligenes eutrophus* (Schlegel and Vollbrecht, 1980).

In conclusion, it appears that the chemolithotrophic capacities of hydrogen bacteria are only of very limited use in natural habitats and are restricted to very special environments or to sporadically suitable conditions. Therefore, it is likely that hydrogen bacteria in nature usually grow heterotrophically or at best mixotrophically.

10. Ecological Niches of Carboxydobacteria

Aerobic chemolithoautotrophic CO-oxidizing bacteria have been isolated from various soil and water samples. They are called carboxydobacteria (Zavarzin and Nozhevnikova, 1977) or carboxydotrophic bacteria (Meyer and Schlegel, 1983; Meyer, 1985). They utilize CO as an energy and carbon source, but are also able to grow like a hydrogen bacterium on H_2, O_2 and CO_2. In addition, they are able to grow heterotrophically and mixotrophically using CO as an additional energy source (Kiessling and Meyer, 1982). However, carboxydobacteria have too high a K_m for CO to account for the oxidation of atmospheric CO in soil and water (see Sections 8.1 and 8.2). Therefore, only habitats with CO concentrations much higher than atmospheric levels can provide an ecological niche for carboxydobacteria.

Most of the possible habitats for carboxydobacteria are identical to those of hydrogen bacteria. Geothermal environments provide CO at similar concentrations as H_2 (Lilley *et al.*, 1982b). The thermophilic carboxydobacteria that have so far been described (Lyons *et al.*, 1984; Krüger and Meyer, 1984) may find a suitable ecological niche in the oxidation of geothermal CO. By analogy to hydrogen bacteria, they may also find a niche in environments with anoxic–oxic interfaces. Relatively high CO concentrations were observed in submerged soils (W. O. Robinson, 1930) and soils with stagnant water (Seiler *et al.*, 1977). High CO concentrations may also emerge from composting materials. In particular in samples from these environments, high rates of CO oxidation were measured (Meyer, 1978). Another habitat with high CO mixing ratios (<10%) is the floating bladder of kelp *(Nereocystis luetkeana)* (Langdon, 1917). So far, the importance of the CO for the kelp and the further fate of the CO are not known.

Other habitats with at least periodically high CO concentrations are the uppermost layers of arid soils, where CO is produced chemically in parallel to the diurnal change of soil temperature, and the illuminated zones of aquatic ecosystems, where CO is photooxidatively produced in parallel to the light intensity (see Sections 4.2–4.4). These habitats are relatively hostile environments because of aridity and radiation stress, but adaptation of microorganisms to these conditions may be possible (Skujins, 1984; Philips and Mitsui, 1982). The CO utilization by carboxydobacteria is relatively restricted, however, since the emerging CO is rapidly diluted with air at the soil–air or water–air interface. Furthermore, the CO oxidase of carboxydobacteria is only synthesized during growth on CO (Cypionka *et al.,* 1980; Meyer and Schlegel, 1983). Therefore, it is uncertain whether CO-utilizing enzymes are induced when the CO is only periodically available in sufficient amounts.

As for hydrogen bacteria, it is likely that carboxydobacteria in nature usually depend on their heterotrophic metabolism rather than on their chemolithotrophic capacities, which may only be useful in special habitats or under sporadically suitable conditions.

ACKNOWLEDGMENTS. I am very grateful to W. Seiler for support and fruitful discussions. In addition, I wish to thank M. Aragno, P. J. Crutzen, A. Holzapfel-Pschorn, O. Meyer, B. Schink, H. Schütz, and F. Slemr, who helped with ideas and comments.

References

Adams, M. W. W., Mortenson, L. E., and Chen, J. S., 1981, Hydrogenase, *Biochim. Biophys. Acta* **594**:105–176.

Anderson, J. P. E., and Domsch, K. H., 1973, Quantification of bacterial and fungal contributions to soil respiration, *Arch. Microbiol.* **93**:113–127.

Aragno, M., and Schlegel, H. G., 1981, The hydrogen-oxidizing bacteria, in: *The Prokaryotes. A Handbook on Habitats, Isolation and Identification of Bacteria* (M. P. Starr, H. Stolp, H. G. Trüper, A. Ballows, and H. G. Schlegel, eds.), Vol. 1, pp. 865–893, Springer, Berlin.

Atlas, R. M., and Bartha, R., 1981, *Microbial Ecology: Fundamentals and Applications,* Addison-Wesley, Reading, Massachusetts.

Bartholomew, G. W., and Alexander, M., 1979, Microbial metabolism of carbon monoxide in culture and in soil, *Appl. Environ. Microbiol.* **37**:932–937.

Bartholomew, G. W., and Alexander, M., 1981, Soil as a sink for atmospheric carbon monoxide, *Science* **212**:1389–1391.

Bartholomew, G. W., and Alexander, M., 1982, Microorganisms responsible for the oxidation of carbon monoxide in soil, *Environ. Sci. Technol.* **16**:300–301.

Bauer, K., Seiler, W., and Giehl, H., 1979, CO-Produktion höherer Pflanzen an natürlichen Standorten. *Z. Pflanzenphysiol.* **94**:219–230.

Bauer, K., Conrad, R., and Seiler, W., 1980, Photooxidative production of carbon monoxide by phototrophic microorganisms, *Biochim. Biophys. Acta* **589**:46–55.

Baxter, R. M., and Carey, J. H., 1983, Evidence for photochemical generation of superoxide ion in humic waters, *Nature* **306**:575–576.

Bell, R. G., 1969, Studies on the decomposition of organic matter in flooded soils, *Soil Biol. Biochem.* **1**:105–116.

Bell, J. M., Williams, E., and Colby, J., 1985, Carbon monoxide oxidoreductases from thermophilic carboxydobacteria, in: *Microbial Gas Metabolism* (R. K. Poole and C. S. Dow, eds.), pp. 153–159, Academic Press, London.

Benson, D. R., Arp, D. J., and Burris, R. H., 1980, Hydrogenase in actinorhizal root nodules and root nodule homogenates, *J. Bacteriol.* **142**:138–144.

Bidwell, R. G. S., and Bebee, G. P., 1974, Carbon monoxide fixation by plants, *Can. J. Bot.* **52**:1841–1847.

Bidwell, R. G. S., and Fraser, D. E., 1972, Carbon monoxide uptake and metabolism by leaves, *Can. J. Bot.* **50**:1435–1439.

Bonjour, F., and Aragno, M., 1984, *Bacillus tusciae*, a new species of thermoacidophilic, facultatively chemolithoautotrophic, hydrogen oxidizing spore former from a geothermal area, *Arch. Microbiol.* **139**:397–401.

Bothe, H., Neuer, G., Kalbe, I., and Eisbrenner, G., 1980, Electron donors and hydrogenase in nitrogen-fixing microorganisms, in: *Nitrogen Fixation* (W. D. P. Stewart and J. R. Gallon, eds.), pp. 83–112, Academic Press, London.

Bowien, B., and Schlegel, H. G., 1981, Physiology and biochemistry of aerobic hydrogen-oxidizing bacteria. *Annu. Rev. Microbiol.* **35**:405–452.

Breznak, J. A., 1982, Intestinal microbiota of termites and other xylophagous insects, *Annu, Rev. Microbiol.* **36**:323–343.

Broecker, W. S., and Peng, T. H., 1974, Gas exchange rates between air and sea, *Tellus* **26**:21–35.

Bullister, J. L., Guinasso, Jr., N. L., and Schink, D. R., 1982, Dissolved hydrogen, carbon monoxide, and methane at the CEPEX site, *J. Geophys. Res.* **87**:2022–2034.

Burke, Jr., R. A., Reid, D. F., Brooks, J. M., and Lavoie, D. M., 1983, Upper water column methane geochemistry in the eastern tropical North Pacific, *Limnol. Oceanogr.* **28**:19–32.

Burns, R. G., 1978, Enzyme activity in soil: some theoretical and practical considerations, in: *Soil Enzymes* (R. G. Burns, ed.), pp. 295–340, Academic Press, London.

Burns, R. G., 1982, Enzyme activity in soil: Location and a possible role in microbial ecology, *Soil Biol. Biochem.* **14**: 423–427.

Burns, R. G., and Hardy, R. W. F., 1975, *Nitrogen Fixation in Bacteria and Higher Plants*, Springer, New York.

Bzdega, T., Karwowska, R., Zuchmantowicz, H., Pawlak, M., Kleczkowski, L., and Nalborczyk, E., 1981, Absorption of carbon monoxide by higher plants. *Polish Ecol. Stud.* **7**:387–399.

Calvert, F., Cloez, S., and Boussingault, M., 1864, Über die Bildung von Kohlenoxyd bei der Einwirkung von Sauerstoff auf pyrogallussaures Kali, *Annl. Chem. Pharmazie (Leipzig)* **130**:248–249.

Chappelle, E. W., 1962, Carbon monoxide oxidation by algae, *Biochim. Biophys. Acta* **62**:45–62.

Choudhry, G. G., 1984, Humic substances. Structural aspects, and photophysical, photochemical and free radical characteristics, in: *The Handbook of Environmental Chemistry* (O. Hutzinger, ed.), Vol. 1C, pp. 1–24, Springer, Berlin.

Conrad, R., 1984, Capacity of aerobic microorganisms to utilize and grow on atmospheric trace gases (H₂, CO, CH₄), in: *Current Perspectives in Microbial Ecology* (M. J. Klug and C. A. Reddy, eds.), pp. 461–467, American Society for Microbiology, Washington, D. C.

Conrad, R., and Seiler, W., 1979a, Field measurements of hydrogen evolution by nitrogen-fixing legumes, *Soil Biol. Biochem.* **11**:689–690.

Conrad, R., and Seiler, W., 1979b, The role of hydrogen bacteria during the decomposition of hydrogen by soil, *FEMS Microbiol. Lett.* **6**: 143–145.

Conrad, R., and Seiler, W., 1980a, Contribution of hydrogen production by biological nitrogen fixation to the global hydrogen budget, *J. Geophys. Res.* **85**:5493–5498.

Conrad, R., and Seiler, W., 1980b, Role of microorganisms in the consumption and production of atmospheric carbon monoxide by soil, *Appl. Environ. Microbiol.* **40**:437–445.

Conrad, R., and Seiler, W., 1980c, Photooxidative production and microbial consumption of carbon monoxide in seawater, *FEMS Microbiol. Lett.* **9**:61–64.

Conrad, R., and Seiler, W., 1981, Decomposition of atmospheric hydrogen by soil microorganisms and soil enzymes, *Soil Biol. Biochem.* **13**:43–49.

Conrad, R., and Seiler, W., 1982a, Arid soils as a source of atmospheric carbon monoxide, *Geophys. Res. Lett.* **9**:1353–1356.

Conrad, R., and Seiler, W., 1982b, Utilization of traces of carbon monoxide by aerobic oligotrophic microorganisms in ocean, lake and soil, *Arch. Microbiol.* **132**:41–46.

Conrad, R., and Seiler, W., 1985a, Influence of temperature, moisture and organic carbon on the flux of H_2 and CO between soil and atmosphere. Field studies in subtropical regions, *J. Geophys. Res.* **90**:5699–6709.

Conrad, R., and Seiler, W., 1985b, Destruction and production rates of carbon monoxide in arid soils under field conditions, in: *Planetary Ecology* (D. E. Caldwell, J. A. Brierley, and C. L. Brierley, eds.). pp. 112–119, Van Nostrand Reinhold, New York.

Conrad, R., and Seiler, W., 1985c, Characteristics of abiological CO formation from soil organic matter, humic acids and phenolic compounds, *Environ. Sci. Technol.* **19**:1165–1169.

Conrad, R., and Seiler, W., 1986a, Influence of the surface layer on the flux of non-conservative trace gases (H_2, CO, CH_4, N_2O) across the ocean–atmosphere boundary layer, *J. Atmos. Chem.*, in press.

Conrad, R., and Seiler, W., 1986b, Exchange of CO and H_2 between ocean and atmosphere, in: *The Role of Air–Sea Exchange in Geochemical Cycling* (P. Buat-Ménard, ed.), pp. 269–282, Reidel, Dordrecht.

Conrad R., and Thauer, R. K., 1983, Carbon monoxide production by *Methanobacterium thermoautotrophicum*, *FEMS Microbiol. Lett.* **20**:229–232.

Conrad, R., Meyer, O., and Seiler, W., 1981, Role of carboxydobacteria in consumption of atmospheric carbon monoxide by soil, *Appl. Environ. Microbiol.* **42**:211–215.

Conrad, R., Seiler, W., Bunse, G., and Giehl, H., 1982, Carbon monoxide in seawater (Atlantic Ocean), *J. Geophys. Res.* **87**:8839–8852.

Conrad, R., Aragno, M., and Seiler, W., 1983a, Production and consumption of hydrogen in a eutrophic lake, *Appl. Environ. Microbiol.* **45**:502–510.

Conrad, R., Aragno, M., and Seiler, W., 1983b, Production and consumption of carbon monoxide in a eutrophic lake, *Limnol. Oceanogr.* **28**:42–49.

Conrad R., Aragno, M., and Seiler, W., 1983c, The inability of hydrogen bacteria to utilize atmospheric hydrogen is due to threshold and affinity for hydrogen, *FEMS Microbiol. Lett.* **18**:207–210.

Conrad, R., Weber, M., and Seiler, W., 1983d, Kinetics and electron transport of soil hydrogenases catalyzing the oxidation of atmospheric hydrogen, *Soil Biol. Biochem.* **15**:167–173.

Conrad, R., Phelps, T. J., and Zeikus, J. G., 1985a, Gas metabolism evidence in support of juxtapositioning between hydrogen producing and methanogenic bacteria in sewage sludge and lake sediments, *Appl. Environ. Microbiol.* **50**:595–601.

Conrad, R., Bonjour, F., and Aragno, M., 1985b, Aerobic and anaerobic microbial consumption of hydrogen in geothermal spring water, *FEMS Microbiol. Lett.* **29**:201–205.

Crozier, T. E., and Yamamoto, S., 1974, Solubility of hydrogen in water, seawater, and NaCl solutions, *J. Chem. Eng. Data,* **19**:242–244.

Crutzen, P. J., 1979, The role of NO and NO$_2$ in the chemistry of the troposphere and stratosphere, *Annu. Rev. Earth Planet. Sci.* **7**:443–472.

Crutzen, P. J., 1982, The global distribution of hydroxyl, in: *Atmospheric Chemisty* (E. D. Goldberg, ed.), pp. 313–328, Springer, Berlin.

Crutzen, P. J., 1983, Atmospheric interactions. Homogeneous gas reactions of C, N, and S containing compounds, in: *The Major Biogeochemical Cycles and Their Interactions* (B. Bolin and R. B. Cook, eds.), pp. 65–114, Wiley, Chichester.

Crutzen, P. J., Delany, A. C., Greenberg, J., Haagenson, P., Heidt, L., Lueb, R., Pollock, W., Seiler, W., Wartburg, A., and Zimmerman, P., 1985, Tropospheric chemical composition measurements in Brazil during dry season, *J. Atmos. Chem.* **2**:233–256.

Cypionka, H., Meyer, O., and Schlegel, H. G., 1980, Physiological characteristics of various species of strains of carboxydobacteria, *Arch. Microbiol.* **127**:301–307.

Dahm, C. N., Baross, J. A., Ward, A. K., Lilley, M. D., and Sedell, J. R., 1983, Initial effects of the Mount St. Helens eruption on nitrogen cycle and related chemical processes in Ryan Lake, *Appl. Environ. Microbiol.* **45**:1633–1645.

Degens, E. T., Reuter, J. H., and Shaw, K. N. F., 1964, Biochemical compounds in offshore California sediments and sea waters, *Geochim. Cosmochim. Acta* **28**:45–66.

Diekert, G., and Ritter, M., 1983, Carbon monoxide fixation into carboxyl group of acetate during growth of *Acetobacterium woodii* on H$_2$ and CO$_2$, *FEMS Microbiol. Lett.* **17**: 299–302.

Diekert, G., Hansch, M., and Conrad, R., 1984, Acetate synthesis from 2 CO$_2$ in acetogenic bacteria: Is carbon monoxide an intermediate?, *Arch. Microbiol.* **138**:224–228.

Dixon, O. D., 1972, Hydrogenase in legume root nodule bacteroids: Occurrence and Properties, *Arch. Microbiol.* **107**:193–201.

Douglas, E., 1967, Carbon monoxide solubilities in sea water, *J. Phys. Chem.* **71**:1931–1933.

Duce, R. A., Mohnen, V. A., Zimmerman, P. R., Grosjean, D., Cautereels, W., Chatfield, R., Jaenicke, R., Ogren, J. A., Pellizzari, E. D., and Wallace, G. T., 1983, Organic material in the global troposphere, *Rev. Geophys. Space Phys.* **21**:921–952.

Duggin, J. A., and Cataldo, D. A., 1985, The rapid oxidation of atmospheric CO to CO$_2$ by soils, *Soil Biol. Biochem.* **17**:469–474.

Egli, T., Lindley, N. D., and Quayle, J. R., 1983, Regulation of enzyme synthesis and variation of residual methanol concentration during carbon-limited growth of *Kloeckera* sp. 2201 on mixtures of methanol and glucose, *J. Gen. Microbiol.* **129**:1269–1281.

Ehhalt, D. H., 1973, On the uptake of tritium by soil water and groundwater, *Water Resources Res.* **9**:1073–1074.

Ehrhardt, M., 1984, Marine gelbstoff, in: *The Handbook of Environmental Chemistry* (O. Hutzinger, ed.), Vol. 1C, pp. 63–77, Springer, Berlin.

Eikmanns, B., Fuchs, G., and Thauer, R. K., 1985, Formation of carbon monoxide from CO$_2$ and H$_2$ by *Methanobacterium thermoautotrophicum, Eur. J. Biochem.* **146**:149–154.

Engel, R. R., Matsen, J. M., Chapman, S. S., and Schwartz, S., 1972, Carbon monoxide production from heme compounds by bacteria, *J. Bacteriol.* **112**:1310–1315.

Engel, R. R., Modler, S., Matsen, J. M., and Petryka, Z. J., 1973, Carbon monoxide production from hydroxocobalamin by bacteria, *Biochim. Biophys. Acta* **313**:150–155.

Evans, H. J., Ruiz-Argüeso, T., Jennings, N., and Hanus, J., 1977, Energy coupling efficiency of symbiotic nitrogen fixation, in: *Genetic Engineering for Nitrogen Fixation* (A. Hollander, ed.), pp. 333–354, Plenum Press, New York.

Fallon, R. D., 1982a, Influences of pH, temperature, and moisture on gaseous tritium uptake in surface soils, *Appl. Environ. Microbiol.* **44**: 171–178.

Fallon, R. D., 1982b, Molecular tritium uptake in southeastern U. S. soils, *Soil Biol. Biochem.* **14**:553–556.

Farquhar, G. J., and Rovers, F. A., 1973, Gas production during refuse decomposition, *Water Air Soil Pollut.* 2:483–495.

Ferenci, T., Ström, T., and Quayle, J. R., 1975, Oxidation of carbon monoxide and methane by *Pseudomonas methanica, J. Gen. Microbiol.* 91:79–91.

Fischer, K., and Lüttge, U., 1978, Light-dependent net production of carbon monoxide by plants, *Nature* 275:740–741.

Fischer, K., and Lüttge, U., 1979, Lichtabhängige CO-Bildung grüner Pflanzen und ihre Bedeutung für den CO-Haushalt der Atmosphäre, *Flora.* 168:121–137.

Fishman, J., and Seiler, W., 1983, Correlative nature of ozone and carbon monoxide in the troposphere: Implications for the tropospheric ozone budget, *J. Geophys. Res.* 88:3662–3670.

Gallon, J. R., 1981, The oxygen sensitivity of nitrogenase: A problem for biochemists and microorganisms, *TIBS* 6:19–23.

Glauser, M., Aragno, M., and Gandolla, M., 1988. Anaerobic digestion of urban wastes: Sewage and organic fraction of garbage, in: *Bioenvironmental Systems* (Vol. 3, D. Wise, ed.), CRC Press, Boca Raton, Florida, in press.

Gohre, K., and Miller, G. C., 1983, Singlet oxygen generation on soil surfaces. *J. Agric. Food Chem.* 31:1104–1108.

Goto, E., Kodama, T., and Minoda, Y., 1978, Growth and taxonomy of thermophilic hydrogen bacteria, *Agric. Biol. Chem.* 42:1305–1308.

Gottschal, J. C., and Kuenen, J. G., 1980, Mixotrophic growth of *Thiobacillus* A2 on acetate and thiosulfate as growth limiting substrates in the chemostat, *Arch. Microbiol.* 126:33–42.

Gunter, B. D., and Musgrave, B. C., 1966, Gas chromatographic measurements of hydrothermal emanations at Yellowstone National Park, *Geochim. Cosmochim. Acta* 30:1175–1189.

Haag, W. R., Hoigne, J., Gassman, E., and Braun, A. M., 1984, Singlet oxygen in surface waters. 2. Quantum yields of its production by some natural humic materials as a function of wavelength, *Chemosphere* 13:641–650.

Hallenbeck, P. C., and Benemann, J. R., 1979, Hydrogen from algae, *Top. Photosynth.* 3:331–364.

Hampson, R. F., 1980, Chemical Kinetic and Photochemical Data Sheets for Atmospheric Research, Report FAA-EE-80-17, National Bureau of Standards, Washington, D. C.

Hanson, R. S., 1980, Ecology and diversity of methylotrophic organisms, *Adv. Appl. Microbiol.* 26:3–39.

Harrison, W. H., and Aiyer, P. A. S., 1913, The gases of swamp soils, *Mem. Dep. Agr. Ind.* 3:65–104.

Hegeman, G., 1980, Oxidation of carbon monoxide by bacteria, *TIBS* 5:256–259.

Heichel, G. H., 1973, Removal of carbon monoxide by field and forest soils, *J. Environ. Qual.* 2:419–423.

Heidt, L. R., Krasnec, J. P., Lueb, R. A., Pollock, W. H., Henry, B. E., and Crutzen, P.J., 1980, Latitudinal distribution of CO and CH_4 over the Pacific, *J. Geophys. Res.* 85:7329–7336.

Herr, F. L., 1984, Dissolved hydrogen in Eurasian arctic waters, *Tellus* 36B:55–66.

Herr, F. L., and Barger, W. R., 1978, Molecular hydrogen in the near surface atmosphere and dissolved in waters of the tropical North Atlantic, *J. Geophys. Res.* 83:6199–6205.

Herr, F. L., Scranton, M. I., and Barger, W. R., 1981, Dissolved hydrogen in the Norwegian Sea: Mesoscale surface variability and deep water distribution, *Deep-Sea Res.* 28:1001–1016.

Herr, F. L., Frank, E. C., Leone, G. M., and Kennicutt, M. C., 1984, Diurnal variability of dissolved molecular hydrogen in the tropical South Atlantic Ocean, *Deep-Sea Res.* **31:**13–20.

Holzapfel-Pschorn, A., Conrad, R., and Seiler, W., 1986, Effects of vegetation on the emission of methane by submerged paddy soil, *Plant Soil* **92:**223–233.

Hu, S. I., Drake, H. L, and Wood, H. G., 1982, Synthesis of acetyl coenzyme A from carbon monoxide, methyltetrahydrofolate, and coenzyme A by enzymes from *Clostridium thermoaceticum, J. Bacteriol.* **149:**440–448.

Ingersoll, R. B., Inman, R. E., and Fisher, W. R., 1974, Soil's potential as a sink for atmospheric carbon monoxide, *Tellus* **26:**151–159.

Inman, R. E., and Ingersoll, R. B., 1971, Note on the uptake of carbon monoxide by soil fungi, *J. Air Pollut. Control Assoc.* **21:**646–647.

Inman, R. E., Ingersoll, R. B., and Levy, E. A., 1971, Soil: A natural sink for carbon monoxide, *Science* **172:**1229–1231.

Jansen, K., Thauer, R. K., Widdel, F., and Fuchs, G., 1984, Carbon assimilation pathways in sulfate reducing bacteria. Formate, carbon dioxide, carbon monoxide, and acetate assimilation by *Desulfovibrio baarsii, Arch. Microbiol.* **138:**257–262.

Jenkinson, D. S., and Ladd, J. N., 1981, Microbial biomass in soil: Measurement and turnover, in: *Soil Biochemistry* (E. A. Paul and J. N. Ladd, eds.), Vol. 5, pp. 415–471, Marcel Dekker, New York.

Jones, R. D., and Morita, R. Y., 1983, Carbon monoxide oxidation by chemolithotrophic ammonium oxidizers, *Can. J. Microbiol.* **29:**1545–1551.

Jones, R. D., Morita, R. Y., and Griffiths, R. P., 1984, Method for estimating *in-situ* chemolithotrophic ammonium oxidation using carbon monoxide oxidation, *Mar. Ecol. Prog. Ser.* **17:**259–269.

Jørgensen, B. B., 1977, Bacterial sulfate reduction within reduced microniches of oxidized marine sediments, *Mar. Biol.* **41:**7–17.

Junge, C., Seiler, W., Bock, R., Greese, K. D., and Radler, F., 1971, Über die CO-Produktion von Mikroorganismen, *Naturwissenschaften* **58:** 362–363.

Junge, C., Seiler, W., Schmidt, U., Bock, R., Greese, K. D., Radler, F., and Rüger, H. J., 1972, Kohlenoxid- und Wasserstoff-Produktion mariner Mikroorganismen im Nährmedium mit synthetischem Seewasser, *Naturwissenschaften* **59:**514–515.

Kawasumi, T., Igarashi, Y., Kodama, T., and Minoda, Y., 1984, *Hydrogenobacter thermophilus* gen. nov. sp. nov., an extremely thermophilic, aerobic, hydrogen-oxidizing bacterium, *Int. J. Syst. Bacteriol.* **34:**5–10.

Kerby, R., Niemczura, W., and Zeikus, J. G., 1983, Single-carbon catabolism in acetogens: Analysis of carbon flow in *Acetobacterium woodii* and *Butyribacterium methylotrophicum* by fermentation and ¹³C nuclear magnetic resonance measurements, *J. Bacteriol.* **155:**1208–1218.

Khalil, M. A. K., and Rasmussen, R. A., 1983, Sources, sinks, and seasonal cycles of atmospheric methane, *J. Geophys. Res.* **88:**5131–5144.

Khalil, M. A. K., and Rasmussen, R. A., 1984, Carbon monoxide in the earth's atmosphere: Increasing trend, *Science* **224:**54–56.

Kiessling, M., and Meyer, O., 1982, Profitable oxidation of carbon monoxide or hydrogen during heterotrophic growth of *Pseudomonas carboxydoflava, FEMS Microbiol. Lett.* **13:**333–338.

Krinsky, N. J., 1978, Non-photosynthetic functions of carotenoids, *Phil. Trans. R. Soc. Lond.* **284B:**581–590.

Kristjansson, J. K., Ingason, A., and Alfredsson, G. A., 1985, Isolation of thermophilic obligately autotrophic hydrogen-oxidizing bacteria, similar to *Hydrogenobacter thermophilus,* from Icelandic hot springs, *Arch. Microbiol.* **140:**321–325.

Krüger, B., and Meyer, O., 1984, Thermophilic *Bacilli* growing with carbon monoxide, *Arch. Microbiol.* **139**:402–408.

Kryukov, V. R., Saveleva, N. D., and Pusheva, M. A., 1983, *Calderobacterium hydrogenophilum* nov. gen., nov. sp., an extreme thermophilic hydrogen bacterium, and its hydrogenase activity, *Mikrobiologija* **52**:781–788.

Krzycki, J. A., Wolkin, R. H., and Zeikus, J. G., 1982, Comparison of unitrophic and mixotrophic substrate metabolism by an acetate-adapted strain of *Methanosarcina barkeri*, *J. Bacteriol.* **149**:247–254.

Kuznetsov, S. I., 1959, *Die Rolle der Mikroorganismen im Stoffkreislauf der Seen*, VEB Deutscher Verlag für Wissenschaften, Berlin.

La Favre, J. S., and Focht, D. D., 1983, Conservation in soil of H_2 liberated from N_2 fixation by Hup⁻ nodules, *Appl. Environ. Microbiol.* **46**:304–311.

Lambert, G. R., and Smith, G. D., 1981, The hydrogen metabolism of cyanobacteria (blue-green algae), *Biol. Rev. Camb. Phil. Soc.* **56**:589–660.

Langdon, S. E., 1917, Carbon monoxide, occurrence free in kelp *(Nereocystis luetkeana), J. Am. Chem. Soc.* **39**:149–156.

Law, A. T., and Button, D. K., 1977, Multiple-carbon-source-limited growth kinetics of a marine coryneform bacterium, *J. Bacteriol.* **129**:115–123.

Lespinat, P. A., and Berlier, Y. M., 1981, The dependence of hydrogen recycling upon nitrogenase activity in *Azospirillum brasilense* Sp. 7, *FEMS Microbiol. Lett.* **10**:127–132.

Li, Y. H., Chin, Y. H., Zhao, H. Y., Zhang, X. J., and Zhou, P. Z., 1980, Survey of hydrogen evolution by leguminoid rhizobia strains, *Wei Sheng Wu Hsueh Pao* **20**:180–184.

Liebl, K. H., and Seiler, W., 1976, CO and H_2 destruction at the soil surface, in: *Production and Utilization of Gases (H_2, CH_4, CO)* (H. G. Schlegel, G. Gottschalk, and N. Pfennig, eds.), pp. 215–229, Goltze, Göttingen.

Lilley, M. D., Baross, J. A., and Gordon, L. I., 1982a, Dissolved hydrogen and methane in Saanich Inlet, British Columbia. *Deep-Sea Res.* **29**:1471–1484.

Lilley, M. D., De Angelis, M. A., and Gordon, L. I., 1982b, CH_4, H_2, CO and N_2O in submarine hydrothermal vent waters, *Nature* **300**:48–49.

Lion, L. W., and Leckie, J. O., 1981, The biogeochemistry of the air–sea interface, *Annu. Rev. Earth Planet. Sci.* **9**:449–486.

Liss, P. S., and Slater, P. G., 1974, Flux of gases across the air–sea interface, *Nature* **247**:181–184.

Loewus, M. W., and Delwiche, C. C., 1963, Carbon monoxide production by algae, *Plant Physiol.* **38**: 371–374.

Logan, J. A., Prather, M. J., Wofsy, S. C., and McElroy, M. B., 1981, Tropospheric chemistry: A global perspective, *J. Geophys. Res.* **86**:7210–7254.

Lovley, D. R., Dwyer, D. F., and Klug, M. J., 1982, Kinetic analysis of competition between sulfate reducers and methanogens for hydrogen in sediments, *Appl. Environ. Microbiol.* **43**:1373–1379.

Lowe, D. C., and Schmidt, U., 1983, Formaldehyde (HCHO) measurements in the nonurban atmosphere, *J. Geophys. Res.* **88**:10844–10858.

Lüttge, U., and Fischer, K., 1980, Light-dependent net carbon monoxide-evolution by C_3 and C_4 plants, *Planta* **149**:59–63.

Lupton, F. S., and Marshall, K. C., 1981, Specific adhesion of bacteria to heterocysts of *Anabaena* sp and its ecological significance, *Appl. Environ. Microbiol.* **42**:1085–1092.

Lupton, F. S., Conrad, R., and Zeikus, J. G., 1984, CO metabolism of *Desulfovibrio vulgaris* strain Madison: Physiological function in absence and presence of exogenous substrate, *FEMS Microbiol. Lett.* **23**: 263–268.

Lyons, C. M., Justin, P., Colby, J., and Williams E., 1984, Isolation, characterization and autotrophic metabolism of a moderately thermophilic carboxydobacterium, *Pseudomonas thermocarboxydovorans* sp. nov., *J. Gen. Microbiol.* **130**:1097–1105.

Malik, K. A., and Schlegel, H. G., 1980, Enrichment and isolation of new nitrogen-fixing hydrogen bacteria, *FEMS Microbiol. Lett.* **8**:101–104.

Marenco, A., and Delaunay, J. C., 1980, Experimental evidence of natural sources of CO from measurements in the troposphere, *J. Geophys. Res.* **85**:5599–5613.

Martens, C. S., 1976, Control of methane sediment–water bubble transport by macroinfaunal irrigation in Cape Lookout Bight, North Carolina, *Science* **192**:998–1000.

Martens, C. S., and Val Klump, J., 1980, Biogeochemical cycling in an organic-rich coastal marine basin. 1. Methane sediment–water exchange processes, *Geochim. Cosmochim. Acta* **44**:471–490.

McFarlane, J. C., Rogers, R. D., and Bradley, Jr., D. V., 1978, Environmental tritium oxidation in surface soil, *Environ. Sci. Technol.* **12**:590–593.

McGee, J. M., Brown, L. R., and Tischer, R. G., 1967, A high temperature hydrogen oxidizing bacterium—*Hydrogenomonas thermophilus* n. sp., *Nature* **214**:715–716.

Meyer, O., 1978, Kohlenmonoxidoxidation und -Assimilation durch das aerobe Wasserstoffbakterium *Pseudomonas carboxydovorans*, Ph. D. Thesis, Göttingen.

Meyer, O., 1985, Metabolism of aerobic carbon monoxide-utilizing bacteria, in: *Microbial Gas Metabolism* (R. K. Poole and C. S. Dow, eds.), pp. 131–151, Academic Press, London.

Meyer, O., and Schlegel, H. G., 1983, Biology of aerobic carbon monoxide-oxidizing bacteria, *Annu. Rev. Microbiol.* **37**:277–310.

Miyahara, S., and Takahashi, H., 1971, Biological CO evolution: Carbon monoxide evolution during autoenzymatic oxidation of phenols, *J. Biochem.* **69**:231–233.

Molongoski, J. J., and Klug, M. J., 1980, Anaerobic metabolism of particulate organic matter in the sediments of a hypereutrophic lake, *Freshwater Biol.* **10**:507–518.

Moortgat, G., and Warneck, P., 1979, CO and H₂ quantum yields in the photodecomposition of formaldehyde in air, *J. Chem. Phys.* **70**:3639–3651.

Nedwell, D. B., 1984, The input and mineralization of organic carbon in anaerobic aquatic sediments, in: *Advances in Microbial Ecology*, Vol. 7 (K. C. Marshall, ed.), pp. 93–131, Plenum Press, New York.

Neitzert, V., and Seiler, W., 1981, Measurement of formaldehyde in clean air, *Geophys. Res. Lett.* **8**:79–82.

Nohrstedt, H. Ø., 1984, Carbon monoxide as an inhibitor of N₂ase activity (C₂H₂) in control measurements of endogenous formation of ethylene by forest soils, *Soil Biol. Biochem.* **16**:19–22.

Norkrans, B., 1980, Surface microlayers in aquatic environments, in: *Advances in Microbial Ecology*, Vol. 4 (M. Alexander, ed.), pp. 51–85, Plenum Press, New York.

Nozhevnikova, A. N., and Yurganov, L. N., 1978, Microbial aspects of regulating the carbon monoxide content of the earth's atmosphere, in: *Advances in Microbial Ecology*, Vol. 2 (M. Alexander, ed.), pp. 203–244, Plenum Press, New York.

Oremland, R. S., 1983, Hydrogen metabolism by decomposing cyanobacterial aggregates in Big Soda Lake, Nevada, *Appl. Environ. Microbiol.* **45**:1519–1525.

Paerl, H. W., 1982, *In situ* hydrogen production and utilization by natural populations of nitrogen-fixing blue-green algae, *Can. J. Bot.* **60**:2542–2546.

Paerl, H. W., 1983, Environmental regulation of hydrogen utilization (tritiated-hydrogen exchange) among natural and laboratory populations of nitrogen and non-nitrogen fixing phytoplankton, *Microb. Ecol.* **9**:79–97.

Pedrosa, F. O., Döbereiner, J., and Yates, M. G., 1980, Hydrogen-dependent growth and autotrophic carbon dioxide fixation in *Derxia*, *J. Gen. Microbiol.* **119**:547–551.

Pedrosa, F. O., Stephan, M., Döbereiner, J., and Yates, M. G., 1982, Hydrogen-uptake hydrogenase activity in nitrogen-fixing *Azospirillum brasilense, J. Gen. Microbiol.* **128:**161–166.

Peiser, G. D., Lizada, C. C., and Yang, S. F., 1982, Dark metabolism of carbon monoxide in lettuce leaf disks, *Plant Physiol.* **70:**397–400.

Peng, T. H., Broecker, W. S., Mathieu, G. G., and Li, Y. H., 1979, Radon evasion rates in the Atlantic and Pacific Oceans as determined during GEOSECS program, *J. Geophys. Res.* **84:**2471–2486.

Pezacka, E., and Wood, H. G., 1984, The synthesis of acetyl-CoA by *Clostridium thermoaceticum* from carbon dioxide, hydrogen, coenzyme A and methyltetrahydrofolate, *Arch. Microbiol.* **137:**63–69.

Philips, E. J., and Mitsui, A., 1982, Light intensity preference and tolerance of aquatic photosynthetic microorganisms, in: *CRC Handbook of Biosolar Resources* (A. Mitsui and C. C. Black, eds.) Vol. 1, pp. 257–308, CRC Press, Boca Raton, Florida.

Popelier, F., Liessens, J., and Verstraete, W., 1985, Soil H_2-uptake in relation to soil properties and rhizobial H_2-production, *Plant Soil* **85:**85–96.

Radler, F., Greese, K. D., Bock, R., and Seiler, W., 1974, Die Bildung von Spuren von Kohlenmonoxid durch *Saccharomyces cerevisiae* und andere Mikroorganismen, *Arch. Microbiol.* **100:**243–252.

Radmer, R. J., and Kok, B., 1979, Rate–temperature curves as an unambiguous indicator of biological activity in soil, *Appl. Environ. Microbiol.* **38:**224–228.

Rasmussen, R. A., and Khalil, M. A. K., 1981, Atmospheric methane (CH_4): Trends and seasonal cyles, *J. Geophys. Res.* **86:**9826–9832.

Rasmussen, R. A., and Khalil, M. A. K., 1983, Global production of methane by termites, *Nature* **301:**700–702.

Robinson, E., Clark, D., and Seiler, W., 1984, The latitudinal distribution of carbon monoxide across the Pacific from California to Antarctica, *J. Atmos, Chem.* **1:**137–150.

Robinson, J. A., and Tiedje, J. M., 1982, Kinetics of hydrogen consumption by rumen fluid, anaerobic digestor sludge, and sediment, *Appl. Environ. Microbiol.* **44:**1374–1384.

Robinson, W. O., 1930, Some chemical phases of submerged soil conditions, *Soil Sci.* **30:**197–217.

Robson, R. L., and Postgate, J. R., 1980, Oxygen and hydrogen in biological nitrogen fixation, *Annu. Rev. Microbiol.* **34:**183–207.

Roelofsen, W., and Akkermans, A. D. L., 1979, Uptake and evolution of hydrogen and reduction of acetylene by root nodules and nodule homogenates of *Alnus glutinosa, Plant Soil* **52:**571–578.

Rudd, J. W. M., and Taylor, C. D., 1980, Methane cycling in aquatic environments, *Adv. Aquat. Microbiol.* **2:**77–150.

Saveleva, N. D., Kryukov, V. R., and Pusheva, M. A., 1982, Obligate thermophilic hydrogen bacteria, *Mikrobiologija* **51:**765–769.

Schink, B., and Zeikus, J. G., 1984, Ecology of aerobic hydrogen-oxidizing bacteria in two freshwater lake ecosystems, *Can. J. Microbiol.* **30:**260–265.

Schink, B., Lupton, F. S., and Zeikus, J. G., 1983, Radioassay for hydrogenase activity in viable cells and documentation of aerobic hydrogen-consuming bacteria living in extreme environments, *Appl. Environ. Microbiol.* **45:**1491–1500.

Schlegel, H. G., 1974, Production, modification, and consumption of atmospheric trace gases by microorganisms, *Tellus* **26:**11–20.

Schlegel, H. G., and Vollbrecht, D., 1980, Formation of the dehydrogenases for lactate, ethanol, and butanediol in the strictly aerobic bacterium *Alcaligenes eutrophus, J.Gen. Microbiol.* **117:**475–481.

Schmidt, U., 1974, Molecular hydrogen in the atmosphere, *Tellus* **26:**78–90.

Schmidt, U., 1978, The latitudinal and vertical distribution of molecular hydrogen in the troposphere, *J. Geophys. Res.* **83**:941–946.

Schmidt, U., 1979, The solubility of carbon monoxide and hydrogen in water and sea-water at partial pressures of about 10^{-5} atmospheres, *Tellus* **31**:68–74.

Schubert, K. R., and Evans, H. J., 1976, Hydrogen evolution: A major factor affecting the efficiency of nitrogen fixation in nodulated symbionts, *Proc. Nat. Acad. Sci. USA* **73**:1207–1211.

Schütz, H., Conrad, R., Goodwin, S. and Seiler, W., 1988, Emission of hydrogen from deep and shallow freshwater environments. *Biogeochemistry,* in press.

Scranton, M. I., 1983, The role of the cyanobacterium *Oscillatoria (Trichodesmium) thiebautii* in the marine hydrogen cycle, *Mar. Ecolg. Progr. Ser.* **11**:79–87.

Scranton, M. I., 1984, Hydrogen cycling in the waters near Bermuda: The role of the nitrogen fixer, *Oscillatoria thiebautii, Deep-Sea Res.* **31**:133–144.

Scranton, M. I., and Farrington, J. W., 1977, Methane production in the waters off Walvis Bay, *J. Geophys. Res.* **82**:4947–4953.

Scranton, M. I., Barger, W. R., and Herr, F. L., 1980, Molecular hydrogen in the urban troposphere: Measurement of seasonal variability, *J. Geophys. Res.* **85**:5575–5580.

Scranton, M. I., Jones, M. M., and Herr, F. L., 1982, Distribution and variability of dissolved hydrogen in the Mediterranean Sea, *J. Mar. Res.* **40**:873–891.

Scranton, M. I., Novelli, P. C., and Loud, P. A., 1984, The distribution and cycling of hydrogen gas in the waters of two anoxic marine environments, *Limnol. Oceanogr.* **29**:993–1003.

Sebacher, D. I., Harriss, R. C., and Bartlett, K. B., 1985, Methane emissions to the atmosphere through aquatic plants, *J. Environ. Qual.* **14**:40–46.

Seiler, W., 1974, The cycle of atmospheric CO, *Tellus* **26**:116–135.

Seiler, W., 1978, The influence of the biosphere and the atmospheric CO and H₂ cycles, in: *Environmental Biogeochemistry and Geomicrobiology* (W. E. Krumbein, ed.), pp. 773–810, Ann Arbor Science Publishers, Ann Arbor, Michigan.

Seiler, W., 1984, Contribution of biological processes to the global budget of CH₄ in the atmosphere, in: *Current Perspectives in Microbial Ecology* (M. J. Klug and C. A. Reddy, eds.), pp. 468–477, American Society for Microbiology, Washington, D.C.

Seiler, W., 1985, Increase of atmospheric methane: Causes and impact on the environment, in: *WMO Special Environmental Report No. 16,* WMO No. 647, pp. 177–203.

Seiler, W., and Conrad, R., 1982, Global carbon monoxide fluxes: Inappropriate measurement procedures, *Science* **216**:761–762.

Seiler, W., and Conrad, R., 1987, Contribution of tropical ecosystems to the global budgets of trace gases, especially CH₄, H₂, CO and N₂O in: *The Geophysiology of Amazonia* (R. E. Dickinson, ed.), pp. 133–162, Wiley, New York.

Seiler, W., and Fishman, J., 1981, The distribution of carbon monoxide and ozone in the free troposphere, *J. Geophys. Res.* **86**:7255–7265.

Seiler, W., and Giehl, H., 1977, Influence of plants on the atmospheric carbon monoxide, *Geophys. Res. Lett.* **4**:329–332.

Seiler, W., and Schmidt, U., 1974, Dissolved nonconservative gases in seawater, in *The Sea* (E. D. Goldberg, ed.), Vol. 5, pp. 219–243, Wiley, New York.

Seiler, W., and Warneck, P., 1972, Decrease of carbon monoxide mixing ratio at the tropopause, *J. Geophys. Res.* **77**:3204–3214.

Seiler, W., and Zankl, H., 1975, Die Spurengase CO und H₂ über München, *Umschau* **75**:735–736.

Seiler, W., and Zankl, H., 1976, Man's impact on the atmospheric carbon monoxide cycle, in: *Environmental Biogeochemistry* (J. O. Nriagu, ed.), Vol. 1, pp. 25–37, Ann Arbor Science Publishers, Ann Arbor, Michigan.

Seiler, W., Liebl, K. H., Stöhr, W. T., and Zakosek, H., 1977, CO- und H_2-Abbau in Böden, *Z. Pflanzenernähr. Bodenkd.* **140**:257–272.

Seiler, W., Giehl, H., and Bunse, G., 1978, The influence of plants on atmospheric carbon monoxide and dinitrogen oxide, *Pure Appl. Geophys.* **116**:439–451.

Seiler, W. Conrad, R., and Scharffe, D., 1984a, Field studies of methane emission from termite nests into the atmosphere and measurements of methane uptake by tropical soils, *J. Atmos. Chem.* **1**:171–186.

Seiler, W., Holzapfel-Pschorn, A., Conrad, R., and Scharffe, D., 1984b, Methane emission from rice paddies, *J. Atmos. Chem.* **1**:241–268.

Seiler, W., Giehl, H., Brunke, E. G., and Halliday, E., 1984c, The seasonality of CO abundance in the Southern Hemisphere, *Tellus* **36**:219–231.

Setser, P. J., Bullister, J. L., Frank, E. C., Guinasso, Jr., N. L., and Schink, D. R., 1982, Relationships between reduced gases, nutrients, and fluorescence in surface waters off Baja California, *Deep-Sea Res.* **29**:1203–1215.

Simpson, F. B., and Burris, R. H., 1984, A nitrogen pressure of 50 atmosphere does not prevent evolution of hydrogen by nitrogenase, *Science* **224**:1095–1097.

Simpson, F. J., Narasimhachari, N., and Westlake, D. W. S., 1963, Degradation of rutin by *Aspergillus flavus.* The carbon monoxide producing system, *Can. J. Microbiol.* **9**:15–25.

Singh, H. B., 1977, Preliminary estimation of the average tropospheric HO concentrations in the Northern and Southern Hemispheres, *Geophys. Res. Lett.* **4**:453–456.

Sjöstrand, T., 1970, Early studies of CO production, *Ann. N. Y. Acad. Sci.* **174**:5–10.

Skujins, J., 1978, History of abiontic soil enzyme research, in: *Soil Enzymes* (R. G. Burns, ed.), pp. 1–49, Academic Press, London.

Skujins, J., 1984, Microbial ecology of desert soil, in: *Advances in Microbial Ecology,* Vol. 7 (K. C. Marshall, ed.), pp. 49–91, Plenum Press, New York.

Spratt, Jr., H. G., and Hubbard, J. S., 1981, Carbon monoxide metabolism in roadside soils, *Appl. Environ. Microbiol.* **41**:1191–1201.

Strayer, R. F., and Tiedje, J. M., 1978, *In situ* methane production in a small, hypereutrophic, hard-water lake: Loss of methane from sediments by vertical diffusion and ebullition, *Limnol. Oceanogr.* **23**:1201–1206.

Swinnerton, J. W., and Lamontagne, R.A., 1974, Carbon monoxide in the South Pacific Ocean, *Tellus* **26**:136–142.

Swinnerton, J. W., Linnenbom, V. J., and Lamontagne, R. A., 1970, Ocean: A natural source of carbon monoxide, *Science* **167**:984–986.

Swinnerton, J. W., Lamontagne, R. A., and Bunt, J. S., 1977, Field Studies of Carbon Monoxide and Light Hydrocarbon Production Related to Natural Biological Processes, Naval Research Laboratory, Washington, D. C., Report 8099, pp. 1–9.

Tenhunen, R., Marver, H. S., and Schmid, R., 1969, Microsomal heme oxygenase. Characterization of the enzyme, *J. Biol. Chem.* **244**:6388–6394.

Trevors, J. T., 1985, Hydrogen consumption in soil, *Plant Soil* **87**:417–422.

Troxler, R. F., 1972, Synthesis of bile pigments in plants. Formation of carbon monoxide and phycocyanobilin in wild-type and mutant strains of the alga, *Cyanidium caldarium, Biochemistry* **11**:4235–4242.

Troxler, R. F., and Dokos, J. M., 1973, Formation of carbon monoxide and bile pigment in red and blue-green algae. *Plant Physiol.* **51**:72–75.

Uratsu, S. K., Keyer, H. H., Weber, D. F., and Lim, S. T., 1982, Hydrogen uptake (HUP) activity of *Rhizobium japonicum* from major U. S. soybean production areas, *Crop Sci.* **22**:600–602.

Vallentyne, J. R., and Whittaker, J. R., 1956, On the presence of free sugars in filtered lake water, *Science* **124**:1026–1027.

Walker, C. C., and Yates, M. G., 1978, The hydrogen cycle in nitrogen-fixing *Azotobacter chroococcum, Biochimie* **60**:225–231.

Wangersky, P. J., 1976, The surface film as a physical environment, *Annu. Rev. Ecol. Syst.* **7**:161–176.

Westlake, D. W. S., Talbot, G., Blakley, E. R., and Simpson, F. J., 1959, Microbial decomposition of rutin, *Can. J. Microbiol.* **5**:621–629.

Wilks, S. S., 1959, Carbon monoxide in green plants. *Science* **129**:964–966.

Williams, P. J. LeB., 1981, Microbial contribution to overall marine plankton metabolism: Direct measurement of respiration, *Oceanol. Acta* **4**:359–364.

Williams, R. T., and Bainbridge, A. E., 1973, Dissolved CO, CH₄, and H₂ in the southern ocean, *J. Geophys. Res.* **78**:2691–2694.

Wilson, D. F., Swinnerton, J. W., and Lamontagne, R. A., 1970, Production of carbon monoxide and gaseous hydrocarbons in seawater: Relation to dissolved organic carbon, *Science* **168**:1577–1579.

Winkler, L. W., 1901, Die Löslichkeit der Gase in Wasser (Dritte Abhandlung), *Ber. Chem. Ges.* **34**:1400–1422.

Wittenberg, J., 1960, The source of carbon monoxide in the float of the Portuguese Man-of War *Physalis physalis, J. Exp. Biol.* **37**:698–705.

Wolff, D. G., and Bidlack, W. R., 1976, The formation of carbon monoxide during peroxidation of microsomal lipids, *Biochem. Biophys. Res. Commun.* **73**:850–857.

Yamane, I., and Sato, K., 1963, Decomposition of organic acids and gas formation in flooded soil, *Soil Sci. Plant Nutr.* **9**:32–36.

Yamane, I., and Sato, K., 1964, Decomposition of glucose and gas formation in flooded soil, *Soil Sci. Plant Nutr.* **10**:127–133.

Yamane, I. and Sato, K., 1967, Effect of temperature on the decomposition of organic substances in flooded soil, *Soil Sci. Plant Nutr.* **13**:94–100.

Yoshida, T., Noguchi, M., and Kikuchi, G., 1982, The step of carbon monoxide liberation in the sequence of heme degradation catalyzed by the reconstituted microsomal heme oxygenase system, *J. Biol. Chem.* **257**:9345–9348.

Zafiriou, O. C., Joussotdubien, J., Zepp, R. G., and Zika, R. G., 1984, Photochemistry of natural waters, *Rev. Environ. Sci. Technol.* **18**:A358–A371.

Zavarzin, G. A., and Nozhevnikova, A. N., 1977, Aerobic carboxydobacteria, *Microb. Ecol.* **3**:305–326.

Zehnder, A. J. B., 1978, Ecology of methane formation, in: *Water Pollution Microbiology* (R. Mitchell, ed.), Vol. 2, pp. 349–376, Wiley, New York.

Zeikus, J. G., 1983, Metabolic communication between biodegradative populations in nature, in: *Microbes in Their Natural Environments* (J. H. Slater, R. Whittenbury, and J. W. T. Wimpenny, eds.), pp. 423–462, Cambridge University Press, Cambridge.

Zimmerman, P. R., Chatfield, R. B., Fishman, J., Crutzen, P. J., and Hanst, P. L., 1978, Estimates on the production of CO and H₂ from the oxidation of hydrocarbon emissions from vegetation, *Geophys. Res. Lett.* **5**:679–682.

Zimmerman, P. R., Greenberg, J. G., Wandiga, S. O., and Crutzen, P. J., 1982, Termites: A potential large source of atmospheric methane, carbon dioxide, and molecular hydrogen, *Science* **218**:563–565.

8

Use of "Specific" Inhibitors in Biogeochemistry and Microbial Ecology

RONALD S. OREMLAND and DOUGLAS G. CAPONE

Inhibitors are like old sports cars: They are fun to play around with, but you should never trust them!

—*Anonymous microbiologist*

1. Introduction

The above statement, although meant to be tongue in cheek, contains an essential truism: all work with inhibitors is inherently suspect. This fact has been known by biochemists for some time. However, use of chemical inhibitors of enzymic systems and membranes continues to be a common approach taken toward unraveling the biochemistry and biophysics of plants, animals, and microorganisms. Various types of "broad-spectrum" biochemical inhibitors (e.g., poisons, respiratory inhibitors, and uncouplers) have been employed by ecologists for many years in order to demonstrate the active participation of microbes in chemical reactions occurring in natural samples (e.g., soils, sediments, and water). In recent years, considerable advances have been made in our understanding of the bio-

Dedicated to Prof. Barrie F. Taylor, our former mentor and bacterial tormentor.

RONALD S. OREMLAND • Water Resources Division, United States Geological Survey, Menlo Park, California 94025. **DOUGLAS G. CAPONE** • Chesapeake Biological Laboratory, Center for Environmental and Estuarine Studies, University of Maryland, Solomons, Maryland 20688-0038.

chemistry of microorganisms of biogeochemical interest. Concurrent with these advances have been the discoveries of novel types of compounds that will block the metabolism of one particular group of microbes, but have little disruptive effect on other physiological types. Thus, the term "specific inhibitor" has been applied to these types of compounds when they are used to probe the functions of mixed populations of microorganisms. These substances provide powerful experimental tools for investigating the activity and function of certain types of microorganisms in natural samples. In addition, some have commercial applications in agriculture and industry. Indeed, efforts to develop pesticides and microbial inhibitors for commercial applications have provided a number of routinely used specific inhibitors. Selective inhibition is not a new approach, however. Selective physical treatments (e.g., differential heating) have been used to study competition for hydrogen between sulfate-reducers and methanogens in marine sediments (ZoBell, 1947). Only more recently has this problem been tackled by use of chemical inhibitors.

The current popularity of employing inhibitors in ecological/biogeochemical studies must be tempered by a knowledge of their limitations. It is our purpose in this review to point out not only the types of inhibitors available and their modes of action, but their potential disruptive effects, as well as cases in which they are ineffective. Thus, while a researcher may think he or she is employing a "silver bullet," in reality it could prove to be either a blank round or a shotgun shell! The "moral" one should draw from this review is that, although inhibitors are convenient tools in microbial ecology, studies that base their findings solely on the results of single inhibition experiments are at best incomplete. All is not hopeless, however, for we shall present a suggested experimental framework by which inhibitor studies can be made more convincing.

Use of inhibitors in ecological studies has been so widespread that it would be impossible to document all the cases in which they have been applied. It is not the intent of this work to establish an exhaustive listing of all conceivable types of inhibitors, their modes of action, target organisms, and limitations. Instead, in this chapter we will primarily emphasize inhibitors used to study important biogeochemical processes.

2. General Considerations

2.1. Definitions, Rationales, and Caveats

Webster's *New World Dictionary* (college edition) defines "inhibitor" as "any substance that slows or prevents a chemical or organic reaction." When this is translated into experimental terms, one sees that, conceptually, several types of inhibition are possible. These include total, partial, delayed, transient, transient/suppressed, and transient/enhanced inhibition, as well as failure to inhibit (Fig. 1). Obviously, the desired effect the experimentalist wishes to achieve is total inhibition. This result is easily interpreted: the target organism is present in the assay system and the inhibitory compound is effective at the concentration applied. However, even this simple case can have complexities that can cloud the results. For example, one should ask whether it was the inhibitor that did the work, or the solvent, impurities, or improper incubation conditions that caused fortuitous and possibly artefactual results. Were the results reproducible? All the other cases of less than total inhibition shown in Fig. 1 present equivocal results, and require further investigation in order to unravel their meaning. Thus, even the absence of any noticeable inhibition does not mean that the target organisms are absent or inactive, even though this could be a desired interpretation. The most positive statement one can logically make is that the results *imply* that the organisms are not present.

An inhibitor may be ineffective for a number of reasons. These include: chemical breakdown during storage, adsorption onto particles, poor dispersion in the assay system, application at too low a concentration, significant chemical disparities between the system assayed and those reported in the literature, as well as the presence of novel target microbes whose biochemistry is somewhat different from those reported in pure culture. Further complications can arise from an increase in resistance to inhibitors associated with mixed populations of bacteria as opposed to the sensitivities reported for pure cultures (Bennett and Bauerle, 1960). The list of possibilities can go on. The point is that further experimentation is required to prove that the target organisms are absent from the system. This can be achieved by several means, such as ensuring that the inhibitor is effective on stock cultures and running dosage–response curves on both the cultures and the environmental samples. It is always extremely important, however, to have other lines of evidence that support the inhibitor results. These should be of the type that substantiate that the target organisms are either present or absent from the assay system (e.g., fluorescent antibody counts, culturing success, enhanced activity with specific substrate addition). In addition, one can use another inhibitor of the process, or chemically manipulate the system in order to achieve the same effect. The cumulative effect of these collateral investigations is that a stronger case is built to support the results obtained with the "specific" inhibitor.

An example of such an approach is the well-studied case of hydrogen consumption by anoxic marine sediments. Uptake of hydrogen by sedi-

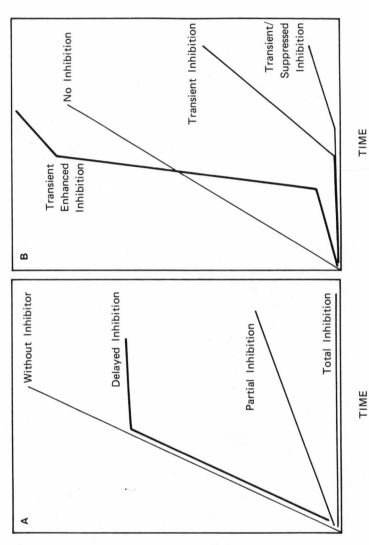

Figure 1. Different possible types of microbial responses to perturbations by an inhibitor.

ment slurries could be due to the activities of either methanogenic or sulfate-respiring bacteria. Hydrogen consumption by tropical marine sediments was abolished by autoclaving, or by the broad-spectrum procaryotic inhibitor chloramphenicol (Oremland, 1976; Oremland and Taylor, 1978). These results demonstrated the bacterial participation in the reaction. Uptake was also inhibited by molybdate (an inhibitor of sulfate-respirers), but not by chloroform (an inhibitor of methanogens), thereby suggesting that sulfate-respirers were responsible (Fig. 2). Further support for the involvement of sulfate-respirers was the fact that sulfate-free slurries were also inhibited (Fig. 2), and that the quantity of hydrogen consumed in the experiments could be accounted for by the quantity of available sulfate ions, but not by the amount of methane formed (Table I). Thus, the conclusions drawn from the "specific" inhibitor studies were reinforced and thereby made more convincing.

Results of inhibitor studies can often have more than one possible interpretation. For example, a pure culture of a methanogenic bacterium can degrade dimethylsulfide to methane and carbon dioxide with a ratio of CH_4/CO_2 of about 3.0 (Kiene et al., 1986). However, addition of ^{14}C-dimethylsulfide to sulfate-rich anoxic sediments results in ratios of ~0.08. Blockage of sulfate reduction by molybdate increases the ratio to ~1.8. Does this mean that sulfate-reducers can also oxidize dimethylsulfide and possibly compete with methanogens for this substrate? Alternatively, because sulfate reduction is the major sink for hydrogen in anoxic, sulfate-rich sediments (e.g., Abram and Nedwell, 1978a; Oremland and Taylor, 1978), inhibition of these organisms will result in increased hydrogen consumption by methanogens. This could cause the conversion of $^{14}CO_2$ produced during methanogenic degradation of ^{14}C-dimethylsulfide to increased $^{14}CH_4$ levels via hydrogen reduction, thereby resulting in an increase in the $^{14}CH_4/^{14}CO_2$ ratio of the molybdate-inhibited sediment. Alternatively, it could be due to a combination of both phenomena.

The other types of inhibition displayed in Fig. 1 are more difficult to explain and are, unfortunately, more common than the literature indicates or admits. Any of the three "transient" responses (Fig. 1B) can be converted into a "total" inhibition by merely shifting the time frame to the left! This practice is not as nefarious as it may sound. Short-term inhibition implies that the target organisms are blocked over a time period during which uninhibited samples can clearly exhibit their activity. It is usually desirable to obtain activity measurements over short incubation periods (if possible), so that experimental conditions closely approximate those in nature. What happens subsequently relative to the controls can be due to numerous causes: growth of mutants unaffected by the inhibitor, degradation of the inhibitor, etc. These defy interpretation without further experimentation.

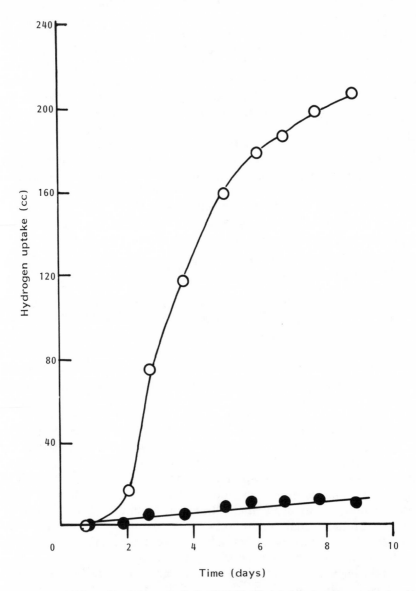

Figure 2. Inhibition of hydrogen uptake by anerobically incubated sediment slurries. (A) (○) Uninhibited control; (●) inhibited with 20 mM molybdate. (B) (○) Uninhibited control; (♦) autoclaved; (●) chloroform; (■) b-fluorolactate; (□) minus sulfate. [Data from Oremland (1976).]

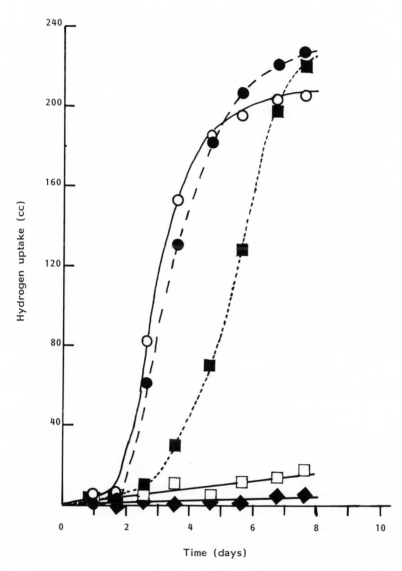

Figure 2.   *(Continued)*

Table I. Hydrogen Uptake and Methane Production by Marine
Sediment Slurries[a]

Condition	Hydrogen uptake[b] (ml)	Methane production[c] (nmole/g dry weight)
Control	218	42 (3)
Minus sulfate	12	67 (18)
Plus molybdate	10	69 (6)

[a]Adapted from Oremland and Taylor (1978), with permission.
[b]Flasks contained 100 ml of slurry and a 100-ml gas phase of H_2. Results represent cumulative H_2 consumption after 213 hr incubation at 26°C. Calculated theoretical uptake was 221 ml. This calculation assumes a sulfate content of seawater of 28 mM, that 4 moles of H_2 is required to reduce 1 mole of sulfate, and a volume temperature correction factor of 10%.
[c]Results represent the mean of three flasks; parentheses indicate one standard deviation.

2.2. "Ideal" and Real Inhibitors

The concept of a "specific" inhibitor obviously implies that the compound blocks only the target organisms, while having no undue influence on the other components of the microbiota or upon the chemical/physical conditions of their milieu. In practice, all inhibitors have certain deficiencies that call into question their specificity and efficacy. Thus, it is of some use to discuss the desired properties that an "ideal" specific inhibitor should have, so that when we examine individual cases, we can point out their shortcomings.

For any given inhibitor, a number of criteria may be described. The ideal inhibitor should have the following properties:

1. It should inhibit only the target organism and not influence the activity of other microbes.
2. It should be effective at low concentrations.
3. It should be water-soluble (to assist dispersion and minimize secondary solvent effects).
4. It should easily penetrate the target cells.
5. It should not degrade rapidly (chemical or biological) or suffer from a high degree of adsorption onto particles, thereby limiting its bioavailability.
6. It should not alter the chemistry of the assay milieu in ways that may influence the activity of nontarget organisms.
7. It should be available in a relatively pure state, and any contaminants should be inocuous.
8. It should not interfere with analytical procedures required during the experiment.

9. It should not produce spurious chemical or biological effects that cloud interpretation or lead to incorrect conclusions.
10. It should be inexpensive and easily obtained.

The fact that this list adds up to ten should not necessarily imply a Mosaic connection! However, like the Ten Commandments, these idealized properties of "specific" inhibitors are never fully complied with. They are nonetheless something to strive for!

2.2.1. General Inhibitors of Growth and Metabolism

Specific inhibitors can be of broad or narrow spectrum, blocking either a wide assortment of physiologically distinct classes or only a particular family of microorganisms.

The modes of action of inhibitors, which are widely varied, directly determine the relative degree of specificity (Franklin and Snow, 1981). Inhibitors whose mode of action involves interaction with major cellular components (e.g., lipid membranes, proteins, DNA) common to a broad range of organisms will likely be broad-spectrum. Broad-spectrum inhibitors are typified by antibiotics, which block protein synthesis, cell wall assembly, or membrane structure (see Section 3.1.2). Heavy metals at high concentrations destroy cross-linked thiol groups and thereby inhibit microbial protein function in general (Gadd and Griffith, 1978; Babich and Stotzky, 1980). Other broad-spectrum inhibitors include general biocides (e.g., phenols, halogens, aldehydes), as well as respiratory cytochrome inhibitors, such as cyanides and azides. The ability to inhibit broadly and effectively microorganisms in general, the opposite end of the spectrum from "specific inhibition," is an important factor in manipulative experiments with regard to controls (Brock, 1978) as well as in commercial/medical applications (Pelczar et al. 1986). However, even some "broad-spectrum" respiratory inhibitors (e.g., azide) have limitations (Rozyccki and Bartha, 1981), and in some instances an approach using multiple antimetabolites is justified (e.g., Paul, 1984).

The inhibition of energy metabolism, be it linked to either autotrophic or heterotrophic processes, provides a somewhat more specific level for inhibition. Broad physiological subgroups can potentially be distinguished by such an approach. A variety of distinct approaches can be used to inhibit aerobic respiration and Krebs cycle functioning [e.g., uncouplers, ionophores, enzyme inhibitors (Ereckinska and Wilson, 1984)]. Similarly, phototrophic energy generation may be disrupted at the level of photosynthetic electron flow (photoautotrophs), inorganic substrate oxidation (chemoautotrophy), or in the "dark" reactions of the Calvin–Benson cycle (Trebst, 1980; Badour, 1978) (see Section 4.1).

The greatest degree of specificity derives from those agents whose interaction is restricted to an enzyme or cofactor unique to a particular class of organism. Narrow-spectrum inhibitors are represented by such compounds as molybdate (Oremland and Taylor, 1978), 2-bromoethanesulfonic acid (BES) (Gunsalus et al., 1978), and nitrapyrin (Goring, 1962). Molybdate uncouples the energy metabolism of sulfate-reducing bacteria at the level of ATP sulfurylase (see Section 3.2). The compound BES is an analogue of a cofactor unique to methanogens and is highly effective in blocking methanogenesis (see Section 3.1). Nitrapyrin is a relatively specific inhibitor of the ammonium oxidase enzyme of nitrification (see Section 3.3). There is a rich biochemical literature on "specific" enzyme inhibitors, which may be of some use in future biogeochemical and microbial ecological studies (e.g., Dawson et al., 1969; Brodbeck, 1980; Jain, 1982; Umezawa, 1982; Katunuma et al., 1983). The medical antimicrobial literature (e.g., Peterson, 1986) and insecticide/pesticide literature (Lal, 1984; Hill and Wright, 1978) also provide a resource for potential specific inhibitors. For instance, a new class of chitin synthetase inhibitors are being developed for application as insecticides (Leighton et al., 1981), which may also be useful in studies of chitin-forming microbes such as fungi (Duran et al., 1979) and certain bacteria (Smucker and Simon, 1986).

2.2.2. Antibiotics

Antibiotics form a class of compounds that readily come to mind in any discussion of microbial inhibitors. They have been applied in studies of microbial ecology and, to a lesser extent, biogeochemistry.

Antibiotics are defined as low-molecular weight microbial metabolites that at low concentrations inhibit the growth of other microorganisms (Lancini and Parenti, 1982). This definition can be expanded to include semisynthetic forms or analogues based on natural examples. Antibiotics generally target a wide range of microorganisms (Table II). In the "antibiotics" literature, the term broad-spectrum generally refers to substances that affect Gram-negative as well as Gram-positive bacteria, whereas narrow-spectrum inhibitors restrict their activity to Gram-positive forms. Gram-negative bacteria generally have lower susceptibility to inhibition by antibiotics, although antibiotics are known that are more effective against Gram-negative forms (Lancini and Parenti, 1982).

In microbial ecological studies, antibiotics have been particularly useful in discerning procaryotic versus eucaryotic contributions to certain processes (Brock, 1961) (see Section 4.2). For example, this approach has been taken to study CO metabolism in soils (Conrad and Seiler, 1980). Theoretically, one may also be able to discern the relative contribution

Table II. Minimum Inhibitory Concentrations (μg/ml) of Representative
Antibiotics for a Variety of Bacteria[a]

Microbe	Penicillin G	Tetracycline	Chloramphenicol
Gram-positive:			
Diplococcus pneumoniae	0.010	0.05–1	1–3
Streptococcus fecalis	1–4	0.1	2–5
Clostridium tetani	0.01–3	0.1–3	1
Corynebacterium diptheriae	0.3–3	0.5–5	0.5–5
Gram-negative:			
Neisseria gonorrhea	0.05	0.1–3	1
Escherichia coli	Uns	0.5–5	2–12
Klebsiella spp.	6	0.5–Uns	0.3–1.5
Pseudomonas aeruginosa	Uns	4–200	4–Uns

[a]Adapted from Lancini and Parentii (1982), with permission. Uns, unsusceptible.

of Gram-negative and Gram-positive bacteria in particular processes and/or environments. For instance, ultrastructural and isolation studies indicate a predominance of Gram-negative forms in the marine water column, with a shift toward Gram-positive bacteria moving into anoxic marine sediments (Sieburth, 1979; Moriarity and Hayward, 1982).

Antibiotics that inhibit eucaryotic (e.g., cycloheximide) or procaryotic (e.g., chloramphenicol) protein synthesis have also been used to determine if lag periods in enzyme activities of natural samples are a result of induction and synthesis of the enzyme during the lag period (e.g., Sutton, 1980; Dicker and Smith, 1980). Comprehensive discussion of the mode of action of particular antibiotics can be found in the volumes by Gottleib and Shaw (1967), Franklin and Snow (1981), Hahn (1983), and Richmond (1969).

3. Examples of Specific Inhibitors: Case Studies

We shall discuss in detail the characteristics of four commonly used inhibitors (2-bromoethanesulfonic acid, molybdate, nitrapyrin, and acetylene) in order to contrast their properties with the idealized compound. We shall discuss other inhibitors more succinctly in the subsequent relevant sections.

3.1. Case 1. 2-Bromoethanesulfonic Acid

The idealized situation is probably most closely approximated by the inhibitor 2-bromoethanesulfonic acid (BES). This compound is a structural analogue of mercaptoethanesulfonic acid, the cofactor known as

"HS-coenzyme M" in methanogenic bacteria (C. D. Taylor and Wolfe, 1974). Coenzyme M (CoM) is associated with the terminal methylation reactions involved in methanogenesis, including the methyl-CoM reductase enzyme complex from which methane is evolved (Gunsalus and Wolfe, 1978, 1980). HS-Coenzyme M can be methylated via a methyltransferase reaction (Shapiro and Wolfe, 1980).

The compound BES, the analogue of CoM, was first employed to study the inhibition of methyl-CoM reductase activity in cell-free extracts of *Methanobacterium thermoautotrophicum* (Gunsalus *et al.*, 1978). Fifty percent inhibition of methane production occurred at $\sim 10^{-5}$ M BES, with total inhibition evident at $\sim 5 \times 10^{-4}$ M. In contrast, chloroethanesulfonic acid (CES) required about tenfold higher concentrations than BES to achieve the same degree of inhibition (Fig. 3). Subsequent investigations revealed that whereas CoM is present in all species of methanogenic bacteria, the compound could not be detected in a broad survey of plant, animal, and other microbial tissues (W. E. Balch and Wolfe, 1979a). In addition, BES was found to be a potent inhibitor of methanogenesis and growth by *Methanobacterium ruminantium* (a rumen species, which requires an exogenous supply of CoM; C. D. Taylor *et al.*, 1974), although there was some ameliorating influence related to the amount of HS-CoM in the medium. At a molar ratio of 15:1 (HS-CoM: BES), no inhibition could be achieved, whereas complete inhibition occurred at a ratio of 1:1. By contrast, an analogue of BES, 3-bromopropanesulfonic acid (BPS), was ineffective as an inhibitor (W. E. Balch and Wolfe, 1979a). Uptake of HS-CoM by whole cells of *M. ruminantium* was completely inhibited by 10^{-6} M BES, whereas 10^{-5} M BPS had no inhibitory effect (W. E. Balch and Wolfe, 1979b). All of these studies were done with methanogens grown upon hydrogen plus carbon dioxide. Another promising candidate as an inhibitor of methyl-CoM reductase is seleno-CoM. However, only very limited work has been done with this compound (L. Daniels, personal communication).

M. R. Smith and Mah (1978) found that growth of *Methanosarcina* strain 227 on acetate could be blocked by 70 μM BES and they suggested that acetate and methanol were dismutated to methane via the methyl-CoM reductase terminal step. This has been confirmed for methanol (Shapiro and Wolfe, 1980; van der Meijden *et al.*, 1983) and trimethylamine (Naumann *et al.*, 1984). Thus, CoM appears to be the terminal methyl carrier proposed by Barker (1956), and presumably all methanogenic substrates must be routed via this step. The BES functions as a competitive inhibitor of the methyl-CoM reductase complex, because reversal of inhibition by cell extracts can be achieved with methyl-CoM, but not HS-CoM (Fig. 4). In addition, BES does not inhibit the methyltransferase system (M. R. Smith, 1983; van der Meijden *et al.*,

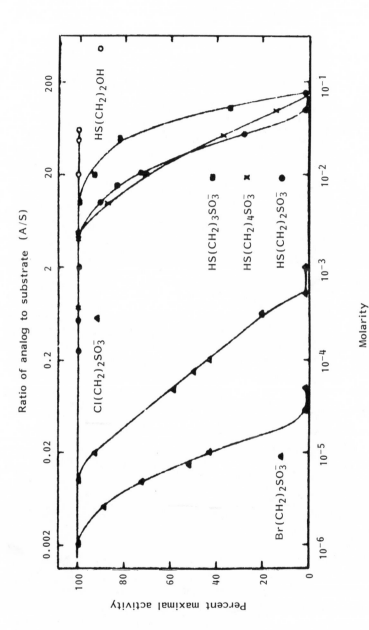

Figure 3. Effect of BES and other analogues of coenzyme M (CoM) on methane formation from CH_3-S-CoM by cell extracts of *Methanobacterium thermoautotrophicum*. [From Gunsalus *et al.*, 1978, with permission.]

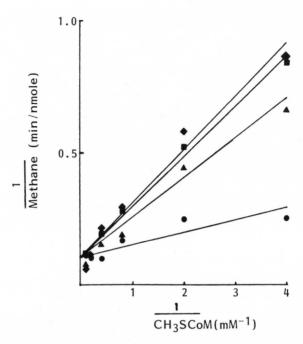

Figure 4. Methylreductase activity as a function of CH_3-S-CoM concentration at different concentrations of BES. Level of BES: (●) none; (▲) 2.5 μM; (■) 5 μM; (◆) 10 μM. [From M. R. Smith (1983), with permission.]

1983). Thus, BES inhibits methanogenesis by all species of methanogens growing on any of the recognized substrates (acetate, hydrogen plus carbon dioxide, methanol, formate, methylated amines), including the recently revealed dimethylsulfide (Kiene *et al.*, 1986). Because CoM is found only in methanogenic bacteria, BES functions as a "specific inhibitor" when used in mixed microbial systems. In support of this concept, BES was found to have no undue influence upon bacterial growth of eight different types of anaerobes, including N_2-fixing strains, while completely preventing diazotrophic growth of three methanogenic species (L. Daniels, personal communication) (Table III).

The compound BES is inexpensive and readily obtained from commerical sources in purified form. Although it functions as a competitive inhibitor, relatively little should be required, since the quantities of CoM in the environment are in the nanomolar range (Mopper and Taylor, 1986). It is also highly water-soluble. Thus, it can easily be added to cultures, sewage sludge, rumen contents, and sediment slurries, and be injected into whole intact sediment cores. The compound is usually effective at blocking methanogenesis in cultures when added at levels in excess of 0.1 mM. It is particularly useful in demonstrating the presence of an active methanogenic biota in unlikely microcosms, such as the dental plaque of primates (Kemp *et al.*, 1983), in estuarine plankton (Oremland, 1979), or in deep sea sediments (Oremland *et al.*, 1982c). It has been employed successfully to demonstrate that methanogens produce ethane (Fig. 5) (Oremland, 1981), are important in the degradation of aromatics (Healy *et al.*, 1980; Grbic-Galic and Young, 1985), utilize trimethylamine and methanol as primary methane precursors in saltmarshes (Fig. 6) (Oremland *et al.*, 1982b), and can attack certain methane precursors added to sediments (Oremland and Polcin, 1982; Dicker and Smith, 1985a).

However, this inhibitor does have some special problems related to the quantity added to assay systems. For example, spontaneous mutants of *Methanosarcina* strain 227 were resistant to 0.24 mM BES when previously exposed to levels of 0.024 mM, and a strain of *Methanobacterium formicicum* was found to be resistant to 0.2 mM BES without any prior exposure (M. R. Smith and Mah, 1981). Resistance to BES appears to be conferred by impermeability to the compound, since cell-free extracts of resistant mutants were susceptible to BES inhibition of methyl-S-CoM reductase (M. R. Smith, 1983). Bouwer and McCarty (1983b) reported that fixed film acetate digestors had only partial (41%) inhibition when exposed to 0.6 mM BES. They speculated that this may have been due to resistant mutants or degradation of BES by the microbial community. In addition, there appears to be differential susceptibility of the methano-

Table III. Effect of 2-Bromoethanesulfonic Acid (BES) on Bacterial Growth of Various Eubacteria and Diazotrophic Methanogens[a]

Organism	Nitrogen source[b]	Generation time (hr)	
		No BES	25 mM BES[c]
Escherichia coli	TYE	2.4	2.3
Lactobacillus casei	TYE	2.0	1.7
Corneybacterium lilium	TYE	2.6	2.1
Eubacterium limosum	TYE	0.7	0.7
Clostridium hystolyticum	TYE	0.7	0.7
Clostridium 3-1	N	4.0	3.5
Clostridium 4-1	N	4.0	4.0
Rhodopseudomonas 5-1	N	6.0	6.0
Methanococcus thermolithotrophicus	N	4.5	NG
Methanospirillum hungatei	N	27.0	NG
Methanobacterium bryantii	N	23.0	NG

[a]Data courtesy of L. Daniels.
[b]TYE, Tryptone plus yeast extract; N, nitrogen gas.
[c]NG, No growth.

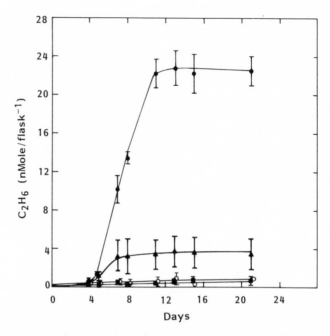

Figure 5. Formation of ethane in anoxic estuarine sediment slurries incubated under H_2. Points indicate the mean and bars the standard deviation of three flasks. (▲) No additions; (●) plus ethyl-S-CoM; (■) plus BES; (○) plus ethyl-S-CoM and BES. Flasks contained 75 ml of slurry and headspace volume was 68 ml.[From Oremland (1981), with permission.]

genic biota of natural ecosystems to different concentrations of BES. Zinder *et al.* (1984) found that while 1 mM BES completely inhibited thermophilic methane formation from acetate in sludge, 50 mM was required to shut off CO_2 reduction. Therefore, the concentration of added BES is an important factor, and it is a good idea to employ relatively high levels (e.g., >10 mM) in order to achieve inhibition. Unfortunately, use of such high concentrations may alter the chemistry of the systems assayed. Even millimolar concentrations, however, are sometimes not infallible at arresting methanogenesis. Oremland *et al.* (1982a) reported that sediment slurries from an alkaline, hypersaline lake did not always respond to BES (1.4–14 mM range), although this may have been due to the unusual chemistry of the system. Methanogenesis in the less saline surface waters was partially inhibited by 37 mM BES, while bottom water was unaffected (Oremland *et al.*, 1988). Likewise, King (1984) reported that >100 mM BES was required to inhibit methanogenesis in whole cores, although this could have been caused by diffusional limitations occurring at the lower concentrations.

It is extremely important to test each individual system to be assayed for an inhibitory response to added BES concentration ranges. This is relatively easy to do in slurries (flushing the gas phase will remove most of the initial methane), but is much harder to achieve with intact or compacted sediments (Kiene and Capone, 1985). Thus, it may not be an easy task to determine the efficacy of BES in sediments having a high methane background. This may require the use of radioisotopes. However, it should be borne in mind that there are several types of potential substrates (e.g., acetate, hydrogen plus carbon dioxide, methylated amines, methanol, dimethylsulfide). Furthermore, successful inhibitor concentrations obtained with slurries do not necessarily apply to compacted sediments (Vogel et al., 1982; King, 1984). In addition, the length of exposure of the systems to BES could be an important factor. For example, King et al. (1983) found that complete inhibition of methanogenesis in sediment slurries required >48 hr incubation.

Use of BES may also influence other microbial processes in unsuspected ways. Addition of 2 and 5 mM BES to saltmarsh sediments caused modest but discernible decreases in the rate of sulfate reduction (16 and 22%, respectively) when compared with sediments lacking inhibitor (Dicker and Smith, 1985a). Although it is impossible to draw conclusions based on these data, one possibility is that sulfate-respiring bacteria may be able to use the sulfonic acid moiety as an electron acceptor in prefer-

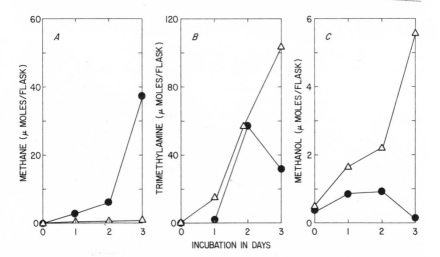

Figure 6. (A) Formation of methane, and accumulation of (B) trimethylamine and (C) methanol in saltmarsh sediment slurries incubated (●) without additions or (△) with BES. Flasks contained 80 ml of slurry under a 58 ml N₂ headspace. [From Oremland et al, (1982b), with permission.]

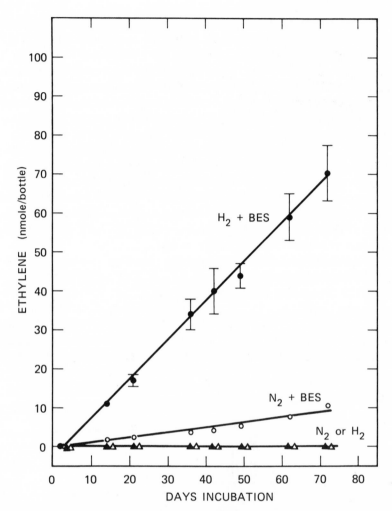

Figure 7. Ethylene levels in the headspace (volume 38 ml) of flasks containing 25 ml of freshwater lake sediment slurries. Flasks were incubated under H_2 or N_2 with or without BES as indicated. Results represent the mean of three flasks, and bars indicate one standard deviation. [R. S. Oremland (unpublished data).]

ence to sulfate. Another "quirk" of BES is that the compound promotes the production of traces of ethylene when added to freshwater (Fig. 7) or estuarine sediments (Oremland, 1981), as well as the anoxic waters from an alkaline, hypersaline lake (R. S. Oremland, unpublished data). Ethylene production is apparently a minor, purely chemical reaction, since autoclaved sediments also formed the gas when supplemented with BES

(Oremland, 1981). However, until this was revealed, it indicated (falsely) that ethylene was an intermediate in the formation of ethane by methanogenic bacteria in anoxic sediments. This particular incorrect explanation was most seductive because ethylene was previously shown to inhibit methanogenesis (Oremland and Taylor, 1975), perhaps by diverting electrons to ethane production. Thus, a misinterpretation of inhibitor results in this case was avoided by a careful examination of alternative explanations and followup experiments.

It should be borne in mind, however, that this particular ethylene-producing quirk of BES can cloud acetylene-reduction (nitrogen-fixation) assays. Because the ability to fix dinitrogen has recently been detected in methanogenic bacteria (Belay *et al.,* 1984; Murray and Zinder, 1984; Bomar *et al.,* 1985), it is probably only a matter of time until someone attempts to determine the significance of methanogenic N_2 fixation in sediments. Use of BES with the acetylene reduction test must be approached with caution, the more so since acetylene is itself an inhibitor of methanogenesis (Oremland and Taylor, 1975) (Section 3.4).

3.2. Case 2. Molybdate and Other Group VI Oxyanions

Interest in blocking the activity of sulfate-respiring bacteria stemmed from the role they play in the corrosion of buried steel-based materials (e.g., pipelines). Thus, initial impetus was economic as opposed to developing tools for ecological research. Economic considerations still foster research in this area (Postgate, 1979). Postgate (1949) first reported inhibition of *Desulfovibrio desulfuricans* by selenate ions. Sulfate reduction was followed by measuring uptake of H_2 by resting cells. Selenate inhibited uptake (~70%), and inhibition could be reversed by addition of sulfite. Although the nature of the inhibition was competitive, partial inhibition occurred at selenate–sulfate ratios of above 0.02, and complete inhibition was observed at ratios above 0.1. Subsequently, Postgate (1952) was able to show that a number of oxyanions stereochemically similar to sulfate (including group VI oxyanions) were competitive inhibitors. Furusaka (1961) demonstrated that transport of sulfate into *D. desulfuricans* was competitively inhibited by selenate ions, and that transport is dependent upon physiological reduction of sulfate.

A biochemical basis (other than transport) also exists for the inhibition of sulfate reduction by group VI oxyanions. The initial step in both assimilatory and dissimilatory sulfate reduction requires the activation of the sulfate ion with ATP and the participation of the enzyme ATP sulfurylase (Robbins and Lipmann, 1958; Wilson and Bandurski, 1958; Peck, 1959). The reaction proceeds as follows:

$$ATP + SO_4^{2-} \rightarrow APS + PPi$$

where APS is the stable adenosine-5′-phosphosulfate intermediate and PPi is pyrophosphate. In sulfate-respiring bacteria, pyrophosphate is rapidly cleaved by the presence of a very active pyrophosphatase, which "pulls" the reaction to the right under anaerobic conditions (Ware and Postgate, 1971). Of the group VI oxyanions, only selenate forms a stable APSe intermediate, whereas molybdate, chromate, and tungstate form putatively unstable analogues (Wilson and Bandurski, 1958). Both assimilatory and dissimilatory sulfate reduction behave similarly at this phase (Peck, 1959). Addition of molybdate to cell-free extracts of *D. desulfuricans* blocked the formation of APS, while stimulating the release of phosphate ions in the presence of added pyrophosphatase (Peck, 1962). This results in the formation of the unstable APMo analogue (in lieu of APS), which breaks down to AMP + molybdate. Without the presence of a stable intermediate like APS, the next sequential enzymatic step, catalyzed by APS reductase, cannot occur:

$$APS + 2e^- \rightarrow AMP + SO_3^{2-}$$

Because sulfate-respiring bacteria must first expend energy (in the form of ATP) in order to activate the sulfate ion, they do not recover this energy until sequential reduction from sulfite to sulfide takes place (Peck, 1960). Thus, molybdate and other group VI oxyanions should rapidly deplete ATP pools in sulfate-respiring bacteria, thereby causing death. Exposure of *D. desulfuricans* and a marine *Desulfovibrio* sp. to group VI oxyanions rapidly lowered ATP levels in cells, thereby giving credence to this hypothesis (B. F. Taylor and Oremland, 1979). Chromate was the most effective ion, followed by molybdate and tungstate (about equal), and finally by selenate (Table IV). The ability of selenate to form APSe while other group VI oxyanions cannot may be the basis for its ineffec-

Table IV. Effect of Oxyanions of Group VI Elements on ATP Levels in Sulfate-Respiring Bacteria[a]

	Desulfovibrio sp.		*Desulfovibrio desulfuricans*	
Addition	ATP (ng/ml)	Percent of control	ATP (ng/ml)	Percent of control
None	897	100	150	100
Molybdate	407	45	32	21
Selenate	744	83	79	53
Tungstate	317	35	82	55
Chromate	41	5	10	7

[a]From B. F. Taylor and Oremland (1979), with permission. Cell suspensions from growing cultures were exposed to 10 mM concentrations of group VI anions for 30 min at 30°C.

Table V. Effect of Oxyanions of Group VI Elements on ATP Levels in *Escherichia coli* (Aerobically and Fermentatively Grown) and Nitrate-Respiring Bacterial Cultures[a]

| | ATP levels as percent of control | | | |
| | *Escherichia coli* | | Denitrifying culture | |
Addition	Aerobic	Anaerobic	Marine	Freshwater
None	100	100	100	100
Molybdate	107	348	98	75
Selenate	83	140	97	106
Tungstate	83	455	93	62
Chromate	80	132	75	59

[a]From B. F. Taylor and Oremland (1979), with permission. Cell suspensions from growing cultures were exposed to 20 mM concentrations of the group VI anions for 30 min at 30°C. The ATP levels (ng/ml) in controls (no group VI anions added) were *E. coli* (aerobic), 827; *E. coli* (anaerobic), 65; marine denitrifying culture, 742; freshwater denitrifying culture, 392.

tiveness at lowering ATP pools. Washed cell suspensions of *Desulfovibrio desulfuricans* can reduce low concentrations (~ 1 μM) of selenate or selenite to selenide, although higher concentrations (~ 1 mM) were inhibitory (Zehr and Oremland, 1987). Other physiological types of bacteria were not affected adversely by group VI oxyanions, and in fact fermentatively grown *Escherichia coli* cultures responded with significant increases (Table V). Thus, it was implied that group VI oxyanions (with the exception of selenate) could serve as effective "specific" inhibitors of bacterial sulfate reduction when added to mixed systems of microorganisms.

Use of specific inhibitors to study the ecology of sulfate reduction in sediments was pioneered by Cappenberg (1974), who employed *b*-fluorolactate in his study of freshwater sediments. The first use of molybdate as a probe for microbial ecosystems was to study the competition for hydrogen between sulfate-respirers and methanogens in tropical marine sediments. Uptake of hydrogen in sediment slurries was caused by sulfate-respirers rather than methanogenic bacteria (Fig. 2). Addition of molybdate to the slurries also increased the rate of methane formation over the uninhibited slurries (Fig. 8), thereby indicating that sulfate-respirers compete with methanogens for the hydrogen present in sediments (Oremland, 1976; Oremland and Taylor, 1978). Other workers also noted this phenomenon without employing molybdate (Winfrey and Zeikus, 1977; Abram and Nedwell, 1978a,b). Abram and Nedwell (1978b) used *b*-fluorolactate to block sulfate-linked hydrogen metabolism in saltmarsh sediments. However, *b*-fluorolactate was found to cause only transient inhibition of hydrogen uptake by tropical marine sediments (Fig. 2B)

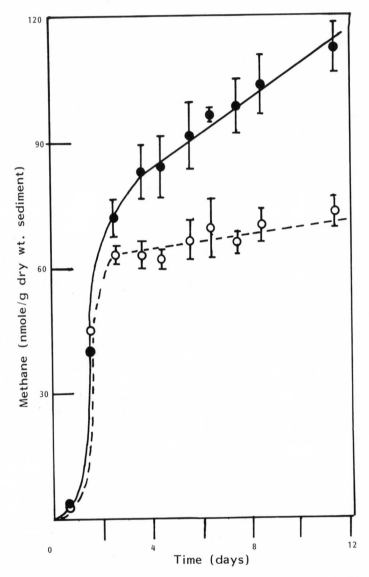

Figure 8. Methane production by marine sediment slurries incubated under N_2 with molybdate (●), or without additions (○). Results represent the mean of three flasks, and bars indicate one standard deviation. [From Oremland (1975).]

(Oremland, 1976), while also causing transient/enhanced kinetics of methane formation (Fig. 9) (Oremland, 1976). These rather frustrating experiences with b-fluorolactate indicated that molybdate was the inhibitor of preference in sediment systems, in that it yielded clear results, which were easy to interpret. Subsequent studies using molybdate in salt-marsh sediments (Nedwell and Banat, 1981; Sørensen et $al.$, 1981; Balba and Nedwell, 1982) confirmed the earlier observations made in tropical marine sediments (Oremland and Taylor, 1978).

Next, molybdate was employed to determine the in $situ$ substrates used by sulfate-reducers in anaerobic sediments. The approaches taken to achieve this consisted of three primary routes. These were: (1) mea-

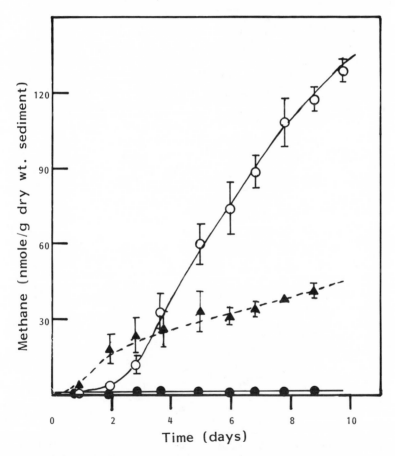

Figure 9. Methane production by marine sediment slurries incubated under H_2 without additions (▲), with b-fluorolactate (O), or with chloroform (●). Results represent the mean of three flasks, and bars indicate one standard deviation. [From Oremland (1975).]

surement of rapid kinetic effects (changes in electrical impedance) elicited by additions of substrates or inhibitors (molybdate) to sediments (Oremland and Silverman, 1979); (2) addition of molybdate to sediments, followed by measurements of the accumulation of fatty acids, hydrogen, and methane (Fig. 10) (Sørensen *et al.*, 1981; Lovely and Klug, 1982; Christensen, 1984; Scranton *et al.*, 1984); and (3) the use of radioisotopes in the form of ^{14}C-labeled substrates or ^{35}S-labeled sulfate in the presence or absence of added molybdate, followed by quantification of the volatile radioactive end products (e.g., ^{35}S-sulfide, ^{14}C-methane, or carbon dioxide). Of these approaches, the third has proven the most popular and has been employed to investigate the ecology of sulfate-respiring bacteria in freshwater (Fig. 11) (R. L. Smith and Klug, 1981; J. G. Jones *et al.*, 1982; Lovley and Klug, 1983; Lovley *et al.*, 1982; Hordijk *et al.*, 1985; J. G. Jones and Simon, 1985; Phelps and Zeikus, 1985) and saline environments (Banat *et al.*, 1981; Balba and Nedwell, 1982; Banat and Nedwell, 1983; King *et al.*, 1983; Winfrey and Ward, 1983; Shaw *et al.*, 1984; Alperin and Reeburgh, 1985; Dicker and Smith, 1985a,b; Iversen *et al.*, 1987; Kiene *et al.*, 1986; Oremland and Zehr, 1986), [see Capone and Kiene (1987) for review].

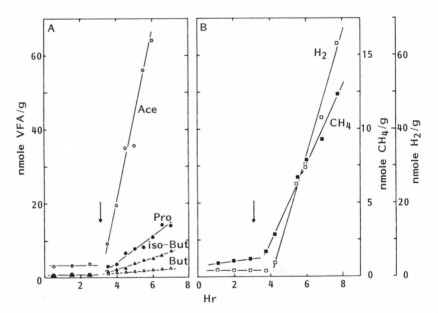

Figure 10. Accumulation of (A) volatile fatty acids and (B) methane and hydrogen in sulfate-containing marine sediments injected with molybdate (arrow). [From Sørensen *et al.* (1981), with permission.]

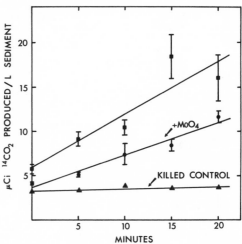

Figure 11. Effect of molybdate on the mineralization of [1-^{14}C] propionate in anaerobic freshwater sediments. Killed controls were poisoned with glutaraldehyde. Points represent the mean of triplicates, and bars indicate one standard deviation. [From R. L. Smith and Klug (1981), with permission.]

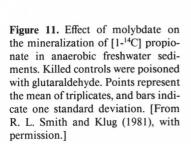

In addition, molybdate has been used to determine that sulfate-reducers (1) are not directly linked to the anaerobic metabolism of acetylene in estuarine sediments (Culbertson *et al.,* 1981), (2) are not directly responsible for respiratory Fe^{3+} reduction in sediments (Sørensen, 1982), (3) are the principal methylators of mercury in anoxic estuarine sediments (Fig. 12) (Compeau and Bartha, 1985, 1987), and (4) are not responsible for the reduction of Mn^{4+} to Mn^{2+} by anaerobic estuarine enrichment cultures (Burdige and Nealson, 1985).

It has also been employed to demonstrate that sulfate-reducers contribute to nitrogen fixation in marine sediments (Table VI) (Capone *et al.,* 1977; Nedwell and Aziz, 1980; Capone, 1982). Most recently it was found that molybdate had no effect on NH_4^+ production in several marine sediments (Jacobson *et al.,*1987), although rates of sulfate reduction and ammonium production are highly correlated in marine muds (Aller and Yingst, 1980; Hines and Lyons, 1982).

Molybdate and other group VI oxyanions are water-soluble and readily available commercially. Because they are competitive inhibitors of sulfate reduction, they must be added at roughly the equivalent level of ambient sulfate dissolved in the system. This poses no difficulties with freshwater systems, in which the sulfate levels are well below 1 mM; however, in brackish and marine systems, 20 mM molybdate is routinely employed. Injection of 2 M sodium molybdate into intact cores was successfully achieved with marine sediments (Christensen, 1984). Some difficulty has been reported with respect to 20 mM molybdate causing inhi-

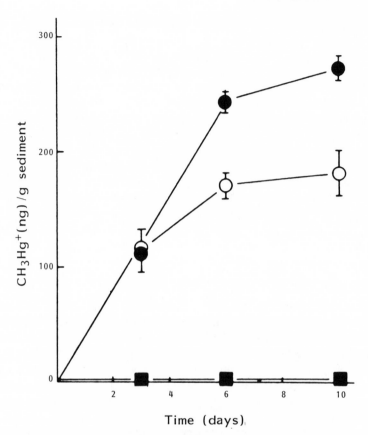

Time (days)

Figure 12. Synthesis of CH_3Hg^+ in anoxic estuarine sediment slurries containing Hg^{2+}. (○) No inhibitors; (●) plus BES; (■) plus molybdate. Results represent the average of duplicates, and bars indicate the range. [From Compeau and Bartha (1985), with permission.]

Table VI. Effect of Glucose, Lactate, and Molybdate on Nitrogen Fixation (Acetylene Reduction) Associated with the Rhizosphere Sediments of *Thalassia testudinum* [a]

Addition	Nitrogen fixation (μg N/g dry weight)			
	6 hr	21 hr	45 hr	121 hr
None	5.5	15	27	98
Glucose, 5 mM	4.6	940	3300	6700
Lactate, 5 mM	4.9	27	650	5200
Molybdate, 20 mM	3.9	10	13	31

[a]From Capone et al. (1977). Assays were performed on sediment slurries (400 ml) under a gas phase of helium plus 10% acetylene. Slurries were incubated in the dark at 26°C with constant reciprocal shaking.

bition of methanogenesis in freshwater systems (J. G. Jones *et al.*, 1982), although ~2 mM presented no difficulties (Lovley and Klug, 1983). Problems also exist in hypersaline waters which contain elevated levels of sulfate (e.g., 50–100 mM). Addition of sodium molybdate to such systems usually causes ion precipitation when added at levels equivalent to ambient sulfate. Use of 20 mM tungstate in anoxic Big Soda Lake water (salinity 89 ppt; sulfate 68 mM) caused only a partial (64%) inhibition of the endogenous rate of sulfate reduction (R. L. Smith and Oremland, 1987).

The efficacy of group VI oxyanions at blocking sulfate reduction in marine sediments has been reported to be molybdate > selenate > tungstate (Banat and Nedwell, 1984). These authors also noted that, whereas inhibition by 20 mM selenate or tungstate could be reversed by doubling the ambient sulfate levels (from 30 mM to 60 mM), inhibition was not similarly reversible with molybdate. The possibility exists, therefore, that molybdate may be a noncompetitive inhibitor of sulfate reduction, although this suggestion has not been confirmed by other investigators. Detailed kinetic studies with sediments and pure cultures are needed to confirm or refute this observation. Use of chromate, although effective at blocking sulfate reduction by pure cultures, is avoided in sediment work due to its strong oxidizing ability and general toxicity to sediment bacteria (Capone *et al.*, 1983; Slater and Capone, 1984).

Molybdate ions suffer a disadvantage in that they form molybdosulfide complexes (e.g., $MoO_2S_2^{2-}$; MoS_4^{2-}), which can bind up the available free sulfide ions (Tonsager and Averill, 1980). This could cause difficulties in sediment experiments because microbes that require sulfide ions (e.g., methanogens) will be at a nutrient disadvantage. The occasional observation of molybdate inhibition of methanogenesis (rather than stimulation) could be caused by this nutritional factor. Wolin and Miller (1980) reported that hydrogenase synthesis by *Ruminococcus albus* was inhibited due to the action of added molybdate ions. Thus, molybdate may have "side effects" that adversely influence some of the other constituents of the microbiota. Another possible adverse effect would be the blockage of assimilatory sulfate-reductases of the non-sulfate-respiring biota, a situation that could conceivably impair their metabolism and growth.

Difficulties also exist when conducting sulfate-reduction assays (^{35}S technique) upon sediments inhibited with molybdate. When these sediments are acidified (to release acid-volatile sulfide), a pronounced blue color results and sulfide release does not occur (Oremland and Silverman, 1979). This means that the effectiveness of the inhibitor cannot be checked in ^{35}S-sulfate reduction assays unless the complex is broken and sulfide released. Banat *et al.* (1981) found that the blue color was caused by a phospomolybdate complex, formation of which was enhanced by the presence of sulfide. They were able to quantitatively recover ^{35}S-sulfide if they added a stronger reducing agent (titanium chloride) just prior to acidification, thereby preventing any sulfide oxidation. An alternative to this procedure is to employ tungstate ions in lieu of molybdate because they do not form such complexes (N. Iversen, personal communication).

Other possible cautions with respect to molybdate include its reduction by microbial (Brierley and Brierley, 1982; A. M. Campbell et al., 1985) or chemical mechanisms (Goodman and Cheshire, 1982). In addition, molybdate will adsorb onto clay surfaces (Phelan and Mattigod, 1984). Thus, chemical and microbiological processes can influence the availability of molybdate ions in an assay system.

Molybdenum is also a trace nutrient required for nitrogenase and nitrate reductase enzymes. Addition of molybdate to systems in which molybdenum is a limiting factor could potentially stimulate nitrogen fixation and/or nitrate reduction. This is the reverse scenario of that proposed by Howarth and Cole (1985), whereby the abundance of sulfate ions in marine environments decreases the availability of molybdenum, thereby limiting nitrogen fixation and nitrogen assimilation by the microbial community. To this end it should be borne in mind that tungstate ions are effective inhibitors of molybdenum-containing enzymes, and have been employed to block nitrate reduction in rumen bacteria (Prins et al., 1980) and dissimilatory reduction of nitrite to ammonia by Citrobacter sp. (M. S. Smith, 1982). Other important enzymes functional in anaerobic bacteria contain group VI ions such as selenium, tungsten, and molybdenum. Growth of Methanococcus vannielii is enhanced by tungstate and selenate (J. B. Jones and Stadtman, 1977), and selenium-containing transfer RNAs (Ching et al., 1984), and hydrogenase (Yamazaki, 1982) occur in this organism. In addition, the formate dehydrogenase of E. coli contains molybdenum and selenium (Enoch and Lester, 1972), whereas that of M. vannielii contains selenium (J. B. Jones and Stadtman, 1981). Although there are many such examples in the literature (e.g., important selenium enzymes are also found in such genera as Selenomonas and Clostridium), the point is that addition of a group VI oxyanion to inhibit sulfate reduction in a sediment system could possibly cause stimulation or inhibition of an associated constituent of the anaerobic microbiota (e.g., methanogens). Because sediments from around the world may contain enormously different levels of available group VI micronutrients, it is difficult to predict whether use of a particular group VI oxyanion as specific inhibitor will enhance, retard, or not influence some other important constituent of the microbiota. Thus, the modest stimulation of denitrification caused by molybdate addition to marine sediment slurries may have been due to molybdenum-deficient nitrate reductases (Slater and Capone, 1984).

3.3. Case 3. Nitrapyrin (N-Serve)

The need for an inhibitor of nitrification grew out of economic necessity related to agriculture. Nitrogenous fertilizers are expensive, and the ability to retard their loss from soils by chemical/biological means is desirable. These fertilizers are usually in the form of ammonia, which is adsorbed onto soil surfaces and therefore tends to be retained in treated soils instead of being washed away. However, bacterial nitrification occurring within the soil oxidizes the ammonia to nitrite and nitrate. This promotes loss of applied nitrogen via increased runoff (due to the lower surface retentive properties of these ions), as well as by enhanced denitrification during episodes of soil wetness, with subsequent loss to the atmosphere as N_2. In the early 1960s, Dow Chemical Company marketed a product known as "N-serve" (nitrapyrin: 2-chloro-6-[trichloromethyl]pyridine), which is employed to prevent chemoautotrophic soil nitrification. This compound does not block "heterotrophic" nitrification (Shattuck and Alexander, 1963; Tate, 1977) and has thus achieved widespread use in agriculture (as well as in microbial ecology) as a "specific" inhibitor of chemoautotrophic soil nitrification (Huber et al., 1977). The end result of use of N-serve in agriculture is to enhance crop productivity while lowering costs (Huber et al., 1969).

Goring (1962) first demonstrated that oxidation of ammonium ions in soil was blocked by nitrapyrin. Subsequently, N. E. R. Campbell and Aleem (1965a) reported that pure cultures of the chemoautotroph *Nitrosomonas* sp., an organism that oxidizes ammonium to nitrite, were inhibited by nitrapyrin. Inhibition was apparently achieved by blocking the copper-containing cytochrome oxidase involved in the enzymatic oxidation of ammonium to hydroxylamine. Although this enzyme contains copper (Anderson, 1965), so do the enzymes involved in the oxidation of hydroxylamine to nitrite (Nicholas et al., 1962), and of nitrite to nitrate oxidation by *Nitrobacter* (N. E. R. Campbell and Aleem, 1965b). Nitrapyrin does not block the oxidation of nitrite to nitrate by *Nitrobacter* (N. E. R. Campbell and Aleem, 1965b). However, thiourea and related compounds will arrest both phases of chemoautotrophic nitrification (*Nitrosomonas* and *Nitrobacter* types), apparently by chelation of the Cu(I) enzymes (Wood et al., 1981). Although it is implied that the basis of nitrapyrin disruption of ammonia oxidation is related to an interaction with the copper component of cytochrome oxidase, its mode of action must be effective against only certain types of copper-containing enzymes.

It is not clear whether nitrapyrin temporarily arrests the growth of nitrifiers or kills them outright. Rodgers et al. (1980) reported that initial inhibition observed in cultures was eventually overcome after several weeks and was not due to adaptation of the organisms. Subsequently, it was shown that the levels of nitrapyrin added to cultures of *Nitrosomas sp.* were critical in determining whether the action was bacteriostatic or bacteriocidal (Rodgers and Ashworth, 1982). Only high concentrations ($\sim$100 μg/ml) had significant bacteriocidal effects.

The reversal of inhibition after prolonged exposure can be due to chemical breakdown of the compound. The hydrolysis product, 6-chloropicolinic acid, did not stop nitrite production by two marine species of *Nitrosomonas* (Table VII) (Salvas and Taylor, 1984), although Powell and Prosser (1985) reported that both nitrapyrin and 6-chloropicolinic acid inhibited ammonium oxidation by *N. europa* over prolonged incubation periods. Apparently, there are considerable differences among strains of ammonia-oxidizing nitrifiers with respect to their sensitivity to nitrapyrin (Belser and Schmidt, 1981). These authors reported that, whereas very high concentrations (1–10 μg/ml) were immediately effective at arresting nitrification in all of the seven strains tested (Fig. 13), 0.2 μg/ml [the level necessary to block *N. europaea* (N. E. R. Campbell and Aleem, 1965a)] was ineffective against five of the seven strains tested. Although 10 μg/ml of nitrapyrin proved capable of blocking nitrification by three strains of *Nitrosomonas,* and this level also inhibited methane oxidation by these organisms, much higher levels ($\sim$100 μg/ml) were required to arrest carbon monoxide oxidation (R. D. Jones and Morita, 1984).

Nitrapyrin suffers the disadvantage of being highly insoluble in water ($\sim$40 mg/liter at 20°C), therefore necessitating the use of organic solvents for application to assay systems. It is soluble in DMSO, ethanol, and acetone, but controls must always be run to assure that these solvents do not

Table VII. Effects of Substituted Pyridine Compounds on Ammonia Oxidation by Autotrophic Nitrifiers[a]

	Nitrite produced[c] (μmole/mg protein per hr)	
Addition[b]	*Nitrosomonas marina*	*Nitrosomonas oceanus*
None	190	197
Pyridine	198	194
Nitrapyrin	6	3
5-Chloropicolinic acid	173	192
Picolinic acid	181	202
2-Chloro-6-methylpyridine	192	198
2-Chloropyridine	190	208

[a]From Salvas and Taylor (1984), with permission.
[b]All additions at a concentration of 10 μM.
[c]Results are the mean of duplicate assays.

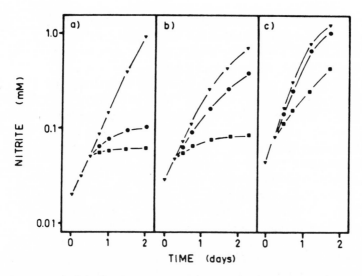

Figure 13. Ammonia oxidation by (a) *Nitrosomonas* strain Tara 7/15, (b) *Nitrospira* strain Sp. 1, and (c) *Nitrosolobus* strain Fargo, as inhibited by nitrapyrin: (▼) control; (●) 0.05 μg/ml; (■) 0.2 μg/ml. [From Belser and Schmidt (1981), with permission.]

themselves perturb the microbiota (including the nitrifiers) in the systems to be assayed. Powell and Prosser (1985) found no inhibitory action of either DMSO or chloroform upon cultures of *N. europaea*. Use of such solvents, however, may interfere with other needed chemical analyses, such as BOD tests (Young, 1983), thereby requiring that the nitrapyrin be added as a dry crystal. Furthermore, use of solvents such as acetone or DMSO provides a carbon source for heterotrophic bacteria. Notton *et al.* (1979) reported that acetone, when added with N-serve, enhanced denitrification of nitrate present in agricultural soils.

The compound does not come in purified form. Reagent grade has a purity of ~99%, whereas technical grade is of lower purity. The main contaminant appears to be a variety of other chlorinated pyridines associated with the synthesis. Thus, an additional purification step (recrystallization, chromatography, etc.) may be needed for certain types of studies in order to rule out the influence of contaminants. Investigators should not employ the "commercial" grade, since it contains emulsifiers and stabilizers needed to meet agricultural requirements.

Nitrapyrin inhibits a variety of important microbial processes in addition to nitrification. Production of methane by marine sediments was blocked by nitrapyrin (~5 μM), whereas 2-chloropyridine as well as other analogues lacking the trichloromethyl moiety had no influence (Salvas and Taylor, 1980). Because methanogenesis is inhibited by chloroform (Bauchop, 1967), the trichloromethyl moiety of nitrapyrin appears to be the basis for inhibition by the latter compound.

Methane-oxidizing bacteria are also blocked by nitrapyrin (Topp and Knowles, 1982, 1984; Salvas and Taylor, 1984). The compound inhibits growth as well as the oxidation of methane, ammonium, and carbon dioxide fixation by methanotrophs. However, nitrapyrin apparently acts only against the methane monooxygenase. Oxidation of other one-carbon substrates (formate, methanol, and formaldehyde) was unaffected. Although methane-monooxygenase is believed to contain copper (Stirling and Dalton, 1979), addition of Cu^{2+} to cell suspensions of *Methylosinus trichosporium* did not alleviate inhibition of methane oxidation by nitrapyrin (Topp and Knowles, 1984). Because such copper additions relieved nitrapyrin inhibition of *Nitrosomonas* (N. E. R. Campbell and Aleem, 1965a), the inhibitor must block the two organisms by a different mechanism.

Salvas and Taylor (1984) found that picolinic acid would block ammonia oxidation by methane-oxidizers, but not by the nitrifiers, and suggested that this compound may prove useful as a diagnostic "tool" to determine if methylotrophs are responsible for significant ammonia oxidation in natural microbial ecosystems. Reports of other inhibitory aspects of nitrapyrin unrelated to nitrification include sulfate reduction (Somville, 1978) and nitrate reduction (denitrification) by a pseudomonad (Henninger and Bollag, 1976). Thus, this particular compound will disrupt the metabolism and activity of other "key players" in the microbiota of soils, sediments, and natural waters and a researcher should be aware of these limitations.

Nitrapyrin is useful for determining if certain observed activities are attributable to chemoautotrophic nitrifiers. For example, nitrous oxide is an intermediate in both denitrification and nitrification, and the emission of this gas from soils could be due to either process. Blackmer *et al.* (1980) found that N-serve-treated soils produced less of this gas than uninhibited samples, thereby concluding that nitrifiers were important in the evolution of this gas.

Nitrapyrin has played an important role in the development of bioassays for measuring the activity of nitrifying bacteria in natural ecosystems (B. F. Taylor, 1983). Because there is no convenient radioisotope of nitrogen, an indirect assay was developed whereby autotrophic fixation of $^{14}CO_2$ into cell material is measured in the presence or absence of the inhibitor (Billen, 1976; Somville, 1978). This basic procedure has been used to study nitrification in a variety of aquatic environments (Table VIII) (e.g., Indrebo *et al.*, 1979; Vincent and Downes, 1981; Hall, 1982; Somville, 1984; Cloern *et al.*, 1983). Other methods used to assess nitrification rates employing N-serve include ammonium accumulation after

Table VIII. Dark Uptake of ^{14}C-Bicarbonate by Bacteria in Water Samples from the Oxic–Anoxic Boundary of Big Soda Lake, Nevada[a]

Addition[b]	Update (dpm)	Percent change
February		
29-m sample, 5-hr incubation None	184; 214	—
N-serve	115; 112	−43
Filtered	13; —	−93
July		
21-m sample, 5-hr incubation None	1241 (135)	—
N-serve	673 (35)	−46
Acetylene	748 (115)	−40
Filtered	79 (—)	−94
October		
21-m sample, 3-hr incubation None	470 (22)	—
N-serve	107 (25)	−77
CH_3F	527 (115)	+12[c]
21-m sample, 7-hr incubation None	1086 (116)	—
N-serve	211 (190)	−80
CH_3F	1017 (209)	−6[c]

[a]From Cloern et al. (1983), with permission. Values are the mean dpm of three replicates, and parentheses indicate one standard deviation; two replicates were run only during February.
[b]N-serve, 22 μM; acetylene, 10 ml; methylfluoride, 2 ml.
[c]Not statistically different from sample not inhibited.

injection of the inhibitor into intact cores (Henriksen, 1980; Hall, 1984) and ^{14}C-carbon monoxide oxidation (R. D. Jones et al., 1984). Experiments that measure the accumulation of nitrite plus nitrate in the presence or absence of nitrapyrin have also been undertaken (Webb and Wiebe, 1975), but in some cases may require too long a time period [e.g., ~30 days (Patrick et al., 1968)], during which the inhibitor can decompose via hydrolysis to the inactive 6-chloropicolinate.

Hydrolysis of N-serve in soils is mainly dependent upon temperature, rather than pH or sorption on colloidal surfaces (Hendrickson and Keeney, 1979a). These authors also caution against using rubber or polyethylene-based containers in studies of this type, due to strong sorption of the nitrapyrin onto these lipophilic materials. This would give the impression that nitrapyrin was hydrolyzed more rapidly than occurs *in situ*. However, sorption of nitrapyrin onto soil surfaces is a major problem with respect to the efficacy of this inhibitor. In addition, loss of the inhibitor occurs via volatilization to the atmosphere (Hendrickson and Keeney, 1979b). In summary, although nitrapyrin is an important chemical for use in agriculture, its employment as a research tool in microbial ecology/biogeochemistry is tempered by several limitations, including nonspecificity of action, insolubility, degradation, and adsorption. Despite these difficulties, this inhibitor continues to enjoy popularity among researchers.

3.4. Case 4. Acetylene

Use of acetylene for assays of nitrogen fixation and denitrification has become widespread over the past 20 years, and has led to many observations concerning the impact of this compound upon various microbial processes. The broader aspects of this subject have been recently reviewed (Payne, 1984). Acetylene inhibits a number of different microbial activities (see this section below), and cannot really be considered a "specific" inhibitor. Nonetheless, it is an extremely useful compound for certain assays, and can be employed as a "backup" inhibitor to confirm the results achieved with more selective compounds or in combination with other, more specific, inhibitors.

Acetylene is highly water-soluble (about one volume of the gas dissolves in one volume of water) and it can be well-dispersed into aqueous assay systems. Headspace equilibration by shaking is a convenient method to assay soils, aqueous samples, or sediment slurries (Capone, 1983), while injections into intact cores can be achieved by saturating water with the gas (Sørensen, 1978). Perfusion of acetylene-saturated water through sediment cores has also been successfully applied to measure nitrogen fixation in sediments (Capone and Carpenter, 1982a). Acetylene can be generated conveniently by reaction of calcium carbide with water. If one is to assay an anaerobic system, care must be taken to eliminate any entrapment of air with the generated acetylene. Formation of acetylene by this mechanism is convenient and therefore amenable for field use. The acetylene formed by carbide reaction is often of much higher purity than that obtained in cylinders from gas vendors. Calcium carbide is easily obtained from commerical scientific chemical companies.

The employment of acetylene to measure nitrogen fixation arose from its ability to act as a substrate analogue for dinitrogen, resulting in the production of ethylene instead of ammonia (Dilworth, 1966). The acetylene-reduction assay was subsequently devised as a method to quantify nitrogenase activity (Stewart et al., 1967; Hardy et al., 1968). This assay has been used extensively to pursue studies on the physiology, biochemistry, and ecology of nitrogen fixation.

Similarly, an assay was devised to quantify denitrification. Observations of N_2O accumulation in soil during exposure to acetylene were first reported by Fedorova et al. (1973). Subsequent studies by Balderston *et al.* (1976) and Yoshinari and Knowles (1976) revealed that the gas inhibits the N_2O-reductase of denitrifying bacteria, thereby causing the accumulation of nitrous oxide. This phenomenon was exploited to devise an assay for denitrification suitable for fieldwork (Yoshinari *et al.*, 1977; Sørensen, 1978; Chan and Knowles, 1979; Kaspar and Tiedje, 1981) and the technique has come into widespread use. Although both the acetylene-reduction and acetylene-blockage assays have limitations with respect to their ability to quantify accurately nitrogen fixation and denitrification, these problems do not present insurmountable obstacles. B. F. Taylor (1983) gives a detailed discussion of the biochemical basis, limitations, and practical aspects of these assays.

Besides serving as an assay tool via its inhibition of nitrogenase and N_2O-reductase, acetylene inhibits a number of other biogeochemically important processes, and is also toxic to certain bacteria. Methanogenesis was reported to be inhibited by acetylene, and inhibition could not be reversed after gassing with hydrogen (Oremland and Taylor, 1975). Methanogenesis in rumen fluid is retarded by acetylene (Elleway *et al.*, 1971) as well as being inhibited in rice paddy soils and by a culture of *Methanosarcina* (Raimbault, 1975). Low levels of acetylene (~ 8 μM) inhibited growth of *Methanospirillum hungatei,* and that of four other species of methanogens, but had no influence upon the growth of three species of *Halobacterium* or four diverse species of eubacteria (Sprott *et al.*, 1982). These authors could not find any adverse influence of acetylene upon various methanogenic enzymes (hydrogenase, methyl-CoM reductase, NADP reductase, or ATP hydrolase), but nonetheless observed a dramatic drop in ATP levels and [63]Ni uptake by *M. hungatei* when exposed to acetylene. They concluded that acetylene disrupted the transmembrane pH gradient, thereby blocking methanogenesis, proton flux, and nickel uptake. However, exposure to acetylene (16 hr) was not necessarily toxic to cells, as indicated by continued viability. Acetylene apparently acts in a similar fashion to gramicidin in that they both inhibit methane formation and cause a drop in intracellular ATP (Jarrell and Sprott, 1983). Acetylene has been used to inhibit methane production from decaying leaves (Tam *et al.*, 1981) and lake sediments (Knowles, 1979). On the other hand, Lipschultz (1981) found that C_2H_2 did not inhibit CH_4 efflux from saltmarsh sediments. However, this may not have reflected active methanogenesis, but rather slow diffusion of gases between the sediments and chamber (Kiene and Capone, 1985), or possibly the inhibition of CH_4 oxidation.

Oxidation of methane by aerobic methylotrophs is also inhibited by acetylene. Methane monooxygenase is a copper-containing enzyme that is inhibited by a number of metal-chelating compounds (Takeda *et al.*, 1976; Stirling and Dalton, 1979). Acetylene readily forms explosive cop-

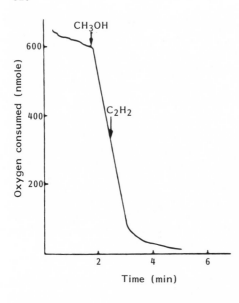

Figure 14. Effect of acetylene on oxygen uptake by cell suspensions of *Methylococcus capsulatus* oxidizing methanol. Methanol and acetylene were added to the suspensions at the times indicated by arrows. [From Dalton and Whittenbury (1976), with permission.]

per and silver acetylides when exposed to these elements (or ions), so its ability to bind with certain metals is well established. Certain methane-oxidizing bacteria are capable of fixing dinitrogen (Coty, 1967); however, the acetylene-reduction assay cannot be employed to assess this activity, because methanotrophs are inhibited by acetylene when growing on methane (Dalton and Whittenbury, 1976). Interestingly, acetylene will not interfere with the oxidation of methanol by whole cells of *Methylococcus capsulatus* (Fig. 14), and a variety of electron donors (methanol, formate, formaldehyde, ethanol, and hydrogen) will support acetylene reduction to ethylene by this diazotrophic organism (Dalton and Whittenbury, 1976).

Some methanotrophs can also oxidize C_2H_4, and this can have added implications for the C_2H_2 reduction assay (de Bont and Mulder, 1974). The consumption of ethylene could result in an underestimate of the actual rate of C_2H_2 reduction, except that C_2H_2 also inhibits C_2H_4 oxidation (de Bont, 1976a,b; de Bont and Mulder, 1976). For these reasons, use of acetylene to assess nitrogen fixation by methanotrophic communities should be discouraged. However, the C_2H_2 reduction assay appears to be valid for pure cultures of methane-oxidizers when methanol is used instead of CH_4 as the growth substrate.

Natural plant and microbial sources of C_2H_4 are known (A. M. Smith, 1976; Cornforth, 1975; Ilag and Curtis, 1968) and certain microorganisms in nature also have the capacity to oxidize C_2H_4 (Cornforth, 1975; de Bont and Mulder, 1974) (see Section 4.5). However, the inhibition of natural C_2H_4 oxidation can pose difficulties for field applications of the C_2H_2 reduction assay. Witty (1979) noted in some soils that the specific activity of $^{14}C_2H_4$ after addition of $^{14}C_2H_2$ was far less than that of the added $^{14}C_2H_2$. This suggested the production of some C_2H_4 that was not linked to $^{14}C_2H_2$ reduction, but was nonetheless stimulated by the presence of C_2H_2. No accumulation of C_2H_4 was noted in "C_2H_2-free" controls. This implied that the C_2H_2 blocked the oxidation of C_2H_4. Cornforth (1975) estimated that ethylene oxidase activity in soil might exceed production by about 50-fold. Hence, inhibition of ethylene oxidation by C_2H_2 could potentially unmask a large natural production of C_2H_4, possibly explaining the observations of Witty (1979). Several workers have further examined this phenomenon, but have not found it to be a significant problem in the C_2H_2 reduction assay as applied to several cultivated (Lethbridge *et al.*, 1982) and grass-inhabited (vanBerkum and Sloger, 1979, 1981) soil systems. Inhibition of ethylene oxidase by C_2H_2 should not be a problem in anoxic systems (Capone, 1987).

Ethylene, the product of C_2H_2 reduction, also acts as an inhibitor of various processes and this could possibly obscure experimental results. Ethylene is a plant hormone and, in modifying plant performance, may affect microbial activities (e.g., nitrogen fixation or denitrification) associated with the plant or its rhizosphere. For instance, concentrations of C_2H_4 of 1 ppm have been shown to reduce photosynthesis in peanuts (Kays and Pallas, 1980). Ethylene, while not as potent as C_2H_2, also inhibits methanogenesis (Oremland and Taylor, 1975 (Section 4.4).

Not surprisingly, acetylene also inhibits nitrification, and because of its aforementioned ability to block methane-monooxygenase as well as methanogenesis, this makes it comparable to nitrapyrin in the types of processes it blocks. Oxidation of ammonia by whole cells of the chemolithotroph *Nitrosomonas europeae* is inhibited by as little as 10^{-4} atm of acetylene (Hynes and Knowles, 1978). As in the case of nitrapyrin, inhibition was centered on ammonia oxygenase ($K_i = 0.25\ \mu M$; Fig. 15), since oxidation of hydroxylamine to nitrite was uninfluenced (Hynes and Knowles, 1982). In addition, these authors reported that heterotrophic nitrification by *Arthrobacter,* as well as chemolithotrophic oxidation of nitrite to nitrate by *Nitrobacter winogradsky,* was not disrupted by acetylene. Hyman and Wood (1985) have recently shown that $^{14}C_2H_2$ specifically binds with and inactivates ("suicide inactivation") ammonium monooxygenase from *Nitrosomonas europaea.* Inhibition of ammonia oxidation by acetylene also occurs in soils (Walter *et al.,* 1979; Mosier, 1980; Berg *et al.,* 1982). This imposes serious limitations on the acetylene-blockage assay for denitrification, because rates will be underestimated if the two processes are closely coupled and the supply of available nitrate plus nitrite is exhausted (Ryden, 1982; Jenkins and Kemp, 1984).

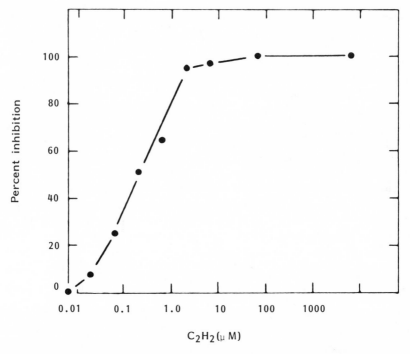

Figure 15. Inhibition by acetylene of ammonia oxidation by cell suspensions of *Nitrosomonas europea* incubated for 6 hr at various levels of acetylene. [From Hynes and Knowles (1982), with permission.]

There has been considerable interest in identifying natural sources and sinks of nitrous oxide (N_2O) because this gas may destroy the atmospheric ozone layer (Delwiche, 1981). Nitrous oxide has been long known to be an intermediate in denitrification (Payne, 1981) and more recently has been recognized to arise as a by-product of nitrification (Nicholas, 1978), assimilatory nitrite reduction (Weathers, 1984), and dissimilatory nitrate reduction (Kaspar and Tiedje, 1981; Dodds and Collins-Thompson, 1985; Samuelsson, 1985). Nitrification-linked N_2O production has been shown to be C_2H_2-sensitive (Hynes and Knowles, 1984). This fact was used to advantage in attempting to assess the relative contributions of nitrifying and denitrifying bacteria to N_2O emissions in natural systems (see Section 3.3). Whereas denitrification-derived N_2O should be stimulated by C_2H_2, in several studies of soils C_2H_2 inhibited N_2O emission (Bremner and Blackmer, 1979; Martikainen, 1985), thereby implicating nitrification as the source. Bremner and Blackmer (1978) had earlier provided corroboration for this by determining that ammonium, rather than nitrate, stimulated N_2O emission, whereas N-serve inhibited this process. Hence, nitrification was the primary source of N_2O (Bremner and Blackmer, 1981; Martikainen, 1985). The recent report of Poth and Focht (1985) complicates this view, since it appears that ammonium-oxidizers may produce N_2O in a truncated denitrification pathway induced during low O_2. The pathway uses NO_2^- as a terminal electron acceptor, with N_2O as the major end product.

There is some indication that acetylene also interferes with dissimilatory reduction of nitrate to ammonia in the bovine rumen (Kaspar and Tiedje, 1981) and with the assimilatory reductase of spinach leaves (Maurino et al., 1983). This latter blockage can be reversed with blue light.

Other reports of acetylene interference with microbial metabolism include the growth of Clostridium pasteurianum under conditions of repressed nitrogenase (Brouzes and Knowles, 1971). Growth and respiration (CO_2 release) by Desulfovibrio desulfuricans and D. gigas were strongly inhibited by an atmosphere of 20% acetylene, although lower amounts caused only partial inhibition (Payne and Grant, 1982). No significant disruption occurred for Desulfotomaculum ruminis, implying that there is a difference among genera of sulfate-respiring bacteria with respect to their susceptibility to acetylene. It is interesting to note that sulfate-linked uptake of molecular hydrogen by marine sediments was uninfluenced by the presence of acetylene (Oremland, 1976). Thus, the general functions of the sulfate-reducing biota may be unaffected by C_2H_2, even though growth of some individual species may be impeded.

Acetylene is a useful probe in studying the production and consumption of molecular hydrogen by diazotrophic bacteria and cyanobacteria. Uptake hydrogenases of Anabaena cylindrica are blocked by acetylene (Daday et al., 1977; Lambert et al., 1979; Houchins and Burris, 1981), as are those of diazotrophic bacteria (L. A. Smith et al., 1976). In addition, C_2H_2 can serve as an electron sink for the hydrogenase associated with the nitrogenase enzyme, thereby limiting the amount of H_2 evolved (Rivera-Oritz and Burris, 1975). In natural ecosystem studies, C_2H_2 was shown to inhibit uptake of $^3H_2$ by a culture of Anabaena circinalis (Fig. 16) as well as by a bloom of Anabaena oscillariodes occurring in the Chowan river (Paerl, 1983). Acetylene was also found to stimulate the dark evolution of H_2 by decomposing cyanobacterial aggregates in Big Soda Lake (Oremland, 1983).

Acetylene can serve as a substrate for the growth of aerobic and anaerobic bacteria. Nocardia rhodochrous grows aerobically on acetylene (Kanner and Bartha, 1979), via an acetaldehyde, acetate, acetyl-CoA pathway (Kanner and Bartha, 1982). Other reports of aerobic growth exist for Rhodococcus A1 (de Bont and Peck, 1980) and for a Bacillus sp. isolated from a freshwater stream (Tam et al., 1983). The latter authors

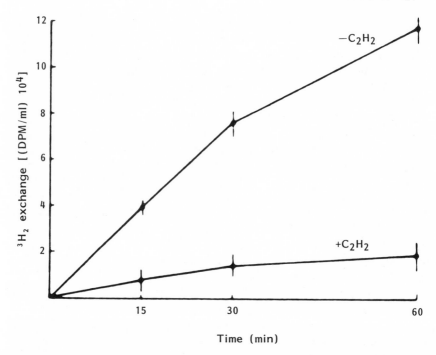

Figure 16. Effect of acetylene on the uptake of $^3H_2$ by *Anaebaena circinalis*. [From Paerl (1983), with permission.]

also reported that aerobically incubated stream sediments could entirely consume added acetylene after 18–21 days. Disappearance of acetylene under anaerobic conditions was reported for rice paddy soils (Watanabe and de Guzman, 1980) and estuarine sediments (Culbertson *et al.,* 1981). Estuarine sediments could consume all the added acetylene within 5 days, with subsequent stimulation of sulfide and methane formation. Acetylene was oxidized anaerobically via an acetate intermediate to carbon dioxide. Although sulfate reduction in slurries was stimulated by the disappearance of acetylene, sulfate-respirers were not the organisms responsible for the metabolism of acetylene. Consumption of acetylene could not be blocked by molybdate or by the absence of sulfate. Enrichment cultures were established that fermented acetylene to ethanol, acetate, and hydrogen via an acetaldehyde intermediate (Fig. 17) (Culbertson, 1983; Culbertson and Oremland, 1983). Schink (1985a) isolated a strict anaerobe, *Pelobacter acetylenicus,* which was capable of fermenting acetylene gas by this reaction. The fact that acetylene can serve as a substrate for the growth of both aerobic and anaerobic bacteria must be taken into account when performing long-term incubations. Bacterial consumption of the gas may lead to a loss of inhibitory function as well as to the release of readily metabolized substrates within the assay milieu.

A final consideration recently pointed out by Hyman and Arp (1987) is that CaC_2-generated and commercial C_2H_2 can have substantial and varied contaminants, which may themselves produce stimulation or inhibition of activities under study. Previous results need to be more closely scrutinized with this observation in mind.

4. Inhibition of Carbon Metabolism

4.1. Autotrophic Processes

The extent of primary carbon input by CO_2 fixation is a prime focus in many biogeochemical studies. It is often the aim to distinguish between autotrophic CO_2 fixation and other possible pathways of CO_2

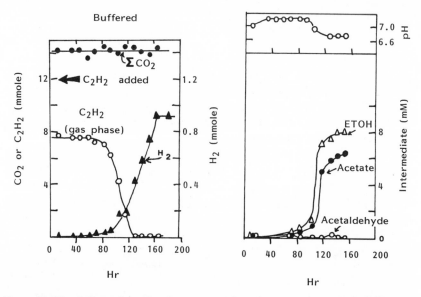

Figure 17. Metabolism of acetylene by an anaerobic enrichment culture obtained from estuarine sediments. Acetylene (total added ∼12 mmole) was dismutated from an acetaldehyde intermediate to equimolar quantities of acetate and ethanol, with hydrogen as a minor product. This experiment was carried out in a bicarbonate-buffered medium, which resulted in only a small drop in pH caused by the accumulation of acetate. When the medium was not buffered, acetate formation caused a drop in pH to ∼5.9, which resulted in the accumulation of more acetaldehyde, with proportionally less ethanol and acetate (not shown). [C. W. Culbertson and R. S. Oremland (unpublished data).]

reduction. Autotrophs effect the reduction and incorporation of CO_2 into organic carbon by the exploitation of either light (photoautotrophy) or chemical (chemoautotrophy) energy, from which they obtain ATP and reductant.

The pentose phosphate pathway ("Calvin–Benson cycle") of CO_2 reduction is common to most autotrophs (McFadden and Purohit, 1978). The key carboxylating enzyme of the Calvin cycle is ribulose bis-phosphate carboxylase (Rubisco) and is distinguished from other carboxylating systems in that the products are two three-carbon (C3) molecules. Other carboxylating enzymes [e.g., phosphoenolpyruvate (PEP) carboxykinase, malic enzyme, and pyruvate carboxylase] are found in nonautotrophic organisms. These yield four carbon (C4) products, which help replenish ("anapleurotic") cellular intermediates, but contribute little to net carbon accumulation for growth (Hochachka and Somero, 1984). PEP carboxylase may effect CO_2 reduction into C4 intermediates in some autotrophs (Glover and Morris, 1979). However, this activity is generally balanced by decarboxylation of C4 compounds and may simply represent a CO_2 transport system. Any of these alternate pathways can contribute to apparent short-term $^{14}CO_2$ reduction measurements, therefore obscuring assessment of "primary productivity." Dark bottle incubations have most often been used to correct for such artefacts in assessment of primary production. However, "primary production" by chemoautotrophic Rubisco activity may thus be subtracted out, thereby limiting the measurement to photoautotrophic activity. On the other hand, PEP carboxylase activity that may not account for any net fixation appears to be positively stimulated by light (Glover and Morris, 1979) and therefore not be accounted for in the dark bottle.

Specific inhibitors of carbon assimilation are known and have been used in enzymological and physiological studies. For instance, Stokes and Walker (1972) reported that D,L-glyceraldehyde was a potent inhibitor of CO_2 assimilation via the Calvin–Benson cycle, probably by inhibiting regeneration of RuBP. Very recently, Yokota and Canvin (1985) have used ^{14}C-carboxypentitol bisphosphate, a substrate analogue of RuBP, to label Rubisco, since it irreversibly binds (and presumably inhibits) this key enzyme. If found to be effective in whole-cell systems, Calvin cycle inhibitors might be fruitfully employed to determine the amount of Rubisco-catalyzed (i.e., autotrophic) CO_2 fixation in natural samples to compare with other procedures for correcting for nonautotrophic CO_2 fixation.

Whereas photoautotrophs are probably the most important carbon-producers on a global scale, chemoautotrophs can be important in more restricted settings, such as marine sediments (Kepkay *et al.*, 1979) or hydrothermal vent regions (Jannasch and Wirsen, 1979). More importantly, chemoautotrophs may play major roles in the cycling of nitrogen (Kaplan, 1983), sulfur (Fenchel and Riedl, 1970), and metals. Inhibitors have played a critical role in the analysis of chemoautotrophic processes in nature. Whereas dark CO_2 fixation assessed with appropriate Rubisco inhibitors may allow more accurate estimation of chemoautotrophic metabolism, finer resolution of particular chemoautotrophic groups requires that electron donor systems be targeted. Inhibition of nitrifiers is specifically discussed in Sections 3.3 and 5.3, inhibition of sulfide oxidation is considered in Section 6.2, while inhibition of metal oxidizing bacteria is discussed in Section 7. Methane-oxidizing bacteria, while not strictly chemoautotrophic, have many similarities with chemoautotrophs, including inhibitor sensitivities (see Section 4.3).

4.1.1. Photoautotrophy

Inhibitors have been used extensively in defining the path of electron flow and ATP generation in photosynthesis (Izawa, 1980; Trebst, 1980). Of the various inhibitors of photosynthetic electron flow, DCMU [3-(3,4-dichlorophenyl)-1,1-dimethylurea] has probably found the most application in ecological studies. DCMU, along with a number of other herbicides, binds to a protein between the primary acceptor molecule of Photosystem II ("Q") and plastoquinone.

DCMU has been proposed for use in control to assess nonphotoautotrophic fixation in primary productivity studies using $^{14}CO_2$ (Legendre et al., 1983). The previous light history has been shown to affect significantly dark CO_2 fixation, since some photoautotrophic CO_2 fixation will continue after transition to the dark. In the presence of DCMU, a more accurate assessment of nonphotosynthetically mediated (abiological adsorption, heterotrophic, and chemoautotrophic) CO_2 fixation may be achieved. Unfortunately, DCMU may inhibit other pathways, thereby tempering its usefulness in identifying photoautotrophic reactions. Legendre et al. (1983) noted inhibition of dark CO_2 fixation by DCMU, which they speculated may be a result of DCMU interaction with ubiquinone, a plastiquinone analogue in the respiratory electron transport system. Interestingly, Li and Dickie (1985a) have found that DCMU inhibits amino acid and glucose uptake in the light or dark.

Another application for DCMU has been in improving fluorometric determination of chlorophyll a extracted from natural samples. By inhibiting linear electron flow between PS II and PS I, DCMU maximizes fluorescence in samples and makes fluorescence yield less dependent on physiological state of the organisms (Slovacek and Hannan, 1977). In the presence of DCMU, chlorophyll a fluorescence provides a more accurate measure of absolute chlorophyll a concentration, and presumably of algal biomass.

The interactions between microalgal symbionts and animal hosts have been probed using DCMU. Calcification rates (as well as photosynthesis) in reef corals (Vandermulen et al., 1972) and in benthic foraminifera (Duguay and Taylor, 1978) are inhibited by DCMU, thus demonstrating the importance of the photoautrophic symbiont.

It is interesting to note that a variety of other herbicides, including the s-triazines, have a similar mode of action to DCMU. This may be due to their common $-C-NH_2$ functional groups (Trebst, 1980). While none of these compounds have been used in experimental ecological studies, the ecotoxicological effects of triazines have been of some concern. Indeed, triazine herbicides were considered as one possible cause for the precipitous decline in submerged aquatic vegetation in the Chesapeake Bay (Stevenson and Confer, 1978; T. W. Jones and Estes, 1984; Orth and Moore, 1983), although recent results indicate other factors may be more important (Macalester et al., 1983).

Numerous functional subgroups exist within the photoautotrophic biota. In aerobic planktonic systems, procaryotic cyanobacteria and eucaryotic microalgae coexist and contribute to overall production. In anaerobic water columns and benthic environments, anoxyphotobacteria may also proliferlate. Various approaches have been used in order to assess the relative contributions of these various subgroups to carbon fixation, as well as other processes, such as N_2 fixation.

Through the use of size fractionation procedures, it has recently been observed that the picoplankton (<2 μm) are important components of the marine phytoplankton community (Li et al., 1983; Platt et al., 1983). This fraction is presumed to contain primarily cyanobacteria (Murphy and Haugen, 1985). However, the observations of Johnson and Sieburth (1982) require that bacteria-sized eucaryotic photoautotrophs (Johnson and Sieburth, 1982) also be considered. Li and Dickie (1985a) reported no effect of cycloheximide on picoplanktonic $^{14}CO_2$ assimilation in the light, indicating in their system a minor role (if any) for eucaryotic photoautotrophs in this size fraction. An alternate possibility, of course, is that eucaryotes were resistant to cycloheximide inhibition at the concentration used.

Among the procaryotic photoheterotrophs, both oxygen-producing and oxygen-tolerant species (oxyphotobacteria, e.g., cyanobacteria and prochlorales) exist, as well as anaerobic organisms (anoxyphotobacteria, e.g., purple sulfur and nonsulfur and green sulfur bacteria). In certain environments, such as near the surface of anaerobic muds, cyanobacteria and anoxyphotobacteria are often found together. Habte and Alexander (1980) have assessed the relative contribution of these photosynthetic procaryotes with respect to nitrogen fixation in flooded rice soils by inhibiting cyanobacteria with propanil (3'4'-dichloropropionanilide), which does not inhibit anoxyphotobacteria. In a study of saltmarsh sediment surfaces, from 79 to 95% of the light-dependent nitrogenase activity was found to be destroyed by propanil treatment (Table IX) (Chemerys, 1983), suggesting a major contribution by cyanobacteria. In this same study, hydrogen production was found to be unaffected by propanil.

Batterton *et al.* (1978) have reported that analine and its methylated derivatives are also potent inhibitors of cyanobacterial growth (comparable to DCMU in potency), but inhibition is not rapidly manifested in short-term rates of photosynthesis, respiration, or nitrogenase activity.

4.2. Heterotrophic Processes

Heterotrophic metabolism in the environment is also of considerable importance. Autotrophic and heterotrophic processes are intimately linked, in that heterotrophy is ultimately dependent on organic carbon produced by autotrophs, whereas autotrophic activity may be limited by inorganic nutrients, which often arise directly from heterotrophic regeneration.

Table IX. Effect of Propanil and Dimethylsulfoxide (DMSO) on Ethylene Production and Hydrogen Evolution by Saltmarsh Sediments Incubated for 24 hr[a]

	Production (nmole/g dry weight)		
	Control	DMSO	Propanil
	Ethylene[b]		
Light:			
9/30/81	66,500	49,600[c]	4,760
2/11/82	337	—	32
4/28/82	6,920	13,400[c]	1,530[c]
6/25/82	2,340	2,350[c]	516
Dark:			
9/30/81	1,080	809[c]	320[c]
	Hydrogen		
Light:			
2/82	2.95	3.43[c]	1.85[c]
6/21/82	1.83	1.84[c]	2.47[c]
5/28/82	3.22	3.23[c]	4.58[c]
Dark:			
2/82	4.88	3.89[c]	2.76[c]

[a]From Chemerys (1983).
[b]From acetylene–reduction assay.
[c]Not statistically different from control.

As autotrophic pathways depend on a variety of mechanisms for generating biochemical energy and reductant, so, too, can heterotrophic metabolism be coupled to a variety of energy conservation mechanisms. These include aerobic respiration, various pathways of anaerobic respiration (e.g., denitrification, sulfate reduction), and fermentation. Unlike autotrophy, however, heterotrophic metabolism employs a wide diversity of carbon substrates and, hence, metabolic pathways. There is considerable physiological diversity among heterotrophs, ranging over the broadest levels of phylogeny (eucaryotes and procaryotes), in terms of oxygen tolerance and use (aerobic, anaerobic, or facultative), general mode of carbon acquisition (osmotrophic, phagotrophic, saphrotrophic), as well as diverse substrate specificities and nutrient requirements.

Various strategies may be employed in inhibitor-based studies of heterotrophic activity, depending on the particular question at hand. For instance, nonmetabolizable substrate analogues such as fluorolactate or fluoroacetate may be used to assess the importance of a particular substrate on total heterotrophic activity or upon a particular heterotrophic process such as sulfate reduction (see Section 6.1). The contribution of a subgroup of heterotrophs to total heterotrophic activity may be considered by examining the effect of a specific inhibitor of that subgroup (e.g., molybdate on sulfate reduction; uncouplers on aerobic respiration) on some general index of overall heterotrophic metabolism (e.g., CO_2 production, ^3H-thymidine uptake, heat flux). Similarly, inhibitors of physiological subgroups may be used to determine the importance of a particular group in the turnover and/or consumption of particular substrates. For instance, Sørensen et al. (1981) (Fig. 10) followed the accumulation of volatile fatty acids in marine sediments by employing molybdate in order to determine the role of sulfate-reducing bacteria in the consumption of these acids. Capone and Kiene (1987) have recently reviewed the extensive literature in sedimentary microbial ecology that has taken advantage of this approach.

Benner et al. (1986) have recently applied a two-pronged approach to define the role of bacteria compared to fungi in the breakdown of lignocelluloses in freshwater and marine wetlands. They obtained generally consistent and corroborating results with a specific inhibitor of fungi (Candicidin) and size fractionation with respect to the metabolism of ^{14}C-labeled lignocelluloses, concluding that bacteria were more important in this process in aquatic systems.

Most recently, attention has been focused on the importance of the "microbial loop" of the marine water column. Recent evidence suggests that heterotrophic bacterioplankton utilize dissolved organic matter equivalent to a large fraction (10–50%) of the indigenous primary production (Fuhrman and Azam, 1980, 1982; Hobbie and Williams, 1984).

Of particular interest have been efforts to identify the nature of the primary participants of this loop and their trophodynamic interrelationships. Antibiotics, in parallel or in combination with other methods, have gained popularity in investigations of the microbial loop.

It has been noted in marine waters that populations of bacteria are strikingly similar among water bodies and relatively invarient with time (Wright and Coffin, 1983; Azam *et al.*, 1983). Given the high estimates for bacterioplankton productivity by various methods, the constancy in their numbers has been interpreted to indicate a dynamic steady state between bacterial production and protozoan grazing. Recently, this hypothesis has been directly tested using antibiotic inhibitors.

Presumably, inhibition of eucaryotic predators should result in an increase in bacterial numbers (biomass) and allow for an independent estimate of bacterial growth rates. Conversely, inhibition of bacterial growth in the presence of grazing should result in a decrease in the bacterial population with time. Indeed, Ramirez and Alexander (1980) have prevented protozoan predation on soil bacteria with eucaryotic inhibitors (cycloheximide and thiram), while McCambridge and McMeekin (1980) have similarly shown that protozoans can account for *E. coli* disappearance from seawater.

Newell *et al.* (1983) developed a method using eucaryotic inhibitors (cycloheximide and thiram), which allowed for estimates of natural bacterial growth rates in seawater. Their results were in reasonable agreement with frequency of dividing cell (FDC) methods for estimating bacterial growth rates *in situ*, but substantially lower than ^{3}H-thymidine incorporation.

An interesting observation has been made by Fuhrman and McManus (1984). They found in marine planktobacterial enrichment cultures (1 vol of 1.0-μm filtered seawater into 9 vol of 0.22-μm filtered seawater) that benzyl-penicillin (100 mg/liter) effectively stopped bacterial cell division and, in the absence of predators (1.0-μm filtration and dilution), bacterial cell numbers remained constant for up to 20 hr. In freshly collected, unfiltered, and undiluted seawater amended with penicillin, bacterial numbers decreased as a result of predation. Predation on penicillin-inhibited bacterial populations could be prevented by the addition of eucaryotic inhibitors (colchicine or cyclohexamide), resulting in a constant concentration of bacteria with time. An unexpected result was that in gently filtered (3-, 1-, or 0.6-μm) samples amended with penicillin, bacteria still disappeared, albeit at a lower rate than in unfiltered samples amended with penicillin, and this disappearance was prevented by the addition of colchicine or cycloheximide (Fig. 18). This was interpreted to indicate that there may be eucaryotic predators small enough to pass these small-pore-size filters.

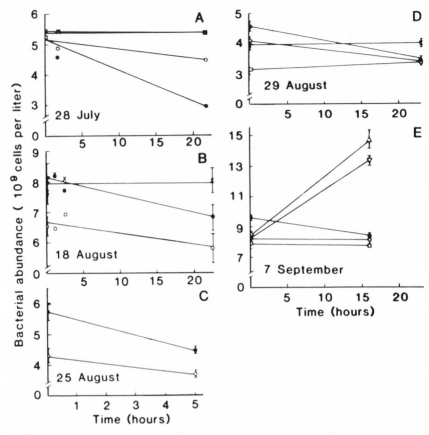

Figure 18. Disappearance of bacteria from unfiltered and prefiltered seawater samples treated with penicillin, and prevention of that disappearance by eucaryote inhibitors: (A) 3-μm filter, cyclohexamide; (B) 0.6-μm filter, cyclohexamide; (C) 0.6-μm filter, no additions; (D) 0.6-μm filter, colchicine; and (E) 1.0-μm filter, cyclohexamide. (●) Unfiltered seawater + penicillin; (×) unfiltered seawater + penicillin + eucaryote inhibitor; (○) filtered seawater + penicillin; (□) filtered seawater + penicillin + eucaryote inhibitor; (▽) filtered seawater only; (△) filtered seawater + cyclohexamide. [From Fuhrman and McManus (1984), with permission.]

Sherr *et al.* (1987) have further developed the "predator inhibition" method, as well as a "prey inhibition" procedure, examining a variety of inhibitors for their efficacy (Table X). A combination of vancomycin (100 mg/liter) and benzyl-penicillin (1 mg/liter) was found to be the most effective inhibitor of procaryotes with the least effect on predators. A combination of cycloheximide (200 mg/liter) and colchicine (100 mg/liter) was the most effective eucaryotic inhibitor with no apparent effect

Table X. Qualitative Results of Assays Evaluating the Effectiveness and Specificity of Procaryotic and Eucaryotic Antibiotics on Bacterioplankton and Heterotrophic Nanoplankton (HNAN)[a]

Antibiotic	Target	Concentration range (mg/ liter)	Effect on growth kinetics Bacterio- plankton	Flagellate	Ciliate	HNAN
Procaryotic						
Chloramphenicol	PS	100	−(C)	−(C)	NA	NA
Cephapirin	CWS	1–100	×	NA	NA	NA
Penicillin	CWS	1–100	−(P, C)	−(P, C)	×	NA
Vancomycin	CWS	100–200	−(P)	×	×	×
Vancomycin + penicillin	CWS	(200) + (1)	−(C)	×	×	×
Eucaryotic						
Thiram	PS	1–100	−(P)	−(C)	−(C)	−(C)
Cyclohexamide	PS	100–200	×	−(C)	−(P)	NA
Demicolcine	MS	0.1–1.0	NA	×	×	×
Colchicine	MS	25–200	×	×	−(P)	NA
Cyclohexamide + colchicine	MS	(200) + (100)	×	−(C)	−(C)	−(C)

[a]From Sherr *et al.* (1987), with permission. PS, Protein synthesis; MS, microtubule synthesis; CWS, cell wall synthesis; −, inhibition; ×, no effect; NA, not assayed; (P), partial; (C), complete; (P, C), partial or complete.

on bacterial growth. The prey inhibition method, in which bacterial growth is inhibited and the disappearance of bacterial cells with time provides an estimate of bacterial mortality (equal to production at steady state) or, alternately, grazing rate on bacterioplankton, yielded the most consistent results. Estimates of bacterial mortality made on natural populations in two seasons with the prey inhibition method were consistent with FDC estimates. The authors suggest that the less consistent results with predator inhibitors (i.e., eucaryotic inhibitors) may be a function of the phasing of the populations of predator and prey, or because of feedback controls of the predators on the productivity of their prey.

Li and Dickie (1985a) have also coupled filtration/fractionation with inhibitors in studies of water-column heterotrophic (as well as autotrophic; see Section 4.1) utilization of amino acids and glucose. They found that while cycloheximide had minor effects on utilization of these compounds, the effect of chloramphenicol was profound and confirmed the domination of bacterioplankton in monomer metabolism. More recently, they reported that the increases in bacterial numbers and activity observed in 0.2-μm filtered seawater are inhibited with penicillin, thereby indicating that some bacteria are able to pass this very small pore size (Li and Dickie, 1985b).

L. Campbell and Carpenter (1986) have applied a procaryotic inhibitor approach (ampicillin) in parallel with predator dilution methods (Landry and Hassett, 1982) to estimate the grazing pressure on the photosynthetic planktobacteria *Synecococcus* spp. They obtained comparable estimates of predator consumption of biomass (37–52%) with each method in nearshore waters. The dilution method was not sensitive enough in more oligotrophic offshore waters, where grazing rates could still be determined by the inhibitor approach.

Fuhrman *et al.* (1986) have recently used antibiotic treatments, along with other procedures, to test the assumptions of the adenine uptake method as a measure for microbial production (both procaryotic and eucaryotic). In a eucaryotic phytoplankton-dominated estuarine system, they found that whereas cycloheximide inhibited the uptake of $^{14}CO_2$ by 62%, it decreased adenine (and thymidine) uptake by only about 10%, indicating that the uptake of these nucleotides is primarily achieved by procaryotes.

Several reports have noted problems with obtaining estimates of bacterial activity, production, or consumption based on inhibitor methods. Yetka and Wiebe (1974) pointed out the inefficiency of antibacterial antibiotics in respiration studies. More recently, Sanders and Porter (1986), working in freshwater Lake Oglethorpe, found inconsistent results in estimates based on penicillin or chloramphenicol inhibition of bacterial growth or cycloheximide or amphotericin B inhibition of eucaryotic predators compared to production estimates based on thymidine incorporation in anaerobic waters and in aerobic waters when bacterial production was low. They attributed this to incomplete inhibition of target populations in general (e.g., incomplete inhibition of predators by cycloheximide will overestimate bacterial production). Several ciliates were shown to continue feeding in the presence of cycloheximide. They attributed the more severe differences noted in anaerobic waters compared to aerobic waters in estimates of rates of bacterial production to the underestimate of bacterial production by thymidine incorporation in anaerobic systems.

G. T. Taylor and Pace (1987) have carried out an extensive and comprehensive series of experiments comparing the effects of various eucaryotic inhibitors at several concentrations and have concluded that none were entirely suitable in assessing bacterial production and/or grazing mortality of bacterioplankton. They attributed the poor performance of this approach to effects of these inhibitors on nontarget phototrophs and indirect effects on bacterioplankton and recommended against the use of such an approach in epipelagic systems with significant phototrophic activity.

4.3. Methane Oxidation

Aerobic oxidation of CH_4 by methanotrophic bacteria proceeds initially via the enzyme CH_4 monooxygenase, which oxidizes CH_4 to CH_3OH (Higgins and Quayle, 1970; Ribbons and Michaelover, 1970; Ferenci, 1974). Methane monooxygenase contains copper (Tonge et al., 1975, 1977) and is susceptible to inhibition by metal-chelating agents (Hubley et al., 1975; Ribbons, 1975; Hou et al., 1979, Stirling and Dalton, 1979). These inhibitors include nitrapyrin and acetylene, which were discussed in detail in Sections 3.3 and 3.4. Hubley et al. (1975) found that whole cells of Methylosinus trichosporium were strongly inhibited by metal-chelating compounds (e.g., thiourea, thiosemicarbazide) as well as by cyanide, azide, and carbon monoxide. Inhibition by these compounds could be relieved by addition of Cu^{2+} ions (Table XI), although in the case of CO, its oxidation occurs in lieu of methane. It is relevant that oxidation of other C1 intermediates (methanol, formate) was not disturbed by the metal-chelating compounds. However, methane monooxygenase can also oxidize ammonia to nitrite (Dalton, 1977), and whole

Table XI. Inhibition of Methane and Methanol Metabolism in Washed Cell Suspensions of Methylosinus trichosporium OB3B by Metal-Binding Compounds[a]

Inhibitor	Concentration (mM)	Rate of disappearance[b] (% control) Methane	Methanol
Thiourea	0.1	0	96
Allythiourea	0.01	28	108
Thioacetamide	1.0	0	100
Diethyldithiocarbamate	0.75	0	90
8-Quinolinol	0.1	0	102
o-Phenylanthroline	0.05	65	90
α,αDipyridyl	0.1	0	92
Ethylxanthate	1.0	0	80
Thiosemicarbazide	0.01	45	100
(Ethylenedinitrilo) tetraacetic acid	1.0	72	76
Carbon monoxide	15%	0	100
Sodium cyanide	0.1	38	100
Histidine	1.0	0	105
Pyridine	1.0	36	108
Imidazole	1.0	0	109
3-Aminotriazole	10.0	27	91

[a]Adapted from Hubley et al. (1975), with permission.
[b]The uninhibited rates of methane and methanol disappearance were 20 and 100 nmole/min per mg dry cell weight, respectively.

cells can produce N_2O from ammonia (Yoshinari, 1984). Presumably these oxidations are susceptible to the same general inhibitors. Methane monooxygenase manifests itself in two forms: soluble and particulate. The particulate form contains copper and is sensitive to metal-chelators, but the soluble form is insensitive to these compounds, and in addition is irreversibly inhibited by copper and other metals (Green *et al.*, 1985). Thus, attempts to inhibit natural populations of methanotrophs using chelators may not prove to be simple. Oxidation of methylamines by methylotrophic bacteria proceeds via several different mechanisms, depending on the type of organism involved [reviewed by Anthony (1982)]. Copper has only been identified with the amine oxidase of *Arthrobacter* P1 (van Vliet-Smits *et al.*, 1981). Fluoroacetate (2–10 μM) inhibited growth of the facultative methylotroph *Pseudomonas* AM1 on methylated amines, but levels as high as 1 mM failed to inhibit growth on non-C1 compounds (Cox and Zatman, 1976). The authors also reported that fluoroacetate did not significantly inhibit growth of bacterium 5H2 on acetate. A maximum of only ~25% inhibition was noted when fluoroacetate levels were as high as 1 mM (substrate acetate concentration 18 mM). Acetate addition (1 mM) to methylamine-growing cultures prevented inhibition by 1 μM fluoroacetate.

Despite the wealth of information on the biochemistry of methane-oxidizing bacteria, little has been done with "specific" inhibitors of the process. One difficulty in examining methane oxidation with "specific" inhibitors is the recognition of the ability of methane-oxidizing and ammonium-oxidizing bacteria to each oxidize methane, ammonium, and, to some degree, carbon monoxide (O'Neill and Wilkenson, 1977; R. D. Jones and Morita, 1983b; R. D. Jones *et al.*, 1984) and to exhibit sen-

Table XII. Oxidation of Selected Methane Derivatives by Whole-Cell Suspensions of *Methylococcus capsulatus* **(Texas)**[a]

Substrate	Rate of oxidation[b] (nmole/min per mg dry weight)
Methane	88
Ethane	15
Propane	6
n-Butane	4
Methylfluoride	3
Bromomethane	50
Methanol	67

[a]From Meyers (1980), with permission.
[b]Oxidation rates were determined by polarographic measurements of oxygen.

sitivity to similar compounds (Topp and Knowles, 1982, 1984; R. D. Jones and Morita, 1984; R. D. Jones *et al.,* 1984) (Sections 3.3., 3.4). As mentioned previously, Salvas and Taylor (1984) suggested the use of picolinic acid to discriminate between ammonia oxidation by methanotrophs versus chemolithotrophic nitrifiers. One promising compound, methylfluoride, has been investigated and found to be a potent inhibitor of oxygen uptake by whole cells and extracts of *Methylococcus capsulatus* (Meyers, 1980) (Table XII). Subsequently, it was shown that methylfluoride also inhibits methane oxidation by methane monooxygenase, as does dimethylether (Meyers, 1982). These two compounds have the advantage of being relatively soluble, and their abundance can be easily measured by flame ionization gas chromatography. Another compound recently revealed as an inhibitor of methane monooxygenase is 1,2-epoxypropane (Habets-Crutzen and de Bont, 1985). This epoxide causes >95% inhibition of methane oxidation, but not methanol oxidation, by cell suspensions of *Methylosinus trichosporium* OB3b.

While there is general interest in the ecology of aerobic methane-oxidizing bacteria in aquatic systems (e.g., Rudd *et al.,* 1974, 1976; Rudd and Hamilton, 1978; Harrits and Hanson, 1980), little work has been done with inhibitors. Lidstrom (1983) reported that acetylene inhibited the uptake of CH_4 by aerobic water from 16 m depth in Framvaren Fjord. Dark CO_2 fixation in Big Soda Lake was blocked by C_2H_2 and N-serve, but not CH_3F (Table VIII). This indicated the involvement of nitrifiers as opposed to methanotrophs. Inhibitors such as CH_3F, dimethylether, and 1,2-epoxypropane should be considered with regard to their usefulness in field studies. Recently, methylfluoride and dimethylether were found to inhibit $^{14}CH_4$ oxidation by endosymbiotic methanotrophic bacteria occurring in the gill tissues of mussels from oceanic hydrocarbon seeps (Table XIII) (Fischer *et al.,* 1987).

Inhibitors have been much more extensively employed in the study of a more controversial phenomenon, namely anaerobic oxidation of methane. This process has been surmised from geochemical data (Barnes and Goldberg, 1976; Reeburgh, 1976; Reeburgh and Heggie, 1977; Martens and Berner, 1977; Devol *et al.,* 1984), and supported by activity measurements (production of $^{14}CO_2$ from $^{14}CH_4$) by natural samples of anoxic waters or sediments (Panganiban *et al.,* 1979; Kosiur and Warford, 1979; Reeburgh, 1980; Iversen and Blackburn, 1981; Devol, 1983; Iversen and Jorgensen, 1985). Interest has focused upon sulfate-respiring bacteria as the causative agents of this process because sulfate is usually the only apparent electron acceptor available in the environments where the process is thought to occur. Davis and Yarbrough (1966) reported low, but discernible activities with *Desulfovibrio desulfuricans.* These low activities were not surprising, since the energy yield associated with methane-

Table XIII. Oxidation of $^{14}CH_4$ by Bacterial
Endosymbionts in Gill Tissues of a
Hydrocarbon Vent Mussel from the Gulf of
Mexico[a]

	dpm/ml gas phase[b]	
Addition	$^{14}CH_4$	$^{14}CO_2$
None	25,159	2740
Methylfluoride	33,415	0
Dimethylether	29,885	0
Boiled	71,182	0

[a]From Fischer et al. (1987).
[b]Flasks contained about 500 mg of fresh gill tissue, 10 ml
of filtered seawater, and an air gas phase of 50 ml. Methylfluoride and dimethylether gases (5 ml each) were
injected into the sealed flasks along with $^{14}CH_4$. Incubation time was 3 hr at 8°C, after which samples were
acidified and gas phases analyzed by radiogaschromato-
graphic procedures.

linked sulfate reduction is poor (Wake et al., 1977). Despite the postulated involvement of sulfate-respiring bacteria in the process, additions
of molybdate (Alperin and Reeburgh, 1985) or tungstate (Iversen et al.,
1987) did not arrest activity in anaerobic samples from Skan Bay or Big
Soda Lake, respectively. However, use of group VI oxyanions to inhibit
methane oxidation by sulfate-respirers is not a straightforward problem.
Addition of molybdate to anoxic systems usually dramatically increases
the methane pool (e.g., Oremland and Taylor, 1978; Sørensen et al., 1981;
Banat et al., 1983). Therefore, a decrease in the quantity of $^{14}CO_2$ produced from $^{14}CH_4$ by molybdate addition to samples may be caused by
isotope dilution with enhanced methane levels rather than blockage of
sulfate-linked methane oxidation (N. Iversen, personal communication).
Anaerobic methane oxidation in water samples was sensitive to the poison formaldehyde, but not to acetylene (Lidstrom, 1983). Anaerobic oxidation was also prevented by filter-sterilization of water samples (Iversen
et al., 1987) and by autoclaving enrichment cultures (Panganiban et al.,
1979), thereby indicating the involvement of microbes in the process.

The possibility that methanogenic bacteria were involved in the process was reported by Zehnder and Brock (1979). Although pure cultures
produced only small quantities of $^{14}CO_2$ from $^{14}CH_4$ and a net consumption of methane was never noted, both production and oxidation of
methane were sensitive to the same levels of BES (Fig. 19). However,
anaerobic methane oxidation in digested sludge and lake sediment samples was much more sensitive to BES than was methane formation (Fig.
20), suggesting perhaps that methanogens convert methane to a soluble

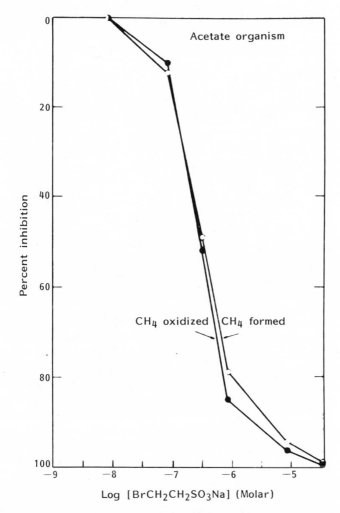

Figure 19. Effect of BES on methane formation and oxidation by an acetate-grown methanogen. [From Zehnder and Brock (1979), with permission.]

intermediate, which is then attacked by another group of anaerobes (Zehnder and Brock, 1980). However, BES failed to inhibit anaerobic methane oxidation in either Skan Bay or Big Soda Lake samples (Alperin and Reeburgh, 1985; Iversen *et al.*, 1987). This implies that methanogens are not involved in the process in these two environments. It is not clear which organisms are responsible for anaerobic oxidation of methane in nature; unfortunately, inhibitor studies have not been particularly helpful.

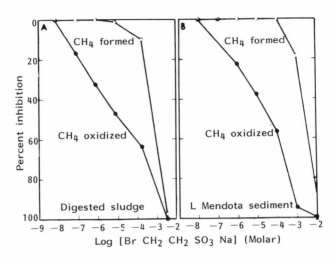

Figure 20. Effect of BES on methane production and oxidation by (A) anaerobic sludge and (B) lake sediments. [From Zehnder and Brock (1980), with permission.]

Anaerobic oxidation of other gaseous hydrocarbons, for instance, acetylene, may help provide clues for the mechanism(s) of methane oxidation. Acetylene is fermented and dismutated to ethanol and acetate via an acetaldehyde intermediate (see Section 3.4; Fig. 17). Although sulfate-linked acetylene oxidation is favorable thermodynamically, pure cultures of sulfate-respirers do not metabolize the gas (Wake *et al.*, 1977). Nonetheless, sulfate reduction was stimulated during acetylene oxidation by estuarine sediment slurries (Culbertson *et al.*, 1981). However, inhibition of sulfate reduction with molybdate failed to block consumption of acetylene by sediments (Culbertson, 1983) (Fig. 21). Sulfate-respirers were apparently oxidizing the products of the initial fermentation/dismutation reactions (H_2, acetate, ethanol) produced by *Pelobacter acetylenicus* (Schink, 1985a) or a related organism(s). An analogous situation could exist for methane, in that methanogens carry out the initial "backreaction" to a soluble product, such as methanol or methane thiol, which is subsequently oxidized to carbon dioxide by sulfate-reducers.

4.4. Methanogenesis

Inhibition of methanogenesis by BES and acetylene has already been discussed at length (Sections 3.1 and 3.4). Chlorinated methane analogues, such as chloroform, carbon tetrachloride, and methylene chloride, were first reported by Bauchop (1967) to inhibit methanogenesis in rumen content samples (Table XIV). Subsequently, Prins *et al.* (1972)

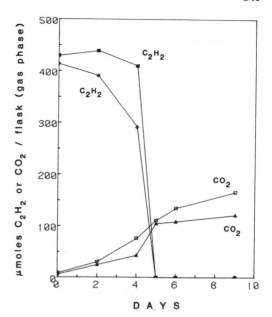

Figure 21. Inability of molybdate to block acetylene consumption by anaerobic estuarine enrichment cultures. Acetylene levels in flasks (◆) without and (■) with molybdate; carbon dioxide levels in flasks (□) without and (▲) with molybdate. Increase in gas-phase carbon dioxide was due to the pH drop associated with acetate formation rather than acetylene oxidation (see Fig. 17). [C. W. Culbertson and R. S. Oremland (unpublished data).]

Table XIV. Effect of Chlorinated Methanes on Methane and Hydrogen Production from Different Substrates[a]

Substrate added	Inhibitor	Concentration (mM)	Gas produced (μmole/ml per hr)		Percent inhibition of CH_4
			Hydrogen	Methane	
None	—	—	0.006	0.848	—
	CH_2Cl_2	15.8	0.012	0.080	91
	$CHCl_3$	0.5	0.007	0	100
	CCl_4	2.0	0.018	0.120	86
Rumen solids	—	—	0	1.72	—
	CH_2Cl_2	15.8	0.127	0	100
	$CHCl_3$	0.5	0.160	0	100
	CCl_4	2.0	0.212	0	100
Formate	—	—	0	2.16	—
	CH_2Cl_2	15.8	0.120	0.06	97
	$CHCl_3$	0.5	0.144	0	100
	CCl_4	2.0	0.200	0.04	98
Hydrogen	—	—	—	1.3	—
	CH_2Cl_2	15.8	—	0	100
	$CHCl_3$	0.5	—	0	100
	CCl_4	2.0	—	0	100

[a]Adapted from Bauchop (1967), with permission.

demonstrated inhibition of pure cultures by micromolar levels of chloroform and carbon tetrachloride. These compounds suffer the disadvantage of usually being added in ethanol to aid in dispersion. This should be borne in mind when examining products of intermediary metabolism (e.g., hydrogen, acetate, etc.) that could be derived from ethanol degradation as opposed to endogenous substrates. Halogenated alkanes (e.g., iodopropane) are known to block the function of corrinoid enzymes in methanogenesis (Wood *et al.,* 1968a; Kenealy and Zeikus, 1981); however, it appears that chloroform also inhibits methyl-Coenzyme M reductase (Gunsalus and Wolfe, 1978). Thus, there may be multiple modes of inhibition by halogenated alkanes. The fact that these compounds can interfere with the functioning of B_{12} enzymes calls into question their specificity for the "target" organism.

Prior to the discovery of BES, chloroform was the most widely used "specific" inhibitor of methanogenesis in ecological/biogeochemical studies. The compound proved effective at blocking methanogenesis in sediments from freshwater (e.g., Cappenberg, 1974; Winfrey and Zeikus, 1979; Lovley and Klug, 1982; J. G. Jones and Simon, 1985) and marine (e.g., Oremland, 1975) locations. It was useful in demonstrating that methylated sulfur compounds can be converted to methane by methanogens present in lake sediments (Zinder and Brock, 1978), an observation that was subsequently confirmed with a pure culture (Kiene *et al.,* 1986). Surprisingly, chloroform, carbon tetrachloride, dichloromethane, as well as chlorinated ethanes and ethenes are all susceptible to anaerobic degradation (Rittman and McCarty, 1980; Bouwer and McCarty, 1983a).

A considerable amount of research has been done upon the influence of the ionophoric drug monensin upon methanogenesis. The interest in this drug stems from economic reasons, because methane formation in the rumen represents an energy waste that might better be shunted to enhance milk and tissue yields of cattle. Monensin addition to populations of mixed rumen bacteria inhibited methanogenesis and increased propionate formation (van Nevel and Demeyer, 1977); however, these authors suggested that the drug did not directly affect the methanogens. Subsequent studies by Chen and Wolin (1979) revealed that the growth of some methanogens was inhibited by monensin (e.g., *Methanobacterium bryantii, M. formicicum, Methanosarcina barkeri*), whereas other methanogens *(Methanobacterium ruminantium)* were not affected. Similar results were obtained with various types of rumen saccharolytic species (e.g., *Ruminococcus albus* and *R. flavefaciens* were inhibited, but *Selenomonas ruminatium* was not). When added to digesting manure, the drug seems to function in a way similar to that of BES; however, there were enough differences to suggest that monensin inhibited both methanogens and nonmethanogens (Wildenauer *et al.,* 1984). Curiously,

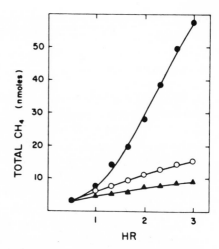

Figure 22. Inhibition of methane production from formate in cell suspensions of *Methanobacterium formicicum* by monensin. (●) No addition; addition of monensin at (O) 1 μg/ml and (▲) 2.5 μg/ml. [From Dellinger and Ferry (1984), with permission.]

monensin strongly inhibits methane production from formate (Fig. 22), but not from $H_2 + CO_2$ by pure cultures of *Methanobacterium formicicum* (Dellinger and Ferry, 1984). It would be interesting to see if this observation could be applied to study formate metabolism in sediments. However, considering that monensin is by no means a "specific" microbial inhibitor, and in addition will disrupt numerous eucaryotic functions (e.g., Pizzey *et al.*, 1983), use of this compound as an ecosystem probe is subject to many possible disadvantages.

A variety of other compounds inhibit methanogenic bacteria, but have not been intensively used in studies of anaerobic ecosystems. Nonetheless, brief mention will be made of these substances. Ethylene inhibits methanogenesis by marine sediments (Oremland and Taylor, 1975), as well as by pure cultures of methanogens (Schink, 1985b). Ethylene does not appear to undergo significant degradation, as does BES, acetylene, or chloroform. In addition, inhibition by ethylene is reversible, methanogenesis being restored upon removal of the gas. Low-potential viologen dyes, such as methylviologen, inhibit methane formation by cultures (Wolin *et al.*, 1964). However, high quantities are needed (~2.5 g/liter) to block activity in freshwater sediments (Pedersen and Sayler, 1981). Nitrogen oxides inhibit methane production by sediments and pure cultures (Balderston and Payne, 1976). Finally, because methanogens have cell walls composed of "pseudomurein" (Kandler and Hippe, 1977), protein subunits (Kandler and Konig, 1978), or heteropolysaccharides (Kandler and Hippe, 1977), they differ from those of eubacteria. Thus, methanogens are not susceptible to a variety of antibiotics that disrupt cell wall synthesis in eubacteria, such as penicillin (Hilpert *et al.*, 1981). Methanogens seem to differ in their susceptibility to such compounds as the ionophore gramicidin (Jarrell and Hamilton, 1985) and the ATP inhibitor N,N'-dicyclohexylcarbodiimide (Sprott and Jarrell, 1982).

4.5. Ethylene Metabolism

Ethylene is a well-known plant growth hormone (Yang, 1974), and, in addition, is the product of acetylene-reduction assay for nitrogen fixation. A variety of soil microorganisms are capable of either producing or consuming this gas. Aerobic ethylene-producing soil bacteria have been isolated that form the gas via metabolism of methionine (Primrose, 1976; Primrose and Dilworth, 1976). Evolution of ethylene by *Escherichia coli* is achieved by the formation of a free soluble intermediate as a degradation product of methionine. The intermediate is chemically similar to 2-keto-4-methylthiobutyric acid and is converted to ethylene by the action of an extracellullar peroxidase. Conversion can be blocked by azide, cyanide, catalase, or anaerobiosis (Primrose, 1977). Production of ethylene by the yeast *Saccharomyces cerevisiae,* however, is not inhibited by anaerobic conditions, but is subject to blockage by cycloheximide (Thomas and Spencer, 1978).

Formation of ethylene in soils, however, seems to be stimulated by anaerobic conditions (K. A. Smith and Russell, 1969; K. A. Smith and Restall, 1971; K. A. Smith and Dowdell, 1974; A. M. Smith, 1976). Sutherland and Cook (1980) found that chloramphenicol and novobiocin inhibited soil ethylene evolution, whereas cycloheximide had no effect (Table XV). These results suggested the importance of anaerobic bacteria over fungi in the tested soils. The authors also found that production was

Table XV. Ethylene Detected in Silt Loam 3 Days after the Soil Was Treated with Chemicals (1 mg/g air-dry soil) and Incubated at 35°C under Nitrogen[a]

Treatment	Ethylene (μg/g soil)[b]
Novobiocin	6.2 (4.1)
Chloramphenicol	9.3 (9.3)
Sodium azide	12.8 (4.6)
Sodium cyanide	14.1 (2.1)
Cycloheximide	18.4 (11.0)
Methionine	20.3 (4.7)
Ethionine	30.1 (2.2)
(Ethylenedinitrilo) tetraacetic acid	32.8 (2.4)
Chlorogenic acid	33.3 (0.8)
Control	21.5 (6.0)

[a]From Sutherland and Cook (1980), with permission.
[b]Values represent the mean of five replicates; parentheses indicate one standard deviation.

partially inhibited by azide and cyanide, and was stimulated by ethionine, EDTA, and chlorogenic acid, but not methionine (Table XV). Evolution from these compounds appears to have proceeded abiogenically, since autoclaved amended soils evolved ethylene. Ethylene production can also be inhibited by high soil salinities (Babiker and Pepper, 1984).

Ethylene is also consumed by certain aerobic soil bacteria (de Bont, 1976a). Ethylene oxidation proceeds via the formation of a free ethylene oxide intermediate, which is then further metabolized via acetyl-CoA and entry into the TCA and glyoxylate cycles (de Bont and Albers, 1976). De Bont and Harder (1978) reported that addition of fluoroacetate to cultures of *Mycobacterium* E20 resulted in the accumulation of ethylene oxide (for ethylene-grown cells) or acetate (for ethane-grown cells). Because fluoroacetate indirectly inhibits the TCA and glyoxylate cycles by causing the formation of fluorocitrate (an inhibitor of aconitate hydratase) (Kun, 1969), the significance of acetyl-CoA in the metabolism of ethylene was underscored. The use of inhibitors to study ethylene oxidation in soils has not been reported. However, the inhibition of natural ethylene oxidation by acetylene can be a potential problem in assays of nitrogen fixation in aerobic soil systems (Witty, 1979) (see Section 3.4).

5. Inhibition of Nitrogen Metabolism

Inhibitors have been used extensively in studies of microbial nitrogen transformations occurring in the environment. The use of C_2H_2 as a substrate analogue in nitrogen fixation, C_2H_2 blockage of nitrous oxide reductase for denitrification, and "selective" inhibition of nitrification by N-serve (see Sections 3.3 and 3.4) have already been discussed in some detail. Other inhibitor approaches have been used for each of these nitrogen transformations, as well as for other microbially catalyzed nitrogen conversions (Table XVI).

5.1. Assimilatory Pathways

In many environments, nitrogen availability often limits productivity (Hardy and Havelka, 1975; Capone and Carpenter, 1982b) and inorganic nitrogen assimilation is likely to be a critical step. Numerous ecological studies have addressed the uptake of various forms of nitrogen as a means of discerning the controls and constraints on primary productivity. Direct chemical analysis of pools and changes in pool size during incubation, as well as ^{15}N isotopic methods [tracer and isotope dilution (e.g., Harrison, 1983)] have been applied to studies of nitrate and/or ammonium uptake. These procedures are inherently insensitive and

Table XVI. Some Inhibitors Used in Nitrogen Transformation Studies

Activity	Inhibitor	Effective concentration	Reference
Assimilation			
Nitrate transport	Chlorate	760 μM (K_m = 62 μM)	W. M. Balch (1987)
Ammonium transport	Methylamine	—	Wheeler (1980)
Glutamine synthetase	MSX[a]	1 mM	Syrett (1981)
Glutamate synthase	Asazerine	1 mM	Syrett (1981)
Nitrogen fixation			
	Acetylene	>0.1 atm	Hardy et al. (1968)
	Acetylene	K_i = 0.052 atm	Rivera-Ortiz and Burris (1975)
	Nitrous oxide	K_i = 0.108 atm	Rivera-Ortiz and Burris (1975)
	Cyclopropene	0.05 atm	McKenna and Huang (1979)
Nitrification			
Ammonium oxidation	N-serve	10 ppm (w/v)	Hall (1984)
	N-serve	20 ppm (w/v)	Henricksen (1980)
	N-serve	10 ppm (w/v)	Jones and Morita (1984)
	Allylthiourea	10 ppm (w/v)	Hall (1984)
	Allylthiourea	100 ppm (w/v)	Hall (1984)
	Carbon monoxide	2 nM	Jones et al. (1984)
	Acetylene	>0.00005 atm	Hynes and Knowles (1978)
	Chlorate	K_i = 2 μM	Hynes and Knowles (1983)
Nitrite oxidation	Chlorate	10 μM	Belser and Mays (1980)
Denitrification			
	Acetylene	0.10 atm	Yoshinari et al. (1977)
	Sulfide	0.3 mM	Sorensen et al. (1980)

[a]L-Methionine-DL-sulfoximine.

tedious. However, specific inhibitors of the assimilatory pathways of nitrogen are known and offer some potential advantages over ^{15}N or pool size methods. These inhibitors have been employed with pure cultures and have not yet been utilized in ecological or geochemical investigations.

In some bacteria, algae, and fungi, nitrate is reduced to NH_4^+ by assimilatory nitrate and nitrite reductases (Falkowski, 1983). Chlorate, an analogue of nitrate, apparently competes with nitrate for transport across the cell membrane and can therefore inhibit nitrate uptake. Chlorate may also be reduced by some assimilatory nitrate reductases to chlorite, which is toxic. The transport of $^{36}ClO_3^-$ has been proposed as a sensitive analogue assay for nitrate assimilation (Tromballa and Broda, 1971; W. M. Balch, manuscript). Similarly, monomethylamine acts as an inhibitor of ammonium uptake (Cresswell and Syrett, 1984). ^{14}C-Methylamine uptake has been used in transport studies as an analogue of ammonium (Wheeler, 1980).

While it was long thought that the glutamate dehydrogenase (GDH) enzyme system was responsible for net ammonium incorporation in ammonium-utilizing organisms, it now appears that incorporation proceeds via the concerted efforts of the glutamine synthetase (GH)/glutamate synthase (GOGAT) pathway (Falkowski, 1983). l-Methionine-DL-sulfoximine (MSX), an analogue of glutamate, the substrate of GS, is a powerful inhibitor of GS (Pace and McDermott, 1952; Syrett, 1981). Numerous studies have used MSX inhibition to confirm that GS rather than GDH was the primary ammonium assimilatory pathway in bacteria (Brenchley, 1973; Moreno-Vivian *et al.,* 1983), cyanobacteria (Flores *et al.,* 1980; Boussiba and Gibson, 1985), and eucaryotic algae (Cresswell and Syrett, 1984; Rigano *et al.,* 1979). Details of the GS regulation have also been addressed using MSX inhibition. Presumably, MSX inhibition in natural samples would extend these pure culture studies in defining the enzymological pathway operating *in situ.* A GOGAT inhibitor, azaserine (a substrate analogue of glutamine), is an inhibitor of GOGAT and has been used in a similar fashion to MSX (Flores *et al.,* 1980; Moreno-Vivan *et al.,* 1983; Elrifi and Turpin, 1986).

Wheeler and Kirchman (1986) have recently examined the uptake of ^{15}N-labeled ammonium and amino acids by marine picoplankters, coupling size fractionation studies to the use of eucaryotic and procaryotic inhibitors. They obtained quite provocative results. Consistent with earlier conclusions, heterotrophic bacteria were found to be most important in the assimilation of amino acids. However, and in contrast to prevailing dogma, heterotrophic bacteria were also found to be an important sink for ammonium, which they apparently compete for with phytoplankton. Surprisingly, eucaryotes were found to be primarily responsible for the regeneration of ammonium.

5.2. Nitrogen Fixation

Nitrogen fixation is the process whereby certain procaryotes utilize molecular nitrogen for cellular nitrogen requirements. It can be quite important on a local scale where nitrogen is limiting, as well as in the global cycling of nitrogen (Postgate, 1982). The capacity of the enzyme responsible for N_2 fixation, nitrogenase, to reduce other low molecular weight, triply bonded compounds other than its natural substrate, N_2, led to the development of the C_2H_2 reduction assay. Despite its limitations (see Section 3.4), the C_2H_2 reduction method is a powerful tool and has allowed for a dramatic increase in our understanding of this process (Postgate, 1982; Payne, 1984).

Acetylene was originally chosen as the alternate substrate of choice for field assays because it is inexpensive, easily generated from CaC_2, highly soluble, and, along with its reduction product C_2H_4, is easily quantified by flame ionization gas chromatography. However, it is also inhibitory to a broad range of organisms besides nitrogen-fixers (see Section 3.4). At the concentrations of C_2H_2 used in the acetylene reduction assay ($\sim 10\%$), C_2H_2 completely displaces the natural substrate, N_2 (Postgate, 1982). Since two electrons are required to reduce C_2H_2 to C_2H_4, whereas six electrons are required for the reduction of dinitrogen to $2NH_4^{2+}$ for each mole of N_2 that would have been reduced, 3 moles of C_2H_4 should be formed from C_2H_2. Under nitrogen-fixing conditions (N_2 as substrate), nitrogenase evolves about 1 mole of H_2 per mole of N_2 reduced, in an apparently wasteful side reaction. This ATP-dependent hydrogenase activity of nitrogenase is inhibited in the presence of C_2H_2 and may account in part for divergences observed between the theoretical (3:1) and observed (often $>3:1$) ratios between C_2H_2 reduction and actual N_2 fixation (usually measured as $^{15}N_2$ uptake) in various natural systems (e.g., Peterson and Burris, 1976; Graham et al., 1980; Capone 1983). Intercalibration of the C_2H_2 reduction method with direct $^{15}N_2$ fixation, while strongly recommended (Burris, 1974), is rarely performed.

The ATP-dependent evolution of hydrogen by nitrogenase is an energetically wasteful process (Postgate, 1982). Interest has been expressed in improving the efficiency of biological nitrogen fixation in agricultural systems either by reducing the extent of nitrogenase-mediated H_2 loss or enhancing the capacity of H_2 recycling through uptake hydrogenases (Evans and Barber, 1977). Postgate (1982) noted that "virtually all diazotrophic systems showed hydrogenase activity but not all micro-organisms which possessed hydrogenase were able to fix dinitrogen." The relationship between nitrogenase, its hydrogenase activity (ATP-dependent H_2 formation), and other hydrogenases has therefore been an area of intense interest, particularly in the cyanobacteria (Bothe, 1982; Lambert and Smith, 1981), and inhibitors have played a critical role in many of these studies. Bothe (1982) recognizes three general classes of hydrogenases. Classical reversible hydrogenases are ATP-independent but generally function in the H_2 formation direction (e.g., clostridial fermentation), and are inhibited by O_2 and CO. Nitrogenase-linked hydrogen evolution is ATP-dependent, CO-insensitive, and inhibited by

C_2H_2. The third class of uptake hydrogenases function in the recycling of H_2 evolved by nitrogenase in aerobic N_2-fixers, such as some cyanobacteria and rhizobia. They are CO-sensitive, but O_2-insensitive. Carbon monoxide inhibition has been used to determine the proportion of nitrogenase activity supported by uptake hydrogenase in cyanobacteria (Bothe *et al.*, 1977a) and *Azospirillum* (Chan *et al.*, 1980). Acetylene may also inhibit uptake hydrogenase in aerobic organisms (L. A. Smith *et al.*, 1976), but not in anaerobes (Bothe *et al.*, 1976). It should be noted that Hyman and Arp (1987) have recently expressed concern about the occasionally high levels of H_2 and C_2H_2 preparations and the conflicting reports of the effect of C_2H_2 on some hydrogenases.

Diazotrophic cyanobacteria apparently have two functionally distinct uptake hydrogenases that can support nitrogenase activity: an aerobic respiratory (Knallgas) type and an anoxic, photodependent system (Peschek, 1979; Bothe *et al.* 1977b). Eisbrenner and Bothe (1979) have provided good evidence for the distinctions between these two systems by using a "specific" inhibitor of the anoxic light-dependent pathway, metronidazole.

Several workers have attempted to determine the extent of H_2-dependent nitrogenase activity in the environment. The cyanobacterium *Oscillatoria* spp. *(Trichodesmium)* is an important diazotroph in the open ocean (Carpenter, 1983a,b) and appears to be closely involved in H_2 cycling in these systems (Scranton, 1984). Saino and Hattori (1982) have found CO to stimulate H_2 production by these organisms, suggesting an active uptake hydrogenase system that may function in efficient recycling of H_2 and oxygen protection. Oremland (1983) found H_2 production by cyanobacterial epiphytes of lake macrophytes to occur only in the dark and to be stimulated by C_2H_2 and inhibited by O_2. Hydrogen evolution was caused by the bacteria rather than the cyanobacteria.

The C_2H_2 reduction method can be used in concert with other inhibitors to provide some unique insights into the ecology of N_2 fixation. As mentioned in Section 3.2, molybdate additions have been shown to severely curtail C_2H_2 reduction in marine sediments (Capone *et al.*, 1977; Nedwell and Aziz, 1980). This suggests that sulfate-respiring bacteria may be important components of the diazotrophic biota in these systems.

Methionine sulfoximine (MSX; see Section 5.1) has also been used to probe the control of nitrogenase in certain systems. This is possible because gene level regulation of nitrogenase by its product, ammonium, occurs via the intervention of glutamine synthetase (or a related gene product of the *gln* gene). Glutamine synthetase is adenylated to a less active form in the presence of ammonium, thereby decreasing ammonium incorporation. The adenylated form of GS (or a related *gln* gene product) also represses nitrogenase synthesis (Postgate, 1982). MSX uncouples NH_4^+ control of nitrogenase, and various studies using MSX have demonstrated that NH_4^- can play an important role in nitrogenase regulation in the various bacterial and cyanobacterial systems examined (Gordon and Brill, 1974; Ramos and Guerrero, 1983; Stewart and Rowell, 1975; Turpin, *et al.* 1984; Hillmer and Fahlbusch, 1979). Whereas activity level control of nitrogenase by ammonium does not appear to be a common property of most nitrogenase systems examined, it does occur in some of the photosynthetic bacteria. MSX is also effective in preventing ammonium "switchoff" in these systems (Yoch and Gotto, 1982). Glutamine, the product of GS, is thought to be the effector molecule in the "switchoff" phenomenon (Arp and Zumft, 1983).

Several studies have recently reported the use of MSX in ecological studies. Capone (1984, 1987; O'Neil and D. G. Capone, in preparation) found MSX to stimulate nitrogenase activity in seagrass and saltmarsh sediments (Fig. 23), suggesting that nitrogenase activity in these systems is suboptimal because of ammonium repression. Yoch and Whiting (1986) have recently examined the kinetics of nitrogenase inhibition in *Spartina alterniflora* sediments and roots and have found evidence for a "switchoff" mechanism (enzyme activity control). MSX at concentrations as low as 25 μM prevented ammonium "switchoff" in these systems.

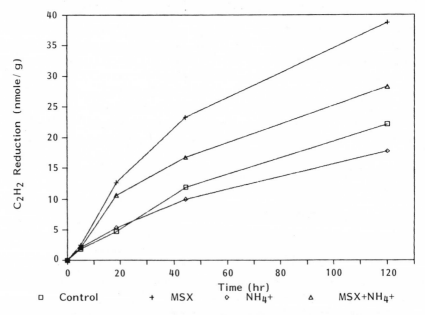

Figure 23. Effect of L-methionine-D,L-sulfoximine (MSX) on nitrogenase activity in saltmarsh sediments. MSX was added to a final concentration of 10 mg/ml; NH_4 to 10 mM. [D. Capone (unpublished data).]

Certain authors have proposed the use of substrates other than C_2H_2 for assessing nitrogenase activity in particular circumstances. For instance, N_2O, which is reduced to N_2 and H_2O by nitrogenase, has been proposed as a suitable substrate for assaying methanotrophic N_2 fixation (Dalton and Wittenbury, 1976), but this may be complicated by production of N_2O from NH_4^+ (Yoshinari, 1984). Cyclopropene, which is reduced to propene and cyclopropane by nitrogenase, has also been suggested as another alternate substrate that may offer certain advantages over C_2H_2 (McKenna and Huang, 1979, Postgate, 1982).

An exciting recent development in N_2-fixation research has been the partial characterization of an alternative nitrogenase known to occur in molybdenum-starved *Azotobacter* spp. (McKenna *et al.*, 1970; Robson *et al.*, 1986). The alternative nitrogenase is a vanadium enzyme, which is also capable of acetylene reduction.

5.3. Nitrification

Nitrification is the oxidation of NH_4^+ to NO_3^-. In nature, it is primarily catalyzed by two specialized groups of chemoautotrophic bacteria, ammonium-oxidizers (e.g., *Nitrosomonas* sp.) and nitrite-oxidizers (e.g., *Nitrobacter* sp.) (Bremner and Blackmer, 1981). In the global nitrogen cycle, nitrification drives the reoxidation of reduced forms of nitrogen (i.e., NH_4^+), and thereby provides substrate for the next step in the cycle, anaerobic reduction of nitrogen oxides by denitrification. The ecological impact of nitrifiers is most evident in the sea, where nitrifiers maintain nitrate as the predominant form of inorganic nitrogen in the water column (Kaplan, 1983). Nitrifiers can have a severe economic impact in agricultural systems by converting fertilizer ammonium to nitrate, which may then be lost from the soil either through denitrification or leaching. In defining the nitrogen cycle of particular systems, it is therefore critical to determine the importance of nitrification. However, as for many aspects of the nitrogen cycle, sensitive (e.g., radioisometric) procedures are not readily available.

Relatively sensitive and convenient methods of measuring rates of nitrification have been developed using "specific" inhibitors of nitrification. These include assays using N-serve, chlorate (ClO_3^-), allylthiourea, and carbon monoxide (CO).

The use of N-serve has already been described in detail (Section 3.3). The simplest application of the N-serve method entails the monitoring of changes in NH_4^+ and NO_3^- pools over time in the presence and absence of N-serve. The N-serve "specifically" targets ammonium oxidation, rather than NO_2^- oxidation. If a steady state existed for nitrate and ammonium concentrations before N-serve addition, the presence of N-result in the buildup of ammonium and decreases in nitrate (Webb and Wiebe, 1975).

The limitations of the N-serve method, as mentioned, include the low water solubility of the inhibitor, its lability, and its nonspecificity. Since it is generally necessary to dissolve N-serve in an organic solvent, multiple level controls are required and solvent–inhibitor interactions need to be considered (Stratton et al., 1982). Acetone is commonly the solvent of choice and it may have secondary, undesired effects in the assay system (L. I. Hendriksen, personal communication). A potential solution is the use of an alternative, water-soluble inhibitor of ammonium oxidation. Hall (1984) and R. D. Jones and Morita (1984) have found that allylthiourea, which is water-soluble, compares favorably with N-serve as an inhibitor of ammonium oxidation. In fact, Hall (1984) suggests that the higher apparent potency of allylthiourea may partially be a result of its greater persistence.

Nutrient chemistry-based nitrification assays using either N-serve or allylthiourea require determination of nitrate and ammonium, the analysis of which can be tedious. Furthermore, nitrate and ammonium often have appreciable environmental backgrounds, which decrease the ability to discern small changes in concentration. Belser and Mays (1980) have described a procedure involving specific inhibition of nitrite oxidation using ClO_3^-. This method monitors levels of nitrite. Nitrite can be detected with greater sensitivity than either ammonium or nitrate, and in addition, background levels of nitrite are usually undetectable. However, Hynes and Knowles (1983) pointed out that whereas ClO_3^- does not appear to inhibit ammonium oxidation (a prerequisite for the success of the assay), nitrite oxidizers can reduce ClO_3^- to ClO_2^-, which strongly inhibits ammonium oxidation. Horrigan and Capone (1985) determined concurrent rates of nitrification and denitrification in nearshore sediments with an isotope dilution method, but were unable to observe any nitrite accumulation in the presence of ClO_3^-. This may have been the result of a higher capacity for nitrate, and presumably nitrite, dissimilation in this system or, alternatively, the buildup of toxic levels of ClO_2^-.

R. D. Jones and Morita (1983a) have recently found that CO can be oxidized to CO_2 by ammonium-oxidizing bacteria. They have proposed CO oxidation as a convenient and very sensitive assay for nitrification (R. D. Jones et al., 1984). The fact that methane-oxidizing bacteria can also perform this oxidation requires that controls be run that account for this. R. D. Jones et al. (1984) used an N-serve-inhibited control to differentiate between the contribution of methane-oxidizers and nitrifiers to CO oxidation. It is interesting to note that N-serve has been found to inhibit methane oxidation (Topp and Knowles, 1982). While the CO oxidation assay appears to be quite sensitive, it is not straightforward to convert from CO oxidation to nitrification.

Finally, Glover (1982) has reported that methylamine is a potent inhibitor of ammonium uptake, oxidation, and CO_2 fixation by nitrifiers.

5.4. Denitrification

Denitrification is the reduction of inorganic oxides of nitrogen (i.e., NO_3^- and NO_2^-) to gaseous end products (N_2O and N_2) by facultatively anaerobic bacteria (Knowles, 1983). It is of importance ecologically and geochemically because it converts nutritionally useful forms of nitrogen to biologically less useful or unavailable forms. In establishing realistic mass balances of nitrogen in the environment, it is critical to evaluate such sinks. Denitrification has been measured by direct $^{15}NO_3^-$ reduction. However, this generally requires substantial additions of nitrate (e.g., Jenkins and Kemp, 1984). Recent modifications by Nishio et al. (1982) have improved this situation, allowing sufficient sensitivity for detecting increases in the $^{15}N/^{14}N$ ratio in N_2 after relatively small absolute increase in the nitrate pool by highly enriched $^{15}NO_3^-$. However, as for all heavy-isotope methods, procedures are tedious and expensive and sample processing is slow.

Direct chromatographic analysis of denitrification end products (N_2 and N_2O) is also possible (Payne, 1973) and has been reported (Seitzinger et al., 1980). Since N_2 is the primary end product, the potential for artefactual results is high, since even minor leakage of N_2 from the ambient air ($\sim 80\%$ N_2) into assay vessels or chromatographic syringes (Pearsall and Bonner, 1980) could be construed as "denitrification."

Through the C_2H_2-mediated inhibition of nitrous oxide reductase, the end products of denitrification are focused into one compound, N_2O. The advantages of the "C_2H_2 block" lie in the fact that the N_2O can be detected at much lower levels (using electron capture) than N_2. Also, ambient N_2O levels are insignificant. Hence, low levels of N_2O production can be discerned over brief periods.

In addition to the general limitations mentioned in Section 3.4 in the use of C_2H_2 (interactions with C_2H_2-inhibited nontarget organisms), other problems are unique to the C_2H_2 blockage method. One serious factor is the apparent loss of C_2H_2 block effectiveness during incubation. This is manifested by the disappearance of N_2O from the assay chamber despite the persistence of C_2H_2 (Yeomans and Beauchamp, 1978; van Raalte and Patriquin, 1979; Slater and Capone, 1984; Oremland et al., 1984). Kaspar (1982) has related the failure of the block to conditions of low nitrate concentration, an observation confirmed by other investigators (Orem-

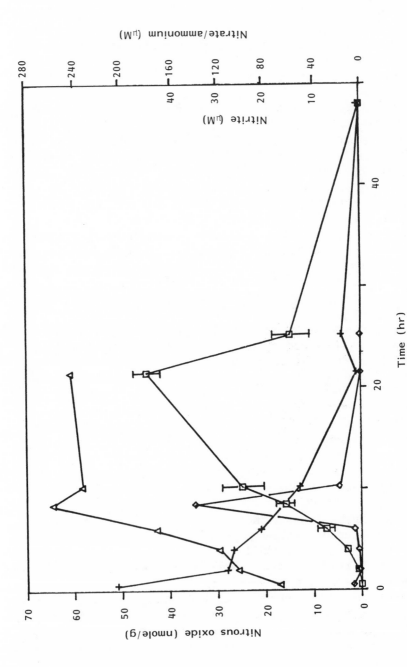

Figure 24. Concentrations of (+) nitrate, (◇) nitrite, (△) ammonium, and (□) nitrous oxide in incubations of saltmarsh sediment in the presence of acetylene. [J. Slater and D. Capone, 1987a, with permission].

land *et al.,* 1984; Slater and Capone, 1987). Nitrous oxide persists while nitrate is still present in saltmarsh sediments, but when nitrate concentrations fall below 2 μM, N_2O disappears (Fig. 24) (Slater and Capone, 1988a,b, submitted). Several investigations have implicated sulfide as the agent causing relief of N_2O reductase blockage by C_2H_2 (Tam and Knowles, 1979; Evans *et al.,* 1985). However, Sørensen *et al.* (1980) have found that sulfide can itself block N_2O reductase activity. Therefore, the mode(s) of action of sulfide in disrupting the C_2H_2-block assay has not yet been clearly defined.

6. Inhibition of Sulfur Metabolism

6.1. Sulfate Reduction

Aspects of the inhibition of sulfate-respiring bacteria by group VI oxyanions were covered in detail in Section 3.2. Saleh *et al.* (1964) compiled a listing of numerous classes of compounds (antibiotics, detergents, dyes, mercurials, phenolics, metal ions, nitro compounds, sulfate analogues, etc.) and their inhibitory efficacy against cultures of sulfate-respiring bacteria. Norqvist and Roffey (1983) tested certain types of inhibitors (nitrate, isothiazolone, molybdate, selenate) against sulfate reduction activity and recoverable populations of sulfate-respirers in sediments. In addition to group VI oxyanions, a variety of other sulfate structural analogues will inhibit sulfate respiration. Postgate (1952) demonstrated noncompetitive inhibition of hydrogen consumption by *Desulfovibrio desulfuricans* exposed to monofluorophosphate. Inhibition could be relieved by addition of sulfite ions. Potassium perchlorate proved an effective inhibitor. However, such compounds as dimethylsulfone, sodium ethylsulfonate, and sulfamic acid had no noticeable influence on cultures. No attempts have been made to pursue ecological studies using the successful versions of these types of compounds.

Fluoridated analogues of the primary substrates of sulfate-respiring bacteria have been employed in several ecological investigations for the purpose of blocking both sulfate reduction and turnover of these compounds in aquatic sediments. The difficulty with using this approach is that one must assume that the compound chosen is the primary substrate of the target organism. In addition, there is some indication that, in certain circumstances, these types of inhibitors can be ineffective due to "swamping" out by the large quantities of substrate available (e.g., Cox and Zatman, 1976). This may have less influence in natural samples, where substrate concentrations are quite low when compared with media. Nonetheless, some success has been achieved with this approach.

Lactate is a common substrate for many species of sulfate-respirers (Postgate, 1979). Cappenberg (1974) reported that lactate turnover, as well as sulfide production in lake sediments, was inhibited by b-fluorolactate. Whereas fluoroacetate did not appear to influence these parameters, it did reduce methanogenic activity. Abram and Nedwell (1978b) reported that use of b-fluorolactate caused inhibition of lactate, ethanol, and hydrogen consumption as well as sulfate reduction in saltmarsh sediments. However, b-fluorolactate caused only transient inhibition of sulfate-linked hydrogen uptake by tropical marine sediments (Fig. 2B), as well as inexplicable transient/enhanced methane kinetics (Fig. 9). The brief popularity of b-fluorolactate was ended by the advent of molybdate as the inhibitor of choice. Use of molybdate also revealed that sulfate-respirers were responsible for the consumption of hydrogen and acetate in sediments (e.g., Sorensen et al., 1981), and that lactate was of lesser significance as an in situ substrate for sulfate reduction. Several studies have focused on the metabolism of acetate in sediments by sulfate-respiring bacteria and have employed fluoroacetate to inhibit sulfate reduction and acetate turnover (Winfrey and Zeikus, 1979; Nedwell, 1982; Winfrey and Ward, 1983; Alperin and Reeburgh, 1985). In addition, fluoroacetate has been used in studies of terminal carbon and electron flows in estuarine sediments (Mountfort et al., 1980).

6.2. Sulfide and Reduced Sulfur Oxidation

An important component of the sulfur cycle in nature is the microbial oxidation of sulfide and other reduced forms of sulfur (Fenchel and Riedl, 1970). A diversity of organisms are active with respect to the oxidative part of the cycle: purple and green sulfur bacteria, thiobacilli, Beggiatoa, and certain cyanobacteria. The distribution of Beggiatoa sp. filaments in a cyanobacteria-covered sediment column was monitored in response to treatment with DCMU, an inhibitor of oxygenic photosynthesis (Moller et al., 1985). However, no reports exist on attempts to inhibit microbial oxidation of reduced sulfur species in natural ecosystems using specific inhibitors. Thus, there is a strong need for specific microbial inhibitors that would help to distinguish between chemical and bacterial oxidation. Unfortunately, such inhibitors have not been reported, and this realm is certainly ripe for future investigation. Useful summaries of relevant topics in this area were written by Kelly (1982). Truper and Fischer (1982), Jørgensen (1982), and Kuenen and Beudeker (1982).

Thiobacilli are subject to inhibitors of terminal oxidases, such as cyanide, azide, and carbon monoxide (Peeters and Aleem, 1970). The pathway of inorganic sulfide oxidation has many enzymes similar to dissimilatory sulfate reduction (Kelly, 1982). It would be of interest to see if compounds such as selenite or molybdite interfere with the functioning of ADP sulfurylase in thiobacilli, since selenate and molybdate disrupt ATP sulfurylase during sulfate reduction. Molybdate did not influence dark CO_2 fixation in the oxic–anoxic interface of Big Soda Lake (Cloern *et al.*, 1983), although there was no way to ascertain if the compound was effective. Various organic compounds, such as fatty acids, pyruvate, and TCA cycle intermediates, inhibit the oxidation of Fe^{2+} and sulfur by *Thiobacillus ferroxidans,* but relatively high levels had to be employed ($\sim 10^{-2}$ M) for effectiveness (Tuttle and Dugan, 1976). Heterotrophic marine bacteria can mimic the chemolithotrophic growth of thiobacilli in that thiosulfate enhances their dark fixation of CO_2 (Tuttle and Jannasch, 1977). Thus, organic amendments that act as inhibitors of lithotrophic reduced sulfur oxidation may not be practical due to the activity of heterotrophs, some of which can oxidize thiosulfate.

7. Inhibition of Metal Metabolism

The involvement of microorganisms in the biogeochemical reactions of metals in nature is a large topic and the reader is referred to the work of Erhlich (1978, 1981) for comprehensive reviews. We will only present a couple of examples in order to give the reader a "feel" for the employment of specific inhibitors in this research area. One of the important questions here concerns the *in situ* significance of metal-active microorganisms. There are numerous examples of microbes with the potential to be important agents in determining the chemical form and availability of metals in nature. Some organisms, for example, *Thiobacillus ferroxidans,* can grow autotrophically upon the oxidation of ferrous ions. However, it is not particularly clear whether or not their abilities (reduction, oxidation, alkylation, dealkylation, precipitation, etc.) are significant with respect to purely chemical kinetics. Unfortunately, as in the case of reduced sulfur oxidation, "specific" inhibitors of microbially catalyzed metal transformations have not yet been "discovered." However, considerable work with broad-spectrum antimicrobial agents and poisons proves the importance of microorganisms under *in situ* conditions. The best example in this area is the case of manganese.

Bacteria can be isolated from marine manganese nodules that have the ability to either oxidize Mn^{2+} (Ehrlich, 1966, 1968; Arcuri and Ehrlich, 1979) or reduce Mn^{4+} (Trimble and Ehrlich, 1968; Ghiorse and Ehrlich, 1976). However, it is not certain if these organisms accelerate these reactions *in situ.* Although there are no known "specific" inhibitors of manganese oxidation or reduction, the organisms involved are susceptible to a variety of respiratory poisons, antibiotics, mercurials, etc. (Ghiorse and Ehrlich, 1976; Rosson and Nealson, 1982; Burdige and Nealson, 1985). Microbial oxidation of Mn^{2+} in lakewater was blocked by filtration (1.2 μm) or ethanol treatment (Chapnick et al., 1982). Uptake of $^{54}Mn^{2+}$ by bacteria in the water column of Saanich Inlet was inhibited by formaldehyde, or a mixture of poisons that included antibiotics and azide (Emerson et al., 1982). The poison mix was tested to determine if complexation with the added manganese "tag" caused artefactual results that falsely implied in situ bacterial oxidation. No significant difficulties were encountered and thus the contribution of bacterial oxidation of manganese in the water column was assessed (Fig. 25). An

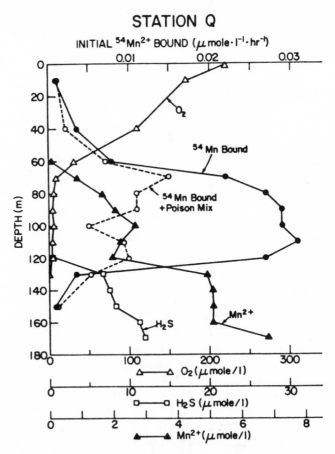

Figure 25. Distributions of oxygen, hydrogen sulfide, and manganese in the water column of Saanich Inlet, British Columbia, and the ^{54}Mn-binding activity in poisoned and unpoisoned water samples incubated for 1 hr. [From Emerson et al, (1982), with permission.]

approach to measuring bacterial manganese oxidation in sediments was devised by Burdige and Kepkay (1983). These investigators used a dialysis "peeper" to follow abiotic and bacterially mediated manganese binding in artificial sediments impregnated with a manganese-oxidizing pseudomonad. Azide-poisoned controls had slower rates of Mn^{2+} deposition than uninhibited samples. This approach could be useful for answering the question of whether or not microorganisms contribute to the formation of sea-floor manganese nodules.

The methylation/demethylation of certain heavy metals (e.g., mercury) is a question of considerable environmental significance. Methylmercury is a toxic and mobile form of this element, which can accumulate in marine food webs at levels lethal to humans. Methylation of mercury occurs in marine sediments (Jensen and Jurnelov, 1969), and extracts of methanogenic bacteria can form methylmercury (Wood et al., 1968b). Because methanogens are present in sediments, it was assumed that they were responsible for formation of methylmercury from Hg^{2+}. However, McBride and Edwards (1977) reported that although cell extracts of methanogens were active in this respect, whole cells were not (Table XVII). Furthermore, addition of chloroform to sewage sludge did not inhibit CH_3Hg^+ formation, even though it blocked methanogenesis (Table XVIII).

Subsequently, Compeau and Bartha (1985, 1987) clearly demonstrated that sulfate-respiring bacteria and not methanogens were responsible for the formation of methylmercury in marine sediments. Molybdate strongly inhibited sediment methylmercury production (Fig. 12), and methylmercury could also be produced by pure cultures of sulfate-respirers. In contrast, addition of BES stimulated methylmercury pro-

Table XVII. Alkylation of Mercury by Methanogenic Bacteria[a]

	Control		Experimental	
	CH_3Hg (μg)	CH_4 ($\mu mole$)	CH_3Hg (μg)	CH_4 ($\mu mole$)
Whole cells				
M. bryantii	0	0	0	9.2
M. thermoautotrophicum	0	0	0	12.3
Cell extracts				
M. bryantii	2.3	0	0.8	6.4
M. thermoautotrophicum	2.3	0	0.9	5.2

[a]From McBride and Edwards (1977). Cell extracts were incubated with CH_3-B_{12}, ATP, H_2, and $^{203}HgCl_2$. Whole cells were incubated with $H_2 + CO_2$ and $^{203}HgCl_2$. Controls were boiled prior to addition of mercury.

Table XVIII. Methylmercury Formation in Sewage Sludge[a]

Treatment	CH_3Hg^+ (μg/liter)	CH_4 (nmole)
None	16.5	140
Aeration	17.3	5
Chloroform	17.5	1
Heat	0	0

[a]From McBride and Edwards (1977).

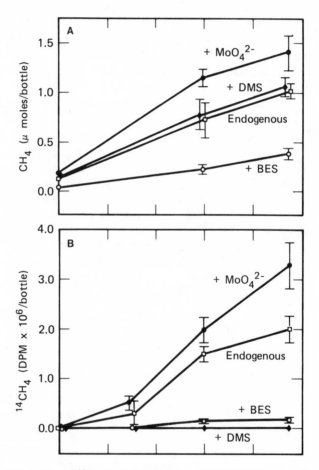

Figure 26. Metabolism of ^{14}C-dimethylselenide by anaerobic saltmarsh sediment slurries incubated without additions, or with 20 mM molybdate, 40 mM BES, or 10 mM DMS as indicated. Production of (A) methane; (B) ^{14}C-methane; (C) ^{14}C-carbon dioxide. [From R. S. Oremland and J. Zehr, 1986.]

duction by sediments (Fig. 12). Although the basis for this was unexplained by the authors, Oremland and Zehr (1986) proposed that certain methylotrophic methanogens can demethylate some types of alkylated metals or metalloids, such as mercury or selenium. These workers used molybdate and BES in order to demonstrate that both methanogens and sulfate-respirers present in sediments could metabolize ^{14}C-dimethylselenide to $^{14}CH_4$ and/or $^{14}CO_2$ (Fig. 26).

8. Summary

The employment of inhibitors in the study of microbe-mediated reactions occurring in the environment has become an accepted scientific approach. The types of inhibitors used cover a wide span of microbial processes, and vary in effectiveness. In this chapter we have attempted to

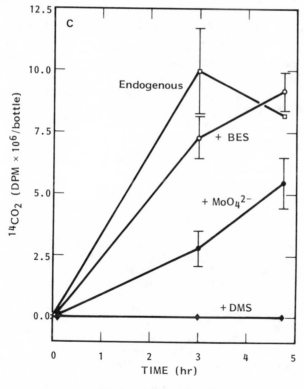

Figure 26. *(Continued)*

outline the types of inhibitors used in biogeochemical and ecological studies and to present enough examples from the literature to give the reader a feel for the "pitfalls" of this approach. The temptation here is that specific inhibitors offer rapid and simple answers to complex questions in microbial ecology. We hope that the reader will be able to take advantage of these benefits, while bearing in mind the limitations of the method. In addition, interesting findings made with inhibitor studies should be followed up by other methods. Thus, although inhibitor results may point to the existence of novel species or metabolic pathways, these findings are never fully accepted by the scientific community until the organisms responsible are isolated. Thus, studies employing inhibitors offer a new dimension in quantitative assessment of the role of certain microbes in nature. However, this approach should supplement rather than eliminate other techniques, including classical isolation of novel species and subsequent physiological investigations.

ACKNOWLEDGMENTS. D. G. C. gratefully acknowledges the support of the National Science Foundation (grants OCE-84-17595, OCE-85-16604, and OCE-85-15886), The Environmental Protection Agency (grant R80-9475-01-0), and the Hudson River Foundation (grants 14-83B-12 and 16-85B-006) for research support during the preparation of this paper.

Invaluable assistance was provided by Robert Ranheim, Judy O'Neil, and Marie Clark, while Jennifer Slater and Jed Fuhrman offered helpful comments. We are grateful to Lacy Daniels for providing unpublished data on the effects of BES. Mention of brand name products does not constitute an endorsement by the U.S. Geological Survey.

References

Abram. J. W., and Nedwell, D. B., 1978a, Inhibition of methanogenesis by sulfate-reducing bacteria competing for transferred hydrogen, *Arch. Microbiol.* **117**:89–92.

Abram, J. W., and Nedwell, D. B., 1978b, Hydrogen as a substrate for methanogenesis and sulfate reduction in anaerobic saltmarsh sediment, *Arch. Microbiol.* **117**:93–97.

Aller, R. C., and Yingst, J. Y., 1980, Relationships between microbial distributions and the anaerobic decomposition of organic matter in surface sediments of Long Island Sound, USA, *Mar. Biol.* **56**:29–42.

Alperin, M. J., and Reeburgh, W. S., 1985, Inhibition experiments on anaerobic methane oxidation, *Appl. Environ. Microbiol.* **50**:940–945.

Anderson, J. H., 1965, Studies on the oxidation of ammonia by *Nitrosomonas, Biochem. J.* **95**:7926–7929.

Anthony, C., 1982, *The Biochemistry of Methylotrophs*, Academic Press, New York.

Arcuri, E. J., and Ehrlich, H. L., 1979, Cytochrome involvement in Mn (II) oxidation by two marine bacteria, *Appl. Environ. Microbiol.* **37**:916–923.

Arp, D. J., and Zumft, W. G., 1983, Methionine-SR-sulfoximine as a probe for the role of glutamine synthetase in nitrogenase switch-off by ammonia and glutamine in *Rhodopseudomonas palustris, Arch. Microbiol.* **134**:17–22.

Azam, F., Fenchel, T., Field, J. G., Gray, J. S., Meyer-Reil, L. A., and Thingstad, F., 1983, The ecological role of water column microbes in the sea, *Mar. Ecol. Prog. Ser.* **10**:257–263.

Babiker, H. M., and Pepper, I. L., 1984, Microbial production of ethylene in desert soils, *Soil Biol. Biochem.* **16**:559–564.

Badour, S. S., 1978, Inhibitors used in studies of algal metabolism, in: *Handbook of Phycological Methods. Physiological and Biochemical Methods* (J. Hellebust and J. Craigie, eds.), pp. 479–488, Cambridge University Press, London.

Balba, M. T., and Nedwell, D. B., 1982, Microbial metabolism of acetate, propionate and butyrate in anoxic sediment from the Colne Point saltmarsh, Essex, U.K., *J. Gen. Microbiol.* **128**:1415–1422.

Balch, W. E., and Wolfe, R. S., 1979a, Specificity and biological distribution of coenzyme M (2-mercaptoethanesulfonic acid), *J. Bacteriol.* **137**:256–263.

Balch, W. E., and Wolfe, R. S., 1979b, Transport of coenzyme M (2-mercaptoethanesulfonic acid) in *Methanobacterium ruminantium, J. Bacteriol.* **137**:264–273.

Balch, W. M., 1987, Studies of nitrate transport of marine phytoplankton using ^{36}Cl-ClO_3 as a transport analog: I. Physiological findings, *J. Phycol.* **23**:107–118.

Balderston, W. L., and Payne, W. J., 1976, Inhibition of methanogenesis in salt marsh sediments and whole-cell suspensions of methanogenic bacteria by nitrogen oxides, *Appl. Environ. Microbiol.* **32**:264–269.

Balderston, W. L., Sherr, B., and Payne, W. J., 1976, Blockage by acetylene of nitrous oxide reduction in *Pseudomonas perfectomarinus, Appl. Environ. Microbiol.* **31**:504–508.

Banat, I. M., and Nedwell, D. B., 1983, Mechanisms of turnover of C_2–C_4 fatty acids in high-sulfate and low-sulfate anaerobic sediments, *FEMS Microbiol. Lett.* **17**:107–110.

Banat, I. M., and Nedwell, D. B., 1984, Inhibition of sulfate reduction in anoxic marine sediments by Group VI anions, *Est. Coast. Shelf Sci.* **18**:361–366.

Banat, I. M., Lindstrom, E. B., Nedwell, D. B., and Balba, M. T., 1981, Evidence for coexistence of two distinct functional groups of sulfate-reducing bacteria in saltmarsh sediment, *Appl. Environ. Microbiol.* **42**:985–992.

Banat, I. M., Nedwell, D. B., and Balba, M. T., 1983, Stimulation of methanogenesis by slurries of saltmarsh sediment after the addition of molybdate to inhibit sulfate-reducing bacteria, *J. Gen. Microbiol.* **129**:123–129.

Barker, H. A., 1956. *Bacterial Fermentations,* Wiley, New York.

Barnes, R. O., and Goldberg, E. D., 1976, Methane production and consumption in anoxic marine sediments, *Geology* **4**:297–300.

Batterson, J., Winters, K., and Van Baalen, C., 1978, Anilines: Selective toxicity to blue-green algae, *Science* **199**:1068–1070.

Bauchop, T., 1967, Inhibition of rumen methanogenesis by methane analogues, *J. Bacteriol.* **94**:171–175.

Belay, N., Sparling, R., and Daniels, L., 1984, Dinitrogen fixation by a thermophilic methanogenic bacterium, *Nature* **312**:286–288.

Belser, L. W., and Mays, E. L., 1980, Specific inhibition of nitrite oxidation by chlorate and its use in assessing nitrification in soils and sediments, *Appl. Environ. Microbiol.* **39**:505–510.

Belser, L. W., and Schmidt, E. L., 1981, Inhibitory effect of nitrapyrin on three genera of ammonia-oxidizing nitrifiers, *Appl. Environ. Microbiol.* **41**:819–821.

Benner, R., Moran, M. A., and Hodson, R. E., 1986, Biogeochemical cycling of ligno-cellulose carbon in marine and freshwater ecosystems: Relative contribution of procaryotes and encaryotes. *Limnol. Oceanogr.* **31**:89–100.

Bennett, E. O., and Bauerle, R. H., 1960, The sensitivities of mixed populations of bacteria to inhibitors, *Aust. J. Biol. Sci.* **13**:142–149.

Berg, P., Klemedtsson, L., and Rosswall, T., 1982, Inhibitory effect of low partial pressures of acetylene on nitrification, *Soil Biol. Biochem.* **14**:301–303.

Billen, G., 1976, Evaluation of nitrifying activity in sediments by dark C-14 bicarbonate incorporation, *Water Res.* **10**:51–57.

Blackmer, A. M., Bremner, J. M., and Schmidt, E. L., 1980, Production of nitrous oxide by ammonia-oxidizing chemoautotrophic microorganisms in soil, *Appl. Environ. Microbiol.* **40**:1060–1066.

Bomar, M., Knoll, K., and Widdel, F., 1985, Fixation of molecular nitrogen by *Methanosarcina barkeri, FEMS Microb. Ecol.* **31**:47–55.

Bothe, H., 1982, Hydrogen production by algae, *Experientia* **38**:59–66.

Bothe, H., Tenniget, J. and Eisbrenner, G., 1977a, The utilization of molecular hydrogen by the blue-green alga *Anabaena cylindrica, Arch. Microbiol.* **114**:43–49.

Bothe, H., Tenniget, J., and Eisbrenner, G., 1977b, The hydrogenase–nitrogenase relationship in the blue-green alga *Anabaena cylindrica, Planta* **33**:237–242.

Boussiba, S., and Gibson, J., 1985, The role of glutamine synthetase activity in ammonium and methylammonium transport in *Anacystis nidulans* R-2, *FEBS Lett.* **180**:13–16.

Bouwer, E. J., and McCarthy, P. L., 1983a, Transformation of 1- and 2-carbon halogenated aliphatic organic compounds under methanogenic conditions, *Appl. Environ. Microbiol.* **45**:1295–1299.

Bouwer, E. J., and McCarthy, P. L., 1983b, Effects of 2-bromethanesulfonic acid and 2-chloroethanesulfonic acid on acetate utilization in a continuous-flow methanogenic fixed-film column, *Appl. Environ. Microbiol.* **45**:1408–1410.

Bremner J. M., and Blackmer, A. M., 1978, Nitrous oxide: Emission from soils during nitrification of fertilizer nitrogen, *Science* **199**:295–296.

Bremner, J. M., and Blackmer, A. M., 1979, Effects of acetylene and soil water content on emission of nitrous oxide from soils, *Nature* **280**:380–381.

Bremner, J. M., and Blackmer, A. M., 1981, Terrestrial nitrification as a source of atmospheric nitrous oxide, in: *Denitrification, Nitrification, and Atmospheric Nitrous Oxide* (C. C. Delwiche, ed.), pp. 151–170, Wiley, New York.

Brenchley, J. E., 1973, Effect of methionine sulfoximine and methionine on glutamate synthesis in *Klebsiella aerogenes, J. Bacteriol.* **114**:666–673.

Brierley, C. L., and Brierley, J. A., 1982, Anaerobic reduction of molybdenum by *Sulfolobus* species, *Zentralbl. Bakteriol. Hyg. I Abt. Orig.* **3**:289–294.

Brock, T. D., 1961, Chloramphenicol, *Bacteriol. Rev.* **25**:32–48.

Brock, T. D., 1978, The poisoned control in biogeochemical investigations, in: *Environmental Biogeochemistry and Geomicrobiology*, Vol. 3 (W. Krumbein, ed.), pp. 717–725, Ann Arbor Science, Ann Arbor, Michigan.

Brodbeck, U., 1980, *Enzyme Inhibitors*, Verlag Chemie, Weinheim.

Brouzes, R., and Knowles, R., 1971, Inhibition of growth of *Clostridium pasteurianum* by acetylene: Implications for nitrogen fixation assay, *Can. J. Microbiol.* **17**:1483–1489.

Burdige, D. J., and Kepkay, P. E., 1983, Determination of bacterial manganese oxidation rates in sediments using an *in situ* dialysis technique. I. Laboratory studies, *Geochim. Cosmochim. Acta* **47**:1907–1916.

Burdige, D. J., and Nealson, K. H., 1985, Microbial manganese reduction by enrichment cultures from coastal marine sediments, *Appl. Environ. Microbiol.* **50**:491–497.

Burris, R. H., 1974, Methodology, in: *The Biology of N₂ Fixation* (A. Quispel, ed.), pp. 9–33, North-Holland, Amsterdam.

Campbell, A. M., del Campillo-Campbell, A., and Villaret, D. B., 1985, Molybdate reduction by *Escherichia coli* K-12 and its chl mutants, *Proc. Natl. Acad. Sci. USA* **82**:227–231.

Campbell, L., and Carpenter, E. J., 1986, Estimating the grazing pressure of heterotrophic nanoplankton on *Synechococcus* spp. using the sea water dilution and selective inhibitor techniques, *Mar. Ecol. Prog. Ser.* **33**:121–129.

Campbell, N. E. R., and Aleem, M. I. H., 1965a, The effect of 2-chloro-6-(trichloromethyl) pyridine on the chemoautotrophic metabolism of nitrifying bacteria. I. Ammonia and hydroxylamine oxidation by *Nitrosomonas, Antonie Leeuwenhoek J. Microbiol. Serol.* **31**:124–136.

Campbell, N. E. R., and Aleem, M. I. H., 1965b, The effect of 2-chloro-6-(trichloromethyl) pyridine on the chemoautotrophic metabolism of nitrifying bacteria. II. Nitrite oxidation by *Nitrobacter, Antonie Leeuwenhoek J. Microbiol. Serol.* **31**:137–144.

Capone, D. G., 1982, Nitrogen fixation (acetylene reduction) by rhizosphere sediments of the eelgrass, *Zostera marina.* L., *Mar. Ecol. Prog. Ser.* **10**:67–75.

Capone, D. G., 1983, Benthic nitrogen fixation, in: *Nitrogen in the Marine Environment* (E. J. Carpenter and D. G. Capone, eds.), pp. 105–137, Academic Press, New York.

Capone, D. G., 1984, Factors controlling nitrogen fixation in marine sediments, in: *Abstracts, 47th Annual Meeting of American Society of Limnol. Oceanogr.,* p. 14.

Capone, D. G., 1987, Benthic nitrogen fixation; Microbiology, physiology and ecology, in: *Nitrogen Cycling in Marine, Coastal Environments* (T. H. Blackburn, J. Sørenson, and T. Roswall, eds.), Wiley, New York, in press.

Capone, D. G., and Carpenter, E. J., 1982a, A perfusion method for assaying microbial activities in estuarine sediments: Applicability to studies of N_2 fixation by C_2H_2 reduction, *Appl. Environ. Microbiol.* **43**:1400–1405.

Capone, D. G., and Carpenter, E. J., 1982b, Nitrogen fixation in the marine environment. *Science* **217**:1140–1142.

Capone, D. G., and Kiene, R. P., 1987, Comparison of microbial dynamics in marine and freshwater sediments: Contrasts in anaerobic carbon catabolism, in: The Comparative Ecology of Freshwater and Marine Ecosystems (S. Nixon, ed.), *Limnol. Oceanogr.* (Special Volume), in press.

Capone, D. G., Oremland, R. S., and Taylor, B. F., 1977, Significance of N_2 fixation to the production of *Thalassia testudinum* communities, in: *Proceedings of the CICAR* [Cooperative Investigations of the Caribbean and Adjacent Regions] *II Symp. Prog. Mar. Res. Caribbean & Adjacent Regions* (H. B. Stewart, Jr., ed.), pp. 71–85, FAO Fisheries Report 200, Rome, Italy.

Capone, D. G., Reese, D. D., and Kiene, R. P., 1983, Effects of metals on methanogenesis, sulfate reduction, carbon dioxide evolution, and microbial biomass in anoxic salt marsh sediments, *Appl. Environ. Microbiol.* **45**:1586–1591.

Cappenberg, T. E., 1974, Interrelations between sulfate-reducing and methane-producing bacteria in bottom deposits of a freshwater lake. II. Inhibition experiments, *Antonie Leeuwenhoek J. Microbiol. Serol.* **40**:297–306.

Carpenter, E. J., 1983a, Nitrogen fixation by marine *Oscillatoria (Trichodesmium)* in the world's oceans, in: *Nitrogen in the Marine Environment* (E. J. Carpenter and D. G. Capone, eds.), pp. 65–104, Academic Press, New York.

Carpenter, E. J., 1983b, Physiology and ecology of marine planktonic *Oscillatoria (Trichodesmium), Mar. Biol. Lett.* **4**:69–85.

Chan, Y., and Knowles, R., 1979, Measurements of denitrification in two freshwater sediments by an *in situ* acetylene inhibition method, *Appl. Environ. Microbiol.* **37**:1067–1072.

Chan, Y. K., Nelson, L. M., and Knowles, R., 1980, Hydrogen metabolism of *Azospirillum brasilense* in nitrogen-free medium, *Can. J. Microbiol.* **26**:1126–1131.

Chapnick, S. D., Moore, W. S., and Nealson, K. H., 1982, Microbially mediated manganese oxidation in a freshwater lake, *Limnol. Oceanogr.* **27**:1004–1014.

Chemerys, R. A., 1983, Nitrogen fixation and hydrogen cycling in salt marsh sediment, M. S. Thesis, State University of New York at Stony Brook.

Chen, M., and Wolin, M. J., 1979, Effect of monensin and lasalocid-sodium on the growth of methanogenic and rumen saccharolytic bacteria, *Appl. Environ. Microbiol.* **38**:72–77.

Ching, W. M., Wittwer, A. J., Tsai, L., and Stadtman, T. C., 1984, Distribution of two selenonucleosides among the selenium-containing tRNAs from *Methanococcus vanielii, Proc. Natl. Acad. Sci. USA* **81**:57–60.

Christensen, D., 1984, Determination of substrates oxidized by sulfate-reduction in intact cores of marine sediments, *Limnol. Oceanogr.* **29**:189–192.

Cloern, J. E., Cole, B. E., and Oremland, R. S., 1983, Autotrophic processes in Big Soda Lake, Nevada, *Limnol. Oceanogr.* **28**:1049–1061.

Compeau, G. C., and Bartha, R., 1985, Sulfate-reducing bacteria: Principal methylators of mercury in anoxic estuarine sediment, *Appl. Environ. Microbiol.* **50**:498–502.

Compeau, G. C., and Bartha, R., 1987, Effect of salinity on mercury-methylating activity of sulfate-reducing bacteria in estuarine sediments, *Appl. Environ. Microbiol.* **53**:261–265.

Conrad, R., and Seiler, W., 1980, Role of microorganisms in the consumption and production of atmospheric carbon monoxide by soil, *Appl. Environ. Microbiol.* **40**:437–445.

Cornforth, I. S., 1975, The persistence of ethylene in aerobic soils, *Plant Soil.* **42**:85-96.

Coty, V. F., 1967, Atmospheric nitrogen fixation by hydrocarbon-utilizing bacteria, *Biotech. Bioeng.* **9**:25–32.

Cox, R. B., and Zatman, L. J., 1976, The effect of fluoroacetate on the growth of the facultative methylotrophs bacterium 5H2 and *Pseudomonas* AM1 and bacterium 5B2, *J. Gen. Microbiol.* **93**:397–400.

Cresswell, R. C., and Syrett, P. J., 1984, Effects of methylammonium and of *l*-methionine-*dl*-sulfoximine on the growth and nitrogen metabolism of *Phaeodactylum tricornutum, Arch. Microbiol.* **139**:67–71.

Culbertson, C. W., 1983, Anaerobic oxidation of acetylene by estuarine sediments and enrichment cultures. M. S. Thesis, San Francisco State University.

Culbertson, C. W., and Oremland, R. S., 1983, Anaerobic growth of the enrichment culture on acetylene gas, in: *Abstracts Third International Symposium on Microbial Ecology,* p. A4.

Culbertson, C. W., Zehnder, A. J. B., and Oremland, R. S., 1981, Anaerobic oxidation of acetylene by estuarine sediments and enrichment cultures, *Appl. Environ. Microbiol.* **41**:396–403.

Daday, A., Platz, R. A., and Smith, G. D., 1977, Anaerobic and aerobic hydrogen gas formation by the blue-green alga *Anabaena cylindrica, Appl. Environ. Microbiol.* **34**:478–483.

Dalton, H., 1977, Ammonia oxidation by the methane oxidizing bacterium *Methylococcus capsulatus* strain Bath, *Arch. Microbiol.* **114**:273–279.

Dalton, H., and Whittenbury, R., 1976, The acetylene reduction technique as an assay for nitrogenase activity in the methane oxidizing bacterium *Methylococcus capsulatus* strain Bath, *Arch. Microbiol.* **109**:147–151.

Davis, J. B., and Yarbrough, H. F., 1966, Anaerobic oxidation of hydrocarbons by *Desulfovibrio desulfuricans, Chem. Geol.* **1**:137–144.

Dawson, R. M. C., Elliott, D. C., Elliott, W. H., and Jones, K. M. (eds.) 1969, *Data for Biochemical Research,* 2nd ed., Oxford University Press, New York.

De Bont, J. A. M., 1976a, Oxidation of ethylene by soil bacteria, *Antonie Leeuwenhoek J. Microbiol. Serol.* **42**:59–71.

De Bont, J. A. M., 1976b, Bacterial degradation of ethylene and the acetylene reduction test, *Can. J. Microbiol.* **22**:1060–1062.

De Bont, J. A. M., and Albers, A. J. M., 1976, Microbial metabolism of ethylene, *Antonie Leeuwenhoek J. Microbiol. Serol.* **42**:73–80.

De Bont, J. A. M., and Harder, W., 1978, Metabolism of ethylene by *Mycobacterium* E 20, *FEMS Microbiol. Lett.* **3**:89–93.

De Bont, J. A. M., and Mulder, E. G., 1974, Nitrogen fixation and cooxidation of ethylene by a methane-utilizing bacterium, *J. Gen. Microbiol.* **83**:113–121.

De Bont, J. A. M., and Mulder, E. G., 1976, Invalidity of the acetylene reduction assay in alkane-utilizing, nitrogen-fixing bacteria, *Appl. Environ. Microbiol.* **31**:640–647.

De Bont, J. A. M., and Peck, M. W., 1980, Metabolism of acetylene by *Rhodococcus* Al, *Arch. Microbiol.* **127**:99–104.

Dellinger, C. A., and Ferry, J. G., 1984, Effect of monensin on growth and methanogenesis of *Methanobacterium formicicum, Appl. Environ. Microbiol.* **48**:680–682.

Delwiche, C. C., 1981, The nitrogen cycle and nitrous oxide, in: *Denitrification, Nitrification and Atmospheric Nitrous Oxide* (C. C. Delwiche, ed.), pp. 1–15, Wiley, New York.

Devol, A. H., 1983, Methane oxidation rates in the anaerobic sediments of Saanich Inlet, *Limnol. Oceanogr.* **28**:738–742.

Devol, A. H., Anderson, J. J., Kuivila, K., and Murray, J. W., 1984, A model for coupled sulfate reduction and methane oxidation in the sediments of Saanich Inlet, *Geochim. Cosmochim. Acta* **48**:993–1004.

Dicker, H. J., and Smith, D. W., 1980, Physiological ecology of acetylene reduction (nitrogen fixation) in a Delaware salt marsh, *Microb. Ecol.* **6**:161–171.

Dicker, H. J., and Smith D. W., 1985a, Effects of organic amendments on sulfate reduction activity, H_2 consumption, and H_2 production in salt marsh sediments, *Microb. Ecol.* **11**:299–315.

Dicker, H. J., and Smith, D. W., 1985b, Metabolism of low molecular weight organic compounds by sulfate-reducing bacteria in a Delaware saltmarsh, *Microb. Ecol.* **11**:317–335.

Dilworth, M. J., 1966, Acetylene-reduction by nitrogen-fixing preparations from *Clostridium pasteurianum, Biochem. Biophys. Acta* **127**:285–294.

Dodds, K. L., and Collins-Thompson, D. L., 1985, Production of N_2O and CO_2 during the reduction of NO_2^- by *Lactobacillus lactis* TS4, *Appl. Environ. Microbiol.* **50**:1550–1552.

Duguay, L. E., and Taylor, D. L., 1978, Primary production and calcification by the soritid foraminifer *Archais angulatus* (Fichtel & Moll), *J. Protozool.* **25**:356.

Duran, A., Cabib, E., and Bowers, B., 1979, Chitin synthetase distribution on the yeast plasma membrane, *Science* **203**:363–365.

Ehrlich, H. L., 1966, Reactions of manganese by bacteria from marine ferromanganese nodules, *Dev. Ind. Microbiol.* **13**:57–65.

Ehrlich, H. L., 1968, Bacteriology of manganese nodules. II. Manganese oxidation by cell-free extracts from a manganese nodule bacterium, *Appl. Microbiol.* **16**:196–202.

Ehrlich, H. L., 1978, Inorganic energy sources for chemolithotrophs and mixotrophic bacteria, *Geomicrobiol. J.* **1**:65–83.

Ehrlich, H. L., 1981, *Geomicrobiology,* Dekker, New York.

Eisbrenner, G., and Bothe, H., 1979, Modes of electron transfer for molecular hydrogen in *Anabaena cylindrica, Arch. Microbiol.* **123**:37–45.

Elleway, R. F., Sabine, J. R., and Nicholas, D. J. D., 1971, Acetylene reduction by rumen microflora, *Arch. Microbiol.* **76**:277–291.

Elrifi, I., and Turpin, D. F., 1986, Nitrate and ammonium induced photosynthetic supression in N-limited *Selenastrum minutum, Plant Physio.* **81**:273–279.

Emerson, S., Kalhorn, S., Jacobs, L., Tebo, B. M., Nealson, K. H., and Rosson, R. A., 1982, Environmental oxidation rate of manganese (II): Bacterial catalysis, *Geochim. Cosmochim. Acta* **46**:1073–1079.

Enoch, H. G., and Lester, R. L., 1972, Effects of molybdate, tungstate, and selenium compounds on formate dehydrogenase and other enzymes in *Escherichia coli, J. Bacteriol.* **110**:1032–1040.

Ereckinska, M., and Wilson, D. F. (eds.), 1984, *Inhibition of Mitochondrial Function,* Pergamon Press, New York.

Evans, H. J., and Barber, L. E., 1977, Biological nitrogen fixation for food and fiber production, *Science* **197**:332–339.

Evans, D. G., Beauchamp, E., and Trevors, J. T., 1985, Sulfide alleviation of the acetylene inhibition of nitrous oxide reduction in soil, *Appl. Environ. Microbiol.* **49**:217–220.

Falkowski, P., 1983, Enzymology of nitrogen assimilation, in *Nitrogen in the Marine Environment* (E. J. Carpenter and D. G. Capone), pp. 809–838, Academic Press, New York.

Fedorova, R. I., Milekhina, E. I., and L'yukhina, I., 1973, Evaluation of the method of "gas metabolism" for detecting extraterrestrial life. Identification of nitrogen-fixing microorganisms, *Izv. Akad. Nauk SSSR Ser. Biol.* **6**:797–806.

Fenchel, T. M., and Riedl, R. J., 1970, The sulfide system: A new biotic community underneath the oxidized layer of marine sand bottoms, *Mar. Biol.* **7**:255–268.

Ferenci, T., 1974, Carbon monoxide-stimulated respiration in methane-utilizing bacteria, *FEBS Lett.* **11**:94–97.

Fischer, C. R., Childress, J. J., Oremland, R. S., and Bidigare, R. R., 1987, The importance of methane and thiosulfate in the metabolism of the bacterial symbionts of two deep sea mussels, *Marine Biol.* **96**:59–72.

Flores, E., Guerro, M. G., and Losada, M., 1980, Short-term ammonium inhibition of nitrate utilization by *Anacystis nidulans* and other cyanobacteria, *Arch. Microbiol.* **128**:137–144.

Franklin, T. J., and Snow, G. A., 1981, *Biochemistry of Antimicrobial Action,* 3rd ed., Chapman and Hall, New York.

Fuhrman, J. A., and Azam, F., 1980, Bacterioplankton secondary production estimates for coastal waters of British Columbia, Antarctica, and California, *Appl. Environ. Micobiol.* **39**:1085–1095.

Fuhrman, J. A., and Azam, F., 1982, Thymidine incorporation as a measure of heterotrophic bacterioplankton production in marine surface waters: Evaluation of field results, *Mar. Biol.* **66**:109–120.

Fuhrman, J. A., and McManus, G. B., 1984, Do bacteria-sized marine eukaryotes consume significant bacterial production?, *Science* **224**:1257–1259.

Fuhrman, J. A., Ducklow, H. W., Kirchman, D. L., Hudak, J., McManus, G. B., and Kramer, J., 1986, Does adenine incorporation into nucleic acids measure total microbial production? *Limnol. Oceanogr.* **31**:627–636.

Furusaka, C., 1961, Sulfate transport and metabolism by *Desulfovibrio desulfuricans, Nature* **192**:427–429.

Gadd, G. M., and Griffiths, A. J., 1978, Microorganisms and heavy metal toxicity, *Microb. Ecol.* **4**:303–317.

Ghiorse, W. C., and Ehrlich, H. L., 1976, Electron transport components of the MnO_2 reductase system and the location of terminal reductase in a marine *Bacillus, Appl. Environ. Microbiol.* **31**:977–985.

Glover, H. E., 1982, Methylamine, an inhibitor of ammonium oxidation and chemoautotrophic growth in the marine, nitrifying bacterium *Nitrosococcus oceanus, Arch. Microbiol.* **132**:37–40.

Glover, H. E., and Morris, I., 1979, Photosynthetic carboxylating enzymes in marine phytoplankton, *Limnol. Oceanogr.* **24**:510–519.

Goodman, B. A., and Cheshire, M. V., 1982, Reduction of molybdate by soil organic matter: EPR evidence for formation of both Mo(V) and Mo(III), *Nature* **299**:618–620.

Gordon, J. K., and Brill, W. J., 1974, Derepression of nitrogenase synthesis in the presence of excess NH_4^+, *Biochem. Biophys. Res. Commun.* **59**:967–971.

Goring, C. A., 1962, Control of nitrification by 2-chloro-6 (trichloromethyl) pyridine, *Soil, Sci.* **93**:211–218.

Gottleib, D., and Shaw, P. D., (eds.), 1967, *Antibiotics,* Vol. 1: *Mechanism of Action,* Springer-Verlag, New York.

Graham, B. M., Hamilton, R. D., and Campbell, N. E. R., 1980, Comparison of the nitrogen-15 uptake and acetylene reduction methods for estimating the rates of nitrogen fixation by freshwater blue-green algae, *Can. J. Fish. Aquat. Sci.* **37**:488–493.

Grbic-Galic, D., and Young, L. Y., 1985, Methane fermentation of ferulate and benzoate: Anaerobic degradation pathways, *Appl. Environ. Microbiol.* **50**:292–297.

Green, J., Prior, S. D., and Dalton, H., 1985, Copper ions as inhibitors of protein C of soluble methane monooxygenase of *Methylococcus capsulatus* (Bath), *Eur. J. Biochem.* **153**:137–144.

Gunsalus, R. P., and Wolfe, R. S., 1978, ATP activation and properties of the methyl coenzyme M reductase system in *Methanobacterium thermoautotrophicum, J. Bacteriol.* **135**:851–857.

Gunsalus, R. P., and Wolfe, R. S., 1980, Methyl coenzyme M reductase from *Methanobacterium thermoautotrophicum.* Resolution and properties of the components, *J. Biol. Chem.* **255**:1891–1895.

Gunsalus, R. P., Roemesser, J. A., and Wolfe, R. S., 1978, Preparation of coenzyme M analogues and their activity in the methyl coenzyme M reductase system of *Methanobacterium thermoautotrophicum, Biochemistry* **17**:2374–2377.

Habets-Crutzen, A. Q. H., and de Bont, J. A. M., 1985, Inactivation of alkene oxidation by epoxides in alkene- and alkane-grown bacteria, *Appl. Microbiol. Biotechnol.* **22**:428–433.

Habte, M., and Alexander, M., 1980, Nitrogen fixation by photosynthetic bacteria in lowland rice culture, *Appl. Environ. Microbiol.* **39**:342–347.

Hahn, F. E. (ed.), 1983, *Antibiotics,* Vol. 6: *Modes and Mechanisms of Microbial Growth Inhibitors,* Springer-Verlag, Berlin.

Hall, G. H., 1982, Apparent and measured rates of nitrification in the hypolimnion of a mesotrophic lake, *Appl. Environ. Microbiol.* **43**:542–547.

Hall, G. H., 1984, Measurement of nitrification rates in lake sediments: Comparison of the nitrification inhibitors nitrapyrin and allylthiourea, *Microb. Ecol.* **10**:25–36.

Hardy, R. W. F., and Havelka, U. D., 1975, Nitrogen fixation research: A key to world food?, *Science.* **188**:633–642.

Hardy, R. W. F., Holsten, R. D., Jackson, E. K., and Burns, R. C., 1968, The acetylene-ethylene assay for N₂ fixation: Laboratory and field evaluation, *Plant Physiol.* **43**:1185–1207.

Harrison, W. G., 1983, Use of isotopes, in: *Nitrogen in the Marine Environment* (E. J. Carpenter and D. G. Capone, eds.), pp. 763–808, Academic Press, New York.

Harrits, S. M., and Hanson, R. S., 1980, Stratification of aerobic methane-oxidizing organisms in Lake Mendota, Madison, Wisconsin, *Limnol. Oceanogr.* **25**:412–421.

Healy, J. B., Young, L. Y., and Reinhard, M., 1980, Methanogenic decomposition of ferulic acid, a model lignin derivative, *App. Environ. Microbiol.* **39**:436–444.

Henninger, N. M., and Bollag, J. M., 1976, Effect of chemicals used as nitrification inhibitors on the denitrification process, *Can. J. Microbiol.* **22**:688.

Hendrickson, L. I., and Keeney, D. R., 1979a, Effect of some physical and chemical factors on the rate of hydrolysis of nitrapyrin (N-serve), *Soil Biol. Biochem.* **11**:47–50.

Hendrickson, L. I., and Keeney, D. R., 1979b, A bioassay to determine the effect of organic matter and pH on the effectiveness of nitrapyrin (N-serve) as a nitrification inhibitor, *Soil Biol. Biochem.* **11**:51–55.

Henriksen, K., 1980, Measurement of *in situ* rates of nitrification in sediment, *Microb. Ecol.* **6**:329–337.

Higgins, I. J., and Quayle, J. R., 1970, Oxygenation of methane by methane grown *Pseudomonas methanica* and *Methylomonas methano-oxidans, Biochem. J.* **118**:201–208.

Hill, I. R., and Wright, S. J. L. (eds.), 1978, *Pesticide Microbiology*, Academic Press, New York.

Hillmer, P., and Fahlbusch K., 1979, Evidence for an involvement of glutamine synthetase in regulation of nitrogenase activity in *Rhodopseudomonas capsulata, Arch Microbiol.* **122**:213–218.

Hilpert, R., Winter, J., Hammes, W., and Kandler, O., 1981, The sensitivity of archaebacteria to antibiotics, *Zentralbl. Bakteriol. Hyg. Abt. Orig.* **2**:11–20.

Hines, M. E., and Lyons, W. B., 1982, Biogeochemistry of nearshore Bermuda sediments. I. Sulfate reduction rates and nutrient generation, *Mar. Ecol. Prog. Ser.* **8**:87–94.

Hobbie, J., and Williams, P. J. LeB. (eds.), 1984, *Heterotrophic Activity in the Sea*, Plenum Press, New York.

Hochachka, P. W., and Somero, G. N., 1984, *Biochemical Adaptation*, Princeton University Press, Princeton, New Jersey.

Hordijk, K. A., Hagenaars, C. P. M. M., and Cappenberg, T. E., 1985, Kinetic studies of bacterial sulfate reduction in freshwater sediments by high-pressure liquid chromatography and microdistillation, *Appl. Environ. Microbiol.* **49**:434–440.

Horrigan, S. G., and Capone, D. G., 1985, Rates of nitrification and nitrate reduction in nearshore marine sediments at near ambient substrate concentrations, *Mar. Chem.* **16**:317–327.

Hou, C. T., Patel, R., Laskin, A. I., and Barnabe, N., 1979, Microbial oxidation of gaseous hydrocarbons: Epoxidation of C_2 to C_4 *n*-alkenes by methylotrophic bacteria, *Appl. Environ. Microbiol.* **38**:127–134.

Houchins, J. P., and Burris, R. H., 1981, Physiological reactions of the reversible hydrogenase from *Anabaena* 7120, *Plant Physiol.* **68**:717–721.

Howarth, R. W., and Cole, J. J., 1985, Molybdenum availability, nitrogen limitation, and phytoplankton growth in natural waters, *Science* **229**:653–655.

Huber, D. M., Murray, G. A., and Crane, J. M., 1969, Inhibition of nitrification as a deterrent to nitrogen loss, *Soil Sci. Soc. Am. J.* **33**:975–976.

Huber, D. M., Warren, H. L., Nelson, D. W., and Tsai, C. Y., 1977, Nitrification inhibitors—New tools for food production, *BioScience* **27**:523–529.

Hubley, J. H., Thomson, A. W., and Wilkinson, J. F., 1975, Specific inhibitors of methane oxidation in *Methylosinus trichosporium, Arch. Microbiol.* **102**:199–202.

Hyman, M. R., and Arp, D., 1987, Quantification and removal of some contaminating gases from acetylene used to study gas-utilizing enzymes and microorganisms, *Appl. Environ. Microbiol.* **53**:298–303.

Hyman, M. R., and Wood, P. M., 1985, Suicidal inactivation and labelling of ammonia mono-oxygenase by acetylene, *Biochem. J.* **227**:719–725.

Hynes, R. K., and Knowles, R., 1978, Inhibition by acetylene of ammonia oxidation in *Nitrosomonas europaea, FEMS Microbiol. Lett.* **4**:319–321.

Hynes, R. K., and Knowles, R., 1982, Effect of acetylene on autotrophic and heterotrophic nitrification, *Can. J. Microbiol.* **28**:334–340.

Hynes, R. K., and Knowles, R., 1983, Inhibition of chemoautotrophic nitrification by sodium chlorate and sodium chlorite: A reexamination, *Appl. Environ. Microbiol.* **45**:1178–1182.

Hynes, R. K., and Knowles, R., 1984, Production of nitrous oxide by *Nitrosomonas europaea:* Effects of acetylene, pH, and oxygen, *Can. J. Microbiol.* **30**:1397–1404.

Ilag, L., and Curtis, R. W., 1968, Production of ethylene by fungi, *Science* **159**:1357.

Indrebo, G., Pengerud, B., and Dundas, I., 1979, Microbial activities in a permanently stratified estuary. II. Microbial activities at the oxic–anoxic interface, *Mar. Biol.* **51**:305–309.

Iversen, N., and Blackburn, T. H., 1981, Seasonal rates of methane oxidation in anoxic marine sediments, *Appl. Environ. Microbiol.* **41**:1295–1300.

Iversen, N., and Jorgensen, B. B., 1985, Anaerobic methane oxidation rates at the sulfate-methane transition in marine sediments from Kattegat and Skagerrat (Denmark), *Limnol. Oceanogr.* **30**:944–955.

Iversen, N., Oremland, R. S., and Klug, M. J., 1987, Big Soda Lake (Nevada). 3. Pelagic methanogenesis and anaerobic methane-oxidation, *Limnol. Oceanogr.* **32**:804–814.

Izawa, S., 1980, Acceptors and donors for chloroplast electron transport, in: *Methods in Enzymology*, Vol. 69 (A. San Pietro, ed.), pp. 413–434, Academic Press, New York.

Jacobson, M. E., Mackin, J. E., and Capone, D. G., 1987, Ammonium production in sediments inhibited with molybdate: Implications for the sources of ammonium in anoxic marine sediments, *Appl. Environ. Microbiol.*, **53**:2435–2439.

Jain, B., 1982, *Handbook of Enzyme Inhibitors*, Wiley, New York.

Jannasch, H. W., and Wirsen, C. O., 1979, Chemosynthetic primary production at east Pacific sea floor spreading centers, *BioScience* **29**:592–598.

Jarrell, K. F., and Hamilton, E. A., 1985, Effect of gramicidin on methanogenesis by various methanogenic bacteria, *Appl. Environ. Microbiol.* **50**:179–182.

Jarrell, K. F., and Sprott, G. D., 1983, The effect of ionophores and metabolic inhibitors on methanogenesis and energy-related properties of *Methanobacterium bryantii*, *Arch. Biochem. Biophys.* **225**:33–41.

Jenkins, M. C., and Kemp, W. M., 1984, The coupling of nitrification and denitrification in two estuarine sediments, *Limnol. Oceanogr.* **29**:598–608.

Jensen, S., and Jurnelov, A., 1969, Biological methylation of mercury in aquatic environments, *Nature* **223**:753–754.

Johnson, P. W., and Sieburth, J. McN., 1982, *In-situ* morphology and occurrence of eucaryotic phototrophs of bacterial size in the picoplankton of estuarine and oceanic waters, *J. Phycol.* **18**:318–327.

Jones, J. B., and Stadtman, T. A., 1977, *Methanococcus vannielii:* Culture and effects of selenium and tungsten on growth, *J. Bacteriol.* **130**:1404–1406.

Jones, J. B., and Stadtman, T. C., 1981, Selenium-dependent and selenium-independent formate dehydrogenases of *Methanococcus vannielii*. Separation of the two forms and characterization of the purified selenium-independent form, *J. Biol. Chem.* **256**:656–663.

Jones, J. G., and Simon, B. M., 1985, Interactions of acetogens and methanogens in anaerobic freshwater sediments, *Appl. Environ. Microbiol.* **49**:944–948.

Jones, J. G., Simon, B. M., and Gardener, S., 1982, Factors affecting methanogenesis and associated anaerobic processes in the sediments of a stratified eutrophic lake, *J. Gen. Microbiol.* **128**:1–11.

Jones, R. D., and Morita, R. Y., 1983a, Carbon monoxide oxidation by chemolithotrophic ammonium oxidizers, *Can. J. Microbiol.* **29**:1545–1551.

Jones, R. D., and Morita, R. Y., 1983b, Methane oxidation by *Nitrosococcus oceanus* and *Nitrosomonas europea*, *Appl. Environ. Microbiol.* **45**:401–410.

Jones, R. D., and Morita, R. Y., 1984, Effect of several nitrification inhibitors on carbon monoxide and methane oxidation by ammonium oxidizers, *Can. J. Microbiol.* **30**:1276–1279.

Jones, R. D., Morita, R. Y., and Griffiths, R. P., 1984, Methods for estimating *in situ* chemolithotrophic ammonium oxidation using carbon monoxide oxidation, *Mar. Ecol. Prog. Ser.* **17**:259–269.

Jones, T. W., and Estes, P. S., 1984, Uptake and phytotoxicity of soil-sorbed atrazine for the submerged aquatic plant, *Potamogeton perfoliatus* L., *Arch. Environ. Contam. Toxicol.* **13**:237–241.

Jørgensen, B. B., 1982, Ecology of the bacteria of the sulfur cycle with special reference to anoxic–oxic interface environments. *Phil. Trans. R. Soc. Lond. B* **298**:543–561.

Kandler, O., and Hippe, H., 1977, Lack of peptidoglycan in the cell walls of *Methanosarcina barkeri, Arch. Microbiol.* **113**:57–60.

Kandler, O., and Konig, H., 1978, Chemical composition of the peptidoglycan-free cell walls of methanogenic bacteria, *Arch. Microbiol.* **118**:141–152.

Kanner, D., and Bartha, R., 1979, Growth of *Nocardia rhodochrous* on acetylene gas, *J. Bacteriol.* **139**:225–230.

Kanner, D., and Bartha, R., 1982, Metabolism of acetylene by *Nocardia rhodochrous, J. Bacteriol.* **150**:989–992.

Kaplan, W. A., 1983, Nitrification, in: *Nitrogen in the Marine Environment* (E. J. Carpenter and D. G. Capone, eds.), pp. 139–190, Academic Press, New York.

Kaspar, H. F., 1982, Denitrification in marine sediments: Measurement of capacity and estimate of *in situ* rate, *Appl. Environ. Microbiol.* **43**:522–527.

Kaspar, H. F., and Tiedje, J. M., 1981, Denitrification and dissimilatory reduction of nitrate and nitrite in the bovine rumen: Nitrous oxide production and effect of acetylene, *Appl. Environ. Microbiol.* **41**:705–709.

Katunuma, N., Umezawa, H., and Holzer, H., 1983, *Proteinase Inhibitors: Medical and Biological Aspects,* Japan Science Society Press, Tokyo.

Kays, S. J., and Pallas, J. E., Jr., 1980, Inhibition of photosynthesis by ethylene, *Nature* **285**:51–52.

Kelly, D. P., 1982, Biochemistry of the chemolithotrophic oxidation of inorganic sulphur, *Phil. Trans. R. Soc. Lond. B* **298**:499–528.

Kemps, C. W., Curtiss, M. A., Robrish, S. A., and Bowen, W. H., 1983, Biogenesis of methane in primate dental plaque, *FEBS Lett.* **155**:61–64.

Kenealy, W., and Zeikus, J. G., 1981, Influence of corrinoid antagonists on methanogen metabolism, *J. Bacteriol.* **146**:133–140.

Kepkay, P. E., Cooke, R. C., and Novitsky, J. A., 1979, Microbial autotrophy: A primary source of organic carbon in marine sediments, *Science* **204**:68–69.

Kiene, R. P., and Capone, D. G., 1985, Degassing of pore water methane during sediment incubations, *Appl. Environ. Microbiol.* **49**:143–147.

Kiene, R. P., and Visscher, P., 1987. Metabolism of the terminal s-methyl group of methionine in anoxic sediments. *Appl. Environ. Microbiol.* **53**:2426–2434.

Kiene, R. P., Oremland, R. S., Catena, A., Miller, L. G., and Capone, D. G., 1986, Metabolism of reduced methylated sulfur compounds in anaerobic sediments and by a pure culture of an estuarine methanogen, *Appl. Environ. Microbiol.* **52**:1037–1045.

King, G. M., 1984, Metabolism of trimethylamine, choline, and glycine betaine by sulfate-reducing and methanogenic bacteria in marine sediments, *Appl. Environ. Microbiol.* **48**:719–725.

King, G. M., Klug, M. J., and Lovely, D. R., 1983, Metabolism of acetate, methanol, and methylated amines in intertidal sediments of Lowes Cove, Maine, *Appl. Environ. Microbiol.* **45**:1848–1853.

Knowles, R., 1979, Denitrification, acetylene reduction and methane metabolism in lake sediment exposed to acetylene, *Appl. Environ. Microbiol.* **38**:486–493.

Knowles, R., 1983, Denitrification, *Microbiol. Rev.* **46**:43–70.

Kosiur, D. R., and Warford, A. L., 1979, Methane production and oxidation in Santa Barbara Basin sediments, *Est. Coast. Mar. Sci.* **8**:379–385.

Kuenen, J. G., and Beudeker, R. F., 1982, Microbiology of thiobacilli and other sulphur-oxidizing autotrophs, mixotrophs and heterotrophs, *Phil. Trans. R. Soc. Lond. B* **298**:473–497.

Kun, E., 1969, Mechanism of action of fluoro analogs of citric acid cycle compounds: An essay on biochemical tissue specificity, in: *Citric Acid Cycle, Control and Compartmentation* (J. M. Lowenstein, ed.), pp. 297–339, Dekker, New York.

Lal, R. (ed.), 1984, *Insecticide Microbiology,* Springer-Verlag, Berlin.

Lambert, G. R., and Smith, G. D., 1981, The hydrogen metabolism of cyanobacteria (bluegreen algae), *Biol. Rev.* **56:**589–660.

Lambert, G. R., Daday, A., and Smith, G. D., 1979, Effects of ammonium ions, oxygen, carbon monoxide, and acetylene on anaerobic and aerobic hydrogen formation by *Anabaena cylindrica* B629, *Appl. Environ. Microbiol.* **38:**521–529.

Lancini, G., and Parenti, F., 1982, *Antibiotics: An Integrated View,* Springer-Verlag, Berlin.

Landry, M. R., and Hassett, R. P., 1982, Estimating the grazing impact of marine microzooplankton, *Mar. Biol.* **67:**282–288.

Legendre, L., Demers, S., Yentsch, C. M., and Yensch, C. S., 1983, The ^{14}C method: Patterns of dark CO_2 fixation and DCMU correction to replace the dark bottle, *Limnol. Oceanogr.* **28:**996–1003.

Leighton, T., Markes, E., and Leighton, F., 1981, Pesticides: Insecticides and fungicides are chitin synthesis inhibitors, *Science* **213:**905–907.

Lethbridge, G., Davison, M. S., and Sparling, G. P., 1982, Critical evaluation of the acetylene reduction test for estimating the activity of nitrogen-fixing bacteria associated with the roots of wheat and barley, *Soil Biol. Biochem.* **14:**27–35.

Li, W. K. W., and Dickie, P. M., 1985a, Metabolic inhibition of size-fractionated marine plankton radiolabeled with amino acids, glucose, bicarbonate, and phosphate in the light and dark, *Microb. Ecol.* **11:**11–24.

Li, W. K. W., and Dickie, P. M., 1985b, Growth of bacteria in seawater filtered through 0.2 μm Nuclepore membranes: Implications for dilution experiments, *Mar. Ecol. Prog. Ser.* **26:**245–252.

Li, W. K. W., Subba Rao, V., Harrison, W. G., Smith, J. C., Cullen, J. J., Irwin, B., and Platt, T., 1983, Autotrophic picoplankton in the tropical ocean, *Science* **219:**292–295.

Lidstrom, M. E., 1983, Methane consumption in Framvaren Fjord, *Limnol. Oceanogr.* **28:**1247–1251.

Lipschultz, F., 1981, Methane release from a brackish intertidal salt-marsh embayment of Chesapeake Bay, Maryland, *Estuaries* **4:**143–145.

Lovley, D. R., and Klug, M. J., 1982, Intermediary metabolism of organic matter in the sediments of a eutrophic lake, *Appl. Environ. Microbiol.* **43:**522–560.

Lovley, D. R., and Klug, M. J., 1983, Sulfate reducers can outcompete methanogens at freshwater sulfate concentrations, *Appl. Environ. Microbiol.* **45:**187–192.

Lovley, D. R., Dwyer, D. F., and Klug, M. J., 1982, Kinetic analysis of competition between sulfate reducers and methanogens for hydrogen in sediments, *Appl. Environ. Microbiol.* **43:**1373–1379.

Macalaster, E. G., Barker, D. A., and Kasper, M. W. (eds.), 1983, *Chesapeake Bay: A Profile of Environmental Change,* U.S. Environmental Protection Agency.

Martens, C. S., and Berner, R. A., 1977, Interstitial water chemistry of anoxic Long Island Sound sediments. I. Dissolved gases, *Limnol. Oceanogr.* **22:**10–25.

Martikainen, P. J., 1985, Nitrous oxide emission associated with autotrophic ammonium oxidation in acid coniferous forest soil, *Appl. Environ. Microbiol.* **50:**1519–1525.

Maurino, S. G., Vargas, M. A., Aparicio, P. J., and Maldonado, J. M., 1983, Blue-light reactivation of spinach nitrate reductase inactivated by acetylene or cyanide, *Physiol. Plant.* **57:**411–416.

McBride, B. C., and Edwards, T. L., 1977, Role of methanogenic bacteria in the alkylation of arsenic and mercury, in: *Biological Implications of Metals in the Environment* (H. Drucker and R. E. Wildung, eds.), pp. 1–17, ERDA Symposium Series 42, NTIS, Springfield, Virginia.

McCambridge, J., and McMeekin, T. A., 1980, Relative effects of bacterial and protozoan predators on survival of *Escherichia coli* in estuarine water samples, *Appl. Environ. Microbiol.* **40**:907–911.

McCarthy, R. E., 1980, Delineation of the mechanism of ATP synthesis in chloroplasts: Use of uncouplers, energy transfer inhibitors, and modifiers of coupling factor, in: *Methods in Enzymology*, Vol. 69 (A. San Pietro, ed.) pp. 719–728, Academic Press, New York.

McFadden, B. A., and Purohit, K., 1978, Chemosynthetic, photosynthetic, and cyanobacterial ribulose bisphosphate carboxylase, in: *Photosynthetic Carbon Assimilation* (W. Siegleman and G. Hine, eds.), pp. 179–207. Plenum Press, New York.

McKenna, C. E., and Huang, C. W., 1979, *In vivo* reduction of cyclopropene by *Azotobacter vinelandii* nitrogenase, *Nature* **280**:609–610.

McKenna, C. E., Benemann, J. R., and Taylor, T. G., 1970, A vanadium containing nitrogenase preparation: Implications for the role of molybdenum in nitrogen fixation, *Biochem. Biophys. Res. Commun.* **41**:1501–1508.

Meyers, A. J., 1980, Evaluation of bromomethane as a suitable analogue in methane oxidation studies, *FEMS Microbiol. Lett.* **9**:297–300.

Meyers, A. J., 1982, Obligate methylotrophy: Evaluation of dimethylether as a C-1 compound, *J. Bacteriol.* **150**:966–968.

Moller, M. M., Nielsen, L. P., and Jørgensen, B. B., 1985, Oxygen responses and mat formation by *Beggiatoa* spp., *Appl. Environ. Microbiol.* **50**:373–382.

Mopper, K., and Taylor, B. F., 1986, Biogeochemical cycling of sulfur: Thiols in coastal marine sediments, in: *Organic Marine Geochemistry* (M. Sohn, ed.), pp. 324–339, American Chemical Society Symposium Series, Washington, D.C.

Moreno-Vivian, C., Cejudo, F. J., Cardenas, J., and Castillo, F., 1983, Ammonia assimilation pathways in *Rhodopseudomonas capsulata* E1F1, *Arch. Microbiol.* **136**:147–151.

Moriarty, D. J. W., and Hayward, A. C., 1982, Ultrastructure of bacteria and the proportion of Gram-negative bacteria in marine sediments, *Microb. Ecol.* **8**:1–14.

Mosier, A. R., 1980, Acetylene inhibition of ammonium oxidation in soil, *Soil Biol. Biochem.* **12**:443–444.

Mountfort, D. O., Asher, R. A., Mays, E. L., and Tiedje, J. M., 1980, Carbon and electron flow in mud and sandflat sediments at Delaware Inlet, Nelson, New Zealand, *Appl. Environ. Microbiol.* **39**:686–694.

Murphy, L. S., and Haugen, E. M., 1985, The distribution and abundance of phototrophic ultraplankton in the North Atlantic, *Limnol. Oceanogr.* **30**:47–58.

Murray, P. A., and Zinder, S. H., 1984, Nitrogen fixation by a methanogenic archaebacterium, *Nature* **312**:284–286.

Naumann, E., Fahlbusch, K., and Gottshalk, G., 1984, Presence of a trimethylamine:HS-coenzyme M methyltransferase in *Methanosarcina barkeri*, *Arch. Microbiol.* **138**:79–83.

Nedwell, D. B., 1982, The cycling of sulfur in marine and freshwater sediments, in: *Sediment Microbiology*, (D. B. Nedwell and C. M. Brown, eds.), pp. 73–106, Academic Press, New York.

Nedwell, D. B., and Aziz, S., 1980, Heterotrophic nitrogen fixation in an intertidal saltmarsh sediment, *Est. Coast. Mar. Sci.* **10**:699–702.

Nedwell, D. B., and Banat, I. M., 1981, Hydrogen as an electron donor for sulfate-reducing bacteria in slurries of salt marsh sediment. *Microb. Ecol.* **7**:305–313.

Newell, S. Y., Sherr, F. B., Sherr, E. B., and Fallon, R. D., 1983, Bacterial response to presence of eukaryote inhibitors in water from a coastal marine environment, *Mar. Environ. Res.* **10**:147–157.

Nicholas, D. J. D., 1978, Intermediary metabolism of nitrifying bacteria, with particular reference to nitrogen, carbon and sulfur compounds, in: *Microbiology—1978* (D. Schlessinger, ed.), pp. 305–309, American Society for Microbiology, Washington, D.C.

Nicholas, D. J. D., Wilson, P. W., Heinen, W., Palmer, G., and Beinert, H., 1962. Use of electron paramagnetic resonance spectroscopy in investigations of functional metal components in micro-organisms, *Nature* **196**:433–436.

Nishio, T., Koike, I., and Hattori, A., 1982, Denitrification nitrogen reduction and oxygen consumption in coastal and estuarine sediments, *Appl. Environ. Microbiol.* **43**:648–653.

Norqvist, A., and Roffey, R., 1983, Alternative method for monitoring the effect of inhibitors on sulfate reduction, *J. Gen. Appl. Microbiol.* **29**:335–344.

Notton, B. A., Watson, E. F., and Hewitt, E. J., 1979, Effects of *N*-serve (2-chloro-6-[trichloromethyl] pyridine) formulations on nitrification and on loss of nitrate in sand culture experiements, *Plant Soil* **51**:1–12.

O'Neill, J. G., and Wilkinson, J. F., 1977, Oxidation of ammonia by methane-oxidizing bacteria and the effects of ammonia on methane oxidation, *J. Gen. Microbiol.* **100**:407–412.

Oremland, R. S., 1975, Methane production in shallow-water, tropical marine sediments, *Appl. Microbiol.* **30**:602–608.

Oremland, R. S., 1976, Studies on the methane cycle in tropical marine sediments, Dissertation, University of Miami, Miami, Florida.

Oremland, R. S., 1979, Methanogenic activity in plankton samples and fish intestines: A mechanism for *in situ* methanogenesis in oceanic surface waters, *Limnol. Oceanogr.* **24**:1136–1141.

Oremland, R. S., 1981, Microbial formation of ethane in anoxic estuarine sediments, *Appl. Environ. Microbiol.* **42**:122–129.

Oremland, R. S., 1983, Hydrogen metabolism by decomposing cyanobacterial aggregates in Big Soda Lake, Nevada, *Appl. Environ. Microbiol.* **45**:1519–1525.

Oremland, R. S., and Polcin, S., 1982, Methanogenesis and sulfate-reduction: Competitive and non-competitive substrates in estuarine sediments, *Appl. Environ. Microbiol.* **44**:1270–1276.

Oremland, R. S., and Silverman, M. P., 1979, Microbial sulfate reduction measured by an automated electrical impedance technique, *Geomicrobiol. J.* **1**:355–372.

Oremland, R. S., and Taylor, B. F., 1975, Inhibition of methanogenesis in marine sediments by acetylene and ethylene: Validity of the acetylene reduction assay for anaerobic microcosms, *Appl. Microbiol.* **30**:707–709.

Oremland, R. S., and Taylor, B. F., 1978, Sulfate reduction and methanogenesis in marine sediments, *Geochim. Cosmochim. Acta* **42**:209–214.

Oremland, R. S., and Zehr, J. P., 1986, Formation of methane and carbon dioxide from dimethylselenide in anoxic sediments and by a methanogenic bacterium, *Appl. Environ. Microbiol.* **52**:1031–1036.

Oremland, R. S., Marsh, L., and Des Marais, D. J., 1982a, Methanogenesis in Big Soda Lake, Nevada: An alkaline, moderately hypersaline desert lake, *Appl. Environ. Microbiol.* **43**:462–468.

Oremland, R. S., Marsh, L. M., and Polcin, S., 1982b, Methane production and simultaneous sulfate reduction in anoxic saltmarsh sediments, *Nature* **296**:143–145.

Oremland, R. S., Culbertson, C. W., and Simoneit, B. R. T., 1982c, Methanogenic activity in sediment from Leg 64, Gulf of California, *Init. Rep. Deep Sea Drilling Project* **64**:759–762.

Oremland, R. S., Umberger, C., Culbertson, C. W., and Smith, R. L., 1984, Denitrification in San Francisco bay intertidal sediments, *Appl. Environ. Microbiol.* **47**:1106–1112.

Oremland, R. S., Cloern, J. E., Sofer, Z., Smith, R. L., Culbertson, C. W., Zehr, J., Miller, L., Cole, B., Harvey, R., Iversen, N., Klug, H., Des Marais, D. J., and Rav, G., 1988, Microbial and biogeochemical processes in Big Soda Lake, Nevada, in: *Lacustrine petroleum source rocks* (K. Kelts and A. Fleet, eds.) Geological Society, London (in press).

Orth, R. J., and Moore, K. A., 1983, Chesapeake Bay: An unprecedented decline in submerged aquatic vegetation, *Science* **222**:51–53.

Pace, J., and McDermott, E., 1952, Methionine sulphoximine and some enzyme systems involving glutamine, *Nature* **169**:415–416.

Paerl, H. W., 1983, Environmental regulation of H_2 utilization ($^3H_2$ exchange) among natural and laboratory populations of N_2 and non-N_2 fixing phytoplankton. *Microb. Ecol.* **9**:79–97.

Panganiban, Jr., A. T., Patt, T. E., Hart, W., and Hanson, R. S., 1979, Oxidation of methane in the absence of oxygen in lake water samples, *Appl. Environ. Microbiol.* **37**:303–309.

Patrick, Jr., W. H., Peterson, F. J., and Turner, F. T., 1968, Nitrification inhibitors for lowland rice, *Soil Sci.* **105**:103–105.

Paul, J. H. 1984. Effects of antimetabolites on the adhesion of an estuarine *Vibrio* sp. to polystyrene, *Appl. Environ. Microbiol.* **48**:924–929.

Payne, W. J., 1973, Gas chromatographic analysis of denitrification by marine bacteria, in: *Estuarine Microbial Ecology* (L. H. Stevenson, ed.), pp. 53–71, University of South Carolina Press, Columbia, South Carolina.

Payne, W. J., 1981, *Denitrification,* Wiley, New York.

Payne, W. J., 1984, Influence of acetylene on microbial and enzymatic assays, *J. Microbiol. Meth.* **2**:117–133.

Payne, W. J., and Grant, M. A., 1982, Influence of acetylene on growth of sulfate-respiring bacteria, *Appl. Environ. Microbiol.* **43**:727–730.

Pearsall, K. A., and Bonner, F. T., 1980, Analysis of dinitrogen–nitrogen oxide mixtures employing direct vacuum line-gas chromatograph injection, *J. Chromatog.* **200**:224–227.

Peck, Jr., H. D., 1959, The ATP-dependent reduction of sulfate with hydrogen in extracts of *Desulfovibrio desulfurican, Proc. Natl. Acad. Sci. USA* **45**:701–708.

Peck, Jr., H. D., 1960, Evidence for oxidative phosphorylation during the reduction of sulfate with hydrogen by *Desulfovibrio desulfuricans, J. Biol. Chem.* **235**:2734–2738.

Peck, Jr., H. D., 1962, The role of adenosine-5'-5-phosphosulfate in the reduction of sulfate to sulfite by *Desulfovibrio desulfuricans, J. Biol. Chem.* **237**:198–203.

Pedersen, D., and Sayler, G. D., 1981, Methanogenesis in freshwater sediments: Inherent variability and effects of environmental contaminants, *Can. J. Microbiol.* **27**:198–205.

Peeters, T., and Aleem, M. I. H., 1970, Oxidation of sulfur compounds and electron transport in *Thiobacillus denitrificans, Arch. Microbiol.* **71**:319–330.

Pelczar, M. J., Jr., Chan, E. C. S., and Krieg, N. R., 1986, *Microbiology,* 5th ed., McGraw-Hill, New York.

Peschek, G. A., 1979, Evidence for two functionally distinct hydrogenases in *Anacystis nidulans, Arch. Microbiol.* **123**:81–92.

Peterson, H. G., 1986, *Antimicrobial Agents Annual,* Elsevier, New York.

Peterson, R. B., and Burris, R. H., 1976, Conversion of acetylene reduction rates to nitrogen fixation rates in natural populations of blue-green algae, *Anal. Biochem.* **73**:404–410.

Phelan, P. J., and Mattigod, S. V., 1984, Adsorption of molybdate anion (MoO_4^{2-}) by sodium-saturated kaolinite, *Clays Clay Minerals* **32**:45–48.

Phelps, T. J., and Zeikus, J. G., 1985, Effect of fall turnover on terminal carbon metabolism in Lake Mendota sediments, *Appl. Environ. Microbiol.* **50**:1285–1291.

Pizzey, J. A., Bennett, F. A., and Jones, G. E., 1983, Monensin inhibits initial spreading of cultured human fibroblasts, *Nature* **305**:315–317.

Platt, T., Subba Rao, V., and Irwin, B., 1983, Photosynthesis of picoplankton in the oligotrophic ocean, *Nature* **301**:702–704.

Postgate, J., 1949, Competitive inhibition of sulfate reduction by selenate, *Nature* **172**:670–671.

Postgate, J. R., 1952, Competitive and non-competitive inhibitors of bacterial sulfate reduction, *J. Gen. Microbiol.* **6**:128–142.

Postgate, J. R., 1979, *The Sulfate-Reducing Bacteria,* Cambridge University Press, Cambridge.

Postgate, J. R., 1982, *The Fundamentals of Nitrogen Fixation,* Cambridge University Press, Cambridge.

Poth, M., and Focht, D. D., 1985, ^{15}N kinetic analysis of N_2O production by *Nitrosomonas europaea:* An examination of nitrifier denitrification, *Appl. Environ. Microbiol.* 49:1134–1141.

Powell, S. J., and Prosser, J. I., 1985, The effects of nitrapyrin and chloropicolinic acid on ammonium oxidation by *Nitrosomonas europaea, FEMS Microbiol. Lett.* 28:51–54.

Primrose, S. B., 1976, Ethylene-forming bacteria from soil and water, *J. Gen. Microbiol.* 97:343–346.

Primrose, S. B., 1977, Evaluation of the role of methional, 2-keto-4-methylthiobutyric acid and peroxidase in ethylene formation by *Escherichia coli, J. Gen. Microbiol.* 98:519–528.

Primrose, S. B., and Dilworth, M. J., 1976, Ethylene production by bacteria, *J. Gen. Microbiol.* 93:177–181.

Prins, R. A., van Nevel, C. J., and Demeyer, D. I., 1972, Pure culture studies of inhibitors for methanogenic bacteria, *Antonie Leeuwenhoek Microbiol. Serol.* 38:281–287.

Prins, R. A., Cline-Thiel, W., Malestein, A., and Counotte, G. H. M., 1980, Inhibition of nitrate reduction in some rumen bacteria by tungstate, *Appl. Environ. Microbiol.* 40:163–165.

Raimbault, M., 1975, Etude d'influence inhibitrice de l'acetylene sur la formation biologique du methane dans un sol riziere, *Ann. Microbiol. Inst. Pasteur* 126A:247–258.

Ramirez, C., and Alexander, M., 1980, Evidence suggesting protozoan predation on *Rhizobium* associated with germinating seeds and in the rhizosphere of beans (*Phaseolus vulgaris* L.), *Appl. Environ. Microbiol.* 40:492–499.

Ramos, J. L., and Guerrero, M. G., 1983, Involvement of ammonium metabolism in the nitrate inhibition of nitrogen fixation in *Anabaena* sp. ATCC 33047, *Arch. Microbiol.* 136:81–83.

Reeburgh, W. S., 1976, Methane consumption in Cariaco Tranch waters and sediments, *Earth Planet. Sci. Lett.* 15:334–337.

Reeburgh, W. S., 1980, Anaerobic methane oxidation: Rate depth distributions in Skan Bay sediments, *Earth Planet. Sci. Lett.* 47:345–352.

Reeburgh, W. S., and Heggie, D. T., 1977, Microbial methane consumption reactions and their effect on methane distributions in freshwater and marine environments, *Limnol. Oceanogr.* 22:1–9.

Ribbons, D. W., 1975, Oxidation of C-1 compounds by particulate fractions from *Methylococcus capsulatus:* Distribution and properties of methane-dependent reduced nicotinamide adenine dinucleotide oxidase (methane hydroxylase), *J. Bacteriol.* 122:1351–1363.

Ribbons, D. W., and Michaelover, J. L., 1970, Methane oxidation by cell-free extracts of *Methylococcus capsulatus, FEBS Lett.* 11:41–44.

Richmond, M. H., 1969, Antimetabolites, antibacterial agents and enzyme inhibitors, in: *Data for Biochemical Research,* 2nd ed. (R. M. C. Dawson, D. C. Elliott, W. H. Elliott, and K. M. Jones, eds.), pp. 335–404, Oxford University Press, New York.

Rigano, C., Rigano, V., Vona, V., and Fuggi, A., 1979, Glutamine synthetase activity, ammonia assimilation and control of nitrate reduction in the unicellular red alga *Cyanidium caldarium, Arch. Microbiol.* 121:117–120.

Rittmann, B. E., and McCarty, P. L., 1980, Utilization of dichloromethane by suspended and fixed-film bacteria, *Appl. Environ. Microbiol.* 39:1225–1226.

Rivera-Ortiz, J. M., and Burris, R. H., 1975, Interactions among substrates and inhibitors of nitrogenase, *J. Bacteriol.* 123:537–545.

Robbins, P. W., and Lipmann, F., 1958, Enzymatic synthesis of adenosine-5'-phosphosulfate, *J. Biol. Chem.* 233:686–690.

Robson, R. L., Eady, R. R., Richardson, T. H., Miller, R. W., Hawkins, M., and Postgate, J. R., 1986, The alternative nitrogenase of *Azotobacter chroococcum* is a vanadium enzyme, *Nature* 322:388–390.

Rodgers, G. A., and Ashworth, J., 1982, Bacteriostatic action of nitrification inhibitors, *Can. J. Microbiol.* 28:1093–1100.

Rodgers, G. A., Ashworth, J., and Walker, N., 1980, Recovery of nitrifier populations from inhibition by nitrapyrin or carbon disulfide, *Zentralbl. Bakteriol. Parasitkd. Infektionskr. Hyg. Abt. 2* 135:477–483.

Rosson, R. A., and Nealson, K. H., 1982, Manganese binding and oxidation by spores of a marine bacillus, *J. Bacteriol.* 151:1027–1034.

Rozyccki, M., and Bartha, R., 1981, Problems associated with the use of azide as an inhibitor of microbial activity in soil, *Appl. Environ. Microbiol.* 41:833–846.

Rudd, J. W., and Hamilton, R. D., 1978, Methane cycling in a eutrophic shield lake and its effects on whole lake metabolism, *Limnol. Oceanogr.* 23:337–348.

Rudd, J. W. M., Hamilton, R. D., and Campbell, N. E. R., 1974, Measurement of microbial oxidation of methane in lakewater, *Limnol. Oceanogr.* 19:519–524.

Rudd, J. W., Fututania, A., Flett, R. J., and Hamilton, R. D., 1976, Factors controlling methane oxidation in shield lakes: The role of nitrogen fixation and oxygen concentration, *Limnol. Oceanogr.* 21:357–364.

Ryden, J. C., 1982, Effects of acetylene on nitrification and denitrification in two soils during incubation with ammonium nitrate, *J. Soil Sci.* 33:263–270.

Saino, T., and Hattori, A., 1982, Aerobic nitrogen fixation by the marine non-heterocystous cyanobacterium *Trichodesmium* (*Oscillatoria*) spp.: Its protective mechanism against oxygen, *Mar. Biol.* 70:251–254.

Saleh, A. M., Macpherson, R., and Miller, J. D. A., 1964, The effect of inhibitors on sulfate reducing bacteria: A compilation, *J. Appl. Bacteriol.* 27:281–293.

Salvas, P. L., and Taylor, B. F., 1980, Blockage of methanogenesis in marine sediments by the nitrification inhibitor 2-chloro-6-(trichloromethyl) pyridine (nitrapyrin or N-serve), *Curr. Microbiol.* 4:305–308.

Salvas. P. L., and Taylor, B. F., 1984, Effect of pyridine compounds on ammonia oxidation by autotrophic nitrifying bacteria and *Methylosinus trichosporium* OB3b, *Curr. Microbiol.* 10:53–56.

Samuelsson, M.-O., 1985, Dissimilatory nitrate reduction to nitrite, nitrous oxide, and ammonium by *Pseudomonas putrefaciens*, *Appl. Environ. Microbiol.* 50:812–815.

Sanders, R. W., and Porter, K. G., 1986, Use of metabolic inhibitors to estimate protozooplankton grazing and bacterial production in a monomictic eutrophic lake with an anaerobic hypolimnion, *Appl. Environ. Microbiol.* 52:101–107.

Schannong Jørgensen, K., Beck Jensen, H., and Sørensen, J., 1984, Nitrous oxide production from nitrification and denitrification in marine sediment at low oxygen concentrations, *Can. J. Microbiol.* 30:1073–1078.

Schink, B., 1985a, Fermentation of acetylene by an obligate anaerobe, *Pelobacter acetylenicus* sp. nov, *Arch. Microbiol.* 142:295–301.

Schink, B., 1985b, Inhibition of methanogenesis by ethylene and other unsaturated hydrocarbons, *FEMS Microbiol. Ecol.* 31:63–68.

Scranton, M. I., 1983, The role of the cyanobacterium *Oscillatoria* (*Trichodesmium*) *thiebautii* in the marine hydrogen cycle, *Mar. Ecol. Prog. Ser.* 11:79–87.

Scranton, M. I., 1984, Hydrogen cycling in the waters near Bermuda: The role of the nitrogen fixer, *Oscillatoria thiebautii, Deep-Sea Res.* 31:133–143.

Scranton, M. I., Novelli, P. C., and Loud, P. A., 1984, The distribution and cycling of hydrogen gas in the waters of two anoxic marine environments, *Limnol. Oceanogr.* **29**:993–1003.

Seitzinger, S., Nixon, S., Pilson, M. E. Q., and Burke, S., 1980, Denitrification and N_2O production in near-shore marine sediments, *Geochim. Cosmochim. Acta* **44**:1853–1860.

Shapiro, S., and Wolfe, R. S., 1980, Methyl-coenzyme M, an intermediate in methanogenic dissimilation of C_1 compounds by *Methanosarcina barkeri, J. Bacteriol.* **141**:728–734.

Shattuck, G. E., and Alexander, M., 1963, A differential inhibitor of nitrifying organisms, *Soil Sci. Soc. Am. Proc.* **27**:600–601.

Shaw, D. G., Alperin, M. J., Reeburgh, W. S., and McIntosh, D. J., 1984, Biogeochemistry of acetate in anoxic sediments of Skan Bay, Alaska, *Geochim. Cosmochim. Acta* **48**:1819–1825.

Sherr, B. F., Sherr, E. B., Andrew, T. L., Fallon, R. D., and Newell, S. Y., 1987, Investigation of the trophic interactions between heterotrophic protozoa and bacterioplankton in estuarine water using selective metabolic inhibitors, *Mar. Ecol. Prog. Ser.,* in press.

Sieburth, J. McN., 1979, *Sea Microbes,* Oxford, London.

Slater, J., and Capone, D. G., 1984, Effect of metals on nitrogen fixation and denitrification in slurries of anoxic saltmarsh sediment, *Mar. Ecol. Prog. Ser.* **18**:89–95.

Slater, J. and Capone, D. G., 1987, Denitrification in aquifer soil and nearshore marine sediments influenced by groundwater nitrate, *Appl. Environ. Microbiol.* **53**:1292–1297.

Slater, J., and Capone, D. G., 1988a, Denitrification by enrichment cultures of bacteria from saltmarsh sediments: Effects of Ni(II) and Cr(VI), manuscript submitted.

Slater, D. G., and Capone, D. G., 1988b, Assessment of denitrification measurement in saltmarsh sediments by acetylene blockage, manuscript submitted.

Slovacek, R. E., and Hannan, P. J., 1977, *In vivo* fluorescence determinations of phytoplankton chlorophyll a, *Limnol. Oceanogr.* **22**:919–924.

Smith, A. M., 1976, Ethylene in soil biology, *Annu. Rev. Phytopathol.* **14**:53–73.

Smith, K. A., and Dowdell, R. J., 1974, Field studies of soil atmosphere. I. Relationships between ethylene, oxygen, soil-moisture content and temperature, *J. Soil Sci.* **25**:217–230.

Smith, K. A., and Restall, S. W. F., 1971, The occurrence of ethylene in anaerobic soil, *J. Soil Sci.* **22**:430–433.

Smith, K. A., and Russell, R. S., 1969, Occurrence of ethylene, and its significance, in anaerobic soil, *Nature* **222:** 769–771.

Smith, L. A., Hills, S., and Yates, M. G., 1976, Inhibition by acetylene of conventional hydrogenase in nitrogen-fixing bacteria, *Nature* **262:**209–210.

Smith, M. R., 1983, Reversal of 2-bromoethanesulfonate inhibition of methanogenesis in *Methanosarcina* sp., *J. Bacteriol.* **156:**516–523.

Smith, M. R., and Mah, R. A., 1978, Growth and methanogenesis by *Methanosarcina* strain 227 on acetate and methanol, *Appl. Environ. Microbiol.* **36:**870–879.

Smith, M. R., and Mah, R. A., 1981, 2-Bromoethanesulfonate: A selective agent for isolating resistant *Methanosarcina* mutants, *Curr. Microbiol.* **6:**321–326.

Smith, M. S., 1982, Dissimilatory reduction of NO_2^- to NH_4^+ and N_2O by a soil *Citrobacter* sp., *Appl. Environ. Microbiol.* **43:**854–860.

Smith, R. L., and Klug, M. J., 1981, Electron donors utilized by sulfate-reducing bacteria in eutrophic lake sediments, *Appl. Environ. Microbiol.* **42:**116–121.

Smith, R. L., and Oremland, R. S., 1987, Big Soda Lake (Nevada). 2. Pelagic sulfate reduction, *Limnol. Oceanogr.,* **32:**794–803.

Smucker, R. A., and Simon, S. L., 1986, Some effects of diflubenzuron on growth and sporogenesis in *Streptomyces* spp., *Appl. Environ. Microbiol.* **51:**25–31.

Somville, M., 1978, A method for the measurement of nitrification rates in water, *Water Res.* **12:**843–848.

Somville, M., 1984, Use of nitrifying activity measurements for describing the effect of salinity on nitrification in the Sheldt Estuary, *Appl. Environ. Microbiol.* **47**:424–426.

Sørensen, J., 1978, Denitrification rates in a marine sediment as measured by the acetylene inhibition technique, *Appl. Environ. Microbiol.* **35**:301–305.

Sørensen, J., 1982, Reduction of ferric iron in anaerobic, marine sediment and interaction with reduction of nitrate and sulfate, *Appl. Environ. Microbiol.* **43**:319–324.

Sørensen, J., Tiedje, J. M., and Firestone, R. B., 1980, Inhibition by sulfide of nitric and nitrous oxide reduction by denitrifying *Pseudomonas fluorescens, Appl. Environ. Microbiol.* **39**:105–108.

Sørensen, J., Christensen, D., and Jørgensen, B. B., 1981, Volatile fatty acids and hydrogen as substrates for sulfate-reducing bacteria in anaerobic marine sediment, *Appl. Environ. Microbiol.* **42**:5–11.

Sprott, G. D., and Jarrell, K. F., 1982, Sensitivity of methanogenic bacteria to dicyclohexylcarbodiimide, *Can. J. Microbiol.* **28**:982–986.

Sprott, G. D., Jarrell, K. F., Shaw, K. M., and Knowles, R., 1982, Acetylene as an inhibitor of methanogenic bacteria, *J. Gen. Microbiol.* **128**:2453–2462.

Stevenson, J. C., and Confer, N. M., 1978, Summary of Available Information on Chesapeake Bay Submerged Vegetation, FWS/OBS-78/66, Fish and Wildlife Service, United States Department of the Interior.

Stewart, W. D. P., and Rowell, P., 1975, Effects of *l*-methionine-*dl*-sulfoximine on the assimilation of newly fixed NH₃, acetylene reduction and heterocyst production in *Anabaena cylindrica, Biochem. Biophys. Res. Commun.* **65**:846–856.

Stewart, W. D. P., Fitzgerald, G. P., and Burris, R. H., 1967, *In situ* studies on N₂ fixation, using the acetylene reduction technique, *Proc. Natl. Acad. Sci. USA* **58**:2071–2078.

Stirling, D. I., and Dalton, H., 1979, Properties of the methane monooxygenase from extracts of *Methylosinus trichosporium* OB3b and evidence for its similarity to the enzyme from *Methylococcus capsulatus* (Bath), *Eur. J. Biochem.* **96**:205–212.

Stokes, D. M., and Walker, D. A., 1972, Photosynthesis by isolated chloroplasts, *Biochem. J.* **128**:1147–1157.

Stratton, G. W., Burrell, R. E., and Corke, C. T., 1982, Technique for identifying and minimizing solvent-pesticide interactions in bioassays, *Arch. Environ. Contam. Toxicol.* **11**:437–445.

Sutherland, J. B., and Cook, R. J., 1980, Effects of chemical and heat treatments on ethylene production in soil, *Soil Biol. Biochem.* **12**:357–362.

Sutton, W. D., 1980, Effects of protein synthesis inhibitors on acetylene reduction activity of lupin root nodules, *Aust. J. Plant Physiol.* **7**:261–270.

Syrett, P. J., 1981, Nitrogen metabolism of microalgae, in: Physiological Basis of Phytoplankton Ecology (T. Platt, ed.), *Can. Bull. Fish. Aquat. Sci. Bull.* **210**:182–210.

Takeda, K., Tezuka, C., Fukuoka, S., and Takahara, Y., 1976, Role of copper ions in methane oxidation by *Methanomonas margaritae, J. Ferment. Technol.* **54**:557–562.

Tam, T. Y., and Knowles, R., 1979, Effects of sulfide and acetylene on nitrous oxide reduction by soil and by *Pseudomonas aeroginosa, Can. J. Microbiol.* **25**:1133–1138.

Tam, T. Y., Mayfield, C. I., and Inniss, W. E., 1981, Nitrogen fixation and methane metabolism in a stream-water system amended with leaf material, *Can. J. Microbiol.* **27**:511–516.

Tam, T. Y., Mayfield, C. I., and Inniss, W. E., 1983, Aerobic acetylene utilization by stream sediment and isolated bacteria, *Curr. Microbiol.* **8**:165–168.

Tate III, R. L., 1977, Nitrification in histosols: A potential role for the heterotrophic nitrifier, *Appl. Environ. Microbiol.* **33**:911–914.

Taylor, B. F., 1983, Assays of microbial nitrogen transformations, in: *Nitrogen in the Marine Environment* (E. J. Carpenter and D. G. Capone, eds.), pp. 809–837, Academic Press, New York.

Taylor, B. F., and Oremland, R. S., 1979, Depletion of adenosine triphosphate in *Desulfovibrio* by oxyanions of group VI elements, *Curr. Microbiol.* **3**:101–103.

Taylor, C. D., and Wolfe, R. S., 1974, Structure and methylation of coenzyme M ($HSCH_2CH_2SO_3$), *J. Biol. Chem.* **249**:4879–4885.

Taylor, C. D., McBride, B. C., Wolfe, R. S., and Bryant, M. P., 1974, Coenzyme M, essential for growth of a rumen strain of *Methanobacterium ruminantium, J. Bacteriol.* **120**:974–975.

Taylor, G. T., and Pace, M. L., 1987, Validity of eucaryote inhibitors for assessing production and grazing mortality of marine bacterioplankton, *Appl. Environ. Microbiol.* **53**:119–128.

Thomas, K. C., and Spencer, M., 1978, Evolution of ethylene by *Saccharomyces cerevisiae* as influenced by the carbon source for growth and the presence of air, *Can. J. Microbiol.* **24**:637–642.

Tonge, G. M., Harrison, D. E. F., Knowles, C. J., and Higgins, I. J., 1975, Properties and partial purification of the methane-oxidizing enzyme system from *Methylosinus trichosporium, FEBS Lett.* **58**:293–299.

Tonge, G. M., Drozd, J. W., and Higgins, I. J., 1977, Energy coupling in *Methylosinus trichosporium, J. Gen. Microbiol.* **99**:229–232.

Tonsager, S. R., and Averill, B. A., 1980, Difficulties in the analysis of acid-labile sulfide in Mo–S and Mo–Fe–S systems, *Anal. Biochem.* **102**:13–15.

Topp, E., and Knowles, R., 1982, Nitrapyrin inhibits the obligate methylotrophs *Methylosinus trichosporium* and *Methylococcus capsulatus, FEMS Microbiol. Lett.* **14**:47–49.

Topp, E., and Knowles, R., 1984, Effects of nitrapyrin [2-chloro-6-(trichloromethyl) pyridine] on the obligate methanotroph *Methylosinus trichosporium* OB3b, *Appl. Environ. Microbiol.* **47**:258–262.

Trebst, A., 1980, Inhibitors in electron flow: Tools for the functional and structural localization of carriers and energy conservation sites, in: *Methods in Enzymology*, (A. SanPietro, ed), pp. 765, Academic Press, New York.

Trimble, R. B., and Ehrlich, H. L., 1968, Bacteriology of manganese nodules. III. Reduction of MnO_2 by two strains of nodule bacteria, *Appl. Microbiol.* **16**:695–702.

Tromballa, H. W., and Broda, E., 1971, Das verhalten von *Chlorella fusca* gegenuber perchlorat und chlorat, *Arch. Microbiol.*. **78**:214–223.

Truper, H. G., and Fischer, U., 1982, Anaerobic oxidation of sulphur compounds as electron donors for bacterial photosynthesis, *Phil. Trans. R. Soc. Lond. B* **298**:529–542.

Turpin, D. H., Edie, S. A., and Canvin, D. T., 1984, *In vivo* nitrogenase regulation by ammonium and methylamine and the effect of MSX on ammonium transport in *Anabaena flos-aquae, Plant Physiol.* **74**:701–704.

Tuttle, J. H., and Dugan, P. R., 1976, Inhibition of growth, iron, and sulfur oxidation in *Thiobacillus ferrooxidans* by simple organic compounds, *Can. J. Microbiol.* **22**:719–730.

Tuttle, J. H., and Jannasch, H. W., 1977, Thiosulfate stimulation of microbial dark assimilation of carbon dioxide in shallow marine waters, *Microb. Ecol.* **4**:9–25.

Umezawa, H., 1982, Low-molecular-weight enzyme inhibitors of microbial origin, *Annu. Rev. Microbiol.* **36**:75–99.

Van Berkum, P., and Sloger, C., 1979, Immediate acetylene reduction by excised grass roots not previously preincubated at low oxygen tensions, *Plant Physiol.* **64**:739–743.

Van Berkum, P., and Sloger, C., 1981, Comparing time course profiles of immediate acetylene reduction by grasses and legumes, *Appl. Environ. Microbiol.* **41**:184–189.

Van der Meijden, P., Heythuysen, H. J., Sliepenbeek, H. T., Houwen, F. P., van der Drift, C., and Vogels, G. D., 1983, Activation and inactivation of methanol: 2-Mercaptoethanesulfonic acid methyltransferase from *Methanosarcina barkeri, J. Bacteriol.* **153**:6–11.

Vandermeulen, J. H., Davis, N. D., and Muscatine, L., 1972, The effect of inhibitors of photosynthesis on zooxanthellae in corals and other marine invertebrates, *Mar. Biol.* **16:**185–191.

Van Nevel, C. J., and Demeyer, D. I., 1977, Effect of monensin on rumen metabolism *in vitro, Appl. Environ. Microbiol.* **34:**251–257.

Van Raalte, C. D., and Patriquin, D. G., 1979, Use of the "acetylene blockage" technique for assaying denitrification in a salt marsh, *Mar. Biol.* **52:**315–320.

van Vliet-Smits, M., Harder, W., and van Dijken, J. P., 1981, Some properties of the amine oxidase of the facultative methylotroph *Arthrobacter* P1, *FEMS Microbiol. Lett.* **11:**31–35.

Vincent, W. F., and Downes, M. T., 1981, Nitrate accumulation in aerobic hypolimnia: Relative importance of benthic and planktonic nitrifiers in an oligotrophic lake, *Appl. Environ. Microbiol.* **42:**565–573.

Vogel, T. M., Oremland, R. S., and Kvenvolden, K. A., 1982, Low temperature formation of hydrocarbon gases in San Francisco Bay sediment (California, U.S.A.), *Chem. Geol.* **37:**289–298.

Wake, L. V., Christopher, R. K., Rickard, P. A. D., Andersen, J. E., and Ralph, B. J., 1977, A thermodynamic assessment of possible substrates for sulfate/reducing bacteria, *Aust. J. Biol. Sci* **30:**115–127.

Walter, H. M., Kenney, D. R., and Fillery, I. R., 1979, Inhibition of nitrification by acetylene, *Soil Sci. Am. J.* **43:**195–196.

Ware, D. A., and Postgate, J. R., 1971, Physiological and chemical properties of a reductant-activated inorganic pyrophosphatase from *Desulfovibrio desulfuricans, J. Gen. Microbiol,* **67:**145–160.

Watanabe, I., and de Guzman, M. R., 1980, Effect of nitrate on acetylene disappearance from anaerobic soil, *Soil Biol. Biochem,* **12:**193–194.

Weathers, P. J., 1984, N_2O evolution by green algae, *Appl. Environ. Microbiol.* **48:**1251–1253.

Webb, K. L., and Wiebe, W. J., 1975, Nitrification on a coral reef, *Can. J. Microbiol.* **21:**1427–1431.

Wheeler, P. A., 1980, Use of methylammonium as an ammonium analogue in nitrogen transport and assimilation studies with *Cyclotella cryptica* (Bascillariophyceae), *J. Phycol.* **16:**328–334.

Wheeler, P. A., and Kirchman, D. L., 1986, Utilization of inorganic and organic nitrogen by bacteria in marine systems, *Limnol. Oceanogr.* **31:**998–1009.

Wildenauer, F. X., Blotevogel, K. H., and Winter, J., 1984, Effect of monensin and 2-bromethanesulfonic acid on fatty acid metabolism and methane production from cattle manure, *Appl. Microbiol. Biotechnol.* **19:**125–130.

Wilson, L. G., and Bandurski, R. S., 1958, Enzymatic reactions involving sulfate, sulfite, selenate, and molybdate, *J. Biol. Chem.* **233:**975–981.

Winfrey, M. R., and Ward, D. M., 1983, Substrates for sulfate reduction and methane production in intertidal sediments, *Appl. Environ. Microbiol.* **45:**193–199.

Winfrey, M. R., and Zeikus, J. G., 1977, Effect of sulfate on carbon and electron flow during microbial methanogenesis in freshwater sediments, *Appl. Environ. Microbiol.* **33:**312–318.

Winfrey, M. R., and Zeikus, J. G., 1979, Anaerobic metabolism of immediate methane precursors in Lake Mendota, *Appl. Environ. Microbiol.* **37:**244–253.

Witty, J. F., 1979, Acetylene reduction assay can overestimate nitrogen/fixation in soil, *Soil Biol. Biochem.* **11:**209–210.

Wolin, M. J., and Miller, T. L., 1980, Molybdate and sulfide inhibit H_2 and increase formate production from glucose by *Ruminococcus albus, Arch. Microbiol.* **124:**137–142.

Wolin, E. A., Wolfe, R. S. and Wilin, M. J., 1964, Viologen dye inhibition of methane formation of *Methanobacillus omelianskii, J. Bacteriol.* **87**:993–998.

Wood, J. M., Kennedy, F. S., and Wolfe, R. S., 1968a, The reaction of multihalogenated hydrocarbons with free and bound vitamin B_{12}, *Biochemistry* **7**:1707–1713.

Wood., J. M., Kennedy, F. S., and Rosen, C. G., 1968b, Synthesis of methyl-mercury compounds by extracts of a methanogenic bacterium, *Nature* **220**:173–174.

Wood, L. B., Hurley, B. J. E., and Matthews, P. J., 1981, Some observation on the biochemistry and inhibition of nitrification, *Water Res.* **15**:543–551.

Wright, R. T., and Coffin, R. B., 1983, Planktonic bacteria in estuaries and coastal waters of northern Massachusetts: Spatial and temporal distributions, *Mar. Ecol. Prog. Ser.* **11**:205–216.

Yamazaki, S., 1982, A selenium-containing hydrogenase from *Methanococcus vannielii, J. Biol. Chem.* **257**:7926–7929.

Yang, S. F., 1974, The biochemistry of ethylene: Biogenesis and metabolism, *Rec. Adv. Phytochem.* **7**:131–164.

Yeomans, J., and Beauchamp, E. G., 1978, Limited inhibition of nitrous oxide reduction in soil in the presence of acetylene, *Soil Biol. Biochem.* **10**:517–519.

Yetka, J. E., and Wiebe, W. J., 1974, Ecological application of antibiotics as respiratory inhibitors of bacterial populations, *Appl. Microbiol.* **28**:1033–1039.

Yoch, D. C., and Gotto, J. W., 1982, Effect of light intensity and inhibitors of nitrogen assimilation on NH_4^+ inhibition of nitrogenase activity in *Rhodopseudomonas rubrum* and *Anabaena* sp., *J. Bacteriol.* **151**:800–806.

Yoch, D. C., and Whiting, G. J., 1986, Evidence for NH_4^+ switch-off regulation of nitrogenase activity by bacteria in salt marsh sediments and roots of the grass *Spartina alterniflora, Appl. Environ. Microbiol.* **51**:143–149.

Yokota, A., and Canvin, D. T., 1985, Ribulose biphosphate carboxylase/oxygenase content determined with [^{14}C] carboxypentitol biphosphate in plants and algae, *Plant Physiol.* **77**:735–739.

Yoshinari, T., 1984, Nitrite and nitrous oxide production by *Methylosinus trichosporium, Can. J. Microbiol.* **31**:139–144.

Yoshinari, T., and Knowles, R., 1976, Acetylene inhibition of nitrous oxide reduction by denitrifying bacteria, *Biochem. Biophys. Res. Commun.* **69**:705–710.

Yoshinari, T, Hynes, R., and Knowles, R., 1977, Acetylene inhibition of nitrous oxide reduction and measurement of denitrification and nitrogen fixation in soil, *Soil Biol. Biochem.* **9**:177–183.

Young, J. C., 1983, Comparison of 3 forms of 2-chloro-6-(trichloromethyl) pyridine as a nitrification inhibitor in BOD tests, *J. Water Pollut. Control Fed.* **55**:415–416.

Zehnder, A. J. B., and Brock, T. D., 1979, Methane formation and methane oxidation by methanogenic bacteria, *J. Bacteriol.* **137**:420–432.

Zehnder, A. J. B., and Brock, T. D., 1980, Anaerobic methane oxidation: Occurrence and ecology, *Appl. Environ. Microbiol.* **39**:194–204.

Zehr, J. P., and Oremland, R. S., 1987, Reduction of selenate to selenide by sulfate-respiring bacteria: Experiments with cell suspensions and estuarine sediments. *Appl. Environ. Microbiol.* **53**:1365–1369.

Zinder, S. H., and Brock, T. D., 1978, Production of methane and carbon dioxide from methane thiol and dimethylsulfide by anaerobic lake sediments, *Nature* **273**:226–228.

Zinder, S. H., Anguish, T., and Cardwell, S. C., 1984, Selective inhibition by 2-bromoethanesulfonate of methanogenesis from acetate in a thermophilic anaerobic digestor, *Appl. Environ. Microbiol*, **47**:1343–1345.

ZoBell, C. E., 1947, Microbial transformations of molecular hydrogen in marine sediments, with particular reference to petroleum, *Am. Assoc. Petrol. Geol. Bull.* **31**:1709–1751.

9

Ecological Significance of Siderophores in Soil

P. BOSSIER, M. HOFTE, and W. VERSTRAETE

1. Introduction

Among the extracellular secondary metabolites, microbial iron-chelating compounds, also called siderophores, have received considerable attention. The ecological interest in these compounds is gradually increasing, especially in terms of the possible function of these compounds in soil. The current increasing interest and research on bacterial siderophores is to a great extent linked to investigations on the inoculation of plant seeds with fluorescent *Pseudomonas* spp. that are considered to produce siderophores counteracting deleterious microorganisms in the root zone. The research on the ecology of fungal siderophores has been focused on the role of the fungal siderophores in the acquisition of iron by plants. Much of the knowledge on siderophores is based on observations *in vitro*. There are, however, considerable differences between the environmental circumstances in soil and in synthetic media. Given these facts, it is of interest to consider the points on which the ecological research on siderophores should focus in order to obtain a better understanding of their role in the soil environment. It is our intention in this chapter to review the ecological significance of siderophores in natural environments such as the soil.

P. BOSSIER, M. HOFTE, and W. VERSTRAETE • Laboratory of Microbial Ecology, University of Ghent, B-9000 Ghent, Belgium.

2. Hydroxamate Siderophores

Siderophores are metabolites with a very high chelating affinity for Fe^{3+} ions and a low affinity for Fe^{2+} ions. In oxygenated environments, these compounds are produced by microorganisms to mobilize Fe^{3+}. Upon uptake of the ferric siderophore, by the aid of a high-affinity siderophore uptake system installed in the cell envelope, the iron is released by reduction to Fe^{2+} (Neilands, 1981).

The bidendate ligand system that chelates the Fe^{3+} ion can be a hydroxamate group, a catechol group, an α-hydroxy acid group, a 2-(2-hydroxyphenyl)-oxazoline group, or a fluorescent quinolinyl chromophore (Fig. 1). The three bidendate ligand systems usually present in siderophores can all be identical, as in most fungal siderophores (three hydroxamate groups). Two different types of bidendate ligand systems may chelate the Fe^{3+} ion, as in the case of arthrobactin (two hydroxamate groups and one citric acid group). Finally, the Fe^{3+} ions can be chelated by three different bidendate ligand systems. This has been found in the fluorescent pseudobactin of *Pseudomonas* B10. In that siderophore, a

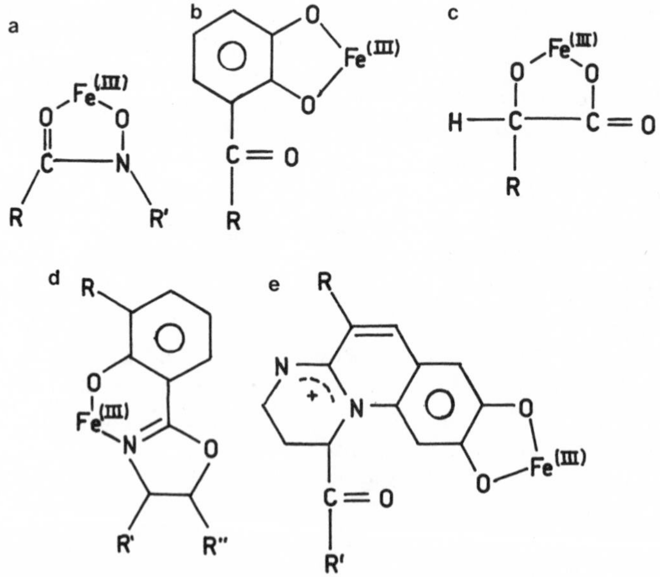

Figure 1. Common bidendate ligand systems in siderophores: (a) hydroxamate, (b) catechol, (c) α-hydroxy acid, (d) 2-(2-hydroxyphenyl)-oxazoline, and (e) fluorescent quinolinyl chromophore.

Table I. Stability Constants of Siderophores with Fe^{3+} Ion

Siderophore	Stability constant,[a] log K	Reference
Ferrichrome	29.1	Raymond et al. (1984)
Coprogen	30.2	Raymond et al. (1984)
Ferrioxamine B	30.5	Raymond et al. (1984)
Fluorescent pigment of a Pseudomonas sp.	32.0	Meyer and Abdallah (1978)

[a]$K = [Fe^{3-n}]/[Fe^{3+}][L^{n-}]$.

hydroxamate group, an α-hydroxy acid, and a fluorescent quinolinyl chromophore are chelating the Fe^{3+}ion. All siderophores are characterized by a very high affinity for Fe^{3+} ions (Table I). Since the majority of soil microorganisms form siderophores containing hydroxamate ligands, in this chapter we deal primarily with this group of siderophores. A remarkable diversity of bacteria, fungi, and even higher organisms appear capable of producing hydroxamate siderophores (Table II).

Little attention has been given up to now to the biochemistry of hydroxamate siderophores. Ornithine auxotrophic mutants of *Neurospora crassa* became siderophore-dependent (Winkelmann, 1979). This indicates that amino acids are intermediates in the production of, for example, hexapeptide siderophores. Of special importance in relation to siderophore biochemistry is the presence of oxygenated N atoms. According to Emery (1966), the hydroxylated N atoms are formed through the direct oxidation with an oxygenase prior to the condensation of the hydroxamate group.

Table II. Examples of Microorganisms Producing Hydroxamate Siderophores

Organism	Siderophore	Reference
Bacteria		
Aerobacter aerogenes	Aerobactin	Neilands (1982)
Arthrobacter pascens	Arthrobactin	Focht and Verstraete (1977)
Pseudomonas B10	Pseudobactin	Teintze et al. (1981)
Actinomyces		
Streptomyces spp.	Ferrioxamine B	Perlman (1965)
Yeast		
Rhodotorula spp.	Rhodotorulic acid	Atkin et al. (1970)
Fungi		
Penicillium spp.	Ferrichrome	Winkelmann (1985)
Penicillium spp.	Coprogen	Zähner et al. (1963)
Dinoflagellates		
Prorocentrum minimum	Prorocentrum	Trick et al. (1983)

This type of reaction is considered to occur in the heterotrophic nitrification pathway. In this pathway many N-oxygenated compounds are formed through the energy-consuming oxygenation of N atoms (Focht and Verstraete, 1977). Some microorganisms even produce nitrite or nitrate via that pathway. Whether N-oxidized amino acids form a pool of products out of which siderophores, as well as other oxygenated organic N compounds, are formed is still an open question and remains to be elaborated. Incidentally, the fact that so many microorganisms are capable of generating oxidized nitrogenous forms has puzzled many biochemists and microbiologists. To our knowledge, up to now no hard data have been produced which, in one way or another, implicate a direct role for the preservation of this heterotrophic nitrification trait in the genome of so many species. A most interesting hypothesis is that for all the heterotrophic nitrifiers, the acquisition of sufficient amounts of iron in an oxic environment is of such crucial importance that it is vital for them to maintain the capacity to generate oxidized nitrogenous compounds such as hydroxamate siderophores.

3. Ecology of Hydroxamate Siderophores

3.1. Fungal Siderophores in Soil

3.1.1. Levels

The concentration of fungal siderophores in soil has been determined by the aid of bioassays. The following assay microorganisms are commonly used: *Arthrobacter* JG-9, a natural siderophore auxotroph (Powell *et al.*, 1980), as well as *Salmonella typhimurium* 7 enb[a] (Akers, 1981), and *Escherichia coli* K12 (RW 193, Ton A) (Powell *et al.*, 1983), two siderophore auxotrophs obtained by mutation. These bioassay microorganisms differ in siderophore growth factor range. The *Escherichia coli* strain (RW 193, Ton A) can use only ferrichrome besides enterobactin as growth factor source (Powell *et al.*, 1983). *Salmonella typhimurium* 7 enb[a] can use more hydroxamate siderophores, such as rhodotorulic acid, ferrichrome, and schizokinen (Akers, 1983b). The exact growth factor range, however, is not know. Finally, *Arthrobacter* JG-9 has a very broad growth factor range. Almost all hydroxamate siderophores (HS) described up to now are known to stimulate the growth of *Arthrobacter* JG-9, with the exception for ferrichrome A. Yet, *Arthrobacter* JG-9 exhibits considerable differences in siderophore growth factor affinity (Burnham and Neilands, 1961). The same can be expected for other bioassay microorganisms.

Considerable differences exist in the designs of the above-mentioned bioassays. For instance, different authors do not use the same siderophore as reference (Table III). It should be noted that most authors extract the siderophores from soil, whereas Bossier and Verstraete (1986a) developed an assay that permits the determination of the HS in the soil directly. Consequently, the results of those bioassays cannot be compared straightforwardly. Despite this diversity in bioassay techniques, the concentration ranges of siderophores in soils reported by several groups of researchers agree quite well (Table III).

The concentrations of siderophores in soil are very low when expressed in $\mu g/kg$ soil. However, when expressed per unit soil microbial biomass, the levels are equal to or somewhat lower than those found in axenic cultures of *Boletus edulis* (Szaniszlo *et al.*, 1981) (Table IV). The major disadvantage of the detection of siderophores by the aid of a bioassay microorganism is the possible nonspecificity of these organisms toward siderophores. An exception should be made for the bioassay developed with *Escherichia coli* K12 (RW 193, Ton A), which is specific for ferrichrome. Because of the low concentrations in soils and the nonspecificity of the bioassay microorganisms, little information is available with regard to the nature of the siderophores present in soil. Based on observations *in vitro,* one can predict that many distinct siderophores will be present in soil. Indeed, for several microorganisms it has already been demonstrated that they produce many different siderophores (Frederick

Table III. Concentration of Siderophores, Mainly of Fungal Origin, in Soils

Bioassay organism	Reference siderophore	Siderophore range in soil ($\mu g/kg$ soil)	Reference
Salmonella typhimurium 7 enb[a]	Rhodoturulic acid	0–31	Akers (1981)
Salmonella typhimurium 7 enb[a]	Rhodoturulic acid	20–51	Akers (1983a)
Salmonella typhimurium 7 enb[a]	Rhodoturulic acid	≤240	Akers (1983b)
Escherichia coli (Ton A, RW 193)	Ferrichrome	53 ≤60	Powell *et al.* (1983)
Arthrobacter JG-9	Ferrioxamine B		Powell *et al.* (1980)
Arthrobacter JG-9	Not indicated	<150	Harrington and Neilands (1982)
Arthrobacter JG-9	Ferrioxamine B	0–132	Bossier and Verstraete (1986a,b)

Table IV. Siderophore Concentration in Relation to Microbial Biomass

Environment	Siderophore concentration (mg/g biomass C)	Reference
Axenic culture		
Boletus edulis (4 days old)[a]		Szaniszlo *et al.* (1981)
0 ppb Fe	2.1	
40 ppb Fe	0.3	
Soils		
Various textures, pH, and	0.04–0.32	Bossier and
organic C levels; $n = 7$	(av. 0.12)	Verstraete (1986b)

[a]Calculation based on the assumption that biomass = (biomass C) × 2.

et al., 1981; Charlang *et al.*, 1981). A better knowledge of the nature of siderophores present in soil would certainly enhance the understanding of their function in the soil environment.

3.1.2. Environmental Influences

The major factor influencing siderophore production is the availability of organic substrates. This has been shown under field conditions as well as in laboratory experiments. In meadows, siderophore concentration was increased with increasing amounts of N fertilizer applied, which in turn stimulated grass production. In the same fields, siderophore concentration increased during the growing season (Bossier and Verstraete, 1986b). These observations are strongly indicative of a rhizosphere effect with respect to siderophore production.

This is in agreement with the results of Powell *et al.* (1982), who found 50-fold higher concentration of siderophores in the rhizosphere soil in respect to the bulk soil. Under laboratory conditions, siderophore production was positively correlated with the amount of organic substrates added to the soil (Bossier and Verstraete, 1986b). Apart from the amount of substrate available for the microorganisms, the nature of the substrate is also important. It was shown that L-ornithine, added to the soil together with a sugar as C source, stimulated siderophore production better than D-ornithine or L-lysine could do (Bossier and Verstraete, 1986b). This was expected, since L-ornithine and not D-ornithine or L-lysine can be recognized as part of the structure of trihydroxamate siderophores of the ferrichrome type. The stimulatory effect of L-ornithine was also found in axenic cultures of *Aspergillus melleus* (Crueger and Zähner, 1968). The ectomycorrhizal fungus *Boletus edulis,* on the other hand, does not respond to L-ornithine by the production of more siderophores (Szaniszlo *et al.*, 1981). The precursor effect of L-ornithine seems to depend upon the microorganism and on the type of siderophores it is producing. Hence, due to differences in the microbial composition of soils, the precursor effect of L-ornithine may vary from soil to soil.

In axenic cultures of microorganisms, an inverse relationship exists between iron concentration and siderophore production. Siderophore production can even be completely inhibited if iron concentration in the culture is too high. Bossier and Verstraete (1986b) also found that in soils siderophore production could be manipulated by influencing the iron availability. Indeed, by adding increasing amounts of EDDA to a soil (up to 1080 ppm), siderophore production was slightly stimulated. It was postulated that, through the addition of EDDA (ethylenediamine-di(o-hydroxyphenylacetic acid)) to that soil, iron became less available, to which the growing biomass responded by producing more siderophores. Yet, in certain soils considerable siderophore production was found even if the pH was as low as 4.61 and EDTA (ethylenediaminetetraacetic acid)-extractable iron as high as 966 ppm (Bossier and Verstraete, 1986b). This strongly indicates that iron extractability is not the only factor determining siderophore production in soil. Some evidence is available showing that factors influencing the mass flow of iron chelated by the siderophore to the growing microbial biomass largely determine the levels of siderophores produced in soil. These factors are mainly the diffusion rate of siderophores in soil and the adsorption of siderophores to the soil–humus complex. Indeed, in soil, siderophore production was found to be inversely correlated with the water activity. Siderophore production upon the addition of C and N sources to the soil was more than three times higher in the dry soil than in the wet soil (Bossier and Verstraete, 1986b). So, it could be that the decreased diffusibility of the siderophores in the dry soil decreased the iron flux to the growing microbial biomass, upon which siderophore production was stimulated. However, there is an alternative explanation for this inverse relationship. In soils with low water activity, fungal growth is favored above bacterial growth, resulting in a higher fungal siderophore production. It must be remembered that the bioassay used by Bossier and Verstraete (1986b) particularly detects fungal siderophores.

The influence of adsorption of siderophores to the soil–humus complex on siderophore production is quite difficult to assess. It was found by Powell *et al.* (1980) that siderophores could still be extracted out of the soil after ten consecutive extractions. This suggests that a considerable reservoir of humus-bound siderophores is present. They also showed that there exists an equilibrium between water dissolved siderophores and humus-bound siderophores. This equilibrium is dependent on soil type and varies with the soil–water ratio. Additionally, Bossier and Verstraete (1986a) showed that in soil suspension cultures the growth of *Arthrobacter* JG-9 in the presence of given amounts of ferrioxamine B varies considerably and is dependent on soil type. Soils high in EDTA-extractable iron promoted the growth of *Arthrobacter* JG-9 better than soils low in extractable iron. It was also noticed that, in soil suspension cultures of soils low in extractable iron, growth of *Arthrobacter* JG-9 was repressed compared to cultures where no such soil was present. Based on this information, it was postulated that siderophores adsorb strongly to the soil–humus complex, thus becoming unavailable for the assay microorganism. However, when the siderophore becomes iron saturated (a process that proceeds faster in a soil high in EDTA-extractable iron), it has less affinity for the soil–humus complex and dissolves in the medium, increasing the availability for ferric siderophores for the siderophore auxotrophic *Arthrobacter* JG-9. Differences may exist, however, between siderophores. The ferrated ferrichrome with a neutral charge appeared to diffuse better in the soil than the positively charged ferrated ferrioxamine B (Reid *et al.*, 1985).

In conclusion, it can be stated that the availability of substrates is the driving force behind siderophore production in soil. The ultimate concentrations found in the soil may depend upon the water activity in the soil, the adsorption capacity for siderophores, the humus content, and the iron extractability.

3.1.3. Role of Fungal Hydroxamate Siderophores

3.1.3a. Iron Supply of the Fungus. Siderophores are obviously produced by fungi as a response to a shortage in the supply of iron. Under oxygenating conditions, Fe^{3+} hardly exists as a soluble ion. As a matter of fact, the amount of soluble Fe^{3+} ions at neutral pH in equilibrium with the so-called soil Fe (the amorphous iron hydroxide form in soil) is only of the order of $10^{-18.3}$ (Lindsay, 1979). Hence, in oxic soils even containing ample levels of total Fe or extractable Fe, the amount of iron available to fungi can be quite strategic.

Iron nutrition by the aid of siderophores is considered to occur according to the active transport system proposed by Neilands (1981). Yet, it differs at the molecular level from fungus to fungus. For instance, *Neurospora crassa* can use many siderophores of the ferrichrome group, except for ferrirubin. Coprogen is taken up as well (Huschka *et al.*, 1985). Based upon the configuration of the peptide backbone at the iron center and the competitivity between the two groups of siderophores for receptor sites, the latter authors conclude that there exist different receptor sites for ferrichrome and coprogen siderophores. However, they have to share the same transport system. It is not known if a multiple-receptor-site system gives a fungus a benefit over other fungi with respect to iron nutrition. *Penicillium parvum*, for example, has receptors for ferrichrome and no receptors for coprogen (Huschka *et al.*, 1985).

3.1.3b. Survival Strategy. It has been a custom to subdivide siderophores into sideramines and sideromycins. The latter exhibit an antibiotic action against Gram-positive as well as some Gram-negative bacteria (Knüsel *et al.*, 1969). In general it has always been difficult to attribute an ecological function to the capacity of certain microorganisms to produce antibiotics. An ecological function may be found in the process of spore germination and emergence. As outlined by Charlang *et al.* (1981), siderophores accumulated in the spores of certain fungi are essential germination factors. In the soil environment, the germinating spore secretes its siderophore in order to provide the growing biomass with iron. In the meantime, however, the majority of the readily biologically available iron in the vicinity of the germinating spore could be chelated by the siderophores. In this way, the availability of iron for other microorganisms can be reduced, thus inhibiting their growth. This could make more nutrients available for the germinating spore. In this way, spore siderophores could have a bacteriostatic or fungistatic activity. Recently a similar function has been outlined for certain antibiotics, such as streptomycin found in the spores of *Streptomyces griseus* (Szabo *et al.*, 1985) and for gramicidin found in the spores of *Bacillus brevis* (Murray *et al.*, 1985). In the case of *Bacillus brevis*, gramicidin could even be an agent controlling the amount of emerging spores of that species. Indeed, it has been shown that germinating spores are sensitive to gramicidin, whereas emerging (development stage subsequent to germination) spores are no longer sensitive. In this way, the available nutrients in the surrounding environment are only used by a limited number of outgrowing spores, increasing their chance to develop and mature and eventually to form spores again.

3.1.3c. Disease Development. Disease development of foliar pathogens can be remarkably influenced by iron. This has been shown for *Colletotrichum musae* on banana fruits, *Colletotrichum lindemuthianum* on *Phaseolus vulgaris*, and *Botrytis cinerea* on *Vicia faba* (Brown and Swinburne, 1981; Slade and Swinburne, 1985a,b). In the case of *Colletotrichum musae*, conidia harvested from iron-deficient media induced lesion formation more rapidly than conidia of iron-replete media. The addition of iron-free siderophores of a *Pseudomonas* sp. to iron-replete conidia also increased disease development. On the contrary, fully chelated siderophores, when put together with the conidia on the host, repressed disease development. The exact mechanism behind these phenomena is not yet totally unraveled. However, it is known that inoculation of the host with iron-replete conidia causes the accumulation of antifungal compounds, phytoalexins (Brown and Swinburne, 1981). It could be that the production of phytoalexin is repressed in the host by iron-deficient conditions. It is likely that during the production of iron-deficient conidia, iron-free siderophores are accumulated in the conidia, and are released from the spore upon germination (Charlang *et al.*, 1981). In iron-replete conidia fewer siderophores could be present, since siderophores are no longer necessary to deliver iron to the germinating conidia. Through the release of iron-free siderophores from the conidia, iron could be scavenged in the immediate vicinity of the conidia, making that element unavailable for the host metabolism. If iron or iron-containing enzymes were involved in the production of phytoalexins, the production of those compounds could be repressed, giving the pathogen the opportunity to form lesions very rapidly. However, caution is warranted, since the hypothesis outlined above remains to be proven. It would be surprising that only iron determines the outcome of the interactions between the host and the pathogen.

   3.1.3d. Iron or Phosphorus Supply of the Plants. Powell *et al.* (1980) first indicated that the concentration of siderophores in soil is high enough to increase the total amount of water-soluble iron in the soil. Based on the work of Lindsay (1979), they concluded that siderophores satisfy all requirements to serve as iron sources for plants or to serve as chemicals increasing the mass flow of iron to the plant roots. Previously, it was shown by Stutz (1964) that iron could be used by tomato plants when supplied as ^{59}Fe-ferrioxamine B. Page (1966) found *Arthrobacter* JG-9-detectable substances in the leaves of several crops, indicating the translocation of siderophores through the plant. It was shown that ferrichrome and ferrichrome A were better iron sources than EDTA for tomato plants in hydroponic solutions (Neilands, 1979). By the aid of an autoradiographic technique Becker *et al.* (1986) could demonstrate an enhanced uptake of Fe in the presence of agrobactin (catecholate siderophore of *Agrobacterium tumefaciens)* in an agar medium. In addition, it was shown by Szaniszlo *et al.* (1981) that the ectomycorrhizal fungus *Boletus edulis* produces siderophores in axenic cultures. The list of ectomycorrhizal fungi producing siderophores has now been extended to nine species (Powell *et al.*, 1982). All those data are strong indications that siderophores present in the soil may be, in one way or another, iron sources for the plant.

   In general, iron acquisition from microbial chelates by plants may occur by three different mechanisms (Szaniszlo *et al.*, 1985): (1) inorganic iron is taken up after the extracellular dissociation of the ferrated chelate; by this mechanism, chelates can only increase the mass flow of iron from iron precipitates to the plant roots (shuttle mechanism); (2) iron is removed directly at receptors on the plasmalemma (direct-removal mechanism); and (3) the intact iron-chelate is taken up (carrier permease mechanism).

Where a plant uses only mechanism 1 to acquire iron from microbial siderophores, it becomes less chlorotic when the concentration of siderophores in nutrient solution is increased, as long as the molar concentration of siderophores does not exceed the total amount of iron. Indeed, by increasing the siderophore concentration, the mass flow of iron from the iron hydroxide precipitates to the plant roots is increased. Where the siderophore concentration is equal to the total iron concentration or exceeds it, the plant will suffer iron deficiency and will develop chlorosis.

Where a plant uses mechanism 2 or 3 to acquire iron from the siderophore, it does not develop chlorosis when the concentration of siderophores exceeds the iron concentration. This is of course only true when the total amount of iron is sufficient for the normal development of the plant. According to Cline *et al.* (1984), it appears possible to differentiate between Fe-efficient and Fe-inefficient plant species or genotypes on the basis of their ability to use siderophores to absorb iron. The Fe-efficient plant species or genotypes respond to Fe-deficiency stress by secreting hydrogen ions and phenolic reductants. This enables the plant to take up iron at a higher rate (Römheld *et al.*, 1982). In addition, a dramatic increase in the activity of a plasma membrane-bound "reductase" can be noticed in such plants (Bienfait *et al.*, 1982, 1983). Whereas Fe-efficient plant species use mechanism 2 or 3 to obtain iron from siderophores, Fe-inefficient plants use only mechanism 1. Additional evidence for the use of mechanism 2 or 3 by Fe-efficient plants is that the inert Cr-ferrichrome could competitively inhibit iron uptake from either inorganic iron or Fe-ferrichrome in an experiment with an Fe-efficient oat (Szaniszlo *et al.*, 1985).

Not all siderophores function equally well as an iron source for plants. Ferric iron in ferrioxamine B is not easily reduced by the plasma membrane-bound reductase found in peanut plants (Römheld and Marschner, 1983). An iron-efficient corn cultivar could obtain iron from ferrichrome when the concentration of the chelate was in excess of the total iron concentration, but not from ferrichrome A or ferrioxamine B under the same conditions (Szaniszlo *et al.*, 1985). This may indicate that the acquisition of iron by plants from microbial siderophores is siderophore-specific.

Plants do not necessarily rely on microbial chelates for their iron acquisition. Monocotyledonous species, especially grasses, seem to have their own high-affinity uptake system of Fe^{3+}. This system is based on the production of phytosiderophores, such as mugineic acid (Sugiura and Nomoto, 1984), and a specific transport system for ferrated phytosiderophores installed in the plasma membrane (Römheld and Marschner, 1986). In this way grasses should be totally independent of microorganisms to mobilize Fe^{3+} in the environment. It remains to be unraveled how important this mechanism is in the soil, since phytosiderophores are easily decomposed by microorganisms (Römheld and Marschner, 1986). However, the latter statement can also be made for microbial siderophores. Microorganisms that can grow on siderophores as single C and N source can easily be isolated (Warren and Neilands, 1965; W. Verstraete, unpublished results). These considerations further emphasize the need for the in-depth study of formation, function, and degradation of siderophores *in situ*.

Our present knowledge about the production and function of microbial siderophores in soil is schematically summarized in Fig. 2. The driving force behind the whole process is the secretion of root exudates by the plant. Rhizosphere microorganisms can use those exudates to grow and produce siderophores. The production of siderophores is possibly stimulated if the plant secretes siderophore precursors (e.g., L-ornithine). Upon release of the siderophore in the soil environment, the siderophore will diffuse away from the microbial cell. In case the siderophores chelate iron, they can diffuse back to the rhizosphere, where they can be used by the rhizomicroorganisms or the plant as an iron source. The exact mechanism by which plants acquire iron from siderophores is not yet clear. The diffusion of the siderophores increases with increasing water activity in the soil. If diffusion is the rate-limiting step, microorganisms can respond by the production of more siderophores. The rate of saturation of the siderophore with iron depends upon iron extractability and upon the humus content of the soil. The latter two parameters are also linked, since soils rich in humus normally contain more EDTA- or DTPA (diethylenetriaminepentaacetic acid)-extractable iron. The delivery of iron to the microorganisms or the plant could be retarded through the sorption capacity of humus for siderophores. In case of ferrioxamine B, it has been shown that the noniron form is adsorbed more effectively by the humus complex than the iron form. Consequently, in a soil poor in extractable iron, the formation of ferric siderophores is limited by iron extractability itself, which in turn increases the adsorption of the siderophore. The overall result is that in a soil poor in extractable iron more siderophores will have to be produced to obtain an equal rate of iron mass flow than in a soil rich in extractable iron. So in the soil, siderophores are produced to increase the mass flow of iron from the soil–humus complex to the site of iron consumption, i.e., the rhizosphere.

The diffusion model presented in Fig. 2 is in part confirmed by the results of Reid *et al.* (1985). They showed that siderophores (ferrioxamine B and ferrichrome) do indeed speed up the diffusion of iron in soil with

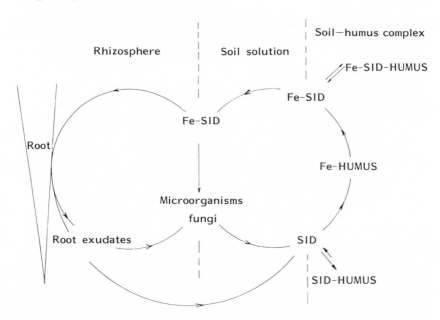

Figure 2. Model of the functioning of fungal and other microbial hydroxamate siderophores (SID) in soil.

respect to water. Siderophores were also more efficient in increasing iron diffusion than Fe chelates with lower stability constants (citrate, oxalate, and EDTA), at least in the soil investigated, with a pH of 7.5.

As mentioned above, plants can use siderophores as a source of iron by reducing the pH in the rhizosphere and by secreting reductants in order to obtain the iron by a reductive release of the Fe^{3+} from the siderophores. Membrane-bound reductases may also be involved in this process. Iron is thus taken up in the ferrous form. A type of reductant often released by the plant roots is a dihydroxyphenolic compound (Olsen *et al.*, 1982). Yet, it appears from the results of El Sayed *et al.* (1986a–c) that, during autotrophic nitrification, quantitatively important amounts of phenolics can become nitrosated and subsequently transformed to aminophenols. Such nitrification-mediated nitrosation reactions could withdraw compounds involved in the reduction of ferric to ferrous ion. It is not clear if this could constitute a quantitatively important mechanism. If we assume that for a crop, 10 kg Fe/ha is taken up, and that, per mole of Fe, a surplus of 10 moles phenolics is excreted and subsequently nitrosated with 1 mole N, we then deal with quantities of ∼20 kg N/ha thus fixed. If Fe^{3+} reductants were nitrosated by the action of autotrophic nitrifiers, plants would remain more dependent on the direct use of siderophores as an iron source. Obviously, this hypothesis is very premature, but points to some unexplored questions. Quite intriguing in this respect is the question of whether iron availability in the rhizosphere could have regulatory impact on the nitrification-mediated nitrosation reactions.

In most soils, the availability of iron is largely governed by Fe(III) oxides and rarely by iron phosphates. The latter appear to be present only in sandy kaolinite clays. Reid *et al.* (1985) present evidence that hydroxamate siderophores, through the mobilization of iron in acid soils containing insoluble Fe phosphate, can increase the level of soluble phosphorus. They further speculate that possibly VAM fungi, which produce hydroxamate siderophores, solubilize solid-phase iron phosphate for the mycorrhizal plants.

3.2. Bacterial Siderophores in Soil

In this section, we deal primarily with the yellow-green fluorescent siderophores or pyoverdines produced by fluorescent *Pseudomonas* species, as these compounds appear to play an important ecological role in soil. The pyoverdines, synthesized by different fluorescent *Pseudomonas* strains, seem to have a common structure with only minor structural differences. Demange *et al.* (1985) elucidated the structures of several pyoverdines and found them to be very similar to pyoverdine Pa, produced by *Pseudomonas aeruginosa,* and pseudobactin, produced by *Pseudomonas B10* (Wendenbaum *et al.,* 1983; Teitze *et al.,* 1981). The latter siderophores show in their turn analogy with pseudobactin 7SR1, produced by a plant-deleterious *Pseudomonas* strain (Yang and Leong, 1984), pyoverdine produced by *Pseudomonas fluorescens* (Philson and Llinas, 1982), and the fluorescent pigment produced by *Pseudomonas syringae* (Torres *et al.,* 1986).

3.2.1. Levels

The presence of pyoverdines in soil has been demonstrated in an indirect way by means of siderophore-negative mutant strains of fluorescent *Pseudomonas* spp. Bakker *et al.* (1986) have demonstrated *in situ* siderophore production by an increased root colonization of sid⁻ mutants in the presence of their wild-type isolates. Apparently, the sid⁻ mutants benefit from the production of the siderophores of the wild-type strain. Unfortunately, no bioassay with a pyoverdine auxotrophic strain has been developed. As a consequence, no data on levels of pyoverdine siderophores in soils are available. Yet, when values of siderophore production by fluorescent pseudomonads in axenic cultures are compared

Table V. Production of Siderophores in Axenic Cultures

Microorganism	Siderophore concentration (mg/g biomass DW)	Reference
Boletus edulis (4-day-old culture)		Szaniszlo *et al.*
0 ppb Fe	4.2	(1981)
40 ppb Fe	0.6	
Pseudomonas sp. (40-hr-old culture)		Meyer and Abdallah
0 ppb Fe	875	(1978)
40 ppb Fe	236	

with the siderophore concentration in axenic cultures of *Boletus edulis,* it becomes clear that fluorescent pseudomonads have a potential for a very high siderophore production (Table V) that warrants further careful examination with regard to its overall ecological significance.

3.2.2. Environmental Influences

The influence of environmental factors on the production of pyoverdine compounds has, to our knowledge, not been studied systematically, certainly not in natural environments.

3.2.3 Role of Siderophores Produced by Fluorescent Pseudomonads

3.2.3a. Iron Supply of Pseudomonads. As for fungal siderophores, it is clear that at least *in vitro* bacterial fluorescent pigments are synthesized and excreted by iron-deficient cells and that these siderophores facilitate iron uptake by the bacterium (Meyer and Hornsperger, 1978). When Fe^{3+} is limiting, an inverse relationship exists between iron content of the medium and the amount of pigment synthesized (Meyer and Abdallah, 1978) (see also Table V). It has been reported that iron starvation induces the synthesis of membrane proteins involved in the uptake of Fe^{3+}–siderophore complexes (Deweger *et al.,* 1986; Hohnadel and Meyer, 1985). All fluorescent *Pseudomonas* strains tested by Hohnadel and Meyer (1985) produced additional outer membrane proteins in case of iron starvation. These proteins appeared to differ from one strain to another. Most strains could only incorporate iron chelated by their own pyoverdine and were not able to incorporate iron chelated by pyoverdine of different origin. *Pseudomonas aeruginosa* (ATCC 15692), *Pseudomonas chlororaphis* (ATCC 9446), and *Pseudomonas fluorescens* (ATCC 13525), however, presented some homology, since each of these strains could incorporate iron chelated by its own pyoverdine, but also, to a lesser extent, by the pyoverdines of the two other strains.

3.2.3b. Iron and Growth Factor Supply of the Plant. There is little evidence that bacterial siderophores serve as a source of iron for plants. Kloepper *et al.* (1980a) found no plant-growth-promoting effect after amending soil with ferric pseudobactin. Becker *et al.* (1985) stated that pseudobactin has even an inhibitory effect on the uptake of iron by *Zea mays* and *Pisum sativum* when these plants were grown under sterile or gnotobiotic conditions on media containing 20 μM $FeCl_3$. Their results showed that pseudobactin competes with plant roots for ferric ions. On the other hand, Jurkevitch *et al.* (1985) found a significant increase of chlorophyll content in peanuts grown in highly calcareous soils after treatment of these soils with the purified iron–pigment complex or with suspensions of fluorescent pseudomonads containing the iron–pigment complex. These results seem to indicate some direct use by the plant of the iron supplemented by the siderophore.

Iswandi (1986) tested 50 *Pseudomonas* strains for their plant-beneficial capacities by cultivating the inoculated plants on acid-washed sterile sand. Sixteen out of the 50 strains tested were beneficial to plant growth. All the beneficial strains were fluorescent pseudomonads. When these 16 strains were tested for their plant-beneficial capacities in plain, nonsterile soil, 9 out of 16 failed to stimulate the plant growth. These strains probably lacked the ability to compete with indigenous soil microorganisms in colonizing the plant root. It is important to note, however, that testing isolates for plant-growth-promoting capacity under gnotobiotic conditions sets apart those strains that have a direct effect on the plant rather than promoting plant growth in an indirect way by protecting the plant against possible deleterious microorganisms. A direct effect on plant growth other than on Fe supply could be caused, for example, by the production of plant growth-promoting substances such as auxins and gibberellins, which are known to be produced by certain fluorescent pseudomonads (Stone and Strominger, 1972; Wehrli and Staehelin, 1971), or production of a phosphate-solubilizing compound, which is reported for a *Pseudomonas putida* strain (Grimes and Mount, 1984). With regard to the auxin hypothesis, a most interesting aspect to study is the possible-growth-stimulating effect of the breakdown products of the pseudomonad hydroxamate siderophores. It is quite conceivable that upon biodegradation of the complex molecules by the producing strain itself or by surrounding soil microbiota such effective metabolites are produced.

3.2.3c. Repression of Competitive Microorganisms. Some fluorescent pseudomonads can exhibit *in vitro* antibiosis against other microorganisms on media low in iron. This antibiosis effect is antagonized by the addition of Fe(III) to the culture medium (Kloepper *et al.*, 1980a). It was also shown that purified pseudobactin, a fluorescent siderophore produced by *Pseudomonas* B10, caused *in vitro* antibiosis against *Gaeuman-*

nomyces graminis and *Fusarium oxysporum* (Kloepper *et al.*, 1980a). Vandenbergh *et al.* (1983) demonstrated that *Pseudomonas putida* PPU3 inhibited the growth of eight fungi tested, whereas the sid⁻ mutant of PPU3 (obtained by NTG mutagenesis) exhibited no antibiotic effect at all. The purified iron-chelating compound of PPU3 inhibited fungal growth, whereas the iron-saturated chelate did not. The results strongly suggest that fluorescent pseudomonads can produce fluorescent siderophores in iron-deficient conditions that *in vitro* are responsible for antibiosis, by depriving other microorganisms of iron. Iswandi (1986) tested the inhibitory effect of eight fluorescent *Pseudomonas* strains isolated from soil on the growth of five fungi on King medium B with and without addition of Fe^{3+} (Table VI). In general, the inhibitory effect of the *Pseudomonas* strains was more pronounced on King medium B without iron. The inhibitory effect was reduced or disappeared completely after amendment of the medium with 5 ppm Fe^{3+}. This suggests that siderophores are at least in part responsible for the inhibitory effect. However, the strains A1K25 and B1K1 inhibited the growth of *Ustilago hordei* more when iron was added, indicating that these strains produce antifungal compounds other than siderophores. Similar phenomena have been described by Hemming *et al.* (1982). It can also be noticed that a certain *Pseudomonas* strain has a strong inhibitory effect on one fungal strain, while not affecting the growth of another.

These observations suggest that each *Pseudomonas* strain has a specific siderophore production and that each fungal strain has a different sensitivity toward these siderophores. Evidence for a possible antibiotic role of fluorescent siderophores in soil is given by the work of Kloepper *et al.* (1980b) and Scher and Baker (1982). Addition of *Pseudomonas* B10 or pseudobactin to soils that were conducive to take-all of wheat (caused by *Gaeumannomyces graminis*) and *Fusarium* wilt (caused by *Fusarium oxysporum*) renders these soils suppressive, while addition of ferric pseudobactin had no effect at all (Kloepper *et al.*, 1980b). Scher and Baker (1982) showed that adding *Pseudomonas putida* or EDDA to soils that are conducive to *Fusarium* wilt pathogens renders these soils suppressive. Competition for Fe seems to be responsible for the suppressiveness in these systems. It is postulated that by adding *Pseudomonas B10* or *Pseudomonas putida* to conducive soils, siderophores are produced that bind Fe making it unavailable for the pathogens.

Most commonly, the role of siderophore production by fluorescent pseudomonads is related to the inhibition of so-called minor pathogens in the root zone. By reducing the deleterious action of these organisms on plant growth (Suslow and Schroth, 1982; Burr and Caesar, 1984) it is conceivable that the *Pseudomonas* species themselves achieve a larger amount of nutrients and energy, because the amount of root exudates is

Table VI. Inhibitory Effect of Fluorescent *Pseudomonas* Strains on the Growth of Five Soil Fungi[a]

Pseudomonas strain	*Gaeumannomyces graminis* f. *tritici*		*Fusarium oxysporum* v. *lini*		*Ustilago hordei*		*Fusarium culmorum*		*Trichoderma* sp.	
	KB	KB + Fe³⁺	KB	KB + Fe³⁺	KB	KB + Fe³⁺	KB	KB + Fe³⁺	KB	KB + Fe³⁺
7NSK2	3.5	2.0^b	3.0	1.5^b	2.6	1.4^b	3.0	1.5^b	6.0	4.0^b
15GK7	3.0	1.5^b	0.5	0.0	0.5	0.0	0.5	0.0	3.0	0.5^b
20FKII3	7.5	6.0^b	6.0	4.0^b	4.0	3.5^b	1.0	1.0	10.0	2.0^b
CNP19	5.0	5.0	8.3	2.7^b	4.0	2.5^b	3.0	1.5^b	4.0	1.0^b
BNP17	3.0	3.0	3.0	0.5	5.1	3.2^b	3.2	1.5^b	0.0	0.0
B1K1	7.0	3.5	4.0	0.7^b	4.0	6.1^b	4.5	2.0^b	5.0	2.0^b
A1K25	0.5	0.5	4.0	0.5^b	8.0	9.7^b	3.2	1.7^b	7.0	4.0^b
ANP15	0.5	0.5	0.0	0.0	7.0	5.2^b	2.5	2.5	0.5	0.0

[a] Iswandi (1986). The values are the average of six measurements.
[b] Significant difference between KB and KB + Fe (5 ppm Fe as Fe citrate) at $t = 0.05$.

proportional to the amount of overall photosynthesis and crop dry matter production. The fact that large quantities of nutrients (1–25% of net photosynthesis) are released into the soil has only recently become established (Kraffczyck et al., 1984; Haider et al., 1985; Haller and Stolp, 1985).

The results in Table VII illustrate that indeed the stimulatory effect of a plant-growth-promoting strain increases with increasing soil microbial activity. The latter can be the result of the addition of organic carbon or of the precropping of the soil before the onset of the experiment. A question that remains to be answered concerns the mechanisms by which siderophores produced by fluorescent pseudomonads counteract possible deleterious microorganisms. It was stated that the deleterious microorganisms are unable to obtain essential quantities of iron for growth in the presence of fluorescent pseudomonads because they either do not produce siderophores, produce comparatively fewer siderophores, or produce siderophores that have less affinity for iron than those of the plant-growth-beneficial pseudomonads (Kloepper et al., 1980a). However, it was demonstrated by Yang and Leong (1984) that siderophore production in deleterious strains often equaled or surpassed that of plant-

Table VII. Effect of Seed Inoculation with the *Pseudomonas* Strain 7NSK2 on the Growth of Barley (cv. Iban) in Soils with Different Levels of Microbial Activity[a]

Soil pretreatment	Soil respiration (mg CO_2 C kg^{-1} soil day^{-1})	Plant growth (g/pot) Blank	Plant growth (g/pot) Inoculated	Increase (%)
Addition of C and N sources				
None	3.00a	0.223	0.249	11.6
66 mg C + 6.6 mg N[b]	3.19b	0.208	0.236	13.5[d]
660 mg C + 66 mg N[b]	4.20c	0.193	0.238	23.3[e]
66 mg C + 6.6 mg N[c]	3.18b	0.215	0.240	11.6[d]
660 mg C+ 66 mg N[c]	4.82d	0.208	0.242	16.3[e]
Precropping period with barley				
None	1.67a	0.437	0.459	5.0
1 month	1.97b	0.394	0.414	5.1
2 months	2.95c	0.360	0.410	16.9
3 months	3.49d	0.350	0.410	18.0[d]
4 months	2.50e	0.335	0.389	16.0

[a]Iswandi (1986). Data followed by the same letter are not significantly different.
[b]C sources: galactose, glucose, arabinose, xylose (each sugar 25% of the total C). N source: NH_4NO_3.
[c]C sources: sugars as in footnote b. N source: L-asparagine and aspartic acid.
[d]Significant at 5%.
[e]Significant at 10%.

growth-promoting strains. The latter authors unraveled the structure of pseudobactin 7SR1, a siderophore from a plant-deleterious *Pseudomonas*, and they found it to be remarkably similar to pseudobactin. A key difference between pseudobactin and pseudobactin 7SR1 was that the fluorescent group of pseudobactin 7SR1 was attached via an ester bond to the peptide, whereas that of pseudobactin was attached via an amide bond. Yang and Leong (1984) stated that for this reason pseudobactin 7SR1 might be more susceptible to hydrolysis than pseudobactin. Such hydrolysis might be expected to diminish the ability of pseudobactin 7SR1 to bind Fe(III) and hence might adversely affect the growth of *Pseudomonas 7SR1* in iron-limiting environments.

Seed inoculation of barley with the plant-beneficial *Pseudomonas* strain 7NSK2 was found by Iswandi (1986) to give rise to a significant decrease of other root microorganisms in the inoculated treatments (Table VIII), indicating that this strain colonizes the roots of barley aggressively and successfully represses competitors.

Schippers *et al.* (1986) reported increased potato plant dry weight and tuber production in pot, as well as in field, experiments after inoculation with fluorescent *Pseudomonas* strains. This only occurred in soils frequently cropped to potato and not in soils from the same field with no history of potato cultivation. Sid^- mutants of the fluorescent *Pseudomonas* (obtained by Tn5 mutagenesis) did not increase potato root development in continuous cropped soil, whereas the wild type did. Their hypothesis is that plant-growth stimulation by *Pseudomonas* isolates is the result of the suppression of the activity of cyanide-producing microbial populations. Siderophore-mediated competition for Fe^{3+} may thus suppress microbial cyanide production in the rhizosphere and diminish effects harmful to roots (Schippers *et al.*, 1986). Ahl *et al.* (1986) studied the suppression of black root rot (caused by *Thielaviopsis basicola*) by a *Pseudomonas fluorescens* strain CHA0 on roots under gnotobiotic conditions and *in vitro*. Under these conditions the strain produced siderophores, cyanic acid, and several antibiotics. All these compounds were able to inhibit the growth of *Thielaviopsis basicola in vitro*. Remarkably, the iron-bound siderophores and not the iron-free siderophores had a strong inhibitory effect on *Thielaviopsis basicola* in liquid media. Ahl *et al.* (1986) try to explain the toxicity of the iron-bound siderophores by stating that siderophores make iron more available and therefore more toxic to the fungus or that the complexed iron siderophores have by themselves an inhibitory effect such as antibiosis. It remains to be proven, however, that the production of siderophores, cyanic acids, or antibiotics is directly involved in disease control in soil.

Table VIII. Effect of Seed Inoculation with the Rhizopseudomonad Strain 7NSK2* on the Microbial Composition of the Roots of the Barley Cultivar Iban, Grown in Soils with Different Levels of Microbial Density and Activity

Pretreatment of soil[a]	Seed inoculation[b]	CFU/g root[c] (dry weight)				
		Total[c]	Rhizopseudomonads	7NSK2*	Fungi	Coliform bacteria
Root (Rhizoplane and endorhizosphere)						
A1	BL	1580×10^5	930×10^5	0	741×10^4	—
	IN	460×10^{5e}	930×10^5	270×10^5	89×10^{4e}	—
A3	BL	$10{,}230 \times 10^5$	950×10^5	0	1122×10^4	—
	IN	1820×10^{5e}	1290×10^{5d}	790×10^5	102×10^{4e}	—
A5	BL	9550×10^5	550×10^5	0	1549×10^4	—
	IN	2890×10^{5e}	760×10^{5d}	540×10^5	12×10^{4c}	—
Endorhizosphere						
A1	BL	11×10^5	7×10^5	0	—	55×10^2
	IN	8×10^{5d}	7×10^5	6×10^5	—	7×10^{2e}
A3	BL	14×10^5	8×10^5	0	—	141×10^2
	IN	6×10^{5e}	3×10^{5e}	4×10^5	—	12×10^{2e}
A5	BL	16×10^5	5×10^5	0	—	42×10^2
	IN	2×10^5	2×10^{5d}	2×10^5	—	14×10^{3e}

[a]A1: No pretreatment of the soil, control; A3: 600 mg C added/kg soil (25% galactose C, 25% glusoce C, 25% arabinase C, 25% xylose C), C/N ratio of the amendment adjusted to 10 with NH_4NO_3. A5: Addition of L-asparagine and aspartic acid (70 mg N/kg soil). Sugars were added to reach a total C amount of the amendments equivalent to 660 mg C/kg soil.
[b]BL, Blank; IN, inoculation.
[c]Total = total bacterial count; —, not determined.
[d]Significantly different from the blank at a level of 10%.
[e]Significantly different at a level of 5%.

3.3. Multiple Hydroxamate Siderophore Production by Fungi and Bacteria

One of the amazing features of siderophore production by fungi is the fact that one species may produce several distinct siderophores. This has been reported, for instance, for *Epicoccus purpurascens* (Frederick *et al.*, 1981), *Boletus edulis* (Szaniszlo *et al.*, 1981), *Neurospora crassa*, *Aspergillus nidulans*, and *Penicillium chrysogenum* (Charlang *et al.*, 1981), and *Aspergillus ochraceous* (Jalal *et al.*, 1984). In the case of *Epicoccus purpurascens*, *Boletus edulis*, and *Aspergillus ochraceous*, up to ten or more different siderophroes could be detected in the culture supernatant. Such a diversity in iron-chelating compounds is without any apparent function, except for their iron-chelating feature.

For bacteria, multiple siderophore production has been reported in axenic cultures of fluorescent pseudomonads. *Pseudomonas* B10 produces pseudobactin and pseudobactin A. The latter component is nonfluorescent and shows only a minor structural difference from pseudobactin. Both ferric pseudobactin and ferric pseudobactin A are equally efficient in transporting iron into *Pseudomonas* B10. Pseudobactin A was produced only in iron-limiting, chemically defined minimal media, whereas pseudobactin production occurred also in nutritionally richer, iron-deficient media (Teintze *et al.*, 1981; Teintze and Leong, 1981).

3.3.1. Sequential Derepression

There are some indications that multiple siderophore production is at least partly due to the special environment created in artificial media. Those media are specially designed to be low in iron and are sometimes treated with 8-hydroxyquinoline or passed over a Chelex 100 column in order to reduce the iron concentration drastically. In rich media low in iron, siderophore production is derepressed, resulting in an overproduction probably both in quantity and nature.

One of the main characteristics of those artificial media is the absence of a precipitated iron form with which siderophores can exchange iron. Such a bulk of precipitated iron is present in soils. The consequence of the presence of mineral iron or humus-bound iron in artificial media for siderophore production was outlined by W. J. Page and Huyer (1984) for *Azotobacter vinelandii*. This bacterium is able to produce three compounds with either low (dihydroxybenzoic acid) or high (azotobactin and azotochelin) affinity for iron. By the addition of insoluble iron sources to iron-limited cultures, it was possible to increase iron availability. The results of W. J. Page and Huyer (1984) provide evidence of a sequential derepression of the siderophore production. Indeed, in the presence of marcasite or humic and fulvic acid, azotobactin and azotochelin production was repressed. Vivianite [$Fe_3(PO_4)_2 \cdot 8H_2O$], olivine, and Fe_3O_4 derepressed azotochelin, whereas azotobactin is still repressed. Hematite, siderite, pyrite, and goethite caused partial derepression of azotobactin and partial to full derepression of azotochelin production. Finally, illite, ilmenite, and micaceous hematite were poor sources of iron, resulting in hyperproduction of both azotochelin and azotobactin. It is not known if azotochelin is a weaker iron-chelating agent than azotobactin. According to Stiefel et al. (1980), the stability constant (log K) of both compounds is 25. However, this does not necessarily mean that they chelate iron equally strongly. The pM values (Raymond et al., 1984), which are the best parameters with which to compare chelates, are to our knowledge not available in the literature, where pM stands for the negative log of free Fe^{3+}-ion concentrations at pH 7.4 when the total concentration of iron is 10^{-6} M and the concentration of the ligand is 10^{-5} M. If this sequential derepression is linked to the iron extractability and hence availability, azotobactin should have a higher pM value than azotochelin.

Similar experiments with fungi that produce multiple hydroxamates are not available. As shown by Charlang et al. (1981) and Jalal et al. (1984), hydroxamate siderophores in iron-deficient culture are produced in sequence. In cultures of Aspergillus ochraceous, ferrichysin was the major siderophore after 4 days of incubation; after 6 days of incubation ferrirubin was the major siderophore. It could well be that this sequential production of siderophores has something to do with the increasing iron stress in the medium due to the consumption of iron by the fungus Aspergillus ochraceous.

In order to obtain more insight into the relationship between iron availability in the soil and siderophore production, similar experiments to the one described by W. J. Page and Huyer (1984) should be performed with a variety of siderophore-producing microorganisms. At the same time, data on the apparent sequential siderophore productions should be gathered. A better simulation of the soil conditions in in vitro cultures could thus yield a better understanding of the ecology of siderophores.

3.3.2. Different Hydroxamate Siderophores Have Different Functions

Charlang et al. (1981) pointed out a possible other function for the different siderophores produced by Neurospora crassa, Aspergillus nidulans, and Penicillium chrysogenum. Typically those fungi produce siderophores that do not appear in the culture supernatant and that are cytoplasmic or cell-membrane bound. *Neurospora crassa* secretes coprogen as an iron-mobilizing agent. Ferricrocin is also produced, but is kept within the cell. *Aspergillus nidulans* secretes ferricrocin, ferrirhodin, and triacetylfusigen (and some uncharacterized siderophores in young cultures). Only ferricrocin and triacetylfusigen are found intracellularly. In the spores of this fungus only ferricrocin could be found. Lastly, *Penicillium chrysogenum* secretes coprogen (and uncharacterized siderophores in young cultures). Ferrichrome was found intracellularly and in spores. Spores of those fungi could be freed from siderophores by exposing them to a solution with low water activity. Since those siderophores are essential germinating factors, spores are no longer able to germinate. Conidial siderophores were able to restore this deficiency. However, not all extracellular siderophores could serve as a germinating factor. Triacetylfusigen, produced by *Aspergillus nidulans,* did not stimulate its siderophore-deficient conidia. Regarded in this way, multiple siderophore production has a physiological and ecological meaning, since distinct siderophores are attributed to separate functions.

Philson and Llinas (1982) reported the production by *Pseudomonas fluorescens* of various forms of pyoverdine and ferribactin besides the production of a purple iron-binding compound with no direct structural relation to ferribactin and pyoverdines. Ferribactin is an iron-binding pigment with a hydroxamate structure, but there is no evidence that it functions in iron transport. Relative to pyoverdines, ferribactin provides weaker iron binding. Ferribactin, unlike the pyoverdines, is not produced until late stages of growth. Philson and Llinas (1982) hypothesize that ferribactin is a precursor of pyoverdine, and only accumulates after the normal metabolism is blocked.

3.3.3. Maximizing Cooperative Interactions

It is well known that microorganisms can use certain siderophores as iron sources that they do not produce themselves (Huschka *et al.,* 1985). It would be interesting, therefore, to know if this feature could form a basis for interaction between microorganisms. Indeed, fungi could be able to select, for example, for certain microorganisms in their environment through the production of certain siderophores. An example of a positive interaction between microorganisms has recently been demonstrated by W. J. Page and Dale (1986). In a medium with insoluble iron sources, *Azotobacter vinelandii* could grow well by producing catechol siderophores that extracted iron from the insoluble sources. On the other hand, *Agrobacterium tumefaciens* could not grow in that medium. The chelating capacity of the catechol agrobactin is apparently not strong enough to mobilize iron. When cultured together, *Agrobacterium vinelan-*

dii was capable of promoting the growth of *Agrobacterium tumefaciens.* In this coculture *Agrobacterium tumefaciens* uses the siderophores of *Azotobacter vinelandii* only as an iron source. *Agrobacterium tumefaciens* did not use the *Azotobacter vinelandii* siderophores directly. So, on the basis of such observations *in vitro,* we can postulate interactions between microorganisms in natural environments via siderophores, by which a given organism, through the production of multiple forms of HS, maximizes its chances of soliciting a cooperative partner. However, these interactions are speculative and remain to be confirmed.

3.3.4. Minimizing Commensalism

Neurospora crassa can use a variety of ferrichrome types of siderophores and also coprogen. The uptake of both types of siderophore, however, is inhibited by ferrirhodin, a siderophore produced by some *Aspergillus* sp. It is not known if an *Aspergillus* sp. is able to inhibit *Neurospora crassa* in this way under natural conditions (for example, the soil environment) and thus reduce, for example, the risk of commensalism on its own siderophore.

Siderophores are relatively simple compounds susceptible to biodegradation. Microorganisms secreting siderophores may be faced with the problem that antagonizing microorganisms are degrading the siderophore before it can be taken up again by the producing microorganism. Warren and Neilands (1965) isolated a *Pseudomonas* sp. capable of using ferrichrome A as a sole C and N source. The importance of such a process in the interactions between microorganisms has not been evaluated thoroughly.

4. Conclusions

1. The capability of forming hydroxamate siderophores by means of heterotrophic nitrification is extremely widespread among microorganisms. The most important environmental factors involved appear to be Fe^{3+} availability and supply of an organic carbon and energy source.

2. There is substantial evidence that in neutral and alkaline oxic soils, the amount of available Fe^{3+} ions is of strategic significance. Siderophores in general and hydroxamate siderophores in particular, both in terms of their chemical characteristics (i.e., complex-forming capability with Fe^{3+}) and their concentration levels in soils, satisfy all requirements for playing an important role in the Fe^{3+} acquisition of both microorganisms and plants.

3. Both fungi and bacteria appear to produce a multitude of hydrox-

amate molecules. A variety of reasons for this are possible. Yet it is our opinion that both soil fungi and bacteria, by means of their siderophores, either directly or indirectly result in optimizing plant growth in order to obtain a maximum amount of energy and carbon supply via the root exudates.

References

Ahl, P., Voisard, C., and Défago, G., 1986, Iron bound-siderophores, cyanic acid and anti-biotics involved in suppression of *Thielaviopsis basicola* by a *Pseudomonas fluorescens* strain, *J. Phytopathol.* **116**:121–134.

Akers, H. A., 1981, The effect of waterlogging on the quantity of microbial iron chelators (siderophores) in soil, *Soil Sci.* **132**:150–152.

Akers, H.A., 1983a, Multiple hydroxamic acid microbial chelators (siderophores) in soils, *Soil Sci.* **135**:156–160.

Akers, H. A., 1983b, Isolation of the siderophore schizokinen from soil of rice field, *Appl. Environ. Microbiol.* **45**:1704–1706.

Atkin, C. L., Neilands, J. B., and Pfaff, H. J., 1970, Rhodotorulic acid from species of *Leucosporidium, Rhodosporidium, Rhodoturula, Sporodia bolus* and *Sporobolomyces* and new alanine-containing ferrichrome from *Cryptococcus melibiosum, J. Bacteriol.* **103**:722–733.

Bakker, P. A. H. M., Weisbeek, P. J., and Schippers, B., 1986, The role of siderophores in plant growth stimulation by fluorescent *Pseudomonas* spp. *Med. Fac. Landbouww. Rijksuniv. Gent.* **51**:1357–1362.

Becker, J. O., Hedges, R. W., and Messens, E., 1985, Inhibitory effect of pseudobactin on the uptake of iron by higher plants. *Appl. Environ. Microbiol.* **49**:1090–1093.

Becker, J. O., Messens, E., and Hedges, R. W., 1986, A convenient autoradiographic technique for the study of uptake of minerals by plants roots and the effects of environmental factors upon the process, *Plant Soil* **92**:299–302.

Bienfait, H. F., Druivenvoorden, J., and Verkerke, W., 1982, Ferric reduction of roots of chlorotic bean plants: Indications for an enzymatic process, *J. Plant Nutr.* **5**:451–457.

Bienfait, H. F., Bino, R. J., Van den Bliek, A. M., Duivenvoorden, J. F., and Fontain, J. M., 1983, Characterisation of ferric reducing activity in roots of Fe deficient *Phaseolus vulgaris, Physiol. Plant* **59**:196–202.

Bossier, P., and Verstraete, W., 1986a, A direct bioassay for the detection of hydroxamate siderophores in soil, *Soil Biol. Biochem.* **18**:481–486.

Bossier, P., and Verstraete, W., 1986b, Ecology of *Arthrobacter* JG9 detectable hydroxamate siderophores in soils, *Soil Biol.Biochem.* **18**:487–492.

Brown, A. E., and Swinburne, T. R., 1981, Influence of iron and iron chelators on formation of progressive lesions of *Colletotrichum musae* on banana fruits, *Trans. Br. Mycol. Soc.* **77**:119–124.

Burnham, B. F., and Neilands, J. A., 1961, Studies of the metabolic function of the ferrichrome compounds, *J. Biol. Biochem.* **236**:554–559.

Burr. T.J., and Caesar, A., 1984, Beneficial plant bacteria, *Crit. Rev. Plant Sci.* **2**(1):1–20.

Charlang, G., Bradford, N., Horowitz, N., and Horowitz, R. M., 1981, Cellular and extracellular siderophores of *Aspergillus nidulans* and *Penicillium chrysogenum, Mol. Cell Biol.* **1**:94–100.

Cline, G. R., Reid, C. P. P., Powell, P. E., and Szaniszlo, P. J., 1984, Effect of a hydroxamate siderophore on iron absorption by sunflower and sorghum, *Plant Physiol.* **76**:36–39.

Crueger, W., and Zähner, H., 1968, Stoffwechselproduktie von Mikroorganismen. 70 Mitteilung. Uber der Einfluss der Kohlenstoffquelle auf die Sideraminebildung von *Aspergillus melleus* Yukawa, *Arch. Mikrobiol.* **63**:376–384.

Demange, P., Wendenbaum, S., Bateman, A., Dell, A., Meyer, J. M., and Abdallah, M. A., 1985, Bacterial siderophores: Structure of pyoverdine and related compounds, Advanced Nato Research Workshop, London (July 1985), Abstract.

Deweger, L. A., Van Boxtel, R., Van der Burg., B., Gruters, R. A., Geels, F. P., Schippers, B., and Lugtenberg, B., 1986, Siderophores and outer membrane proteins of antagonistic, plant growth stimulating root-colonizing *Pseudomonas* spp, *J. Bacteriol.* **165**:585–594.

El Sayed, A., Verhé, R., Proot, M., Sandra, P., and Verstraete, W., 1986a, Binding of nitrite-N on polyphenols during nitrification, *Plant Soil* **94**:369–382.

El Sayed, A., Vandenabeele, J., and Verstraete, W., 1986b, Nitrification and organic nitrogen formation in soils, *Plant Soil* **94**:383–400.

El Sayed, A., Van Cleemput, O., and Verstraete, W., 1986c, Nitrification mediated nitrogen immobilization in soils, *Plant Soil* **94**:401–440.

Emery, T., 1966, Initial steps in the biosynthesis of ferrichrome: Incorporation of δ-N-hydroxyornithine and δ-N-acetyl δ-N-hydroxyornithine, *Biochemistry* **5**:3694–3701.

Focht, D. D., and Verstraete, W., 1977, Biochemical ecology of nitrification and denitrification, in: *Advances in Microbial Ecology*, Vol. 1 (M. Alexander, ed.) pp. 135–214. Plenum Press. New York.

Frederick, C. B., Szaniszlo, P. J., Vickrey, P. E., Bentley, M. D., and Shive, W., 1981, Production and isolation of siderophores from the soil fungus *Epicoccum purpurascens*, *Biochemistry* **20**:2432–2436.

Grimes, H. D., and Mount, M. S., 1984, Influence of *Pseudomonas putida* in nodulation of *Phaseolus vulgaris*, *Soil Biol. Biochem.* **16**:27–30.

Haider, K., Mosier, A., and Heinemeyer, O., 1985, Phytotron experiments to evaluate the effect of growing plants on denitrification. *Soil Sci. Soc. Am. J.* **49**:636–641.

Haller, T., and Stolp, H., 1985, Quantitative estimation of root exudation of maize plants, *Plant Soil* **86**:207–216.

Harrington, G. J., and Neilands, J. B., 1982, Isolation and characterisation of dimerumic acid from *Verticillium dahliae*, *J. Plant Nutr.* **5**:675–682.

Hemming, B. C., Orser, C., Jacobs, D. L., Sands, D. C., and Strobel, G. A., 1982, The effects of iron on microbial antagonism by fluorescent Pseudomonads, *J. Plant. Nutr.* **5**:683–702.

Hohnadel, D., and Meyer, J. M., 1985, Pyoverdine-facilitated iron uptake among fluorescent pseudomonads, Advanced Nato Research Workshop, London (July 1985), Abstract.

Huschka, H., Naegeli, H. V., Leuenberger-Ryf, H., Keller-Schierlein, W., and Winkelmann, G., 1985, Evidence for a common siderophore transport system but different siderophore receptors in *Neurospora crassa*, *J. Bacteriol.* **162**:715–721.

Iswandi, A., 1986, Seed inoculation with *Pseudomonas* spp., Ph.D. Thesis, State University of Gent, Faculty of Agricultural Sciences, Belgium.

Jalal, M. A. F., Mocharla, R., Barnes, C. L., Hossain, M. B., Powell, D. G., Eng-Wilmot, D. L., Grayson, S. L., Benson, B. A., and Van der Helm, D., 1984, Extracellular siderophores from *Aspergillus ochraceous*, *J. Bacteriol.* **158**:683–688.

Jurkevitch, E., Hadar, Y., and Chen, Y., 1985, The effect of *Pseudomonas* siderophores on iron nutrition of peanuts. Advanced Nato Research Workshop, London (July 1985), Abstract.

Kloepper, J. W., Leong, L., Teintze, M., and Schroth, M. N., 1980a, Enhanced plant growth by siderophores produced by PGPR, *Nature* **286**:885–886.

412   P. Bossier *et al.*

Kloepper, J. W., Leong, J., Teintze, M., and Schroth, M. N., 1980b, *Pseudomonas* sidero-phores: A mechanism explaining disease suppressive soils, *Curr. Microbiol.* **4**:317–320.

Knüsel, F., Schiess, B., and Zimmermann, W., 1969, The influence exerted by sideromycins on poly-U-directed incorporation of phenylalanine in the S-30 fraction of *Staphylococcus aureus*, *Arch. Mikrobiol.* **68**:99–106.

Kraffczyck, J., Trolldenier, G., and Beringer, H., 1984, Soluble root exudates of maize: Influence of potassium supply and rhizosphere microorganisms, *Soil Biol. Biochem.* **16**:315–322.

Lindsay, W. L., 1979, *Chemical Equilibria in Soils*, Wiley-Interscience, New York.

Meyer, J. M., and Abdallah, M. A., 1978, The fluorescent pigment of *Pseudomonas fluorescens:* Biosynthesis, purification and physico-chemical properties, *J. Gen. Microbiol.* **107**:321–331.

Murray, T., Lazaridis, I., and Seddon, B., 1985, Germination of spores of *Bacillus brevis* and inhibition by gramicidin S: A strategem for survival, *Lett. Appl. Microbiol.* **1**:63–65.

Neilands, J. B., 1979, Biomedical and environmental significance of siderophores, in: *Trace Metals in Health and Disease* (N. Kharash, ed.) pp. 27–41, Raven Press, New York.

Neilands, J. B., 1981, Iron absorption and transport in microorganisms, *Annu. Rev. Nutr.* **1**:27–46.

Neilands, J. B., 1982, Iron envelope proteins, *Annu. Rev. Microbiol.* **36**:285–309.

Olsen, R. A., Brown, J. C., Bennett, J. H., and Blume, D., 1982, Reduction of Fe^{3+} as it relates to Fe chlorosis, *J. Plant Nutr.* **5**:433–447.

Page, E. R., 1966, Sideramines in plants and their possible role in iron metabolism, *Biochem. J.* **100**:34.

Page, W. J., and Dale, P. L., 1986, Stimulation of *Agrobacterium tumefaciens* growth by *Azotobacter vinelandii* ferrisiderophores, *Appl. Environ. Microbiol.* **51**:451–454.

Page, W. J., and Huyer, M., 1984, Derepression of the *Azotobacter vinelandii* siderophore system, using iron-containing minerals to limit iron repletion. *J. Bacteriol.* **158**:496–502.

Perlman, D., 1965, Microbial production of metal-organic compounds and complexes, in: *Advances in Applied Microbiology*, Vol. 7 (W. W. Umbreit, ed.), pp. 103–138, Academic Press, New York.

Philson, S. B., and Llinas, M., 1982, Siderochromes from *Pseudomonas fluorescens*. I. Isolation and characterisation, *J. Biol. Chem.* **257**: 8081–8085.

Powell, P. E., Cline, G. R., Reid, C. P. P., and Szaniszlo, P. J., 1980, Occurrence of hydroxamate siderophore iron chelators in soils, *Nature* **287**:833–834.

Powell, P. E., Szaniszlo, P. J., Cline, G. R., and Reid, C. P. P., 1982, Hydroxamate siderophores in the iron nutrition of plants, *J. Plant Nutr.* **5**:653–673.

Powell, P. E., Szaniszlo, P. J., and Reid, C. P. P., 1983, Confirmation of occurrence of hydroxamate siderophores in soil by a novel *Escherichia coli* bioassay, *Appl. Environ. Microbiol.* **46**:1080–1083.

Raymond, K. N., Mueller, G., and Matzanke, B. F., 1984, Complexation of iron by siderophores. A review of their solution and structural chemistry and biological function, *Top. Curr. Chem.* **123**:49–102.

Reid, R. K., Reid, C. P. P., and Szaniszlo, P. J., 1985, Effects of synthetic and microbially producted chelates on the diffusion of iron and phosphorus to a simulated root in soil, *Biol. Fertil. Soils* **1**:42–45.

Römheld, V., and Marschner, H., 1983, Mechanism of iron uptake by peanut plants, 1. Fe reduction, chelate splitting and release of phenolics, *Plant Physiol.* **71**:949-954.

Römheld, V., and Marschner, H., 1986, Evidence for a specific uptake system for iron phytosiderophores in roots of grasses. *Plant Physiol.* **80**:175–180.

Römheld, V., Marschner, H., and Kramer, D., 1982, Responses of Fe-deficiency in roots of "Fe-efficient" plant species, *J. Plant Nutr.* **5**: 489–499.

Scher, F. M., and Baker, R., 1982, Effect of *Pseudomonas putida* and a synthetic iron chelator on induction of soil suppressiveness to *Fusarium* wilt pathogens, *Phytopathology* **72**:1567–1573.

Schippers, B., Bakker, P. A. H. M., Bakker, A. W., Weisbeek, P. J., and Lutgenberg, B., 1986, Plant growth inhibiting and stimulating rhizosphere microorganisms, in: *Microbial Communities in Soil* (V. Jensen, A. Kjoller, and L. H. Sorensen, eds.), pp. 35–49, Elsevier, London.

Slade, S. J., and Swinburne, T. R., 1985a, Infection development of *Colletotrichum lindemuthianum* race β on resistant and susceptible cultivars of *Phaseolus vulgaris* affected by a bacterial siderophore, Advanced Nato Research Workshop, London (July 1985), Abstract.

Slade, S. J., and Swinburne, T. R., 1985b, Phytoalexin accumulation elicited abiotically in *Vicia faba* reduced by a bacterial siderophore, Advanced Nato Research Workshop, London (July 1985), Abstract.

Stiefel, E. I., Burgess, B. K., Wherland, S., Newton, W. E., Corbin, J. L., and Watt, G. D., 1980, *Azotobacter vinelandii* biochemistry: $H_2(D_2)N_2$ relationships of nitrogenase and some aspects of iron metabolism in: *Nitrogen Fixation*, Vol. 1 (W. E. Newton and W. H. OrmeJohnson, eds.), pp. 221–222, University Park Press, Baltimore, Maryland.

Stone, K. J., and Stominger, J. L., 1972, Inhibition of sterol biosynthesis by bacitracin, *Proc. Natl. Acad. Sci. USA* **69**:1287.

Stutz, E., 1964, Aufnahme von ferrioxamine B durch Tomatenpflanzen, *Experimentia* **20**:430–431.

Sugiura, Y., and Nomoto, K., 1984, Phytosiderophores: Structures and properties of mugineic acids and their metal complexes, *Structure Bonding* **58**:107–135.

Suslow, T. V., and Schroth, M. N., 1982, Role of deleterious rhizobacteria as minor pathogens in reducing crop growth, *Phytopathology* **72**:111–115.

Szabo, I., Benedek, A., and Barabas, G., 1985, Possible role of streptomycin released from spore cell wall of *Streptomyces griseus, Appl. Environ. Microbiol.* **50**:438–440.

Szaniszlo, P. J., Powell, P. E., Reid, C. P. P., and Cline, G. R., 1981, Production of hydroxamate siderophore iron chelators by ectomycorrhizal fungi, *Mycologia* **73**:1158–1175.

Szaniszlo, P. J., Tai, S. C., Crowley, D. E., and Reid, C. P. P., 1985, Mechanisms of iron acquisition from hydroxamate siderophores by two monocot plant species, Advanced Nato Research Workshop, London (July 1985), Abstract.

Teintze, M., and Leong, J., 1981, Structure of pseudobactin A, a second siderophore from plant growth promoting *Pseudomonas* B10, *Biochemistry* **20**:6457–6462.

Teintze, M., Hossain, M. D., Barnes, C. L., Leong, J., and Van den Helm, D., 1981, Structure of ferric pseudobactin, a siderophore from a plant growth promoting *Pseudomonas, Biochemistry* **20**:6446–6457.

Torres, L., Perez-Ortin, J. E., Tordera, V., and Beltran, J. P., 1986, Isolation and characterization of an Fe(III)-chelating compound produced by *Pseudomonas syringae, Appl. Environ. Microbiol.* **52**:157–160.

Trick, C. G., Andersen, R. J., Gillam, A., and Harrison, P. J., 1983, Prorocentrin: an extracellular siderophore produced by the marine dinoflagellate, *Science* **219**:306–308.

Vandenbergh, P. A., Gonzalez, C. F., Wright, A. M., and Kunka, B. S., 1983, Iron-chelating compounds produced by soil pseudomonads: Correlation with fungal growth inhibition, *Appl. Environ. Microbiol.* **46**:128–132.

Warren, R. A. J., and Neilands, J. B., 1965, Mechanism of microbial catabolism of ferrichrome A, *J. Biol. Chem.* **240**:2055–2058.

Wehrli, W., and Staehelin, M., 1971, Actions of the rifamycin, *Bacteriol. Rev.* **35**:290–309.

Wendenbaum, S., Demange, P., Dell, A., Meyer, J. M., and Abdallah, M. A., 1983, The structure of pyoverdine Pa, the siderophore of *Pseudomonas aeruginosa, Tetrahedron Lett.* **24:**4877–4880.

Winkelmann, G., 1979, Surface iron polymers and hydroxy acids. A model of iron supply in sideramine-free fungi, *Arch. Mikrobiol.* **121:**43–51.

Winkelmann, G., 1985, Specificity of siderophore iron uptake by fungi, in: *The Biological Chemistry of Iron* (H. B. Dunford, D. Dolphin, K. N. Raymond, and L. Sieker, eds.), pp. 107–116, D. Reidel, Dordrecht.

Yang, C. C., and Leong, J., 1984, Structure of Pseudobactin 7SR1, a siderophore from a plant deleterious *Pseudomonas, Biochemistry* **23:**3534–3540.

Zähner, H., Keller-Schierlein, W., Hutter, R., Hess-Leisinger, K., and Deer, A., 1963, Stoff-wechselprodukte van Mikroorganismen: 40. Mitteilung. Sideramine aus *Aspergillaceen, Arch. Mikrobiol.* **45:**119–135.

10

Bacteria and Chromium in Marine Sediments

MARGARET W. LOUTIT, JACQUELINE AISLABIE,
PHILIP BREMER, and CHRISTOPHER PILLIDGE

1. Introduction

Chromium (Cr), atomic number 24 and mass 52.01, is one of the most widely used metals in industry (Stern, 1982; Kimbell and Panulas, 1984; Moore and Ramamoorthy, 1984) and its use is increasing (Papp, 1983). Since many Cr-containing effluents are discharged into bodies of water (Moore and Ramamoorthy, 1984), the possibility that the discharged Cr interacts with the biota immediately or subsequently has to be considered. From the literature there is little evidence that consideration has been given to this possibility (Moore and Ramamoorthy, 1984).

Studies on the interaction of any of the 45 biologically important heavy metals (Duxbury, 1986) and the aquatic biota have concentrated on the macrobiota and have involved the assessment of toxic effects usually using LD_{50} values following the addition of soluble forms of the metals to laboratory tanks (Eisler and Hennekey, 1977; Phillips, 1980; Ahsanullah, 1982; Taylor *et al.*, 1985). Rarely has attention centered on the requirement of the aquatic biota for these metals (Kuwabara 1981, 1982) and only recently has consideration been given to the effect of sublethal

MARGARET W. LOUTIT and PHILIP BREMER • Microbiology Department, University of Otago, Dunedin, New Zealand. JACQUELINE AISLABIE • Department of Biology, University of Louisville, Louisville, Kentucky 40292. CHRISTOPHER PILLIDGE • Department of Microbiology, University of Maryland, College Park, Maryland 20742.

concentrations of soluble metals on the physiology or reproductive capacity or on the young of aquatic organisms (Calabrese *et al.*, 1973; Brkovic-Popovic and Popovic, 1977; Martin *et al.*, 1981).

Studies on the effects of Cr have followed the same pattern. The effects of soluble Cr, both Cr(VI) and Cr(III), on the aquatic macrobiota have in the main assessed the toxicity of the element (Mearns and Young, 1977; Oshida and Word, 1982; Frey *et al.*, 1983), as have the few studies on Cr and aquatic microorganisms (Albright and Wilson, 1974). The possibility that Cr is an essential element for marine microorganisms or that microorganisms in marine sediments are involved in its transformation has received little attention (Smillie *et al.*, 1981) in spite of the fact that the highest concentrations of Cr have been detected in sediments (Aston and Chester, 1976), particularly around point discharges (Mayer and Fink, 1980; Smillie *et al.*, 1981; Duedall *et al.*, 1983). Since bacteria are known to be abundant in sediments (Rheinheimer, 1980), even those polluted by metal-containing effluents (Austin *et al.*, 1977), the interaction of Cr and bacteria warrants attention.

In this paper we consider bacterial interaction with Cr in marine sediments, and the implications of this interaction.

2. Sources of Chromium in Sediment

Chromium enters marine habitats from natural and anthropogenic sources (NAS, 1974; Cary, 1982; Duedall *et al.*, 1983; Buat-Menard, 1984; Moore and Ramamoorthy, 1984).

2.1. Natural Sources

The Cr in the earth's surface enters the marine environment from streams, rivers, runoff following rain, land subsidence, wind-blown dust, volcanic debris (Moore and Ramamoorthy, 1984), and deep-sea vents (Jeandel and Minster, 1984). The quantity of Cr entering from these natural sources is now exceeded in certain areas by that entering due to human activities (Leland *et al.*, 1978; Cary, 1982; Duedall *et al.*, 1983; Forstner, 1984; Moore and Ramamoorthy, 1984).

2.2. Anthropogenic Sources

The increase in anthropogenic sources of Cr is due to increasing urbanization and industrialization, with the result that large volumes of

Cr-containing effluents are discharged directly or indirectly into the marine environment. Discharges from tanneries, steelworks, chrome-plating and paint factories, and smelters create point sources of elevated Cr concentrations in the sediments and water column around the discharge point and for varying distances out from it (Duedall *et al.*, 1983; Moore and Ramamoorthy, 1984).

The high volume of sewage discharged by some cities also contributes significant Cr to the marine environment (Mearns and Young, 1977; Jan and Young, 1978). Where several highly industrialized areas or large urban areas are in close proximity the extent of pollution of the inshore area may be considerable (Forstner and Wittmann, 1981; Duedall *et al.*, 1983).

2.3. Deposition of Chromium into Sediments

The Cr in effluents entering bodies of water can be deposited or remain in the water column, either suspended or in solution, depending on the form of the Cr in the effluent, the composition of both the effluent and the receiving waters, and the condition of these (Cranston and Murray, 1980; Forstner and Wittmann, 1981; Katz and Kaplan, 1981; Schulz-Baldes *et al.*, 1983). For example, Cr in the soluble (VI) form in an effluent with little accompanying suspended matter may be carried considerable distances from the point of entry to the marine environment (Osterberg *et al.*, 1965; Cutshall *et al.*, 1966). At the other extreme, Cr(III) in effluents from tanneries with high Cr and high organic matter is deposited rapidly into marine sediments (Katz and Kaplan, 1981; Smillie *et al.*, 1981).

Where Cr deposition does occur, the elevated Cr concentrations in the sediments are detectable within a relatively short distance from the discharge point (Mearns and Young, 1977; Johnson *et al.*, 1981; Smillie *et al.*, 1981; Duedall *et al.*, 1983). The area affected, however, varies. Mearns and Young (1977) reported concentrations up to 1317 μg Cr/g dry weight in sediments off the California coast adjacent to the discharge points for sewage and industrial wastes, and a considerable area some 8 km in length was affected. Hershelman *et al.*, (1981), in a later study of the same area, found that the top 5 cm of sediment was affected over a more extensive area. Much higher concentrations have been reported (Johnson *et al.*, 1981; Smillie *et al.*, 1981) for sediments in Otago Harbour, New Zealand, receiving tannery effluent. The concentration reached was 3000 μg Cr/g dry weight sediment, but the area affected was small and the Cr concentration decreased to 100 μg Cr/g dry weight sediment within 100 m of the effluent entry point.

3. Speciation, Distribution, and Partitioning of Chromium in Marine Sediments

3.1. Speciation

Chromium exists in nature in a variety of compounds, and can occur in oxidation states from -2 to $+6$, although only the (III) and (VI) forms are important biologically (Mertz, 1969). Considerable attention has been paid to the forms of Cr existing in seawater; in the open ocean Cr(VI) is dominant (Curl *et al.*, 1965; Cutshall *et al.*, 1966; Fukai, 1967; Elderfield, 1970; Schroeder and Lee, 1975; Cranston and Murray, 1978; Jan and Young, 1978; Zhou *et al.*, 1979; Osaki *et al.*, 1980; Nakayama *et al.*, 1981a; Van der Weijden and Reith, 1982) whereas nearshore, most Cr is in the (III) form (Elderfield, 1970; Zhou *et al.*, 1979; Jenkins, 1982). Little information is available on the speciation of Cr in marine sediments because attention has tended to focus on how much of the Cr is deposited and the total Cr concentration in the sediments after deposition. Loring (1979), however, has pointed out that total element concentrations are of little value in determining the availability of heavy metals to the biota, so it becomes important to know something of the speciation of the Cr in sediments.

Where speciation in sediments has been studied, Cr(III) is reported to dominate (Curl *et al.*, 1965; Elderfield, 1970; Schroeder and Lee, 1975; Zhou *et al.*, 1979; Nakayama *et al.*, 1981a). Although Cr(VI) has been reported in freshwater sediments (Luli *et al.*, 1983), no such report has appeared for marine sediments. Whether Cr interacts with Mn or Fe, resulting in a change in form in the sediments, as has been shown in seawater (Nakayama *et al.*, 1981b), has yet to be established. The lack of information on speciation may be due in part to the exacting techniques required to obtain reliable results (Cranston and Murray, 1978; Nakayama *et al.*, 1981c).

Although a number of countries have nearshore sediments with elevated Cr concentrations, in only a few cases do we know anything about the speciation or the partitioning of the Cr in them (Katz and Kaplan, 1981). Until recently, it has been a matter of conjecture whether the Cr remains in the sediment once deposited. The fact that Cr does not form sulfides, but can form oxides and hydroxides (Curl *et al.*, 1965; Schroeder and Lee, 1975), would seem to support the contention that, once deposited, the Cr could be unavailable to the biota. However, recent studies (see Section 5.3) suggest that this may not always be the case.

In 1977 we began a study to determine if bacteria and Cr interact in the sediments of Sawyers Bay, Otago Harbour (Fig. 1), which had received untreated tannery effluent for 100 years. The Cr in the sediment

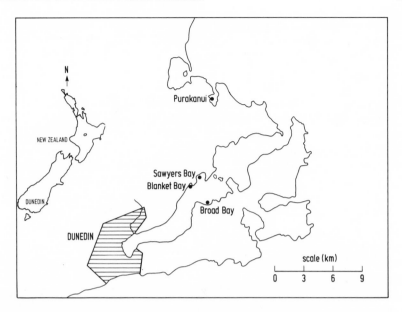

Figure 1. Maps showing location of sampling area in Otago Harbour, New Zealand.

was found to be in the Cr(III) form (Smillie, 1980), and the very low concentration of Cr(VI) that was detectable in the effluent at that time was apparently reduced to Cr(III) in the sediment (Smillie, 1980).

3.2. Distribution with Depth

Most studies on Cr in sediments have involved examination of samples collected from the top few centimeters (Pankow *et al.*, 1977; Loring, 1979), but Capuzzo and Anderson (1973) examined a series of cores from different depths in the Great Bay estuary of New Hampshire as part of a study on sedimentation rates. The area was subject to considerable pollution by industrial wastes containing Cr and the results indicated that the highest concentrations of Cr (221 ppm) were in the top few centimeters at all sites sampled. Further, in a critical review of the extensive work on the fate of heavy metals entering the sea off the California coast, Katz and Kaplan (1981) reported that the highest concentrations of Cr were found in the surface layers of the sediment, but that elevated Cr concentrations were detectable in even the deepest layers of cores taken from sediments close to the effluent entry points.

In a series of cores taken over a period from the Cr-polluted sediment in Otago Harbour, we found that Cr concentration through the sediment profile was variable (Fig. 2). The finding that the concentration of

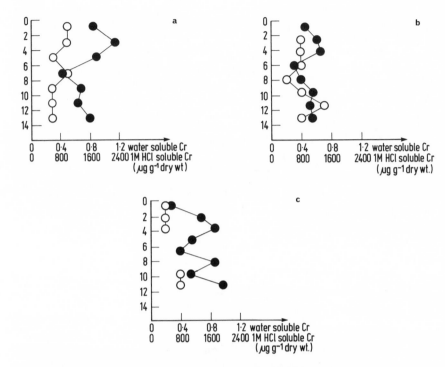

Figure 2. Concentration in water of (○) soluble and (●) total Cr (extractable with 1 M HCl) in sediments from Sawyers Bay, Otago Harbour, collected in (a) April, (b) June, and (c) July.

water-soluble Cr in the sediment was extremely low (Fig. 2) seemed to support the postulate that once deposited into the sediment, Cr, in this case Cr(III), does not readily diffuse into the overlying water; concentration of soluble Cr in the overlying water was between 1 and 10 ng Cr/ml (Pillidge, 1985).

3.3. Partitioning of Chromium in Sediment

Before discharge into the harbor the untreated effluent, when analyzed in 1978 (Smillie, 1980), contained up to 41 μg Cr/ml and most of the Cr was associated with an insoluble floc, which had an organic matter content of 36%. The remaining percentage was in the form of oxides and hydroxides. At that time, Cr concentration in the sediment was up to 3000 μg Cr/g dry weight of sediment near the outfall. At a later date the tannery began treatment of the effluent to remove the Cr. Five years after treatment began, however, the Cr concentration in the sediment remains at approximately 2000 μg/g dry weight.

When the sediment was analyzed by the method of Gupta and Chen (1975), the greatest concentration of Cr was associated with the organic fraction (Table I), with only 4% of the Cr associated with oxides and hydroxides. Capuzzo and Anderson (1973) and Mayer and Frank (1980) also reported a correlation between Cr and organic matter in their studies on sediments polluted with tannery wastes. Similarly, Lindau and Hossner (1982), in work on natural and experimental marshes, found Cr associated with the organic fraction.

The methods we used to investigate the partitioning of the Cr gave only broad groupings and such methodology is not suitable for obtaining an accurate balance sheet or an indication of the movement of Cr from one complex to another with time. Rao and Sastri (1982) reviewed methods for detecting Cr in natural waters and concluded that there was a need for more sensitive methods to allow the determination of very low concentrations of Cr. The determination of whether, for example, Cr is bound in low concentrations to fulvic acids (Schnitzer and Kerndorff, 1981), or to other organic complexes as determined by Douglas *et al.* (1986), should assist in this problem. The possibility that the Cr binds with organic acids produced by bacteria or complexes with other organic

Table I. Chromium Partitioning between Different Geochemical Fractions of Sediment Samples Obtained Using the Method of Gupta and Chen (1976)

	Cr (μg/g dry weight)	
Extractant	Oxidized sediment (0–1 cm)	Reduced sediment (10–12 cm)
1. Deoxygenated deionized water (soluble solids)	0.07	0.30
2. Deoxygenated 1 M ammonium acetate ions	0.23	0.90
3. 1 M acetic acids (carbonates, some oxides)	38.0[a]	280.0[a]
	40.0[a]	310.0[a]
4. 0.1 M NH$_2$OH, HCl in 0.01 M HNO$_3$ (easily reducible phase)	17.0	103.0
	20.0	91.0
5. 30% H$_2$O$_2$ digestion plus extraction with 1 M ammonium acetate in 6% HNO$_3$ (organics, sulfides)	1050.0	6100.0
	880.0	7060.0
6. 2% Na$_2$S$_2$O$_8$ in 8% Na citrate (moderately reducible phase)	50.0	94.0
	31.0	108.0
7. Residual: HNO$_3$/HF digestion (lithogenous fraction)	50.0	56.0
	48.0	53.0
Sum	1210.0	6630.0
	1020.0	7670.0

[a]Duplicate samples.

molecules (Sterritt and Lester, 1980b) during the degradation of the organic matter discharged from the tannery needs to be investigated.

A point of interest to emerge during our attempts to understand the partitioning of the Cr was the very low concentration of soluble Cr detected and, similarly, the very low concentration of Cr in the interstitial and the overlying water (Pillidge, 1985). Lindau and Hossner (1982) also reported low exchangeable Cr and low soluble Cr in their studies on salt marsh sediments.

4. Bacteria in Chromium Polluted Sediments

4.1. Bacteria in Polluted Sediments

In general, higher numbers of bacteria are found in the top layers of sediments close to shore due to the presence of higher concentrations of organic matter (ZoBell, 1946; Walker and Colwell, 1975). Changes in tides, climate, salinity, and freshwater input (ZoBell, 1973; Sieburth, 1979; Nedwell and Brown, 1982) must have an influence on the physical and chemical environment of these sediment bacteria. The physiological versatility of sediment bacteria, commented on by Hauxhurst *et al.* (1980, 1981), apparently explains their continued presence in near-shore, heavily polluted sediments (Goulder *et al.,* 1980). Even where heavy metals are present, bacteria survive and grow, although some activities, such as methanogenesis, sulfate reduction, and carbon dioxide evolution may be reversibly or irreversibly affected by their presence (Capone *et al.,* 1983). Whether nitrifying bacteria are affected in some marine sediments by metals, as in activated sludge (Fargo and Fleming, 1977) and soils (James and Bartlett, 1984), is not yet established. The ability of bacteria to survive and grow in metal-polluted sediments has been explained as due either to an increase in the numbers of metal-tolerant bacteria (Nelson and Colwell, 1975; Kurata *et al.,* 1977; Austin *et al.,* 1977; Traxler and Wood, 1981), which could be due to plasmids mediating resistance, or to the resident bacterial population adapting to the increase in metal (Timoney *et al.,* 1978; Aislabie and Loutit, 1984). Such an adaption could be mediated by increased polysaccharide production (Aislabie and Loutit, 1986) or some other metabolic activity of the bacteria. While bacteria clearly exist in metal-polluted sediments, in only a few cases has their interaction with metals been studied. The interaction of Hg, As, Cd, and Pb with bacteria has received most attention (Jernelov and Martin, 1975; Summers and Silver, 1978; Trevors *et al.,* 1986). Little attention has focused on Cr.

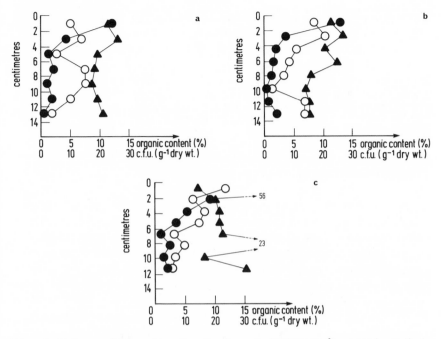

Figure 3. Estimates of (●) aerobic colony-forming units (CFU $\times$ 10^{-5}) and (○) facultatively anaerobic bacteria (CFU $\times$ 10^{-4}) and (▲) organic matter (% dry weight) in sediment cores taken from Sawyers Bay in (a) April, (b) June, and (c) July.

4.2. Distribution of Bacteria in Chromium Polluted Sediments

Results of estimations of viable aerobic bacteria in core samples from the polluted sediments in Otago Harbour and the organic matter content of the sediment are given in Fig. 3. The results of pH and Eh estimations made at the same time and on the same samples are given in Fig. 4.

The highest numbers of aerobic heterotrophs were found in the top 6 cm of sediment (Fig. 3), which was also where the highest concentration of Cr was found (Fig. 2), suggesting that the bacteria were able to survive in this Cr-rich environment. Numbers of facultatively anaerobic bacteria were variable through the sediment profile, but the greatest numbers were also in the top 6 cm (Pillidge, 1985).

The aerobic heterotrophic bacterial populations in the top 1 cm of sediment from the polluted sediment were also compared with those in samples from a nearby site with comparable organic matter content but with low Cr (Aislabie and Loutit, 1984). The results showed that although

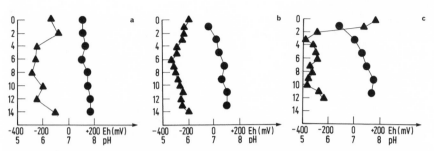

Figure 4. Estimates of (●) pH and (▲) Eh (mV) in sediment cores from Sawyers Bay taken in (a) April, (b) June, and (c) July.

the numbers of bacteria were similar, as were the types present, the percentages of these types in the communities at the two sites differed. There was also a seasonal variation in the composition of the communities. The highest numbers of bacteria were reported when the organic matter content at both sites was highest, supporting the reported correlation between bacterial numbers and organic matter concentration (ZoBell, 1946; Walker and Colwell, 1975).

4.3. Types of Bacteria in Chromium Polluted Sediments

Studies on the types of bacteria in sediments indicate that Gram-negative strains are commonly found (Sieburth, 1967; Austin *et al.,* 1977; Rheinheimer, 1980), but Gram-positive types may also be present in marine sediments (Boeye *et al.,* 1975; Timoney *et al.,* 1978; Sieburth, 1979). Gram-positive strains made up to between 32 and 90% of the population in our study of the Otago Harbour sediments, depending on the time of sampling (Aislabie and Loutit, 1984). The high proportion of Gram-positive organisms may reflect the fact that both the polluted and the control sediments are coastal and subject to run-in from streams.

The identification of the sediment isolates proved difficult, in part because they were metabolically inert (Aislabie and Loutit, 1984), but mainly because the isolates were morphologically atypical, as has been reported in other studies (Wood, 1967; Hodgkiss and Shewan, 1968; Sieburth, 1979). Nevertheless, grouping of the isolates was possible and the populations differed in composition at the two sites (Aislabie and Loutit, 1984). The differences were thought to be due in part to the presence of the Cr(III) and this postulate was investigated by attempting to determine the tolerance of the bacteria to Cr.

It is perhaps worth pointing out the difficulties we had in trying to determine the Cr tolerance of the isolates. It had been our intention to use a chemically defined liquid medium, because of the reports that metals bind to certain organic components in media (Ramamoorthy and Kushner, 1975; Washington et al., 1978; Sterritt and Lester, 1980a; Duxbury, 1981), making at least a proportion of the metal unavailable to the bacteria. We wished to ensure, if possible, that all the Cr added would be available to the organisms. The use of a chemically defined liquid medium was not possible because, when a medium was eventually found that supported the growth of the sediment isolates, the addition of Cr(III) salts resulted in precipitation of the Cr (Aislabie, 1984). Den Dooren de Jong (1971) observed such a precipitation when $CrCl_3$ was added to solid, defined medium, as had Baldry et al. (1977) when using chemically defined media to study the toxicity of metals to bacteria. The deposition of the Cr in defined medium paralleled the problem of the binding of metals to organic components in a medium. In both cases there was an inability to assess how much of the Cr was available to the bacteria. Our attempts to study the tolerance of isolates to Cr, then, had to be carried out in media containing some organic components and, because our isolates grew more rapidly on solid medium, this was used in our experiments. Using this particular medium, it became apparent that some Cr was available to the bacteria, since analysis of the cells showed they concentrated Cr, depending on the isolate used (Aislabie, 1984; Aislabie and Loutit, 1986). A further complication in studying the effect of metals on bacteria is that addition of metal salts to a medium may lower the pH. Unless this is recognized, it is difficult to know whether an observed effect is due to a change in pH or the presence of metal ions.

Using a plate count medium (Difco) made up in estuarine salts solution (Austin et al., 1977), we found that more isolates from the Cr-polluted sediment were tolerant to Cr than were those from the sediment not subject to Cr pollution (Aislabie and Loutit, 1984). The observed tolerance for isolates from the Cr-polluted sediment varied from 5 μg/ml of Cr(III) in January to 10 μ/ml in June (Aislabie, 1984). This variation may be explained by the seasonal variation in the composition of the bacterial communities (Aislabie and Loutit, 1984). Considering the high Cr concentration in the sediment, the tolerance of the sediment isolates to Cr was expected to be higher. Their low tolerance to Cr(III) seemed to support the contention that little of the Cr in the sediment was biologically available. The isolates were also tested for their tolerance to Cr(VI) (Aislabie, 1984), which was found to be toxic over a range of concentrations. Since Cr(VI) is reported to be considerably more toxic than Cr(III) (Baldry et al., 1977; Petrilli and de Flora, 1977; Ross et al., 1981), the results were not unexpected.

An observation of some interest was the apparent increase in numbers of bacteria that could be isolated from Cr-polluted sediment if sublethal concentrations of Cr(III) were incorporated in the medium (Aislabie, 1984). This increased efficiency of isolation of marine bacteria in the presence of metals has been reported by others (Goyne and Jones, 1973; Traxler and Wood, 1981), but the reason for the stimulation is not known. Aislabie (1984) showed that the bacteria did not require the Cr for growth above the trace amounts present in the unsupplemented medium, and suggested that the apparent increase in numbers isolated was due to increased production of polysaccharide in response to the Cr, making the colonies easier to detect on the supplemented plates.

Polysaccharide production in response to the presence of Cr is dealt with in Section 5.2 as a mechanism of tolerance. In attempting to find an explanation for the tolerance of a greater proportion of the isolates from the polluted sediment to Cr(III), the production of extracellular polysaccharide must be considered as a likely possibility.

5. Bacteria and Chromium Interaction

Heavy metals are known to be essential for the growth of microorganisms (Diekert *et al.*, 1981), but above certain concentrations they are toxic (Petrilli and de Flora, 1977; Babich and Stotzky, 1980, 1982; Babich *et al.*, 1982). From a variety of habitats subject to heavy metal pollution it has been possible to obtain isolates and to study some of their tolerance mechanisms (Horitsu *et al.*, 1978; Luli *et al.*, 1983; Nordgren *et al.*, 1983; Haefeli *et al.*, 1984). Microorganisms and heavy metal interaction has been reviewed by several authors (Jernelov and Martin, 1975; Summers and Silver, 1978; Gadd and Griffiths, 1978; Sterritt and Lester, 1980b; Foster, 1983), but information on Cr is sparse, particularly for marine sediment bacteria.

5.1. Chromium as an Essential and a Toxic Element

Most studies on Cr and bacteria have concentrated on the toxicity of the element and have been carried out in pure culture (Albright *et al.*, 1972; Baldry *et al.*, 1977; Petrilli and de Flora, 1977; Ross *et al.*, 1981; McFeters *et al.*, 1983; Ajmal *et al.*, 1984; Thompson and Watling, 1984). Although marine sediment bacteria were not used in the above studies, our experiments with bacteria isolated from Otago Harbour sediment polluted with tannery effluent gave similar findings. The toxicity of the Cr depended on the speciation and the concentration of the element (Aislabie, 1984).

Hexavalent Cr was more toxic than trivalent Cr, as has been reported for *Klebsiella aerogenes* (Baldry *et al.*, 1977) and for soil bacteria (Ross *et al.*, 1981), including nitrifiers (James and Bartlett, 1984). In assessing the effect of the Cr(III), however, it must be remembered that Cr(III) may precipitate in the test medium at neutral pH (Den Dooren de Jong, 1971; Baldry *et al.*, 1977; Ross *et al.*, 1981) and it is not known if Cr(III) and Cr(VI) bind differentially to components in the medium. Chromium (VI) has been shown to affect the metabolism of *Desulfovibrio* in pure culture (Taylor and Oremland, 1979), and Capone *et al.* (1983) showed that sulfate reduction was inhibited in salt marsh sediments.

The form of the Cr also influenced its effect on the isolates we tested. Potassium salts were more toxic than sodium salts and dichromate was more toxic than chromate when added to media containing sediment bacteria (Aislabie, 1984). Capone *et al.* (1983) reported that biomass, CO_2 evolution, and sulfate reduction in marsh sediments were inhibited by $K_2Cr_2O_7$.

5.2. Tolerance and Resistance of Bacteria to Chromium

An arbitrary distinction is made between the terms tolerance and resistance. Resistance is used where the mechanism is proven to be mediated by a plasmid or due to an enzyme acting directly on the Cr. Tolerance is used where some metabolic activity allows the organism to survive and grow in the presence of Cr. Such metabolic activity would occur during the growth of the organism irrespective of the presence of Cr, and its effect on the Cr can be said to be indirect.

In only a few instances has resistance to Cr by bacteria been reported, and then only for Cr(VI) (Summers and Jacoby, 1978; Bopp *et al.*, 1983). There are no reports of resistance to Cr(III). A number of papers use the term resistance, but no evidence has been produced for plasmid control or for the presence of a constitutive enzyme in those bacteria studied from freshwater sediments (Simon-Pujol *et al.*, 1979; Luli *et al.*, 1983). On the other hand Horitsu *et al.* (1983) discuss the tolerance of sewage bacteria to Cr when in fact a genetic basis for the process is apparent. In some papers the terms tolerance and resistance are used interchangeably (Baldry *et al.*, 1977), perhaps because, as Trevors *et al.* (1985) pointed out, concentrations to distinguish metal-sensitive and metal-resistant bacteria have not been defined. Duxbury (1981) has proposed an approach for determining tolerance of soil bacteria to metals in an attempt to overcome this problem. He plated soil dilutions on a medium containing different concentrations of metals. The numbers of bacteria were plotted against the concentration of metal in the plating medium. When the data were subjected to curve-fitting analysis each curve could be described by an exponential function, and by extrapolation it was possible to distinguish tolerant and nontolerant bacteria within the conditions of the experiment.

Whereas a number of factors in sediments will undoubtedly modify the effects of Cr on the microorganisms (Babich and Stotzky, 1980, 1981, 1982, 1983), as shown for metals in pure culture studies, the possibility must not be ignored that the microorganisms themselves may modify the effects of metals, rendering them less toxic to their cells. Such modifications constitute mechanisms of tolerance. Thus, Cr(VI) can be reduced to Cr(III) by H_2S produced by sediment bacteria (Smillie *et al.,* 1981) and Cr(III) can be bound to extracellular polysaccharide produced by sediment bacteria (Aislabie and Loutit, 1986); both mechanisms allow tolerance to Cr. Production of polysaccharide has been postulated as a necessary adjunct to survival (Costerton *et al.,* 1981) by conditioning the environment around the cell (Geesey, 1982) and polysaccharide production by sediment bacteria is in line with these suggestions. In the case of certain isolates, polysaccharide production may be stimulated by Cr, and if Cr is removed, polysaccharide production is reduced (Aislabie and Loutit, 1986).

Other mechanisms of tolerance to heavy metals that have been described include the production of chelating agents, such as organic compounds (Zajic, 1969; Sterritt and Lester, 1980a), chemotaxis for marine bacteria (Young and Mitchell, 1973), and morphological changes. For the latter, most reports concern fungi (Gadd, 1981; Mowll and Gadd, 1984), although Den Dooren de Jong (1971) studied morphological changes in *Azotobacter* strains and Dubinina (1976) in *Leptothrix*.

5.3. Mobilization of Chromium in Sediments by Bacteria

The low concentrations of Cr detected in the interstitial water in the polluted Otago Harbour sediment and in the overlying water (Pillidge, 1985) suggest that either little mobilization of Cr occurs or that any mobilized Cr is rapidly rebound to some other sediment component. In attempting to establish what might be happening in the polluted sediment, a series of experiments was conducted with units constructed from boiling tubes in which Cr-containing sediment or different forms of Cr were incorporated in agar in the base of the tubes. The amount of Cr that could be detected in the medium overlying the agar was determined after incubation in the presence or absence of bacterial cells (Pillidge, 1985; Loutit and Pillidge, 1987). Bacteria incubated under anaerobic conditions were found to solubilize Cr from sediment and from Cr oxides held in the agar base. The release was greater when hydrated Cr oxide was incorporated in the agar than when Cr-polluted sediment was present (Pillidge, 1985). Some chemical release of Cr occurred, but the presence of some strains of bacteria enhanced the solubilization of Cr. Stimulation of the bacteria by the addition of glucose led to an increased release of Cr (Pillidge, 1985).

An observation of interest was that cells either grown in a sac of dialysis tubing suspended in liquid plate count medium (Difco) made up in estuarine salts (Austin *et al.,* 1977) or grown unconfined did not concentrate Cr, unlike cells grown on the surface of solid medium containing Cr (Loutit and Pillidge, 1987). The explanation for this apparent contradiction lies in the fact that the very small amount of Cr released by bacterial activity in the boiling tube experiments immediately binds to the organic components in the plate count medium and is unavailable to the bacterial cells (Pillidge, 1985).

From these and other observations we can piece together a picture of the interactions of Cr and bacteria in the tannery-polluted sediment in Otago Harbour.

5.4. What Really Happens in the Polluted Sediment?

The Cr in the sediment is in the form of Cr(III); any Cr(VI) in the effluent at discharge has been reduced by bacterially produced H_2S (Smillie *et al.,* 1981) or by other reducing substances. The Cr is associated mainly with the organic matter in the sediment and is not present in any appreciable quantity as oxides or hydroxides (less than 4%) (Table I), in spite of the fact that the discharge originally contained a high proportion of oxides and hydroxides (Smillie, 1980). There is some indication that the Cr in the sediment can be mobilized by bacterial action (Pillidge, 1985; Loutit and Pillidge, 1987), and the Cr associated with the sediment organic matter was found to be less easily rendered soluble by bacterial action than when it existed as oxides and hydroxides (Pillidge, 1985). A fraction of the organic matter with which the Cr is associated will be bacterial cells and their extracellular polysaccharides. Bacterial cells are found to be abundant in the polluted sediments (Aislabie and Loutit, 1984; Pillidge, 1985) and many produced copious quantities of polysaccharide to which Cr(III) binds (Aislabie and Loutit, 1986).

The concentration of soluble Cr was low in the interstitial water and in the overlying water (Smillie, 1980; Pillidge, 1985), an observation also made by Lu and Chen (1977) for polluted sediments incubated under oxidized and reduced conditions.

If Cr can be solubilized by bacterial action and yet the concentration in the interstitial water and the overlying water remains very low, then what happens to the released Cr? Some of Pillidge's experiments seem to hold the answer (Pillidge, 1985; Loutit and Pillidge, 1987). The Cr released binds rapidly to other components, be they bacterial polysaccharide (Aislabie and Loutit, 1986) or other organic compounds, perhaps similar to those in the medium Pillidge used. Because Cr has been reported to bind to humic and fulvic acids and other organic components in sediments (Knezevic and Chen, 1977; Schnitzer and Kerndorff, 1981), it is possible that Cr solubilized by organic acids produced by bacteria under anaerobic conditions binds to any of a number of compounds. No preferential order of binding has been established between Cr and the variety of compounds present in sediments.

If this is the picture for a sediment polluted with tannery effluent, then do the same events occur in sediments polluted with other Cr-containing effluents? It seems possible they do, provided organic matter is present either in the effluent or from some other source. Should the effluent be very low in organic matter and the concentration of Cr(VI) be so high that it has an adverse effect on the sediment bacterial community, reduction to Cr(III) may not occur. In freshwater sediments, Cr(VI) has been reported to be the predominant form (Luli *et al.*, 1983), but we are not aware of such a situation for marine sediments.

The variation in the bacterial numbers with season (Aislabie and Loutit, 1984) may regulate the amount of organic acids produced and, hence, the amount of Cr solubilized. Similarly, the availability of fermentable substrate must influence the quantity of acid produced and, hence, the amount of Cr solubilized. The presence of free organic acids in the sediment may be transitory, as they could be rapidly used by other microorganisms.

It is possible that the release of minute quantities of Cr occurs at microsites in the sediments, only to be bound to the extracellular layers or products of bacterial cells. If this is so, the status quo appears to have been maintained; that is, the Cr is still bound in the sediment. If the Cr is still in the sediment and it is not released to the overlying water, then what is the interest in these sediment processes?

6. Implications of Bacterial–Chromium Interaction in Sediments

The interest is that animals collected from the area polluted with tannery effluents have elevated Cr concentrations (Aislabie, 1984; C. Pillidge, unpublished results; Aislabie and Loutit, 1986), which, in view of the extremely low Cr concentrations in the overlying and interstitial water, makes it unlikely that the concentration in the animals derives from this source. Thus, we have postulated that some of the Cr is from ingested food and not from soluble Cr. The postulate has been proven for at least one animal (Bremer and Loutit, 1986a) fed sediment bacteria and their polysaccharides to which Cr was bound. Significantly, the observation that bacterial polysaccharide with Cr bound to it is more resistant to degradation by other sediment bacteria and that the form of the polysaccharide changes when Cr binds to it (Bremer and Loutit, 1986b) may mean that the polysaccharide persists in the sediment for a longer period, making it available to grazing animals in detrital food chains for a longer time. We believe that the activity of bacteria at microsites in marine sediments solubilizes Cr, which then binds to bacterial polysaccharides, which in turn are ingested by animals living in or on the surface of the sediment. Bacteria are thus a vehicle by which Cr enters food chains.

The effect of the entry of this Cr to the food chain has yet to be fully investigated. Chromium (III) is not as toxic as Cr(VI) (Mertz, 1969), but the suggestion has been made that certain skin ulcerations in fish may be associated with high sediment Cr (McDermott et al., 1976). The possibility exists that ingestion of Cr(III) affects the physiology or reproductive capacity, or the embryo or larval stages, of some animals, as has been demonstrated for other metals (Martin et al., 1981), and that Cr(III) can be oxidized to Cr(VI) somewhere in the animal.

7. Conclusions

When considering the fate of heavy metals in sediments the effects of microorganisms should not be ignored. Chromium behaves differently from many of the heavy metals that are discharged into the marine environment; in particular, it does not form sulfides and little Cr leaches out into the interstitial water and into the overlying water. The fact that in sediments the Cr is bound to organic matter or is present as oxides or hydroxides does not mean that it is unavailable for entry to the food chain. Bacterial production of acids to solubilize Cr and the ability of many sediment bacteria to produce polysaccharides to which Cr binds ensures that, when the next trophic level ingests the bacteria and its surrounding polysaccharide or even the polysaccharide alone, Cr enters the food chain. The subsequent effects of this entry are still largely unknown. Whether organisms that form the food of higher trophic levels are affected needs to be assessed. Considering that vast quantities of Cr are being discharged into the aquatic environment and that much of this Cr is deposited into sediments, the role of bacteria in facilitating the entry of even a small proportion of this Cr into the food chain cannot be ignored.

References

Ahsanullah, M., 1982, Acute toxicity of chromium, mercury, molybdenum and nickel to the amphipod *Allorchestes compressa, Aust. J. Mar. Freshwater Res.* 33:465–474.

Aislabie, J., 1984, Aerobic heterotrophic bacteria in a marine sediment polluted with chromium, Ph.D. Thesis, University of Otago, Dunedin, New Zealand.

Aislabie, J., and Loutit, M. W., 1984, The effect of effluent high in chromium on marine sediment aerobic heterotrophic bacteria, *Mar. Environ. Res.* **13**:69–79.

Aislabie, J., and Loutit, M. W., 1986, Accumulation of Cr(III) by bacteria isolated from polluted sediment, *Mar. Environ. Res.* **20**:221–232.

Ajmal, M., Nomani, A. A., and Ahmad, A., 1984, Acute toxicity of chrome electroplating wastes to microorganisms: Adsorption of chromate and chromium (VI) on a mixture of clay and sand, *Water Air Soil Pollut.* **23**:119–127.

Albright, L. J., and Wilson, E. M., 1974, Sub-lethal effects of several metallic salts–organic compound combinations upon the heterotrophic microflora of a natural water, *Water Res.* **8**:101–105.

Albright, L. J., Wentworth, J. S., and Wilson, E. M., 1972, Technique for measuring metallic salt effects upon the indigenous heterotrophic microflora of a natural water, *Water Res.* **6**:1589–1596.

Aston, S. R., and Chester, R., 1976, Estuarine sedimentary processes, in: *Estuarine Chemistry* (J. D. Burton and P. S. Liss, ed.), pp. 37–53, Academic Press, London.

Austin, B., Allen, D. A., Mills, A. L., and Colwell, R. R., 1977, Numerical taxonomy of heavy metal-tolerant bacteria isolated from an estuary, *Can. J. Microbiol.* **23**:1433–1447.

Babich, H., and Stotzky, G., 1980, Environmental factors that influence the toxicity of heavy metal and gaseous pollutants to microorganisms, *CRC Crit. Rev. Microbiol.* **8**:99–145.

Babich, H., and Stotzky, G., 1981, Manganese toxicity to fungi: Influence of pH, *Bull. Environ. Contam. Toxicol.* **27**:474–480.

Babich, H., and Stotzky, G., 1982, Nickel toxicity to fungi: Influence of environmental factors, *Ecotoxicol. Environ. Safety* **6**:577–589.

Babich, H., and Stotzky, G., 1983, Nickel toxicity to estuarine/marine fungi and its amelioration by magnesium in sea water, *Water Air Soil Pollut.* **19**:193–202.

Babich, H., Schiffenbauer, M., and Stotzky, G., 1982, Comparative toxicity of trivalent and hexavalent chromium to fungi, *Bull. Environ. Contam. Toxicol.* **28**:452–459.

Baldry, M. G. C., Hogarth, D. S., and Dean A. C. R., 1977, Chromium and copper sensitivity and tolerance in *Klebsiella aerogenes, Microbios Lett.,* **4**:7–16.

Boeye, A., Wayenbergh, M., and Aerts, M., 1975, Density and composition of heterotrophic bacterial populations in north sea sediment, *Mar. Biol.* **32**:263–270.

Bopp, L. H., Chakrabarty, A. M., and Ehrlich, H. L., 1983, Chromate resistance plasmid in *Pseudomonas fluorescens, J. Bacteriol.* **155**:1105–1109.

Bremer, P. J., and Loutit, M. W., 1986a, Bacterial polysaccharide as a vehicle for the entry of Cr(III) to a food chain, *Mar. Environ. Res.* **20**:235–248.

Bremer, P. J., and Loutit, M. W., 1986b, The effect of Cr(III) on the form and degradability of a polysaccharide produced by a bacterium isolated from a marine sediment, *Mar. Environ. Res.* **20**:249–260.

Brkovic-Popovic, I., and Popovic, M., 1977, Effects of heavy metals on survival and respiration rate of tubificid worms: Part II—Effects on respiration rate, *Environ. Pollut.* **13**:93–98.

Buat-Menard, P. E., 1984, Fluxes of metals through the atmosphere and oceans, in: *Changing Metal Cycles and Human Health* (J. O. Nriagu, ed.), pp. 43–69, Springer-Verlag, New York.

Calabrese, A., Collier, R. W., Nelson, D. A., and MacInnes, J. R., 1973, The toxicity of heavy metals to embryos of the American oyster *Crassostrea virginica, Mar. Biol.* **18**:162–166.

Capone, D. G., Reese, D. D., and Kiene, R. P., 1983, Effects of metals on methanogenesis, sulfate reduction, carbon dioxide evolution, and microbial biomass in anoxic salt marsh sediments, *Appl. Environ. Microbiol.* **45**:1586–1591.

Capuzzo, J. D., and Anderson, F. E., 1973, The use of modern chromium accumulations to determine estuarine sedimentation rates, *Mar. Geol.* **14**:225–235.

Cary, E. E., 1982, Chromium in air, soil and natural waters, in *Biological and Environmental Aspects of Chromium* (S. Langard, ed.), pp. 49–64. Elsevier Biomedical Press, New York.

Costerton, J. W., Irwin, R. T., and Cheng, K. Y., 1981, The bacterial glycocalyx in nature and disease, *Annu. Rev. Microbiol.* **35**:299–324.

Cranston, R. E., and Murray, J. W., 1978, The determination of chromium species in natural waters, *Anal. Chim. Acta.* **99**:275–282.

Cranston, R. E., and Murray, J. W., 1980, Chromium species in the Columbia River and Estuary, *Limnol. Oceanogr.* **25**:1104–1112.

Curl, H., Cutshall, N., and Osterberg, C., 1965, Uptake of chromium (III) by particles in sea-water, *Nature* **205**:275–276.

Cutshall, N., Johnson, V., and Osterberg, C., 1966, Chromium-51 in sea water: Chemistry, *Science* **152**:202–203.

Den Dooren De Jong, L. E., 1971, Tolerance of *Azotobacter* for metallic and non-metallic ions, *Antonie Leeuwenhoek* **37**:119–124.

Diekert, G., Konheiser, U., Piechulla, K., and Thauer, R. K., 1981, Nickel requirement and factor F430 content of methanogenic bacteria, *J. Bacteriol.* **148**:459–464.

Douglas, G. S., Mills, G. L., and Quinn, J. G., 1986, Organic copper and chromium complexes in the interstitial waters of Narragansett Bay (Rhode Island, USA) sediments, *Mar. Chem.* **19**:161–174.

Dubinina, G. A., 1976, Ecology of freshwater iron bacteria, *Biol. Bull. Acad. Sci. USSR* **3**:473–488.

Duedall, I. W., Ketchum, B. H., Park, P. K., and Kester, D. R., 1983, in: *Global Inputs, Characteristics, and Fates of Ocean-Dumped Industrial and Sewage Wastes: An Overview* (I. W. Duedall, P. K. Park, B. H. Ketchum, and D. R. Kester, eds.), pp. 3–46, Wiley, New York.

Duxbury, T., 1981, Toxicity of heavy metals to soil bacteria, *FEMS Microbiol. Lett.* **11**:217–220.

Duxbury, T., 1986, Microbes and heavy metals: An ecological review, *Microb. Sci.,* **3**:330–333.

Eisler, R., and Hennekey, R. J., 1977, Acute toxicities of Cd^{2+}, Cr^{6+}, Hg^{2+}, Ni^{2+} and Zn^{2+} to estuarine microfauna, *Arch. Environ. Contam. Toxicol.* **6**:315–323.

Elderfield, H., 1970, Chromium speciation in seawater, *Earth Planet. Sci. Lett.* **9**:10–16.

Fargo, L. L., and Fleming, R. W., 1977, Effects of chromate and cadmium on most probable number estimates of nitrifying bacteria in activated sludge, *Bull. Environ. Contam. Toxicol.* **18**:350–354.

Forstner, U., 1984, Metal pollution of terrestrial waters, in: *Changing Metal Cycles and Human Health* (J. O. Nriagu, ed.), pp. 71–94, Springer-Verlag, New York.

Forstner, U., and Wittmann, G. T. W., 1981, *Metal Pollution in the Aquatic Environment,* Springer-Verlag, New York.

Foster, T. J., 1983, Plasmid determined resistance to antimicrobial drugs and toxic metal ions in bacteria, *Microbiol. Rev.* **47**:361–409.

Frey, B. E., Riedel, G. F., Bass, A. E., and Small, L. F., 1983, Sensitivity of estuarine phytoplankton to hexavalent chromium, *Est. Coast. Shelf Sci.* **17**:181–187.

Fukai, R., 1967, Valency state of chromium in sea water, *Nature* **213**:901–902.

Gadd, G. M., 1981, Mechanisms implicated in the ecological success of polymorphic fungi in metal polluted habitats, *Sci. Technol. Lett.* **2**:531–536.

Gadd, G. M., and Griffiths, A. J., 1978, Microorganisms and heavy metal toxicity, *Microb. Ecol.* **4**:303–317.

Geesey, G. C., 1982, Microbial exopolymer: Ecological and economic considerations, *ASM News* **48**:9–14.

Goulder, R., Blanchard, A. S., Sanderson, P. L., and Wright, B., 1980, Relationships between heterotrophic bacteria and pollution in an industrialized estuary, *Water Res.* **14**:591–601.

Goyne, E. R., and Jones, G. E., 1973, An ecological survey of the open ocean and estuarine microbial populations II. The oligodynamic effect of Ni on marine bacteria, in: *Marine Ecology* (H. L. Stevenson, ed.), pp. 243–257, University of South Carolina Press.

Gupta, S. K., and Chen, K. Y., 1975, Partitioning of trace metals in selective chemical fractions on nearshore sediments, *Environ. Lett.* **10**:129–158.

Haefeli, C., Franklin, C., and Hardy, K., 1984, Plasmid determined silver resistance in *Pseudomonas stutzeri* isolated from a silver mine, *J. Bacteriol.* **158**:389–392.

Hauxhurst, J. D., Krichevsky, M. I., and Atlas, R. M. 1980, Numerical taxonomy of bacteria from the Gulf of Alaska, *J. Gen. Microbiol.* **120**:131–148.

Hauxhurst, J. D., Kaneko, T., and Atlas, R. M. 1981, Characteristics of bacterial communities in the Gulf of Alaska, *Microb. Ecol.* **7**:167–182.

Hershelman, G. P., Schafer, H. A., Jan. T. K., and Young, D. R., 1981, Metals in marine sediments near a large California Municipal outfall, *Mar. Pollut. Bull.* **12**:131–134.

Hodgkiss, W., and Shewan, J. M., 1968, Problems and modern principles in taxonomy of marine bacteria, in: *Advances in Microbiology of the Sea*, Vol. I (M. R. Droop and E. J. Fergusson Wood, eds.), pp. 127–166, Academic Press, London.

Horitsu, H., Nishida, H., Kato, H., and Tomoyeda, M., 1978, Isolation of potassium chromate tolerant bacterium and chromate uptake by the bacterium, *Agric. Biol. Chem.* **42**:2037–2043.

Horitsu, H., Futo, S., Ozawa, K., and Kawai, K., 1983, Comparison of characteristics of hexavalent chromium-tolerant bacterium, *Pseudomonas ambigua* G-1, and its hexavalent chromium-sensitive mutant, *Agric. Biol. Chem.* **47**:2907–2908.

James, B. R., and Bartlett, R. J., 1984, Nitrification in soil suspensions treated with chromium (III, VI) salts or tannery wastes, *Soil Biol. Biochem.* **16**:293–295.

Jan, T.-K., and Young, D. R., 1978, Chromium speciation in municipal wastewater and seawater, *J. Water Pollut. Control. Fed.* **50**:2327–2336.

Jeandel, C., and Minster, J. F., 1984, Isotope dilution measurements of inorganic chromium (III) and total chromium in seawater, *Mar. Chem.* **14**:347–364.

Jenkins, S. H., 1982, Chromium (VI) reduction in sea water, *Mar. Pollut. Bull.* **13**:77–78.

Jernelov, A., and Martin, A., 1975, Ecological implications of metal metabolism by microorganisms, *Annu. Rev. Microbiol.* **29**:61–77.

Johnson, I., Flower, N., and Loutit, M. W., 1981, Contribution of periphytic bacteria to the concentration of chromium in the crab *Helice crassa, Microb. Ecol.* **7**:245–252.

Katz, A., and Kaplan, I. R., 1981, Heavy metals behavior in coastal sediments of Southern California: A critical review and synthesis, *Mar. Chem.* **10**:261–299.

Kimbell, C. L., and Panulas, J., 1984, Minerals in the world economy, in: *Minerals Yearbook, 1982*, Vol. III *Area Reports: International*, pp. 1–35, Bureau of Mines, U. S. Department of the Interior, Washington, D. C.

Knezevic, M. Z., and Chen, K. Y., 1977, Organometallic interactions in recent marine sediments, in: *Chemistry of Marine Sediments* (T. F. Yen, ed.), pp. 231–241, Ann Arbor Scientific, Ann Arbor, Michigan.

Kurata, A., Yoshida, Y., Kadota, H., and Taguchi, F., 1977, Distribution of Ni tolerant bacteria in water and sediments of the sea of Aso, *Bull. Jpn. Soc. Sci. Fish.* **43**:1203–1208.

Kuwabara, J. S., 1981, Gametophytic growth by *Macrocystis pyrifera* (Phaeophyta) in response to various iron and zinc concentrations, *J. Phycol.* **17**:417–419.

Kuwabara, J. S., 1982, Micronutrients and kelp cultures: Evidence for cobalt and manganese deficiency in southern California deep sea water, *Science* **216**:1218–1221.

Leland, H. V., Luoma, S. N., Elder, J. F., and Wilkes, D. J., 1978, Heavy metals and related trace elements, *J. Water Pollut. Control Fed.* **50**:1469–1514.

Lindau, C. W., and Hossner, L. R., 1982, Sediment fraction of copper, nickel, zinc, chromium, molybdenum and iron in 1 experimental and 3 natural marshes, *J. Environ. Qual.* **11**:540–545.

Loring, D. H., 1979, Geochemistry of cobalt, nickel, chromium, and vanadium in the sediments of the estuary and open Gulf of St. Lawrence, *Can. J. Earth Sci.* **16**:1196–1209.

Loutit, M. W., and Pillidge, C. J., 1987, Sediment bacteria and mobilization of Cr (III), in: *Proceedings of the 4th International Congress of Microbial Ecology, Ljubljana, Yugoslavia, August 1986*, in press.

Lu, J. C. S., and Chen, K. Y., 1977, Migration of trace metals in interfaces of seawater and polluted surficial sediments, *Environ. Sci. Technol.* **11**:174–182.

Luli, G. W., Talnagi, J. W., Strohl, W. R., and Pfister, R. M., 1983, Hexavalent chromium-resistant bacteria isolated from river sediments, *Appl. Environ. Microbiol.* **46**:846–854.

Martin, M., Osborn, K. E., Billig, P., and Glickstein, N., 1981, Toxicities of ten metals to *Crassostrea gigas* and *Mytilus edulis* embryos and *Cancer magister* larvae, *Mar. Pollut. Bull.* **12**:305–308.

Mayer, L. M., and Fink, L. K., 1980, Granulometric dependence of chromium accumulation in estuarine sediments, *Mar. Est. Coast. Mar. Sci.* **11**:491–503.

McDermott, D. J., Alexander, G. V., Young, D. R., and Mearns, A. J., 1976, Metal contamination of flatfish around a large submarine outfall, *J. Water Pollut. Control Fed.* **48**:1913–1917.

McFeters, G. A., Bond, P. J., Olson, S. B., and Tchan, Y. T., 1983, A comparison of microbial bioassays for the detection of aquatic toxicants, *Water Res.* **17**:1757–1762.

Mearns, A. J., and Young, D. R., 1977, Chromium in the southern Californian environment, in: *Pollutant Effects on Marine Organisms* (C. S. Giam, ed.), pp. 125–142, Lexington Books, D. C. Heath and Company, Lexington.

Mertz, W., 1969, Chromium occurrence and function in biological systems, *Physiol. Rev.* **49**:163–172.

Moore, J. W., and Ramamoorthy, S., 1984, Chromium, in: *Heavy Metals in Natural Waters: Applied Monitoring and Impact Assessment* (J. W. Moore and S. Ramamoorthy, eds.), pp. 58–73, Springer-Verlag, New York.

Mowll, J. L., and Gadd, G. M., 1984, Cadmium uptake by *Aureobasidium pullulans, J. Gen. Microbiol.* **130**:279–284.

Nakayama, E., Kuwamoto, T., Tokoro, H., and Fujinaga, T., 1981a, Chemical speciation of chromium in seawater: Part 3. The determination of chromium species, *Anal. Chim. Acta* **131**:247–254.

Nakayama, E., Kuwamoto, T., Tsurubo, A., and Fujinaga, T., 1981b, Chemical speciation of chromium in seawater: Part 2. Effects of manganese oxide and reducible organic materials on the redox processes of chromium, *Anal. Chim. Acta* **130**:401–404.

Nakayama, E., Tokoro, H., Kuwamoto, T., and Fujinaga, T., 1981c, Dissolved state of chromium in seawater, *Nature,* **290**:768–770.

NAS, 1974, Medical and Biological Effects of Environmental Pollutants, Chromium, Committee on the Biologic Effects of Atmospheric Pollutants, Medical Sciences National Research Council, Washington, D. C.

Nedwell, D. B., and Brown, C. M., 1982, *Sediment Microbiology,* Society of Microbiology, Academic Press, London.

Nelson, D. J., and Colwell, R. R., 1975, The ecology of mercury resistant bacteria in Chesapeake Bay, *Microb Ecol.* **1**:191–218.

Nordgren, A., Baath, E., and Soderstrom, B., 1983, Microfungi and microbial activity along a heavy metal gradient, *Appl. Environ. Microbiol.* **45**:1829–1837.

Osaki, S., Osaki, T., Nishino, K., and Takashima, Y., 1980, Oxidation and reduction of chromium in natural water I. Oxidation rate of chromium (III) by oxygen in the presence of Mn (II), *Nippon Kaguku Kaishi* **5**:711–716.

Oshida, P. S., and Word, L. S., 1982, Bioaccumulation of chromium and its effects on reproduction in *Neanthes arenaceodentata* (Polychaeta), *Mar. Environ. Res.* **7**:167–174.

Osterberg, C., Cutshall, N., and Cronin, J., 1965, Chromium-51 as a radioactive tracer of Columbia River water at sea, *Science* **150**:1585–1587.

Pankow, J. F., Leta, D. P., Lin, J. W., Ohl, S. E. Shum, W. P., and Janauer, G. E., 1977, Analysis for chromium traces in the aquatic ecosystem. II. A study of Cr(III) and Cr(VI) in the Susquehanna River basin of New York and Pennsylvania, *Sci. Total Environ.* **7**:17–26.

Papp, J. F., 1983, Chromium, in: *Mineral Commodity Profiles*, pp. 1–18, United States Department of the Interior, Bureau of Mines, Washington, D. C.

Petrilli, F. L., and de Flora, S., 1977, Toxicity and mutagenicity of hexavalent chromium on *Salmonella typhimurium, Appl. Environ. Microbiol.*, **33**:805–809.

Phillips, D. J. H., 1980, Biological indicators: A retrospective summary, in: *Quantitative Aquatic Biological Indicators: Their Use to Monitor Trace Metal and Organochlorine Pollution* (D. J. H. Phillips, ed.), pp. 377–411, Applied Science Publishers, London.

Pillidge, C. J., 1985, Bacterial mobilization of chromium (III) in a polluted marine sediment, Ph. D. Thesis, University of Otago, Dunedin, New Zealand.

Ramamoorthy, S., and Kushner, D. J., 1975, Binding of mercuric and other metal ions by microbial growth media, *Microb Ecol.* **2**:162–176.

Rao, V. M., and Sastri, M. N., 1982, Determination of chromium in natural waters—A review, *J. Sci. Ind. Res.* **41**:607–615.

Rheinheimer, G., 1980, *Aquatic Microbiology*, 2nd Ed., Wiley, Chichester.

Ross, D. S. Sjogren, R. E., and Bartlett, R. J., 1981, Behaviour of chromium in soils: IV. Toxicity to microorganisms, *J. Environ. Qual.* **10**:145–148.

Schnitzer, M., and Kerndorff, H., 1981, Reactions of fulvic acid with metal ions, *Water Air Soil Pollut.* **15**:97–108.

Schroeder, D. C., and Lee, G. F., 1975, Potential transformations of chromium in natural waters, *Water Air Soil Pollut.* **4**:355–365.

Schulz-Baldes, M., Rehm, E., and Farke, H., 1983, Field experiments on the fate of lead and chromium in an intertidal benthic mesocosm, the Bremerhaven Caisson, *Mar. Biol.* **75**:307–318.

Sieburth, J. M. N., 1967, Seasonal selection of estuarine bacteria by water temperature, *J. Exp. Mar. Biol. Ecol.* **1**:98–121.

Sieburth, J. M. N., 1979, *Sea Microbes*, Oxford University Press, New York.

Simon-Pujol, M. D., Marques, A. M., Ribera, M., and Congregado, F., 1979, Drug resistance of chromium tolerant Gram-negative bacteria isolated from a river, *Microbios Lett.* **7**:139–144.

Smillie, R. H., 1980, Metals in wastewater, Ph. D. Thesis, University of Otago, Dunedin, New Zealand.

Thompson, G. A., and Watling, R. J., 1984, A simple method for the determination of bacterial resistance to metals, *Bull. Environ. Contam. Toxicol.* **31**:705–711.

Timoney, J. F., Port, J., Giles, J., and Spanier, J., 1978, Heavy-metal and antibiotic resistance in the bacterial flora of sediments of New York Bight, *Appl. Environ. Microbiol.* **36**:465–472.

Traxler, R. W., and Wood, E. M., 1981, Multiple metal tolerance of bacterial isolates, *Dev. Ind. Microbiol.* **22**:521–528.

Trevors, J. T., Oddie, K. M., and Belliveau, B. H., 1985, Metal resistance in bacteria, *FEMS Microbiol. Rev.* **32**:39–54.

Trevors, J. T. Stratton, G. W., and Gadd, G. M., 1986, Cadmium, transport, resistance and toxicity in bacteria, algae and fungi, *Can. J. Microbiol.* **32**:447–464.

Van der Weijden, C. H., and Reith, M., 1982, Chromium (III)–chromium (VI) interconversions in seawater, *Mar. Chem.* **11**:565–572.

Walker, J. D., and Colwell, R. R., 1975, Factors affecting enumeration and isolation of *Actinomycetes* from Chesapeake Bay and South East Atlantic Ocean sediments, *Mar. Biol.* **30**:193–201.

Washington, J. A. I. I., Snyder, R. J., Kohner, P. C., Curtis, G., Wilt, S. E., Ilstrup, D. M., and McCall, J. T., 1978, Effect of cation content of agar on the activity of gentamicin, jobramycin and amikacin against *Pseudomonas aeruginosa, J. Infect. Dis. 1* **137**:103–111.

Wood, E. F. G., 1967, *Marine Microbial Ecology,* Reinhold, New York.

Young, L. Y., and Mitchell, R., 1973, Negative chemotaxis of marine bacteria to toxic chemicals, *Appl. Environ. Microbiol.* **25**:972–976.

Zajic, J. E., 1969, *Microbial Biogeochemistry,* Academic Press, New York.

Zhou, J., Wanying, C., Ming, K., Wang, L., Yuting, W., and Kueichu, L., 1979, Marine geochemistry I. The valence state of chromium in sea water and the sea water–sediment chromium interchange, Paper presented at International Association on Physical Sciences of the Ocean—Symposium on Marine Pollution Transfer Processes, Canberra, Australia.

ZoBell, C. E., 1946, *Marine Microbiology, A Monograph on Hydrobacteriology,* Chronica Botanica, Waltham, Mass.

ZoBell, C. E., 1973, *Microbial and Environmental Transitions in Estuaries,* Belle W. Baruch Coastal Research Institute, University of South Carolina Press, Columbia, South Carolina.

Index